DAIRY CATTLE SCIENCE

FOURTH EDITION

Howard D. Tyler
Iowa State University

M. E. Ensminger
Late, of Iowa State University

PEARSON
Prentice
Hall

Upper Saddle River, New Jersey
Columbus, Ohio

Library of Congress Cataloging-in-Publication Data
Tyler, Howard.
Dairy cattle science / Howard Tyler, M.E. Ensminger.—4th ed.
p. cm.
Rev ed. of: Dairy cattle science / by M.E. Ensminger. 1993.
ISBN 0-13-113412-4 (casebound)
1. Dairy cattle. 2. Dairy farming. I. Ensminger, M. Eugene. II. Ensminger, M. Eugene.
Dairy cattle science. III. Title.
SF208.T95 2006
636.2′142—dc22
2005043003

Executive Editor: Debbie Yarnell
Assistant Editor: Maria Rego
Production Editor: Alexandria Benedicto Wolf
Production Coordinator: Carlisle Publishers Services
Design Coordinator: Diane Ernsberger
Cover Designer: Jim Hunter
Cover art: Index Stock
Production Manager: Matt Ottenweller
Marketing Manager: Jimmy Stephens

This book was set in New Baskerville by Carlisle Communications, Ltd. It was printed and bound by R.R. Donnelley & Sons. The cover was printed by Phoenix Color Corp.

Pearson Education Ltd.   Pearson Education Australia Pty. Limited
Pearson Education Singapore Pte. Ltd.   Pearson Education North Asia Ltd.
Pearson Education Canada, Ltd.   Pearson Educacin de Mexico, S.A. de C.V.
Pearson Education—Japan   Pearson Education Malaysia Pte. Ltd.

10 9 8 7 6 5 4 3 2 1
ISBN 0-13-113412-4

This edition marks the end of an era. Marion Eugene Ensminger passed away July 5, 1998, at ninety years of age. Dr. Ensminger was president of Agriservices Foundation, a nonprofit foundation serving world agriculture. Throughout his life, he served animal agriculture in various positions, including staff positions at the University of Massachusetts, the University of Minnesota, and Washington State University. He received numerous awards and honors throughout his extensive and distinguished career.

Dr. Ensminger founded the International Ag-Tech Schools, which he directed for more than fifty years. He directed schools, lectured, and/or conducted seminars in seventy countries; he authored or coauthored twenty-two books, which are published in several languages and used all over the world.

Dr. E. often stated that he considered the whole world as his classroom; he waived all royalties on the foreign editions of his books to help educate people around the globe. His was a life that was truly dedicated to agriculture, education, and improving the world.

CONTENTS

PREFACE

The dairy industry is rapidly evolving throughout the world. To be competitive in today's industry still requires a healthy balance between possessing a thorough understanding of the scientific principles of dairy science and the practical art of implementing those principles. While the base of scientific knowledge in dairy science is rapidly expanding and increasing in complexity, the importance of integrating and implementing the scientific principles has not diminished. Integrating concepts from different disciplines is critical to assessing management decisions or solving herd problems. The economic implications of these decisions have increased greatly because the industry has consolidated and shifted toward fewer large farms. Dairy science students face greater challenges and need more extensive preparation than ever before; we hope that this text will contribute to their preparation as they meet this challenge.

The Ensminger series of textbooks has been a fundamental source of information in agricultural education for over fifty years. *Dairy Cattle Science* is a relatively young book in this series; the first edition was published in 1971. The first three editions were solely the work of Dr. Ensminger; this edition is the first to include a coauthor. Readers who are familiar with Dr. Ensminger's style will be pleased to see that it is still very much in evidence in this fourth edition. The influence of Dr. E. will no doubt reverberate throughout *Dairy Cattle Science* for many editions to come.

Numerous changes, however, have been implemented in this edition. The structure of the text has been extensively reorganized to build a foundation of basic concepts before they are integrated into management systems. The management sections have been greatly expanded and represent the heart of the new edition. These sections will enhance your understanding of the complex interactions among the many disciplines that comprise dairy science. Extensive use of pictures, diagrams, and illustrations make this text "classroom ready," as well as a valuable reference beyond the classroom. The traditional chapters have been reorganized into much shorter and more focused, topic-based sections for easy access. This streamlined approach and organization should ultimately provide a more manageable text for classroom use.

ACKNOWLEDGEMENTS

As with any project of this magnitude, many people have contributed in a significant way. First, I am indebted to Olivia Kilian, who went far beyond the call of duty and through the course of many long days and nights made exceptional contributions to this book. Dana Boeck and Nathan Klein provided the elegant artwork that transforms difficult-to-explain concepts into readily understandable diagrams and illustrations. Brianna Klein spent long hours updating the information and making sure the factual content was current.

My editor, Debbie Yarnell kept this project on track despite numerous setbacks and obstacles. My current and former bosses in the Department of Animal Science at Iowa State, Maynard Hogberg, Susan Lamont, and Dennis Marple, provided encouragement and access to resources that made the entire project possible.

Valerie Heffernan of Carlisle Publishers Services managed the transformation of my rough draft into a finished product. In addition, throughout this revision, my greatest critics and strongest supporters have been my students, both undergraduate and graduate. In particular, Carrie Hammer and Josie Booth spent many hours critiquing my words and providing invaluable suggestions regarding what constitutes a "textbook worth buying and keeping."

All of my current and former dairy colleagues have contributed ideas and advice, most notably Leo Timms, Lee Kilmer, Doug Kenealy, and Marj Faust. Additionally, a large number of associations, companies, and individuals graciously donated pictures and figures that have greatly enhanced the quality of this textbook. They are acknowledged individually at the appropriate places throughout the book. Lawrance Fox, Washington State University, and Peter Spike, Ohio State University, provided excellent reviews.

Finally, I also want to acknowledge the sacrifice and support of my family (Kris, Tracy, Nic, John, Drew and Wil) throughout the long process of revision and publication.**Other books in the Ensminger series available from Prentice Hall/Pearson Education:**

Swine Science
Poultry Science
Equine Science

ADDITIONAL RESOURCES

ONLINE SUPPLEMENTS ACCOMPANYING THE TEXT

An online Instructor's Manual and TestGen are also available to instructors through the Tyler catalog page at **www.prenhall.com.** Instructors can search for a text by author, title, ISBN, or by selecting the appropriate discipline from the pull down menu at the top of the catalog home page. To access supplementary materials online, instructors need to request an instructor access code. Go to www.prenhall.com, click the **Instructor Resource Center** link, and then click **Register Today** for an instructor access code. Within 48 hours after registering you will receive a confirming e-mail including an instructor access code. Once you have received your code, go to the site and log on for full instructions on downloading the materials you wish to use.

AGRICULTURE SUPERSITE

This site is a free on-line resource center for both students and instructors in the Agricultural field. Located at http://www.prenhall.com/agsite, students will find additional study questions, job search links, photo galleries, PowerPoints, *The New York Times* eThemes archive, and other agricultural-related links.

Instructors will find a complete listing of Prentice Hall's agriculture titles, as well as instructor supplements supporting Prentice Hall agriculture textbooks available for immediate download. Please contact your Prentice Hall sales representative for password information.

THE NEW YORK TIMES THEMES OF THE TIMES FOR AGRICULTURE

Taken directly from the pages of *The New York Times*, these carefully edited collections of articles offer students insight into the hottest issues facing the industry today. These free supplements can be packaged along with the text.

AGRIBOOKS: A CUSTOM PUBLISHING PROGRAM FOR AGRICULTURE

Just can't find the textbook that fits *your* class? Here is your chance to create your own ideal book by mixing and matching chapters from Prentice Hall's agriculture textbooks. Up to 20 percent of your custom book can be your own writing or come from outside sources. Visit us at **http://www.prenhall .com/agribooks.**

SafariX: TEXTBOOKS ONLINE

SafariX Textbooks Online™ is an exciting new choice for students looking to save money on required or recommended academic textbooks. As an alternative to purchasing the print textbook, students can subscribe to the same content online and save up to 50 percent off the suggested list price of the print text. With a SafariX WebBook, students can search the text, make notes online, print out reading assignments that incorporate lecture notes, and bookmark important passages for later review. For more information, visit **www.safari.com.**

1

The Dairy Industry

Topic 1
History of the Dairy Industry

Topic 2
Breeds of Dairy Cattle

Topic 3
Dairy Industry Overview

Topic 4
Dairy Product Processing

Topic 5
Milk-Marketing and Pricing Systems

Topic 6
Employee Management

1

History of the Dairy Industry

- To understand the history and development of the dairy industry.
- To understand the uses of cattle in ancient times.

Under natural conditions, wild mammals produce only enough milk for their offspring. Long before recorded history, however, people domesticated milk-producing animals and began selecting them for higher production. In addition to cows, buffalo, and goats, people have also used ewes, mares, camel, reindeer, and other mammals for producing milk for human consumption in different parts of the world (Figure 1.1). The importance of the cow in milk production is attested to by her well-earned designation as the foster mother of the human race (Figure 1.2).

Figure 1.1 Because milk provides many essential nutrients in a single, highly digestible food, milk cows are highly valued in many developing nations. *(Courtesy of FAO, Rome)*

TAXONOMY OF DAIRY CATTLE

Domesticated cattle belong to the family *Bovidae*, which includes ruminants with hollow horns. Members of this family possess one or more specialized foregut compartments for fermenting feeds, and they chew their cuds. In addition to cattle or oxen, the family *Bovidae* (and the subfamily *Bovinae*) includes the true buffalo, bison, musk-ox, banteng, gaur, gayal, yak, and zebu.

Although there are several taxonomy systems utilized by scientists, Linnaean taxonomy is the most commonly used taxonomy system for classifying living things. The following outline shows the basic position of the domesticated cow as described by the classical binomial nomenclature of Linnaean taxonomy:

Kingdom *Animalia:* Animals collectively; the animal kingdom.

Phylum *Chordata:* One of approximately twenty-one phyla of the animal kingdom in which there is either a backbone (in the vertebrates) or the rudiment of a backbone (in the chorda).

Figure 1.2 The image of dairy cows on pasture has dominated the public's perception of the dairy industry. *(Courtesy of USDA Soil Conservation Service)*

Class *Mammalia:* Mammals or warm-blooded, hairy animals that produce live young and suckle them for a variable period on a secretion from the mammary glands.

Order *Artiodactyla:* Even-toed, hoofed mammals.

Family *Bovidae:* Ruminants having polycotyledonary placenta; hollow, nondeciduous, up-branched horns; and nearly universal presence of a gallbladder.

Genus *Bos:* Ruminant quadrupeds, including wild and domestic cattle, distinguished by a stout body and hollow, curved horns standing out laterally from the skull.

Species *Bos taurus:* *Bos taurus* includes the ancestors of the European cattle and of the majority of the cattle found in the United States.

HISTORY AND DEVELOPMENT OF THE DAIRY INDUSTRY

Cattle are arguably the most important of all the domesticated animals and, next to the dog, the most ancient of these animals. There are well over 100 million milk cattle in the world. The word *cattle* most likely originated from the Roman word *chattel,* meaning "possession." This is not surprising; in Rome, a person's wealth was often computed in terms of cattle possession. The fact that the ownership of cattle implied wealth is further attested to by the fact that the earliest known coins bear an ox head, and the Roman word *pecunia* for "money" (preserved in our adjective *pecuniary*) was derived from the Latin word *pecus,* meaning "cattle." Cattle were probably first domesticated in Europe and Asia during the New Stone Age. Modern cattle descended from these ancient ancestors. All major dairy breeds in the United States belong to the species *Bos taurus* (Figure 1.3).

Figure 1.3 The seasonal nature of milk availability has been linked historically to the availability and quality of pasture growth. *(Courtesy of Iowa State University)*

USE OF CATTLE IN ANCIENT TIMES

Like other animals, cattle were first hunted and used as a source of food and other materials. As civilization advanced and people turned to tillage of the soil, cattle were probably first domesticated because of their projected value for draft purposes. Eventually populations increased in density, feed became more abundant, and cattle became more plentiful. People adopted a more settled life and began selecting animals that possessed the increasingly desired qualities of rapid growth, fat storage, and milk production. Using cattle as a source of milk dates back to the beginnings of civilization. Cows were being milked as early as 9000 B.C. Prehistoric drawings from the Sahara (circa 8000 B.C.) feature picture stories of cattle. A mosaic frieze (3500 B.C.) found at the temple of Ur near Babylon depicts a dairy scene that includes milk containers and strainers. Sanskrit writings, thousands of years old, relate that milk was one of the most essential of all foods. Hippocrates recommended milk as a medicine five centuries before Christ.

In contrast with the great importance of cattle in western Asia and Europe in both ancient and modern times, cattle were never very highly valued in China, Japan, or Korea. The people of these countries historically have not consumed much beef, milk, butter, or cheese. In India and Pakistan, however, cattle play an important role as a food source; for many, they also retain a great religious significance.

In medieval times, the small amount of cow's milk that was available was produced primarily during the grazing season. In fact, more goat's milk than cow's milk was consumed in liquid form. Even in the thirteenth century, when farming methods had improved, one writer indicated that three cows could be expected to produce only 3.5 pounds of butter per week. Most cow's milk was used in making cheese.

HISTORY OF THE DAIRY INDUSTRY IN THE UNITED STATES

Table 1.1 lists the dates of important developments in the U.S. dairy industry. When Christopher Columbus first arrived in America, there were no cows on the American continent. On his second voyage in 1493, he brought four calves and two heifers to the West Indies. Later, Coronado brought about 150 head of cattle to North America during his expedition to Mexico. These cattle were the ancestors of the Longhorns of the Southwest.

The first cows brought to the Jamestown colony arrived in 1611. The Pilgrims did not bring any cows with them on the Mayflower, which arrived at Plymouth, Massachusetts (about 40 miles south of Boston), in 1620. As a result, the lack of milk is said

TABLE 1.1   Milestones of Milk History in the United States

Year	Milestone
1611	Cows arrive in the Jamestown colony.
1624	Cows reach Plymouth colony.
1841	First regular shipment of milk by rail: Orange County to New York City.
1856	Pasteur began his pasteurization experiments.
	Gail Borden received first patent on condensed milk from both the United States and England.
1878	Continuous centrifugal cream separator invented by Dr. Gustav De Laval.
1884	Milk bottle invented by Dr. Hervey D. Thatcher, Potsdam, New York.
1885	Automatic bottle filler and capper patented.
1890	Tuberculin testing of dairy herds introduced.
	Test for fat content of milk and cream perfected by Dr. S. M. Babcock.
1890	Sherman Anti-Trust Act establishes federal antimonopoly policy.
1892	Certified milk originated by Dr. Henry L. Coit in Essex County, New Jersey.
1895	Commercial pasteurizing machines introduced.
	Thistle milking machine introduces intermittent pulsation.
1908	First compulsory pasteurization applies to all milk sold in Chicago except that from tuberculin-tested cows.
1911	Automatic rotary bottle filler and capper perfected.
1914	Tank trucks first used for transporting milk.
1919	Homogenized milk sold successfully in Torrington, Connecticut.
1922	Capper-Volsted Act codifies agricultural cooperatives.
1932	Practical methods developed for increasing vitamin D in milk.
	First plastic-coated paper milk cartons introduced commercially.
1937	Agricultural Marketing Agreement Act establishes federal milk-marketing orders.
1938	First farm bulk tanks for milk began to replace milk cans.
1942	Every-other-day milk delivery started (initially as a war conservation measure).
1946	National School Lunch Act signed by President Truman.
	Vacuum pasteurization method perfected.
1948	Ultrahigh temperature pasteurization is introduced.
1949	Agricultural Adjustment Act establishes dairy support price at $3.14/cwt.
1950	Milk vending machines win place in distribution.
1955	Flavor-control equipment for milk is introduced commercially.
1964	Plastic milk container introduced commercially.
1968	Electronic testing for milk is introduced commercially, marking the official acceptance of the process.
1974	Nutrition labeling of fluid milk products begins.
1980	American Dairy Association launches the nationwide introduction of the "REAL"® Seal dairy symbol.
1981	Ultrahigh temperature (UHT) milk gains national recognition.
1982	Creation of National Dairy Promotion and Research Board.
1988	Lower-fat dairy products gain widespread acceptance.
	Lowfat and skim milk sales combined exceed whole milk sales for first time.
1993	Mandatory animal drug-residue testing program established.
1994	Recombinant bovine somatotropin (rBST) approved for commercial use in the United States.
	Nutrition Labeling and Education Act requires mandatory nutrition labeling.
1995	Launch of processor-funded milk mustache advertising campaign.
2000	Federal milk-marketing orders reformed.
	Component pricing introduced.

Source: Milk Facts 2002 Edition, p. 4.

to have contributed to the high death rate of the colonists, particularly of the children. It would be four years before the colony received its first cows; three heifers and a bull were delivered by the Charity when it landed in 1624.

Throughout the colonial period and until the middle of the nineteenth century, dairying was limited to a few cows cared for by the family. A farm generally had one or two cows to supply fresh milk.

The availability of milk was largely seasonal, with the peak coming in the spring and early summer, when pastures were lush. Urban populations in this period received almost no fluid milk because of its highly perishable nature, lack of rapid transportation, and lack of adequate storage facilities. Milk in urban areas was either shipped in from neighboring rural areas or produced by local cows fed distillers' by-products. In either case, milk was not subjected to health or quality-control regulations. Consequently, it was often diluted with water and of low quality. Surpluses of milk in this period were converted to cheese to prevent spoilage.

The middle of the nineteenth century proved to be the turning point for the dairy industry. New techniques in handling, storage, and processing of milk were developed at a rapid pace. The government recognized the value of agricultural research in the later half of the nineteenth century and provided the needed funds to establish agricultural colleges (the Land Grant Act) and to foster agricultural research (the Hatch Act).

The early years of the twentieth century ushered in an era where the quality of food was given considerable attention. Laws regulating the quality and preparation of food were passed, and new techniques were developed that improved the safety, acceptability, and nutritive value of milk.

In 1949, researchers developed the ability to freeze bull semen. This discovery led to a rapid expansion in the use of artificial insemination that revolutionized the dairy industry. Since that time, numerous other techniques have increased the yields of dairy cattle and improved the handling and quality control of milk and milk products. These techniques include flavor-control equipment, adaptation of computers to record production records and more accurately predict genetic merit, new milk containers, electronic testing for milk fat, embryo transfer, and the application of biotechnology.

SUMMARY

- Domesticated cattle belong to the family *Bovidae.*
- Early cattle possessed qualities including rapid growth, fat storage, and milk production.
- Cattle are important in western Asia and Europe, and they possess religious significance in India.
- Cattle were never highly valued in China, Japan, or Korea.
- City dwellers obtained milk from two sources: neighboring rural areas or local cows.

QUESTIONS

1. Give the zoological scheme showing the position of the domesticated cow.
2. What is the ancient ancestor of all U.S. dairy breeds?
3. Where and when were the first cows brought to the United States? Who brought them?
4. What year was artificial insemination introduced for dairy cattle?

ADDITIONAL RESOURCES

Books

Becker, R. B. *Dairy Cattle Breeds.* Gainesville, FL: University of Florida Press, 1973.

Porter, A. R., J. A. Sims, and C. F. Foreman. *Dairy Cattle in American Agriculture.* Ames, IA: Iowa State University Press, 1965.

Articles

Plourd, R. "The Drive for Efficiency." *Dairy Herd Management* 37 (March 2000).

Sanders, A. H. "The Taurine World." *National Geographic* (December 1925): 591–710.

Internet

Linnaean Taxonomy: http://encyclopedia.thefreedictionary.com/Linnaean%20taxonomy.

Ng, J. 2001. "Bos taurus," Animal Diversity Web: http://animaldiversity.ummz.umich.edu/site/accounts/information/Bos_taurus.html.

Nowak, R. Walker's Mammals of the World Online 5.1: http://www.press.jhu.edu/books/walkers_mammals_of_the_world/artiodactyla/artiodactyla.bovidae.bos.html.

2
Breeds of Dairy Cattle

OBJECTIVES

- To characterize the six major breeds of dairy cattle.
- To describe the physical characteristics of each major breed.

Although all dairy cattle belong to the same genus and species (*Bos taurus*), considerable genetic variation exists among the breeds of cattle that have been developed throughout history. There are many breeds of cattle worldwide; however, only six major breeds of dairy cattle are utilized primarily by commercial dairy producers in the United States (Figure 2.1 and Tables 2.1 and 2.2). Other minor breeds exist as well. Although they are rarely maintained as commercial milk-producing cattle in modern herds, they are important from a historical perspective and for maintaining the genetic diversity of the dairy cattle population. They each have distinctive characteristics that define the breed. The historical development and defining characteristics, including both

physical characteristics and performance characteristics, of each breed (major and minor breeds) are discussed below. (See also Tables 2.3 and 2.4.)

AYRSHIRES

The Ayrshire breed traces back two to three centuries to native cows from the county of Ayr in southwestern Scotland. But the first true Ayrshires were not reported until the late 1700s. The first Ayrshires imported into the United States arrived in 1822, and the U.S. version of the Ayrshire Herd Book was established 1859. It listed 217 cows and 79 bulls under 130 different ownerships. The Ayrshire Breeders Association was formed in 1886. Ayrshire numbers peaked at over 25,000 registered cattle in 1951. Only 7,000 Ayrshires are on Dairy Herd Improvement Association (DHIA) test currently, and their numbers have decreased at the rate of 10 percent per year in recent years.

Ayrshires are various shades of reddish brown and white or mahogany and white (Figures 2.2 and 2.3). The mahogany color is sometimes expressed as nearly black on many bulls. Brindle and roan

AYRSHIRE COW BROWN SWISS COW GUERNSEY COW

HOLSTEIN-FRIESIAN COW JERSEY COW MILKING SHORTHORN COW

Figure 2.1 The physical characteristics of the six major breeds. *(Courtesy of M. E. Ensminger)*

TABLE 2.1   Breed Enrollment in Dairy Herd Improvement Test Plans (2004)

Ayrshire		Brown Swiss		Guernsey		Holstein		Jersey		Milking Shorthorn		Mixed	
Herds	Cows	Herds	Cows	Herds	Cows	Herds	Cows	Herds	Cows	Herds	Cows	Herds	Cows
140	7,033	313	14,769	223	9,589	22,831	3,732,424	1,228	163,473	50	2,515	1,420	139,512

Source: Modified from USDA-AIPL: http://www.aipl.arsusda.gov/publish/dhi/current/lacavx.html.

TABLE 2.2   Trends in Breed Registration Numbers (1985–2000)

Year	Holstein	Jersey	Brown Swiss	Guernsey	Ayrshire
1985	394,506	65,357	11,974	25,106	11,120
1990	395,906	54,213	12,133	13,930	8,020
1995	329,948	63,399	10,799	7,387	6,456
2000	317,567	63,766	10,648	6,151	6,046

Source: American Jersey Cattle Club: http://www.usjersey.com/.

TABLE 2.3   Breed Averages of Dairy Herds Enrolled in DHI (2003)

Breed	Herds	Cow Years	Cow Years per Herd	Milk (lb)	Fat (%)	Fat (lb)	Protein (%)	Protein (lb)
Ayrshire	143	7,518	53	15,468	3.88	601	3.17	491
Brown Swiss	309	15,430	50	17,737	4.04	716	3.36	595
Guernsey	226	10,234	45	14,693	4.53	665	3.36	494
Holstein	24,219	3,965,426	164	21,671	3.65	795	3.05	664
Jersey	1,225	179,158	146	15,751	4.62	728	3.57	563
Milking Shorthorn	45	2,593	58	14,206	3.67	522	3.12	443
Mixed	1,539	148,966	97	17,495	3.87	682	3.18	560
All breeds	27,739	4,331,143	156	21,236	3.69	787	3.07	655

Source: Modified from USDA-AIPL: http://www.aipl.arsusda.gov/publish/dhi/current/.

TABLE 2.4   Milk Production in Pounds

Breed	Calving Year									
	1980	1990	1995	1996	1997	1998	1999	2000	2001	2002
Ayrshire	13,144	14,799	15,684	15,647	16,611	16,932	17,424	17,389	17,878	17,880
Brown Swiss	14,172	16,250	17,493	17,818	18,782	19,386	20,148	20,300	20,694	20,869
Guernsey	11,666	13,297	14,051	14,177	15,132	15,546	15,963	16,043	16,410	16,398
Holstein	17,566	20,178	21,618	21,926	23,143	23,675	24,380	24,517	24,889	24,996
Jersey	11,437	13,407	14,842	15,097	15,838	16,294	16,940	17,038	17,472	17,663
Milking Shorthorn	11,560	14,011	15,341	15,464	16,007	16,109	16,704	16,548	17,008	17,144

Source: Modified from USDA-AIPL: http://www.aipl.arsusda.gov/publish/dhi/current/lacavx.html.

rarely occur, but these colors are not discriminated against by registry restriction. The most distinctive characteristic of the breed throughout most of their development is their horns. The horns curve outward, then up and back, often exceeding a foot in length. In the past, breeders trained the horns to grow to a more graceful appearance (Figure 2.3). Mature size for Ayrshire cows is about 54 inches at the shoulder and 1,200 pounds body weight. They are known as excellent grazers and browsers, with strong, easy-to-raise calves. They were initially developed and are still noted for strong, well-formed udders and excellent body conformation, both of which contribute to their longevity. They produce average volumes of milk in relation to the other breeds, with moderate levels of fat and protein.

Figure 2.2   An Ayrshire cow. *(Courtesy of Susie Vaughan)*

Figure 2.3   An Ayrshire cow. Note the unique and characteristic appearance of the horns. *(Courtesy of National Dairy Council)*

Figure 2.4   A Brown Swiss cow. *(Courtesy of Susie Vaughan)*

Figure 2.5   A Brown Swiss bull. *(Courtesy of Genex)*

tal cow numbers to the Holstein. In the United States, there are approximately 15,000 Brown Swiss on DHIA test, and their numbers have declined at a rate of nearly 8 percent per year.

BROWN SWISS

Brown Swiss (known as Braunvieh in their native Switzerland) are the oldest breed of dairy cattle. They were developed in the mountainous areas of the Swiss Alps as a triple-purpose breed: for milk, meat, and draft. They were first imported to the United States in 1869, and the Brown Swiss Association of America was organized in 1880.

Brown Swiss cattle are either dark brown or gray and solid in color (Figures 2.4 and 2.5). They are slightly smaller in size and weight than the Holstein, with mature cows weighing approximately 1,400 pounds. They produce milk with higher protein content than does the Holstein. They are large-boned, structurally sound cattle with a reputation as excellent grazers. Worldwide, they are second in to-

DEVON

The Devon is an extremely old breed and was originally developed in Devonshire, England, as a triple-purpose cow (for milk, meat, and draft). Devon cows are related to the Hereford and Sussex breeds. Devons were first imported into the United States in 1817, and the American Devon Herd Book Association was organized in 1860. In 1952, the American Devon Cattle Club moved toward the development of a more beef-type animal, although a small group of breeders formed a separate dairy association to ensure survival of the original genetics.

Milking Devons are medium-size cattle that are known as a hardy, long-lived breed. In color, they range from light red to chestnut to dark, rich red

with dappled markings. They may have some white on the tail switch and udder. They have distinctive light-colored horns that curve upward or inward and are dark at the tips.

DEXTER

Dexter cattle were developed as a triple-purpose breed in southern Ireland and are closely related to the Kerry. They are mostly black, but they may have some white color in the area of the udder. Occasionally, they may be red or dun in color. Mature cows weigh only 600 to 800 pounds, and they have very short legs; their height at maturity is typically less than 42 inches. They were first imported into the United States in 1905. The American Dexter Cattle Association was founded in 1911 and handles breed business. Currently, only about 850 Dexter cattle are registered in the United States.

DUTCH BELTED

Dutch Belted were first described in 1750 in northeastern Switzerland and in Austria, but they were reportedly developed in Holland by the nobility specifically for their distinctive white belt around an otherwise coal black body (Figures 2.6 and 2.7). They were imported to the United States in 1838. They are probably ancestrally related to Holsteins. The Dutch Belted Cattle Association was formed in 1886 in America. Fewer than 200 Dutch Belted cattle are registered in the United States. Cows average about 1,150 pounds at maturity. Dutch Belted cattle are fine-boned cattle with a reputation for having little calving difficulty.

FRENCH-CANADIAN

The French-Canadian is a minor breed descended from the same stock as the Guernsey and Jersey breeds. They are also known as Canadiennes, Black Canadians, or simply Canadians. The French-Canadian Herd Book was established in 1886. The Canadienne Cattle Breeding Association was formed in 1895. Only a few hundred French-Canadian Cattle are registered in North America at this time. These cattle are light-boned and reach a mature size of only 1,000 to 1,100 pounds. Although they are born light-colored, they become dark brown to black as they mature, with lighter udders, stomach, and chest.

GUERNSEY

The Guernsey is one of the Channel Island breeds (along with the Jersey). The breed was quite similar to the Jersey early in its development, but it was se-

Figure 2.6   A Dutch Belted cow. *(Courtesy of Iowa State University)*

Figure 2.7   A Dutch Belted bull. *(Courtesy of Iowa State University)*

lected for the unique characteristics that now define the breed. The first documented importation into the United States occurred in 1830. The American Guernsey Cattle Club was founded in 1877. Although over 100,000 Guernseys were registered in the early 1950s, currently approximately only 10,000 Guernseys are on DHIA test in the United States, and their numbers are rapidly decreasing at the rate of nearly 15 percent per year.

Guernseys are tan and white in color and intermediate in size and weight between Holsteins and Jerseys. Mature cows weigh about 1,100 pounds (Figures 2.8 and 2.9). They have a reputation as good grazers and easy calvers. They are quite docile and adaptable. They produce milk with a high percentage of fat and protein, as well as a high concentration of beta-carotene, proving a yellowish tint to their milk and their carcass fat.

Figure 2.8 A Guernsey cow. *(Courtesy of Susie Vaughan)*

Figure 2.10 A Holstein cow. *(Courtesy of USDA)*

Figure 2.9 A Guernsey bull. *(Courtesy of Genex)*

Figure 2.11 A Holstein bull. *(Courtesy of Genex)*

HOLSTEIN-FRIESIAN

This breed originated in the province of Friesland in Holland and is one of the oldest dairy breeds. Historical references trace the breed origins back at least 2,000 years. They were first imported into the United States in 1795, and the Holstein Herd Book Association was organized in 1873. In 1878, the Dutch Friesian Association was formed. The compound name originated from the merger of the Holstein Breeders Association with the Dutch Friesian Association in 1885, but these cows are commonly known as Holsteins in the United States. The association is currently known as the Holstein Association.

Holstein cows can be either black and white or red and white (Figures 2.10 and 2.11). Currently, approximately 3.9 million Holsteins are on DHIA test in the United States, comprising about 90 percent of the total dairy cow population. Their numbers are decreasing at the rate of 1 percent per year. Holsteins have the highest breed average for milk production. Mature cows average approximately

1,500 pounds in body weight and over 57 inches at the shoulder. As a breed, Holsteins produce the greatest volume of milk among all the dairy breeds.

JERSEY

The Jersey is one of the Channel Island breeds (along with the Guernsey). These islands lie roughly 9 miles off the coast of France in the entrance to the English Channel. Jerseys are among the smallest of the major dairy breeds (the Kerry and Dexter are smaller), and they are ancestrally related to the Brown Swiss, Devon, and Kerry, but not to Holsteins. Early in their development, they were commonly called Alderney cattle. The first Jerseys were imported to America in 1850. The American Jersey Cattle Club was formed in 1868 to handle breed business.

Jerseys range in color from light gray to dark fawn to nearly black (Figures 2.12 and 2.13). Cows reach a mature weight of about 950 pounds and are characterized by their dished faces (Figure 2.14). They are relatively heat tolerant in comparison to

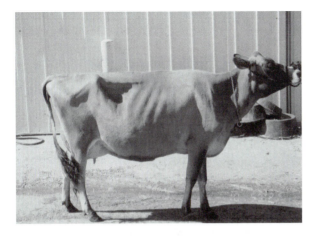

Figure 2.12   A Jersey cow at the World Dairy Expo, Madison, Wisconsin. *(Courtesy of Susie Vaughan)*

Figure 2.13   A Jersey bull. *(Courtesy of Genex)*

Figure 2.14   A Jersey cow. Note the dished appearance of the face. *(Courtesy of USDA-ARS)*

other breeds and have a reputation as excellent grazers that produce milk with high fat and protein content. They are also noted for their ability to utilize forage efficiently. Approximately 160,000 Jerseys are on DHIA test in the United States, and their numbers continue to increase at a rate of over 2 percent per year.

KERRY

The Kerry is the second smallest of all dairy breeds (next to the Dexter). They originated in Kerry County in southern Ireland. They were first imported into the United States in 1818, but by the 1930s, the breed was almost nonexistent. The Kerry is a fine-boned breed, and they reach a mature weight of only 850 pounds. They are known for their longevity and for producing milk with high fat content. Only a few hundred Kerry cattle exist worldwide.

MILKING SHORTHORN

The Milking Shorthorn was developed as a dual-purpose breed (for milk and meat). It was originally known as the Durham, derived from its origins in northeastern England (Durham, Yorkshire, and Northumberland counties). The breed origins trace back to Roman times, but the development of the modern Shorthorn occurred in the late 1700s. The original American Shorthorn Breeders Association developed an auxiliary association, the Dairy Shorthorn Breeders Association, in 1910. In 1915, a separate group formed the American Milking Shorthorn Breeders Association to help develop the dairy strains of the breed. This organization was later discontinued, but the Milking Shorthorn Society was formed in 1920 to promote the development of the Milking Shorthorn. The American Milking Shorthorn Society was the final incarnation of the society and was incorporated in 1948.

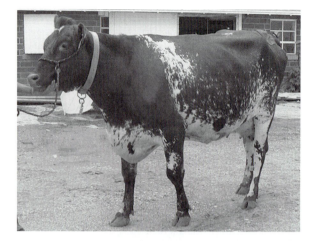

Figure 2.15 A Milking Shorthorn cow. *(Courtesy of Mark Kirkpatrick)*

Figure 2.16 A Milking Shorthorn cow at the World Dairy Expo. *(Courtesy of Susie Vaughan)*

Milking Shorthorn cattle can be red, white, red and white, or roan in color (Figures 2.15 and 2.16). They reach a mature weight of about 1,350 pounds. Milking Shorthorns are known for their structural soundness and longevity. They are also reported to have outstanding feed efficiency in comparison to other breeds. Their calves are born easily and grow efficiently; those calves not raised for breeding purposes produce highly graded carcasses. Canadian Milking Shorthorns, Dairy Shorthorns, and the Australian Illawarra are all based predominately on Milking Shorthorn ancestry. Currently, the Milking Shorthorn has the lowest number of registered animals among the six major breeds. Less than 0.5 percent of U.S. dairy cattle are Milking Shorthorns.

SUMMARY

- Six major breeds of dairy cattle exist in the United States.
- There are breed differences in milk, protein, and butterfat production.
- The place of origin, color, head shape, and other distinguishing characteristics can be used to identify each breed.

QUESTIONS

1. List the six major breeds of dairy cattle in the United States.
2. Which breed is highest in milk production? In butterfat production?
3. Which breed was first imported to the United States?
4. What is the place of origin of the Brown Swiss? Of the Milking Shorthorn?

ADDITIONAL RESOURCES

Books

Becker, R. B. *Dairy Cattle Breeds.* Gainesville, FL: University of Florida Press, 1973.

Schmidt, G. H., L. D. Van Vleck, and M. F. Hutjens. *Principles of Dairy Science,* 2nd ed. Englewood Cliffs, NJ: Prentice Hall, 1988.

Article

Franck, R. "Wanted: Jersey Cows." *Dairy Herd Management* 36 (March 1999): 28–32.

Internet

American Dexter Cattle Association: http://www.dextercattle.org/.

American Livestock Breeds Conservancy: http://www.albc-usa.org/.

American Milking Shorthorn Society: http://www.milkingshorthorn.bravepages.com/about.html.

American Jersey Cattle Association: http://www.usjersey.com/.

Ayrshire Breeders Association: http://usayrshire.com/.

Brown Swiss Association: http://www.brownswissusa.com/.

Breeds of Livestock, North American Dairy Breeds: http://www.ansi.okstate.edu/breeds/cattle/.

Canadian Cattle Breeders Association: http://www.clrc.on.ca/canadien.html.

Holstein Association USA: http://www.holsteinusa.com/.

Milking Devon Cattle Association: http://www.milkingdevons.org/.

Red and White Cattle Association: http://www.redandwhitecattle.com/.

The American Guernsey Association: http://www.usguernsey.com/.

3

Dairy Industry Overview

OBJECTIVES

- To describe the advantages and disadvantages of managing a dairy production enterprise.
- To describe the changes in the dairy industry over the last century.

Cattle furnish most of the milk of the world; other species, such as buffalo, goats, and sheep, provide less than 10 percent of the world's current milk supply. Table 3.1 shows the productivity of cows in the principal milk-producing countries of the world. The United States ranks fourth in the number of milk cows and second in average milk production per cow, behind Japan. Some countries produce more dairy products than they can use and become exporters of those products, whereas others are importers. New Zealand has low milk production costs and accounts for about 20 percent of the total world dairy exports. By contrast, the United Kingdom is an especially heavy importer of butter, cheese, and condensed milk. Many factors

determine the level of development of the dairy industry in different countries, including the dietary preferences of the people, adaptation of the country to dairying, relative size of urban and rural populations, and extent and effectiveness of dairy research and education.

DAIRYING IN THE UNITED STATES

Dairy production is an especially demanding profession, but it also presents some advantages compared to raising other livestock on the farm. These advantages include the following:

1. The dairy industry is stable. Total milk production varies little from year to year relative to the output of most other agricultural products; the change is often less than 1 percent and is usually not more than 2 percent.
2. The dairy cow is an amazingly efficient producer of human food. A cow producing 25,000 pounds of milk per year supplies as much food nutrients as are produced by five 1,250-pound steers, and she is still available for more productive years.

TABLE 3.1   Ten Leading Countries in Milk Cows on Farms and in Milk Production per Cow

Country	Number of Milk Cows on Farms (1,000 Head)	Country	Average Milk Production per Cow (pounds)
India	35,900	**Japan**	18,845
Brazil	16,045	**United States**	18,139
Russia	12,500	**Netherlands**	17,328
United States	9,115	**Sweden**	17,328
Mexico	6,800	**Canada**	16,010
Ukraine	5,375	**Denmark**	15,952
Germany	4,564	**United Kingdom**	14,061
France	4,412	**Germany**	13,470
New Zealand	3,269	**Spain**	12,544
Poland	3,047	**France**	12,430

Source: Milk Industry Foundation, from data provided by USDA.

3. Dairying provides a steady source of income. A grain farmer, fruit producer, or vegetable grower receive income only when products are sold, usually once per year. Likewise, beef cow-calf producers secure most of their income when the calf crop is marketed, again, generally once per year. On the other hand, a dairy producer receives a milk check at frequent intervals (biweekly or monthly) throughout the year.
4. Dairying provides steady employment for hired labor. Many types of agricultural work are highly seasonal, and the labor force must be increased and decreased accordingly, particularly during harvest. In the dairy enterprise, however, fairly uniform labor needs exist throughout the year, which makes it possible to keep better employees on a permanent basis.
5. Dairy cows use a large amount of unsalable forages. Each year, considerable amounts of roughage are produced on the farms and ranches of America that would have little value if they were not used by dairy cattle and other ruminants. Also, much of the rolling land on which such feeds are grown is unsuited to the production of grain or other crops.
6. Soil fertility is maintained. By returning the manure to the land, the fertility and physical condition of the soil are preserved.

Some factors make dairy production a challenging profession to undertake:

1. Considerable capital is required. The investment per cow for dairy farming is at least double what is normally required in investment per cow in a beef cow-calf enterprise.
2. Successful dairy management requires specialized education and training. A top manager must be knowledgeable in business administration, labor management, animal physiology, genetics, and nutrition. He or she also requires the skills to meld this knowledge into a smooth functioning, efficient production unit.
3. The number of regulatory programs can be overwhelming. Producers must follow the federal, state, and local regulatory programs applicable to the operation. No matter how important the objectives of these regulatory programs, compliance takes considerable planning and time.
4. Dairying is labor intensive and provides little free time. Cows must be milked regularly; owners or managers interested in vacation time, short work weeks, or even short days should not enter into a one-worker dairy enterprise.

5. Hourly returns are low. Hourly returns to dairy farmers have, on the average, been below returns in many other types of farming, and they are well below the average for all U.S. manufacturing industries.

Through the years, dairy farms in the United States have become larger and more specialized. The practices of separating cream, making butter, processing market milk, growing and mixing concentrates, and keeping bulls have largely disappeared from the average dairy farm. Dairy producers tend to buy more of their feed and replacements instead of investing in more land. A few produce part or all of their forage requirements. Some create agreements with specialized heifer-raising operations. Dairy farmers are increasingly dependent on the selective adoption of new technologies to remain competitive.

Tables 3.2 and 3.3 describe trends in the leading dairy states. Major population centers that provide a large market for milk are major factors in determining the intensity of dairying in an area. Also, the suitability of the land, climate, and the availability of feed all exert a considerable influence. These factors have been a driving force toward the tremendous growth of the dairy industry in the western states over the last twenty-five years. Wisconsin is the leading dairy state in the United States; it has the most milk cows on farms. Many dairy farms in the major dairy areas, especially the Lake States, Corn Belt, and Northeast, continue to be operated as smaller, more diversified enterprises. Several states, especially those of the Southwest, have highly intensified dairy farms due to the limitations of arable land and the cost of irrigation. The location of dairy farm centers is also driven to a large extent by the availability of processors. The dairy-processing industry has undergone marked change in recent decades, including fewer but larger plants, increased importance of producer cooperatives, and regional shifts precipitated by population shifts and shifts in milk production in excess of fluid sales.

Several dramatic changes have occurred in the dairy industry over the last several decades. These changes are expected to continue in the future:

1. **Decline in dairy farms and cow numbers; increase in herd size.** The number of farms reporting dairy cows has declined sharply and will, in all likelihood, continue to decline in the future. During the period 1950 to 2000, the number of dairy farms declined by over 90 percent, but herd size increased by over

TABLE 3.2   Top Five Milk-Producing States, 1909–2001

Year	Rank				
	1	*2*	*3*	*4*	*5*
1909	New York	Wisconsin	Iowa	Pennsylvania	Minnesota
1919	Wisconsin	New York	Minnesota	Pennsylvania	Ohio
1930	Wisconsin	Minnesota	New York	Iowa	Illinois
1940	Wisconsin	Minnesota	New York	Iowa	Illinois
1950	Wisconsin	New York	Minnesota	Iowa	California
1960	Wisconsin	New York	Minnesota	California	Pennsylvania
1970	Wisconsin	New York	Minnesota	California	Pennsylvania
1980	Wisconsin	California	New York	Minnesota	Pennsylvania
1990	Wisconsin	California	New York	Minnesota	Pennsylvania
2000	California	Wisconsin	New York	Pennsylvania	Minnesota
2001	California	Wisconsin	New York	Pennsylvania	Minnesota

Sources: USDA, National Agricultural Statistics Service. Milk Facts, 2002 Edition, p. 15.

TABLE 3.3   Milk Production in the Five Largest Dairy States, in Millions of Pounds, 1960–2001

	1960	*1970*	*1980*	*1990*	*2001*
California	8,059	9,457	13,577	20,947	32,251
Wisconsin	17,780	18,435	22,380	24,187	22,199
New York	10,171	10,341	10,974	11,067	11,778
Pennsylvania	6,878	7,124	8,496	10,014	10,849
Minnesota	10,272	9,636	9,535	10,030	8,812
Top five	53,160	54,993	64,962	76,245	85,889
Top ten	77,312	74,762	84,803	98,683	116,675
Top twenty	99,535	95,357	106,708	123,831	143,121
Total for the United States	123,109	117,007	128,406	147,720	165,336

Sources: USDA, National Agricultural Statistics Service. Milk Facts, 2002 Edition, p. 14.

1,000 percent. During the forty-year period between 1960 to 2000, the number of milk cows in the United States declined from 17,500,000 to 9,200,000. The number of farms with over 200 cows is dramatically increasing in the United States (Table 3.4).

These changes are driven by changes in the costs of dairying. Tight margins have forced producers to spread expenses over more animal units to maintain returns on investment. Feed has traditionally been the largest single expenditure in dairying; it accounts for 45 to 65 percent of the cost of milk production. One commonly used indicator of the relative prosperity of dairy producers is the milk:feed price ratio. This ratio gives an indication of how much return producers are receiving for their feed inputs. To turn a profit, producers must maintain a favorable milk:feed price ratio. The ratio fluctuates from season to season and from year to year, primarily in response to the availability and cost of feed and the demand for milk, although it does not indicate how energy, labor, and other managerial costs have increased in recent years. Thus, producers have increased size and increased the use of technology to reduce labor demands.

Starting a new dairy requires the investment of large amounts of capital. Land, buildings, silos, machinery, milking equipment, and the animals themselves all add up to a sizable investment. In addition to the cost of facilities and animals, depreciation and repairs of the physical plant, interest on loans, insurance, and taxes must be considered in the operating expenses. The U.S. Department of Agriculture estimates that the average dairy

TABLE 3.4   Averages of Dairy Herd Improvement (DHI) Cow Herds, by Herd Size, 2003

Herd Size (Cow Years per Herd)	Records		Milk		Fat		Protein		
	Herds	Cow Years	(lbs)	(%)	(lbs)	% Reporting	(%)	(lbs)	% Reporting
5–24	993	16,490	16,805	4.03	677	100	3.19	537	100
25–49	6,430	254,284	18,736	3.82	716	100	3.07	575	100
50–74	7,183	437,776	19,736	3.79	749	99	3.06	605	99
75–99	3,872	332,015	20,159	3.76	759	99	3.07	620	99
100–149	3,661	440,599	20,120	3.75	757	97	3.07	620	97
150–199	1,502	258,860	20,234	3.75	768	96	3.08	628	96
200–299	1,433	348,617	20,766	3.73	782	95	3.08	644	95
300–399	706	244,636	20,971	3.69	786	94	3.07	655	91
400–499	421	187,751	21,591	3.64	798	91	3.06	676	85
500–749	606	369,955	22,447	3.63	824	94	3.07	697	90
750–999	315	271,302	22,529	3.61	819	96	3.06	700	88
1,000–1,499	302	369,253	22,717	3.62	827	96	3.07	702	87
1,500–1,999	137	236,276	22,647	3.65	829	99	3.09	702	89
2,000–2,999	108	260,263	22,925	3.59	824	99	3.07	707	90
3,000–3,999	37	126,233	23,084	3.56	822	100	3.07	705	92
4,000+	33	176,833	22,849	3.59	819	95	3.10	706	93
All herds	27,739	4,331,143	21,236	3.69	787	97	3.07	655	93

Source: Modified from USDA-AIPL: http://www.aipl.arsusda.gov/publish/dhi/current/.

cow is worth $1,207.00 and the average cost to produce 100 pounds of milk is $11.84. Other expenses are inherent in all dairy operations, including bedding, veterinary bills, utilities, general operational supplies, breeding fees, DHIA expenses, and transportation costs. Increased size and efficiency of farms has been driven by tremendous increases in the costs of labor, building materials, and utilities, coupled with relatively static feed costs and milk prices. The same standard of living provided by sixty cows in 1970 now requires more than 200 cows (see Figure 3.1).

2. **Increase in production per cow.** The number of milk cows has decreased dramatically to the current population of around 9 million cows. However, the average annual production per cow in the United States has steadily increased from 3,138 pounds in 1920 to over 18,000 pounds per cow in 2001 (Table 3.5). As a result, the total amount of milk produced has increased by about 30 percent in the last forty years, despite declining cow populations. The increase in average production per cow can be attributed to improved breeding, feeding, disease control, and management. Improved genetics accounts for much of this increase, especially since the development of technol-

Figure 3.1   A small dairy farm located in western Maryland. *(Courtesy of USDA-ARS)*

ogy for freezing semen. In addition, the extensive utilization of formalized record-keeping systems such as Dairy Herd Improvement (DHI) by producers has allowed rapid improvement in many aspects of dairy herd management. The cows enrolled in dairy herd

TABLE 3.5   **U.S. Milk Production per Cow**
and Number of Milk Cows

Year	Milk Production per Cow (Pounds)	Number of Milk Cows
1920	3,138	21,455,000
1930	4,508	23,032,000
1940	4,625	24,940,000
1950	5,314	23,853,000
1960	7,002	19,527,000
1970	9,385	12,483,000
1980	11,891	10,799,000
1990	14,642	10,127,000
2000	18,201	9,206,000

Sources: USDA, National Agricultural Statistics Service.
Milk Facts, 2002 Edition, p. 12.

improvement associations produced more milk and fat than those not enrolled. In addition, degree of diversification affects level of production. On a typical dairy farm, the business may consist of one or more of the following enterprises: milking herd, raising heifers for replacement purposes, production of forage, and production of cash crops. The producers who have the greatest intensity of focus on the milking herd tend to have cows that produce the greatest amounts of milk.

SUMMARY

- Special advantages of dairy production include business stability, food production efficiency, steady income and employment, use of unsalable roughage, and maintenance of soil fertility.
- The disadvantages of dairying include the considerable capital required, overwhelming regulatory programs, work confinement, and low hourly returns.
- Factors determining present dairy industry development include dietary preferences, adaptation of the country, size of urban and rural populations, and dairy research and education.
- Changes in the dairy industry include a decline in farms and cow numbers with an increase in herd size and an increase in production per cow.

QUESTIONS

1. Where does the United States rank according to the number of milk cows? According to milk production?
2. What has been one major factor in determining the increased intensity of dairying?
3. Which is the leading dairy state?
4. What has traditionally been the largest single expenditure in dairying?
5. Explain the milk:feed price ratio.
6. What are some of the start-up costs of a new dairy?

ADDITIONAL RESOURCES

Books

Becker, R. B. *Dairy Cattle Breeds.* Gainesville, FL: University of Florida Press, 1973.

Porter, A. R., J. A. Sims, and C. F. Foreman. *Dairy Cattle in American Agriculture.* Ames, IA: Iowa State University Press, 1965.

Articles

Plourd, R. "The Drive for Efficiency." *Dairy Herd Management* 37 (March 2000).

Sanders, A. H. "The Taurine World." *National Geographic* (December 1925): 591–710.

Internet

Linnaean Taxonomy: http://encyclopedia.thefreedictionary.com/Linnaean%20taxonomy.

Ng, J. 2001. "Bos taurus," Animal Diversity Web: http://animaldiversity.ummz.umich.edu/site/accounts/information/Bos_taurus.html.

Nowak, R. Walker's Mammals of the World Online 5.1: http://www.press.jhu.edu/books/walkers_mammals_of_the_world/artiodactyla/artiodactyla.bovidae.bos.html.

4

Dairy Product Processing

OBJECTIVES

- To describe manufactured dairy products.
- To provide an overview of the factors affecting the movement and distribution of fluid milk.

The profitability of the dairy industry depends on maintaining a strong market for dairy products (Figure 4.1). The factors that affect processor profitability include the regional availability of milk, costs of processing, and consumer market for the final product (Tables 4.1 and 4.2). Manufactured dairy products utilized over 60 percent of U.S. milk production. The production of manufacturing grade milk is primarily centered in the Midwest and the Great Lakes area. Thus, with the exception of ice cream making, the processing of manufactured dairy products (butter, nonfat dry milk powder, cheese, evaporated and condensed milk, and other products) is concentrated near those areas of production.

Since the 1940s, the number of milk manufacturing plants in the United States has been decreasing, while the output per plant has been increasing. During the last half of the twentieth century, the number of plants decreased by over 90 percent, while the plants became ten times larger. The reduction in the number of plants has contributed to increased regionalization of the industry (Tables 4.3 and 4.4). The costs associated with the transportation of raw milk from the farm to the plant dictates to a great extent the distance from the plant that dairy farms can be sited and still be profitable (Figure 4.2).

There are three common ways to assess the profitability of processing firms: net return on sales, return on assets, and return on net worth. Return on sales averages about 2 percent. Also, earnings on assets and earnings on net worth are typically low. The volume and efficiency of operations make it possible for the fluid milk industry to operate on these comparatively small margins.

Figure 4.1 The young milk drinker, by Aved. *(Courtesy of M. E. Ensminger)*

FLUID MILK

Milk producers sell most of their product in the form of whole milk (Figure 4.3). Dairy farmers once marketed a considerable amount of their product as farm-separated cream. In 1940, farm-separated cream accounted for 38 percent of the total product marketed. Today, marketing of cream is almost nonexistent.

Milk moves from the farm to the consumer in the following three stages:

1. Assembly and transportation from farms to processing plants (Figure 4.4).
2. Processing and packaging or manufacturing into various dairy products (Figure 4.5).
3. Distribution of packaged milk and manufactured milk products to consumers (Figure 4.6).

TABLE 4.1 Fluid Milk Industry Statistics, 1992–2000*

	1992 (Census)†	1997 (Census)	2000 (ASM)‡
Companies	525.0	405.0	NA
All establishments (*plants*)			
Total	746.0	612.0	NA
With twenty or more employees	506.0	435.0	NA
All employees			
Number (*1,000*)	63.4	57.6	58.1
Payroll (*million dollars*)	1,841.0	1,898.0	2,067.9
Production workers			
Number (*1,000*)	32.5	30.0	30.3
Hours (*Million*)	70.8	64.8	64.4
Total compensation (*million dollars*)	984.0	2,467.8	2,633.2
Value added by manufacturer (*million dollars*)	25,966.0	6,285.0	7,781.0
Cost of materials (*million dollars*)	15,974.0	15,719.0	15,947.0
Value of shipments (*million dollars*)	21,927.0	21,995.0	23,699.0
New capital expenditures (*million dollars*)	363.0	423.0	543.0

*Revised. NA = not available. Note: Latest data available. The "fluid milk industry" is defined by the Bureau of the Census as being comprised of manufacturing establishments for which the value of the shipments of fluid milk, cream, and related products are both primary and secondary to the industry.
†Economic Census of Manufacturers.
‡Annual Survey of Manufacturers covering sample of establishments.
Sources: Bureau of Census. Milk Facts 2002 Edition, p. 24.

TABLE 4.2 Supply and Utilization, Total Milk, in Billions of Pounds, 1996–2001*

	1996	1997	1998	1999	2000†	2001‡
Production	154.3	156.6	157.4	162.7	167.5	165.3
Farm use	1.4	1.4	1.4	1.4	1.3	1.3
Marketing	152.8	155.2	156.1	161.3	166.2	164.1
Beginning commercial stocks	4.1	4.7	4.9	5.3	6.1	6.9
Imports	2.9	2.7	4.6	4.8	4.4	5.7
Total supply	157.8	162.6	165.6	171.4	176.8	176.7
Ending commercial stocks	4.7	4.9	5.3	6.1	6.9	7.0
Net removals	0.1	1.1	0.4	0.3	0.8	0.2
Commercial disappearance—actual	155.0	156.6	159.9	164.9	169.1	169.5
Percentage change from previous year	0.1	1.0	2.1	3.2	2.5	0.2

*Milk equivalent, milk fat basis.
†Revised.
‡Preliminary.
Sources: USDA, Economic Research Service. Milk Facts 2002 Edition, p. 28.

Producers also market their milk as Grade A milk or as manufacturing milk (Grade B).

Once milk is processed, it must be marketed rather rapidly because of its highly perishable nature. When consumers have a bad experience with milk, it is difficult to get them back into the habit of drinking milk. Hence, it is imperative that marketed fluid milk is consistently high quality and possesses both desirable taste and attractive packaging. Milk is distributed in several ways, including home delivery, retail stores, and vending machines and other forms of self-service. Today, the vast majority of milk is sold in retail stores mostly because new technology in processing increases the storage life of milk and because of the increased costs involved with home delivery (Figure 4.7).

Many retail stores sell their own brand of milk at a lower price than the regional or national brands, but they continue to offer other regional or national competitive brands to provide customers with a wider choice. Through the use of private brands, the store has considerably more control over pricing and merchandising than it does with outside brands.

TABLE 4.3   Top Ten States for Production of Butter, Cheese, and Ice Cream, 2001*

Rank	Butter (1,000 lbs)		Cheese† (1,000 lbs)		Ice Cream‡ (1,000 gal)	
1	California	341,101	Wisconsin	2,133,156	California§	180,352
2	Wisconsin	324,675	California	1,619,391	Indiana	111,958
3	Washington	102,110	New York	702,752	Pennsylvania	79,242
4	Pennsylvania	74,378	Minnesota	594,085	Texas‖	62,450
5	New York	28,087	Idaho	569,419	Ohio	54,427
6			Pennsylvania	366,972	New York	46,943
7			Iowa	263,552	Minnesota‖	44,609
8			South Dakota	146,897	Missouri	43,363
9			Washington	146,715	Florida	39,132
10			Ohio	146,184	Washington	24,950
All other states		366,450		1,174,707		700,451#
Total for the United States		1,236,801		8,129,094		1,387,877

*State figures not shown when less than three plants reported or when individual plant data might be disclosed; included in total for "All other states."
†All types of cheese except cottage cheese and lowfat cottage cheese.
‡Data includes regular and lowfat ice cream.
§Includes freezer-made milk drink.
‖Does not include lowfat ice cream because fewer than three plants reported or individual plant operations could not be disclosed.
#Individual state data on nonfat ice cream production not available. State data reflects only those products reported by the U.S. Department of Agriculture (USDA) for each state.
Sources: USDA, National Agricultural Statistics Service. Milk Facts, 2002 Edition, p. 18.

TABLE 4.4   Top Ten States for Production of Cottage Cheese and Nonfat Dry Milk, 2001*

Rank	Creamed Cottage Cheese (1,000 lbs)		Lowfat Cottage Cheese (1,000 lbs)		Nonfat Dry Milk for Human Use (1,000 lbs)	
1	New York	58,812	New York	88,661	California	694,481
2	Illinois	36,796	California	67,220	Washington	187,346
3	California	31,690	Ohio	29,434	Idaho	88,162
4	Ohio	19,447	Illinois	28,572		
5	Kentucky	17,227	Wisconsin	12,165		
6	Iowa	14,562	Iowa	12,027		
7	Indiana	14,478	Kentucky	9,451		
8	Pennsylvania	12,996	Pennsylvania	7,399		
9	Wisconsin	12,374	Indiana	5,928		
10	Missouri	11,423	Oregon	3,265		
All other states		118,917		104,430		443,810
Total for the United States		371,525		370,884		1,413,799

*State figures not shown when less than three plants reported or when plant data might be disclosed; included in total for "All other states."
Sources: USDA, National Agricultural Statistics Service. Milk Facts, 2002 Edition, p. 19.

Figure 4.2   Milk haulers deliver their product to the processing plant. *(Courtesy of Mark Kirkpatrick)*

Figure 4.3 A high-speed gallon filler. *(Courtesy of Iowa State University)*

Figure 4.4 Bulk milk transporters unload milk at the processing plant. *(Courtesy of Mark Kirkpatrick)*

Figure 4.5 Milk can be processed into a large variety of delicious dairy products. *(Courtesy of United Dairy Industry Association)*

Figure 4.6 After manufacturing, cheese is distributed to retailers for sale to consumers. *(Courtesy of Mark Kirkpatrick)*

Figure 4.7 Although increasingly rare, home delivery of dairy products is still available in some areas. *(Courtesy of Mark Kirkpatrick)*

Another type of retail outlet is the milk store operated by a large dairy, which demonstrates integration from production through processing and marketing. As dairy operations become larger and more integrated in the future, the number of these specialty markets will likely increase.

Over one-third of the milk marketed by dairy farmers today is consumed in fluid form (Table 4.5). Fluid milk is retailed as pasteurized milk, homogenized milk, fortified milk (vitamin D), skim milk, flavored milk (whole milk with flavor added), or flavored milk drink (skim milk with flavor added).

BUTTER

Butter is made from cream. As marketed, it consists of about 80 percent milk fat. The remainder is water, salt, and traces of other substances. It takes over 20 pounds of milk to produce each pound of butter. One of the major factors affecting butter consumption has been the development of margarine and increasing health concerns about high intakes of

TABLE 4.5 Per Capita Sales of Fluid Milk Items, in Pounds, 1980–2001

Year	Plain Whole Milk	Reduced and Lowfat Milk	Nonfat Milk	Flavored Milk and Milk Drinks	Buttermilk	Total
1980	137.5	70.1	11.6	10.0	4.1	233.3
1981	132.5	72.6	11.3	9.3	4.0	229.7
1982	126.7	73.5	10.6	8.6	4.1	223.5
1983	123.5	75.4	10.6	9.1	4.3	222.9
1984*	119.6	78.6	11.6	9.8	4.3	223.9
1985	116.7	83.3	12.6	9.7	4.4	226.7
1986	110.1	88.1	13.5	9.9	4.2	225.8
1987*	105.8	88.3	14.1	10.1	4.3	222.5
1988*	100.9	90.9	16.3	10.0	4.1	222.2
1989*	92.5	96.2	20.3	9.6	3.7	222.3
1990*	85.5	98.2	22.8	9.4	3.5	219.4
1991*	82.1	98.9	23.7	9.5	3.4	217.6
1992*	78.7	98.3	24.8	9.5	3.1	215.5
1993*	74.9	95.5	26.3	9.5	3.0	209.2
1994*	73.1	94.0	28.2	9.7	2.9	207.8
1995*	70.1	90.9	31.4	9.8	2.8	204.9
1996*	69.4	89.1	32.9	10.2	2.6	204.3
1997*	67.5	86.9	33.5	10.4	2.5	200.9
1998*	65.8	85.0	33.4	11.0	2.5	197.6
1999*	66.2	84.5	32.2	11.5	2.4	196.8
2000*	65.4	83.8	29.9	11.8	2.2	193.1
2001†	63.7	82.9	28.9	12.4	2.1	189.5

*Revised.
†Preliminary.
Source: USDA, Economic Research Service.

animal fats. As a result, per capita consumption of margarine surpassed butter in 1957, and currently about twice as much margarine is consumed than butter.

CASEIN

Casein, which is the major protein of milk, is found only in milk. It is obtained by acid or rennet coagulation of defatted milk. Casein contains a minimum of 80 percent crude protein and gives milk its white color. Casein is used as an ingredient in coffee whitener and whipped toppings, in baked goods, and as a source of protein in the manufacture of meat analogs and in the protein supplementation of some meat products. In 1987, the United States imported 108,136 metric tons of casein, which far exceeded imports of dried milk (1,301 metric tons) and of butter (905 metric tons).

CHEESE

Cheese is made by exposing milk to specific bacterial fermentation, or treating milk with enzymes (rennet), or using both methods to coagulate (curdle)

Figure 4.8 A vat containing curds and whey early in the cheese-making process. *(Courtesy of Mark Kirkpatrick)*

some of the proteins (Figure 4.8). The end result of the curdling process is the production of curds and whey (the liquid remainder) (Figure 4.9). It takes approximately 10 pounds of milk to produce each pound of cheese (Figures 4.10 and 4.11).

Milk is processed into many different kinds of cheese (Figure 4.12). Some are made from whole

Figure 4.9 Cheese curds are created through mechanical stirring. *(Courtesy of Mark Kirkpatrick)*

Figure 4.11 It takes 10 pounds of milk to produce 1 pound of cheese. *(Courtesy of USDA-ARS)*

Figure 4.10 Cheese products are packaged prior to shipping. *(Courtesy of Mark Kirkpatrick)*

Figure 4.12 Milk can be processed into many kinds of cheeses. *(Courtesy of United Dairy Association)*

milk, others from milk that has had part of the fat removed, and still others from skim milk. American types of cheese (cheddar, colby, washed curd, stirred curd, monterey, and jack) make up 60 percent of the nation's cheese output. The most important variety produced from skim milk is cottage cheese. Other important types of cheese are Italian (mostly soft varieties), swiss, muenster, brick, blue, and processed cheese. Over 30 percent of all the milk used in manufactured dairy products is processed into cheese (not including cottage cheese). The rising popularity of pizza in the United States accounts for much of the increase in cheese production and consumption in recent years (Figure 4.13).

Figure 4.13 The continuing increase in mozzarella cheese consumption is directly tied to increases in pizza consumption. *(Courtesy of United Dairy Industry Association)*

CONDENSED AND EVAPORATED MILKS

The primary products within this category are evaporated milk and condensed milk packed in cans for consumer use and condensed whole and skim milk shipped in bulk. Condensed and evaporated milk are manufactured by removing a major portion of the water from the whole milk in a machine called a vacuum pan. Condensed milk is further treated by the addition of large amounts of sugar.

Concentrating the fat portion of milk yields cream. Prior to the advent of the cream separator, this was accomplished by gravity separation. Today, milk is passed through a cream separator. Whipping cream contains about 40 percent fat; coffee or table cream, 18 to 20 percent; and half-and-half, 12 percent.

DRIED MILK (WHOLE MILK, SKIM MILK, AND WHEY)

Among the dried milk products produced from milk are nonfat dried milk (skim milk) for both human food and animal feed, dried whey for both human food and animal feed, and dried whole milk. Approximately 1 million pounds of each of these products are produced in the United States each year.

ICE CREAM AND SIMILAR FROZEN DESSERTS

Currently, 99 percent of all frozen desserts in the United States consist of ice cream, ice milk, sherbet, and mellorine (made with a vegetable fat base). Ice cream must be at least 10 percent milk fat (Figure 4.14), but other frozen products are lower in fat content. Other frozen desserts include frozen custard, frozen malted milk, artificially sweetened ice cream and ice milk, and water ices.

CULTURED MILK PRODUCTS

Numerous cultured and acidified milk products are sold as specialty dairy products. Among such products are yogurt, cultured buttermilk, cultured sour cream, and acidophilus milk. These products, especially yogurt and frozen yogurt (Figure 4.15), have become increasingly important as an outlet for the milk produced on dairy farms.

Figure 4.14   Premium ice creams may contain as much as 16 percent milk fat. *(Courtesy of United Dairy Industry Association)*

Figure 4.15   Frozen yogurt was introduced in the 1960s, and consumption has increased continually since then. *(Courtesy of United Dairy Industry Association)*

SUMMARY

- Dairy products include fluid milk, butter, cheese, condensed and evaporated milk, and frozen and other cultured products.
- Milk moves from the farm to the consumer in three stages: assembly and transport from farms to plants, processing and packaging or manufacturing products, and distribution to consumers.
- Stores can offer brands of milk processed by their own plants, by their competitors, or by specialty markets where the milk is sold by a large dairy operation.

QUESTIONS

1. What are three ways of gauging the profitability of processing farms?
2. What is the major protein of milk?
3. How is cheese made?
4. Besides fluid milk, which dairy product is utilized to the greatest extent?

ADDITIONAL RESOURCES

Articles

Bailey, K. W. "Dairy Processing." *Veterinary Clinicians of North American Food Animal Practices* 19 (July 2003): 295–317.

Kapoor, R., P. Lehtola, and L. E. Metzger. "Comparison of Pilot-Scale and Rapid Visco Analyzer Process Cheese Manufacture." *Journal of Dairy Science* 87 (September 2004): 2813–2821.

Internet

got milk: http://www.gotmilk.com.

The Basics of Making Cheese: http://www.efr.hw.ac.uk/SDA/cheese2.html.

5

Milk-Marketing and Pricing Systems

OBJECTIVES

- To list the factors influencing the price of raw, unprocessed products.
- To describe government controls in stabilizing the milk market and pricing.
- To describe the purpose of cooperatives.
- To list factors affecting milk imports and exports.

Milk production is seasonal in nature, and the economic principles of supply and demand dictate that price declines when milk availability increases (Figure 5.1). Milk production peaks in May and June and reaches its yearly low in November. Typically, milk price trends are a month behind milk production, with the lower yearly prices occurring in April to June and the highest in November or December. Knowledgeable dairy producers and dairy manufacturers follow market reports to project future trends in supply and demand, and they plan their programs accordingly.

MILK PRICING

Chaotic conditions in milk marketing, resulting from the breakdown of private controls and the serious economic plight of farmers during the depression years of the early 1930s, brought requests from organized producers and distributors for government control. Out of this crisis evolved two

Figure 5.1   A bulk milk hauler picks up a load of milk at a dairy farm. *(Courtesy of Mark Kirkpatrick)*

forms of government controls: those established by the federal government and those established by state governments. Both were designed to bring more stability into the marketing of milk. Today, federal and state agencies, directly or indirectly, still affect the pricing of milk marketed by dairy farmers in the United States.

Federal Milk-Marketing Orders

Federal milk-marketing orders were established and are administered by the secretary of agriculture under acts of Congress passed in 1933 and 1937. They are legal instruments, and they are very complex. Stated in simple terms, however, these orders are designed to stabilize the marketing of fluid milk and to assist farmers in negotiating with distributors for the sale of their milk. Prices paid to farmers are controlled, but there is no direct control of retail prices.

Federal orders are not concerned with sanitary regulations. They regulate the handling and pricing of about 80 percent of the Grade A milk; only Grade A milk is regulated by federal milk-marketing orders. Most of the remaining 20 percent of Grade A milk is regulated under state regulations. Each of the market orders has a market administrator and provisions for setting minimum farm prices and regulating transactions between farmers and milk dealers in their area. Prices in other Grade A markets are influenced by prices established under federal orders or state control programs.

Cooperatives

Cooperatives are groups of individual farmers, ranchers, and manufacturers or businesses that have similar interests that work together in marketing, shipping, and related activities to sell their product more efficiently. Because of the challenges inherent in dealing individually with a large number of producers, cooperatives were organized. These cooperative associations are of two general types:

1. Bargaining associations that make all business arrangements without actually handling any milk.
2. Associations that process and distribute milk or collect it for fluid use.

About 75 percent of the total deliveries of milk to plants and dealers in the United States are handled by cooperatives. In the last few decades, a large number of small cooperatives have merged together to form regional organizations that offer a more powerful position for bargaining and more efficient use of marketing channels. Cooperatives allow the dairy producers in a given area to integrate the various aspects of marketing. Procurement, assembling, marketing, and routing of milk are all handled by the cooperative. Thus, producers that are members of a cooperative are assured of a market for their milk.

Standards and Grades

The U.S. Department of Agriculture is responsible for the development of standards and grades for milk and dairy products. Over 90 percent of the nation's milk supply is Grade A, and about 45 percent of all Grade A milk that is sold is used for fluid milk products (beverage milk). Milk used for fluid products is designated (by federal orders) as Class I. Most orders have two other classes: Class II includes milk used for soft products, including fluid cream, ice cream, cottage cheese, and yogurt; Class III includes milk used for hard products, including cheese, butter, and nonfat dry milk.

Milk Pricing

The federal government has established a goal of eliminating government control of milk pricing. Currently, thirty-two federal milk-marketing orders are still used to assist in stabilizing milk prices, with the greatest impact in those areas of the country where Class I utilization of milk accounts for more than 50 percent of the total milk produced (Table 5.1). In 1995, the Minnesota-Wisconsin price was replaced by the basic formula price (BFP). The BFP is determined by the value of manufacturing grade milk that is utilized for butter, nonfat dry milk, and cheese in Minnesota and Wisconsin. The production of cheese is far greater than that of the other commodities; therefore, changes in the price of cheese have a great impact on the BFP.

As in the past, the price producers actually receive for their milk is far more complicated. Processors pay producers a blend price for milk based on the classified value of the products they produce. Prices received by producers are further complicated by Class I differentials, which are incentives to move fluid milk from a region of high milk production to a milk-deficit region and are unique to each federal milk-marketing order; quality premiums and/or reblended prices paid by milk-marketing co-ops; and the California make allowance, which was established to solidify

TABLE 5.1   Milk Cows, Milk Production, and Income by State, 2001

	Number of Milk Cows (Thousands)*	Milk per Cow (lbs.)	Milk Production (Million lbs.)	Sold to Plants and Dealers All Milk (Million lbs.)	Sold to Plants and Dealers Percentage Grade A†	Sold to Plants and Dealers Average Price ($ per Cwt)	Farm Cash Receipts from Milk and Cream Total Received (Million $)‡	Farm Cash Receipts from Milk and Cream Percentage of All Farm Receipts§
Alabama	21	14,286	300	297	100	16.90	49	1.5
Alaska	1	13,055	14	14	100	20.50	2	4.8
Arizona	140	20,679	2,895	2,882	100	14.70	359	15.7
Arkansas	35	12,343	432	419	100	16.10	64	1.3
California	1,590	20,913	33,251	33,215	99	13.94	3,707	14.3
Colorado	91	21,648	1,970	1,944	100	14.80	223	4.8
Connecticut	25	18,240	456	452	100	16.20	67	13.5
Delaware	9	16,778	151	150	100	16.10	20	2.7
Florida	153	15,758	2,411	2,409	100	17.80	384	5.4
Georgia	86	16,640	1,431	1,420	100	15.90	183	3.7
Hawaii	8	14,107	106	104	100	25.50	28	5.3
Idaho	366	21,194	7,757	7,724	99	13.50	762	21.3
Illinois	116	17,414	2,020	2,004	98	15.00	255	3.6
Indiana	153	16,732	2,560	2,533	96	16.10	309	6.7
Iowa	210	18,024	3,785	3,744	97	14.70	455	4.2
Kansas	93	17,312	1,610	1,599	99	14.50	177	2.2
Kentucky	128	12,969	1,660	1,628	99	16.20	229	6.8
Louisiana	54	11,704	632	620	100	16.00	95	5.2
Maine	38	17,211	654	649	100	16.30	93	18.4
Maryland	82	15,780	1,294	1,284	100	16.20	181	12.3
Massachusetts	21	17,048	358	354	100	16.20	52	13.1
Michigan	303	19,323	5,855	5,800	99	15.20	729	18.9
Minnesota	510	17,278	8,812	8,707	95	14.90	1,127	15.0
Mississippi	35	14,200	497	495	100	16.10	75	2.6
Missouri	145	13,441	1,949	1,921	96	14.90	270	5.9
Montana	19	18,211	346	337	100	15.10	42	2.3
Nebraska	72	16,056	1,156	1,144	97	14.60	145	1.6
Nevada	25	19,400	485	479	100	13.00	52	13.4
New Hampshire	18	17,944	323	320	100	13.00	43	27.7
New Jersey	14	16,643	233	230	100	16.10	32	3.9
New Mexico	268	20,750	5,561	5,504	100	14.80	644	30.9
New York	672	17,527	11,778	11,651	100	15.80	1,544	49.2
North Carolina	67	17,373	1,164	1,147	100	17.10	174	2.4
North Dakota	46	14,000	644	632	73	14.20	76	2.8
Ohio	260	16,612	4,319	4,294	92	15.20	559	12.7
Oklahoma	89	14,528	1,293	1,279	100	15.90	174	4.1
Oregon	95	18,074	1,717	1,709	99	15.50	207	6.8
Pennsylvania	599	18,112	10,849	10,794	99	19.60	1,521	37.7
Rhode Island	1	16,571	23	23	100	16.40	4	8.2
South Carolina	21	17,476	367	354	100	16.50	52	3.5
South Dakota	99	15,960	1,580	1,566	93	15.10	207	5.5
Tennessee	92	14,511	1,335	1,330	99	16.20	193	9.7
Texas	352	15,689	5,099	5,076	100	15.80	766	5.7
Utah	93	17,581	1,635	1,610	96	14.70	186	18.4
Vermont	153	17,431	2,667	2,647	100	15.80	367	72.3
Virginia	118	15,898	1,876	1,863	100	17.00	279	12.2
Washington	247	22,324	5,514	5,484	100	15.30	711	14.0
West Virginia	16	15,563	249	246	100	15.80	35	8.9
Wisconsin	1,292	17,182	22,199	21,914	95	14.80	2,688	51.1
Wyoming	5	14,000	63	62	77	14.20	8	0.9
Total for the United States	**9,115**	**18,139**	**165,336**	**164,072**	**98**	**15.05**	**20,608**	**10.6**

*Average number of farms during year, including dry cows but excluding heifers not yet fresh. Total may not add due to rounding.
†Percentage of milk sold to plants and dealers that is approved by health authorities.
‡Based on preliminary estimate by U.S. Department of Agriculture (USDA).
§Based on 2000 data.

Sources: USDA, National Agricultural Statistics Service. Milk Facts 2000, pp. 16–17.

the manufacturing capacity of a geographically isolated state.

In addition, multiple-component pricing breaks down the BFP to determine values for individual components of milk. Thirteen federal milk-marketing orders, representing nearly 60 percent of all milk produced under federal orders, currently use variations of multiple-component pricing to value milk. All component-pricing systems in use pay producers per pound of fat produced, and several systems pay per pound of protein produced. This system is beneficial to processors, especially in areas where large quantities of Class III products are produced; it is also beneficial to producers marketing high-solids milk.

Export markets continue to expand, especially in Mexico and the Far East. Recent estimates suggest that about 1 percent of the price received for milk in the United States is due to export sales. The recent reductions in subsidies paid by governments of the European Community (EC) to dairy producers in those countries that were required by the General Agreement and Tariffs and Trade (GATT) should effectively increase world prices and make the United States more competitive in the international market.

Changes occurring over the last few years have led to increased volatility in mailbox prices for producers. Reductions in price-support programs and commodity-purchasing programs have contributed most to this volatility, and most experts agree that the remaining changes planned over the next few years will not create further volatility in milk pricing. New opportunities in risk management and forward-contracting programs allow progressive producers to maintain profit margins. Producers utilizing futures contracts to lock in prices can easily reduce price volatility by 50 percent or more. Clearly, milk-marketing strategies are moving from government control to the responsibility of the individual producer. This change requires more educated marketing decisions by producers seeking to retain a competitive edge in the dairy industry.

Marketing is that all-important end of the line; it gives point and purpose to all that has gone before. Successful milk producers must understand milk markets and pricing systems, along with the factors affecting them, if they are to take full advantage of their financial opportunities. In the future, increasing competition will likely force marginal dairy operations, in all probability, out of business. Thus, it is imperative that milk producers be aware of all the factors that influence the price that they receive for their raw, unprocessed product.

TABLE 5.2 International Dairy Prices, (dollars per metric ton) 1995–2001[1]

Year		Butter	Cheese	Nonfat Dry Milk
1995	January	1,600	1,950	1,825
	July	2,225	2,280	2,200
1996	January	2,175	2,400	2,150
	July	1,650	2,425	1,875
1997	January	1,613	2,425	1,913
	July	1,675	2,400	1,650
1998	January	1,925	2,350	1,600
	July	1,875	2,225	1,413
1999	January	1,738	2,000	1,400
	July	1,338	1,900	1,225
2000	January	1,350	1,800	1,488
	July	1,275	1,875	2,075
2001	January	1,300	2,000	2,200
	July	1,438	2,250	2,063

Sources: USDA, Agricultural Marketing Service. Milk Facts 2002 Edition, p. 62.
[1]Freight on board Northern Europe, midpoint of range.

IMPORTS AND EXPORTS

Imports of several dairy products, including several types of cheese, butter, butter oil, butterfat mixtures, ice cream, frozen cream, nonfat dry milk, dried buttermilk and whey, evaporated milk, condensed milk, chocolate crumb, and animal feed with milk solids, are restricted by specific import quotas. Although not formally restricted, certain other dairy products may be limited by agreement between the United States and the exporting country. As long as domestic prices are above world prices (Table 5.2) and world supplies are ample, exporting countries will look to the United States as a possible market. As a result, import pressure will persist (Table 5.3). Yet it is expected that imports of many commodities will continue to be limited by quotas.

Again, because domestic prices are above world prices, exports of dairy products are rather small. U.S. exports of dairy products on a milk-equivalent basis amount to only about 1 percent of the total U.S. milk supply. The three main US dairy product exports, ranked in descending order of tonnage, are nonfat dry milk, cheese, and butter. The exports of nonfat dry milk are greater than the exports of all other dairy products combined.

Exports of dairy products will continue to be influenced by the availability of surplus products and foreign policy. A more active role in meeting food deficiencies in the less-developed countries of the world could increase total demand for dairy products and thus create a demand for export products from the United States (Table 5.4).

TABLE 5.3   World Trade in Dairy Products (Imports), 1995–2000*

	1995	*1996*	*1997†*	*1998†*	*1999†*	*2000‡*
Butter/butter oil						
World	846	788	869	765	880	900
European Union	72	96	92	88	105	103
Russia	246	126	190	83	179	226
Algeria	22	NA	NA	NA	NA	NA
Egypt	49	50	38	40	43	45
Morocco	22	28	16	16	NA	NA
Mexico	20	19	25	27	34	25
Iran	17	27	10	10	NA	NA
Jordan	22	15	15	15	3	NA
United States	1	5	13	27	15	13
Other countries	375	432	476	469	512	500
Nonfat dry milk						
World	1,190	950	1,050	950	1,000	950
European Union	43	61	74	65	73	78
Russia	0	NA	NA	31	109	30
Japan	87	75	73	57	57	52
Philippines	104	79	98	76	87	96
Brazil	54	34	41	42	46	41
Mexico	107	127	133	93	110	105
Algeria	108	55	79	87	71	70
Other countries	687	519	552	499	447	478
Whole milk powder						
World	1,165	1,100	1,225	1,200	1,140	NA
Russia	0	NA	NA	35	35	40
Algeria	75	78	91	105	108	110
Brazil	217	116	105	133	147	125
Mexico	30	30	30	46	45	45
Venezuela	66	66	56	60	67	60
Saudi Arabia	63	69	63	64	NA	NA
Malaysia	60	62	65	50	NA	NA
Singapore	30	26	26	20	NA	NA
Philippines	36	42	52	47	37	58
Other countries	588	611	737	640	701	NA
Cheese						
World	1,036	1,080	1,150	1,100	1,100	1,150
European Union	84	97	109	127	146	147
Switzerland	30	31	31	31	31	31
Russia	85	110	195	85	60	90
United States	157	152	140	163	198	192
Brazil	89	34	29	24	20	18
Egypt	17	15	16	14	16	15
Japan	153	164	168	178	181	197
Iran	26	NA	NA	NA	NA	NA
Saudi Arabia	58	74	52	52	64	NA
Australia	27	33	32	31	33	38
Other countries	310	370	378	395	411	422

*NA = not available.
†Revised.
‡Preliminary.
Sources: International Dairy Federation. Milk Facts 2002 Edition, p. 69.

TABLE 5.4   World Trade in Dairy Products, Exports in Thousands of Metric Tons, 1995–2001*

	1995	1996	1997†	1998†	1999†	2000‡	2001‡
Butter/butter oil							
World	787	761	754	770	781	850	850
European Union	216	189	219	162	169	173	170
United States	38	21	21	11	5	16	NA
Australia	58	64	100	106	118	125	139
New Zealand	235	237	314	317	277	336	330
Other countries	214	250	250	100	174	212	200
Nonfat dry milk							
World	1,151	958	1,079	951	1,162	1,140	1,000
European Union	376	227	282	175	272	256	180
Canada	NA	45	30	31	41	29	NA
United States	59	22	92	111	183	84	80
Australia	188	168	205	199	243	244	235
New Zealand	138	127	183	166	174	172	170
Other countries	3,902	369	286	269	248	255	475
Whole milk powder							
World	1,284	1,154	1,302	1,392	1,420	1,430	1,450
European Union	596	540	571	588	576	573	550
United States	64	16	27	20	18	25	15
Australia	93	93	109	110	139	169	190
New Zealand	318	278	341	359	362	393	435
Other countries	213	282	316	412	475	170	260
Cheese							
World	1,094	1,140	1,225	1,203	1,350	1,450	1,480
European Union	528	517	511	448	398	457	450
Switzerland	64	62	61	56	60	58	NA
Australia	112	123	138	167	202	212	218
New Zealand	153	173	236	232	240	249	250
United States	33	36	37	37	38	47	NA
Other countries	204	229	242	263	413	427	562
Casein							
World	160	155	160	175	185	190	NA
European Union	65	62	58	59	61	79	NA
Poland	4	2	6	8	9	9	NA
Australia	7	5	6	7	9	9	NA
United States	0	0	4	7	5	NA	NA
New Zealand	77	75	81	94	98	103	NA
Other countries	8	11	5	1	10	NA	NA

*NA = not available.
†Revised.
‡Preliminary.
Sources: International Dairy Federation. Milk Facts 2002 Edition, p. 68.

SUMMARY

- Two forms of government regulate milk pricing, bring stability into marketing, and assist farmers in negotiating with distributors: the federal government and the state governments.
- Some cooperatives are designed to help dairy producers either make business arrangements without handling any milk or process and distribute milk for fluid use.
- Marketing gives point and purpose. To be successful, producers must understand milk markets, pricing systems, and the factors affecting both.
- Imports are restricted by specific quotas, and U.S. exports are rather few in number.

QUESTIONS

1. During which months does milk production peak?
2. To which class does fluid milk belong?
3. Grade B milk is used for what purpose(s)?
4. Name three main U.S. dairy export products.

ADDITIONAL RESOURCES

Articles

Dalton, T. J., G. K. Criner, and J. Halloran. "Fluid Milk Processing Costs: Current State and Comparisons." *Journal of Dairy Science* 85 (April 2002): 984–981.

Macmillan, K. L., and A. H. Kirton. "Impact of Exporting Dependence on Livestock Production Systems, Industry Structure, and Research" (review). *Journal of Animal Science* 75 (February 1997): 522–532.

Internet

Milk Money: How Dairy Cooperative Impact Farm-Level Milk Prices: http://www.wisc.edu/uwcc/info/farmer/pre2001/milk.html.

Milk Pricing in the United States: http://www.ers.usda.gov/publications/aib761/aib761.pdf.

Opening Up Global Dairy Trade: http://www.aae.wisc.edu/www/pub/dairyland/rd6.pdf.

6

Employee Management

OBJECTIVES

- To stress the importance of employee management for a successful dairy business.
- To list the essential features of an incentive plan.
- To consider the importance of insurance for a dairy operation.

Whether managing a small family farm with a single paid employee or owning a multimillion dollar diversified dairy conglomerate with hundreds of employees, the underlying principles of successful people management are the same. Management entails accomplishing work goals through the labor and skills of others. The means of accomplishing this objective is what distinguishes an effective manager from an ineffective one.

People management skills have become more important as dairy expansion has dominated the industry. Many smaller dairy operations have been successfully managed by one person: the owner-operator. To position their facility for the future, many of these people have explored opportunities to expand their facilities. Following expansion, they often find that they are no longer managing cattle but instead are primarily managing people. Despite tremendous knowledge of dairy cattle, managers who lack people skills or motivational skills may find that they cannot extract the same level of performance from their herd. Successful managers must study and understand the needs and reasons for the behavior of their employees with the same zeal that they studied the needs and reasons for the behavior of their cows.

Four major ingredients are essential to success in the dairy business: good cattle, good feeding, good management, and good records. A manager can make or break any dairy enterprise. Effective managers must have the ability to establish and maintain effective relationships with employees and prospective employees. It is important that workers know to whom they are responsible and for what they are responsible. The bigger and the more complex the operation, the more important this information becomes. It should be written down in an organization chart (Figure 6.1).

On operations where multiple employees do the same jobs (milking, feeding, etc.), it is especially crucial that every employee performs these tasks in exactly the same way. Cows thrive on routine, and any break in routine imparts stress to the cow and affects her performance. The easiest way to ensure this consistency is to create written protocols or standard operating procedures for every procedure on the operation. In that way, every employee understands exactly what is expected. These protocols should be written as simply as possible and in as much detail as possible to avoid any misinterpretation or misunderstanding. Every employee should have a copy of the protocols for any job he or she might perform. In addition, laminated copies of protocols should be posted in areas where the jobs are actually performed. For example, milking protocols should be posted in the parlor, and feed mixing protocols should be posted on mixer wagons.

Miscommunication between the manager and employee regarding job expectations is the primary cause of employee dissatisfaction, ranking well above salary issues. All communications should be discussed first in person and then followed with a written summary of the discussion. Managers and employees who rely strictly on face-to-face communications often walk away with different interpretations of their meeting or may have different memories of the discussion at a later date. Written follow-ups prevent, or at least minimize, these potential sources of job friction. Conversely, managers who rely strictly on written communication are often perceived as distant and uncaring by their employees. A combination of communication methods is the best approach to maintaining a stress-free workplace.

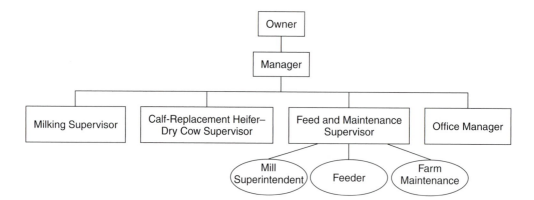

Figure 6.1   A simplified example of a dairy management flowchart. *(Courtesy of M. E. Ensminger)*

Even in situations where communication is excellent, employee performance sometimes does not meet expectations. Good managers should assess the reasons for this poor performance rather than simply dismissing the employee. In cases where employee's have a positive attitude and effort is put forth, the employee may simply be in a position that doesn't fit his or her skills or personality. Occasionally, changing job responsibilities can put an employee in a better position to utilize strengths. The ability to evaluate employees' strengths and weaknesses and to place people in situations where they are most likely to be successful is the primary trait separating good managers from poor managers.

Surveys of dairy employees have indicated that the most important factor affecting job satisfaction is whether or not they feel they are valued as an asset to the operation. This is more important than salary, work hours, and work conditions. Therefore, a management priority is to create a team environment where every employee has the opportunity for input. This doesn't mean that employees should be allowed to do their jobs any way they want; it simply means that they must feel that their input is valued and that the successes of the operation are, in part, their successes. It doesn't matter if the manager is the smartest dairy producer in the world and the dairy is the most successful operation in the world; the employees will not feel part of that success if they don't feel like they are valued employees. There is an old adage that employees must "know how much you care before they care how much you know," and this holds true whether you are managing a single person or managing a corporation with hundreds of employees.

Labor costs average 20 percent of the total cost of milk production on average, and the agricultural workforce situation is going to become more difficult in the years ahead. An incentive basis makes hired help partners in profit and can provide a strong motivation for employees to treat the operation as their own. Many manufacturers have long operated on an incentive basis. Executives are frequently accorded stock-option privileges, through which they prosper as the business prospers. Laborers may receive bonuses based on piecework or quotas (number of units, pounds produced). Also, most factory workers get overtime pay and have group insurance and a retirement plan. A few industries have a true profit-sharing arrangement based on net profits, a specified percentage of which is divided among employees. No two systems are alike; yet each is designed to pay more for labor, provided labor improves production and efficiency. In this way, both owners and employees benefit from better performance. Family-owned and family-operated farms have a built-in incentive basis; there is pride of ownership, and all members of the family are fully cognizant that they prosper as the business prospers. Many different incentive plans can be used. One is not best for all operations. The incentive basis chosen should be tailored to fit the specific operation with consideration given to the size of the operation, extent of the owner's supervision, present and projected productivity levels, mechanization, and other factors.

Normally, we think of incentives as monetary in nature, as direct payments or bonuses for extra production or efficiency. However, there are other ways of encouraging employees to do a better job. The latter are known as indirect incentives, and they include housing allowances; the use of the farm truck or car; utility allowances; vacation time with pay; time off; sick leave; group health insurance; job security; the opportunity for self-improvement; the right to invest in the business; paid trips to short courses, shows, or conventions; and year-end bonus for staying all year. These indirect incentives are critical to the success of a dairy operation. Employees cite mutual respect between themselves and their employers and provision of reasonable amounts of both responsibility and authority as the primary factors leading to job satisfaction (Figure 6.2).

Figure 6.2 Effective managers inspire outstanding employee performance rather than inspiring fear of failure. *(Courtesy of Dana Boeck)*

Figure 6.3 Managers and employees should work together as a team to set farm goals. *(Courtesy of Iowa State University)*

INCENTIVE PAY

After reaching a decision to operate on an incentive basis, it is necessary to calculate how much to pay. Here are some guidelines that may be helpful:

1. Pay a base, guaranteed salary and then add the incentive pay above this amount.
2. Determine the upper limit of the total stipend (the base salary plus incentive).
3. Check the plan on paper to see how it would have worked out in past years based on the records and how it will work out as goals are achieved.

Owners should always start with a simple plan; a change can be made to a more inclusive and sophisticated plan with experience. Regardless of the incentive plan adopted for a specific operation, it should encompass the following essential features:

1. It must be fair to both employer and employees.
2. It must compensate for extra performance rather than substitute for a reasonable base salary and other considerations (house, utilities, and certain provisions).
3. It must be as simple, direct, and easy to understand as possible.
4. It should compensate all members of the team. For example, without the cooperation of the milkers, no dairy farm incentive program will succeed (Figure 6.3).
5. It must be in writing to avoid misunderstanding. For example, if some production-sharing plan is used in a market milk operation, it should stipulate the ration (or who is responsible for ration formulation).

6. It is preferable, although not essential, that workers receive incentive payments at frequent intervals, rather than annually, and immediately after accomplishing the incentive-laden goal.
7. It should give the employee a certain amount of responsibility from which she or he will benefit through the incentive arrangement.
8. It must be backed up by good records; otherwise, there is nothing on which to base incentive payments.

LIABILITY INSURANCE AND WORKERS' COMPENSATION INSURANCE

Most farmers are in a financial position that leaves them vulnerable to damage suits. The number of damage suits each year is increasing at an almost alarming rate, and astronomical damages are being claimed. Roughly 95 percent of the court cases involving injury result in damages being awarded. Several types of liability insurance offer a safeguard against liability suits brought as a result of injury suffered by another person or damage to his or her property. Comprehensive personal liability insurance protects farm operators who are sued for alleged damages suffered from an accident involving their property or family. The kinds of situations from which a claim might arise are quite broad, including suits for injuries caused by animals, equipment, or personal acts.

Both workers' compensation and employer's liability insurance protect farmers against claims or court awards resulting from injury to hired help. Workers' compensation usually costs slightly more than straight employer's liability insurance, but it carries more benefits to the worker. An injured employee must prove negligence by the employer before the company will pay a claim under employer's liability insurance, whereas workers' compensation benefits are established by state law and settlements are made by the insurance company without regard to whose negligence caused the injury. Conditions governing participation in workers' compensation insurance vary among the states.

SUMMARY

- Management entails accomplishing work goals through the labor and skills of others.
- People management skills have become more important as dairy expansion has dominated the industry.
- Successful managers must study and understand the needs and reasons for the behavior of their employees with the same zeal that they studied the needs and reasons for the behavior of their cows.
- Effective managers must have the ability to establish and maintain effective relationships with employees and prospective employees.
- The easiest way to ensure routine is to create written protocols for every procedure on the operation.
- Written follow-ups to oral communications prevent, or at least minimize, potential sources of job friction.
- Several types of liability insurance offer a safeguard against liability suits brought as a result of injury suffered by another person or damage to his or her property.

QUESTIONS

1. What are the four major ingredients that are essential to success in the dairy business?
2. On operations where multiple employees do the same jobs, who should receive a copy of the protocols and where should they be placed?
3. What is the primary cause of employee dissatisfaction?
4. What are three indirect incentives that a dairy operation may use to encourage employees to do a better job?
5. What are three helpful guidelines used to determine how much to pay an employee?
6. What are five essential features that an incentive plan should encompass?
7. How does comprehensive personal liability insurance protect farm operators?

ADDITIONAL RESOURCES

Articles

Waller, J. A. "Injuries to Farmers and Farm Families in a Dairy State." *Journal of Occupational Medicine* 34 (April 1992): 414–421.

Wolf, C. A. "The Economics of Dairy Production" (review). *Veterinary Clinicians of North America Food Animal Practices* 19 (July 2003): 271–293.

Internet

Agricultural Labor Management: Employee Incentive Pay in Dairies: http://www.cnr.berkeley.edu/ucce50/ag-labor/7dairy/7dairy.htm.

Managing Dairy Labor: http://ianrpubs.unl.edu/dairy/g1064.htm.

2

Concepts in Genetic Improvement

7

Dairy Records and Programs

OBJECTIVES

- To identify the various methods of animal identification.
- To describe record systems, correction factors, terminology, and registry association programs.
- To describe the limitations of records.

Genetic progress in the dairy industry has proceeded at a remarkable rate since the development of technology for freezing semen. The ability to introduce the best genetics into any herd at a reasonable price has made the dairy industry the model for the systematic improvement of a species. Genetic progress depends on several factors: accurate identification of animals, the ability to identify superior animals accurately, and the ability to utilize those animals most effectively. The decisions that are made can be no better than the information used to make them. A thorough, accurate record-keeping system is the foundation on which decisions can be based with confidence.

IDENTIFICATION

A critical component of any dairy record program or any program for genetic improvement is the accurate identification of each animal in the herd. Daily management decisions concerning breeding, feeding, selection, calving, and culling depend on identification of animals. Proper identification is also necessary for keeping records of cows on official production testing programs and for registration of purebred cattle in the breed registry associations.

Identification of animals by appearance (size, color patterns, or other distinguishing marks) is usually satisfactory in a small herd of fewer than fifty animals; however, some permanent system of iden-

tification is essential in larger herds. Also, positive identification of registered animals is necessary. For registration, the broken-colored breeds (Ayrshires, Guernseys, and Holsteins) require sketches of color markings or photographs, while the Jersey and Brown Swiss breeds require permanent ear tattoos. These identifications are listed on the registration papers. Other methods of identification include ear, leg, and tail tags; neck chains; hide brands; transponders or other electronic identification methods; brisket tags; and marking paint. Each of the main methods of identification will be discussed briefly.

Ear Tags

Ear tags are the most widely used means of identifying dairy animals, especially grade cattle. They are made of steel, aluminum, nylon, or plastic (Figure 7.1). Ear tags are attached easily, and the plastic and nylon tags do not cut through the ears as easily as do the sharper metal varieties.

Hide Brands

When applied properly, hide brands are permanent and easily read. The three most common methods of applying hide brands are:

1. **Freeze branding.** This method makes use of a super-chilled (by dry ice or liquid nitrogen) copper branding iron (Figure 7.2), which is applied to the closely clipped surface for about 20 seconds, thereby depigmenting the hair follicles. After the procedure, the hair grows out white. When done properly, this method is painless and permanent, and there is no hide damage. On white cattle, deliberate overbranding (30 seconds or more) produces a bald brand useful for identification after clipping.

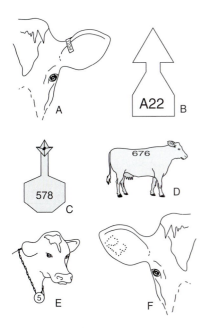

Figure 7.3 Hot branding is still used for permanent identification of animals on some dairies. *(Courtesy of Howard Tyler)*

Figure 7.1 Identification methods: A, metal ear tags; B, blank ear tags; C, prenumbered ear tags; D, brands; E, neck chains; and F, tattoos. *(Courtesy of M. E. Ensminger)*

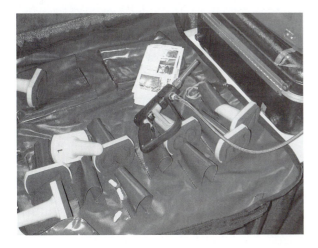

Figure 7.2 A freeze brand is a permanent method of animal identification. *(Courtesy of Leo Timms)*.

2. **Hot iron.** This method is the traditional brand of the west. Irons are heated to a temperature that burns sufficiently deep to make the scab peel but does not leave deep scar tissue (Figure 7.3). The proper temperature of the hot iron is indicated by a yellowish color. Branding is accomplished by placing the heated branding iron firmly against the body area to mark and by not allowing it to slip for the few seconds when the hide is burned. The branding iron should be kept free from dirt and adhering hair at all times. When electric-

ity is available, the electric iron may be used; it keeps an even temperature. If used properly, it makes a clear, uniform brand.

3. **Branding fluids.** Branding fluids, which are less widely used in making hide brands, consist of caustic material applied by means of a cold iron. Best results are obtained if the area is first clipped. The chemical method of producing hide brands is slower than the hot iron and the results are generally less satisfactory, particularly if the operator is inexperienced with the method. In addition, the resulting brand is less permanent.

A good hide brand, regardless of how it is applied, is one that is easily read, cannot be easily changed or tampered with, and interferes with the circulation as little as possible.

Neck Chains or Straps

Neck chains or straps are the most common temporary identification of dairy cattle. Occasionally, they may be lost, but this is not particularly serious if the caretaker replaces each one that is lost immediately, without allowing several losses to accumulate before taking action. In rare instances, an animal will hang itself by the chain. Neck chains or straps must be adjusted as young animals grow or as animals change in condition.

Tattoos

Most purebred dairy cattle registry associations require that registered animals be individually tattooed. This method of marking consists of piercing the skin with instruments equipped with needle points that form letters or numbers; indelible ink is then rubbed onto the freshly pierced area. The tattooing instrument should be disinfected carefully between each operation. A major disadvantage of tattooing as the sole means of identification is that cattle must be restrained so that anyone can read tattoo numbers. Even then, tattoos are difficult to decipher on dark-skinned animals.

Electronic Devices

Various electronic devices are in different stages of research and development. They include the following:

1. **Radio transmitter in the reticulum.** The animal swallows a small radio transmitter enclosed in a ¾ in. × 2½ in. plastic capsule, which lodges in the reticulum. From there, it transmits a coded number when signaled by a receiving unit. The transmitter can be retrieved at slaughter and reused.

2. **Transponder.** The transponder is a new technology. It can be used on livestock or machines for identification, tracking, and theft recovery. On dairy cows, it can be used to identify each individual cow for a grain feeder and in the milking parlor. The transponder consists of an electromagnetic coil and microchip in a glass capsule, which varies in size from about the size a grain of rice to much larger. The transponder has no power source of its own. A reader emits a magnetic field that activates the transponder so that it transmits its code number (Figure 7.4). The transponder may be implanted just below the skin of the animal or on a cow's neckstrap (Figure 7.5). Although not currently used, transponders may replace other methods of animal identification in the future.

3. **Radio frequency identification (RFID).** In response to increased concerns regarding foreign disease outbreaks, such as foot and mouth disease or bovine spongiform encepholopathy, there has been a concerted effort to create a national identification system that would electronically track dairy cattle from birth through slaughter. The goal of the National Animal Identification System is to have the capability to identify all animals and premises that had direct contact with a foreign animal disease (FAD) within 48 hours after discovery. The system currently being tested integrates both a premises identification system and an individual animal numbering system. Initially, implementation of the new system will be on a voluntary basis; however, the intent is to eventually implement a

Figure 7.4 As cows enter the milking parlor, an identification reader transmits cow identification to electronic milk-monitoring devices at each stall. *(Courtesy of Howard Tyler)*

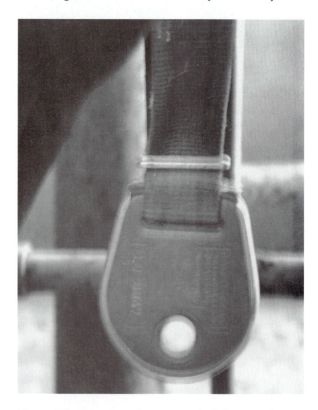

Figure 7.5 A transponder on a neck chain transmits cow identification information to a parlor reader. *(Courtesy of Howard Tyler)*

mandatory system for animal tracking. The system incorporates a RFID eartag that is placed in the left ear of each animal and a RFID handheld reader that connects directly to a personal computer.

RECORD SYSTEMS

Production records are the most important management tool on the dairy farm (Figure 7.6). The primary purpose of dairy records is to give the producer detailed information on individual cows, which can then be used to make daily management decisions; allot concentrates to cows; breed, dry off, and cull cows; and provide treatments as necessary. Records are also essential to evaluate the status of the dairy herd. A summation of the records on a regular basis allows the producer to determine the strengths, weaknesses, and profitability of the operation. Such summations make possible an informed evaluation of past management practices and long-range planning for the years ahead. The desirable characteristics of a dairy record-keeping system are that it be simple, complete, accurate, up to date, and understandable and that it require a minimum amount of time to maintain.

The National Cooperative Dairy Herd Improvement (DHI) program is a voluntary cooperative effort to improve the level and efficiency of milk production and increase dairy profits. It involves milk producers, local and state DHI organizations, extension services of land grant colleges and universities, and the U.S. Department of Agriculture (USDA). The U.S. Department of Agriculture aids in conducting and distributing the results of the sire evaluation phase of the DHI program. It also coordinates, furnishes materials, provides statistical information, analyzes data, and researches various aspects of the program.

State and local dairy herd improvement associations (DHIAs) conduct the program among producers, working through the Cooperative Extension Service in cooperation with the Federal Extension Service and Animal Science Research Service of the USDA. Dairy cattle record-keeping plans may be either official or unofficial, with alternate choices under each grouping. Approximately 30 percent, of cows enrolled in DHI programs are enrolled in unofficial programs. These less expensive options require less supervisor oversight and (or) less milk sample testing.

Official DHI testing includes the Standard Dairy Herd Improvement Association (DHIA) and the Dairy Herd Improvement Registry

Figure 7.6 Dairy records are an invaluable resource for cow and herd evaluation. *(Courtesy of Leo Timms)*

(DHIR). Because both programs are official, a supervisor tests herds four to twelve times annually. Records from both programs are used in proving dairy sires.

Dairy Herd Improvement Association (DHIA)

This program, first adopted in 1926, is the most complete of all dairy production and record plans. Approximately half of the cows in the United States on production test are part on this program (Tables 7.1 and 7.2). Both registered and grade cows can be enrolled. In this program, a supervisor or tester employed by the local or state testing association visits the herd one day each month (Figure 7.7). The tester identifies all cows in the herd, weighs and takes representative samples of the milk from all animals in the herd for two consecutive milkings (three milkings for herds on three-times-daily milking), and then combines the milk samples and sends them to a central testing laboratory for analyses of components such as butterfat, protein, and somatic cell count (SCC) (Figure 7.8). Records are obtained on individual cows based on monthly and cumulative records for milk, fat, and protein; for amount of feed, cost of feed, and income over feed cost; and for breeding dates, calving dates, dry dates, and other factors affecting productivity. In some testing associations, somatic cell counts or results of the California Mastitis Test (CMT) are provided as an aid in monitoring udder health (Table 7.3). All of this information is fed into a computer, which is programmed to provide monthly summaries of both individual cows and the whole herd. This information is provided to the producer.

TABLE 7.1   Dairy Cow Enrollment in DHI, by Test Plan, as of January 1, 2004

Plan Tag	Herds	Cows	Percentage of DHI Cows	Plan Tag	Herds	Cows	Percentage of DHI Cows
DHI	2,125	641,286	15.8	DHI-OS-MO-AP	48	9,754	0.2
DHI-AP-T	1,632	400,795	9.8	DHI-OS-ACT	17	2,544	0.1
DHI-APCS	1,064	356,353	8.8	DHI-OS-AP-ACT	5	519	0.0
DHIR	1,159	146,495	3.6	DHI-comm	25	19,131	0.5
DHIR-AP-T	826	129,757	3.2	DHI-comm-AP	112	57,287	1.4
DHIR-APCS	244	73,338	1.8	DHI-comm-MO	4	3,419	0.1
DHIR-AP	1,445	149,638	3.7	DHI-SS	27	6,615	0.2
DHI-AP	11,104	1,478,539	36.3	DHI-SS-AP	477	100,088	2.5
DHI-MO	54	18,113	0.4	DHI-SS-APCS	34	15,683	0.4
DHI-MO-AP	79	22,072	0.5	DHI-SS-MO	18	4,286	0.1
DHI-OS	1,536	105,104	2.6	DHI-SS-MO-AP	64	18,815	0.5
DHI-OS-AP	3,691	253,364	6.2	DHI-basic	126	8,017	0.2
DHI-OS-APCS	254	32,495	0.8	Other DHI	3	755	0.0
DHI-OS-MO	63	16,837	0.4	All plans	26,236	4,071,099	100.0

DHIA TESTING PLANS

Code Plan Description

00 DHI The DHI technician visits the farm for each milking in a 24-hour period. All milkings are weighed and sampled each test day.

01 DHI-AP-T (AM-PM with Time Monitor). For herds milked twice daily, milk weights and samples are obtained at the AM or PM milking, alternating each test period. For herds milking all cows three times daily, two consecutive milkings are weighed, one or two (optional) sampled. Weighted and sampled milkings are rotated among the milkings on subsequent test days. Appropriate adjustment factors are applied to each cow's production to determine the 24-hour total. Herd must have a time monitor installed.

02 DHI-APCS For herds milked twice daily, milk weights are obtained at both AM and PM milkings, but samples are taken at the AM or PM milking alternating on consecutive test days. For herds in which all cows are milked three times daily, milk weights are obtained from all three milkings, but samples are taken from one or two (optional) milkings that are rotated among all three milkings on consecutive test days. Appropriate adjustments factors are applied to determine the component percentage.

20 DHIR (Dairy Herd Improvement Registry). Application for DHIR testing must be made by the dairyman to the national breed registry organization. The herd cannot be enrolled on DHIR until the breed organization issues a permit for the herd. Same protocol as Code 01 with additional rules imposed by breed associations.

21 DHIR-AP-T (Dairy Herd Improvement Registry AM-PM with Time Monitor). This test is the same as Code 01 except the herd must be enrolled in DHIR with the national breed registry organization as in test code 20.

22 DHIR-APCS (Dairy Herd Improvement Registry with Alternate AM-PM Component Sampling). This test is the same as Code 02 except the herd is enrolled in DHIR with the national breed registry organization as in test code 20.

23 DHIR-AP (Dairy Herd Improvement Registry AM-PM without Time Monitor). This test is the same as Code 01 except a time monitor is not required, and the herd must be enrolled in DHIR with the national breed registry organization as in test code 20. Code 23 may not be accepted by all breed associations.

31 DHI-AP (AM-PM without Time Monitor). This test is the same as Code 01 except a time monitor is not required. Instead, test day/milk shipped relation must be monitored, and Record Standards data must be reported.

33 DHI-MO (Milk Only). This test is the same as Code 00 without component sampling.

34 DHI-MO-AP (Milk Only AM-PM). This test is the same as Code 01 without component sampling.

40 DHI-OS (Owner Sampler). Milk weights and samples are obtained from each milking in a 24-hour period. The dairy farmer or someone other than the technician is responsible for taking the milk weights and samples.

41 DHI-OS-AP (Owner Sampler AM-PM). This test is the same as Code 31 except the dairy farmer or someone other than the technician is responsible for taking the milk weights and samples. Records are not used in USDA sire summaries.

42 DHI-OS-APCS (Owner Sampler with alternate AM-PM Component Sampling). This test is the same as Code 02 except the dairy farmer or someone other than the technician is responsible for taking the milk weights and samples.

43 DHI-OS-MO (Owner Sampler Milk Only). The dairy farmer or someone other than the technician is responsible for taking milk weights for all milkings in a 24-hour period. No samples are taken.

44 DHI-OS-MO-AP (Owner Sampler Milk Only AM-PM). The dairy farmer or someone other than the technician takes milk weights for one milking in a 24-hour period. No samples are taken.

45 OS-BA (Owner Sampler with Breed Average fat test). The dairy farmer or someone other than the technician takes milk weights for all milkings in a 24-hour period. No samples are taken, and the breed average fat test is used for each individual cow.

46 OS-AP-PT (Owner Sampler AM-PM with Plant fat test). This test is the same as Code 44 except the herd's plant fat test is used for each milking cow in the herd.

47 OS-TC (Owner Sampler with Tri-monthly Component tests). The dairy farmer or someone other than the technician takes milk weights for all milkings in a 24-hour period. Samples are taken from all milkings every third test day.

70 DHI-SS (Supervisor Sampled). The DHI technician weights and samples all milkings each test day. One or more of the NCDHIP Rules are not followed.

71 DHI-SS-AP (Supervisor Sampled AM-PM). The technician takes milk weights and samples from one milking only. The milking weighed and sampled is rotated each test period. One or more of the NCDHIP Rules are not followed.

72 DHI-SS APCS (Supervisor Sampled with alternate AM-PM Component Sampling). The technician obtains milk weights from each milking in a 24-hour period, but samples only one. The milking sampled is rotated each test period.

73 DHI-SS-MO (Supervisor Sampled Milk Only). The technician obtains milk weights from each milking in a 24-hour period, but takes no samples. One or more of the NCDHIP Rules are not followed.

74 DHI-SS-MO-AP (Supervisor Sampled Milk only AM-PM). The technician obtains milk weights from one milking only. No samples are taken. One or more of the NCDHIP Rules are not followed.

80 DHI-basic A family of testing plans offering basic management information. These are operated differently in each DHIA, with data processed either at a DRPC or by the DHIA on an in-house computer.

Source: Modified from USDA-AIPL: http://www.aipl.arsusda.gov/publish/dhi/current/lacavx.html.

TABLE 7.2 Dairy Herd Testing Characteristics, 2003

Herds	Cows	Test Days/ Herd Year	Supervised (%)	Sampled (%)	All Milkings Weighed (%)	All Milkings Sampled (%)	Where > 1 Day Recorded (%)	Three-Times-Daily Milking (%)
26,253	4,114,037	11.4	87.9	95.0	35.8	22.7	5.5	26.6

Figure 7.7 DHIA testers record individual cow milk production and collect milk samples for laboratory evaluation. *(Courtesy of Mark Kirkpatrick)*

Figure 7.8 Preserved individual milk samples are packaged and ready for shipping to an approved DHIA testing laboratory. *(Courtesy of Mark Kirkpatrick)*

TERMINOLOGY

In addition to the monthly record provided by a testing program, it is important for determining genetic progress to have lifetime records on each individual cow. This should provide complete identification of the individual animal, individual lifetime lactation summaries, breeding records, calving records, and health and veterinary records. Production records are used for cow evaluation. The terminology relevant to this evaluation is discussed as follows:

TABLE 7.3 Herd and Cow Participation in Protein Reporting and Mastitis Screening, 2004

	Protein	Mastitis
Herds	25,576	25,255
Animals	3,762,512	3,777,834

Source: Modified from USDA-AIPL: http://www.aipl.arsusda.gov/publish/dhi/current/.

Lactation record: A lactation record is simply the total pounds of milk and fat (and sometimes protein) that a cow produced from the date she calved until her production was terminated by a dry date or other reason (Figure 7.9).

Mature equivalent record: This formula standardizes lactation records to 305 days in length, twice-a-day milking, an average environmental month of calving, and a mature age (five to eight years). The resulting standardized lactation is called a 305-2×-ME record. This standardization increases heritability estimates and predictive values associated with lactation records.

Herdmates: Herdmates are defined as all cows of the same breed that calved in the same herd during the same year, season, or general period of time. Therefore, herdmates are all cows producing under the same environmental conditions, assuming there is no preferential treatment given to any individual cows.

Yield deviation: This value compares a cow's 305-2×-ME with the average of her herdmates. This figure may be positive or negative, depending on whether the cow outproduced her herdmates or not.

CORRECTION FACTORS

Good culling decisions depend on the ability to compare accurately the performance of individuals or groups of animals. To accomplish this, all records must be adjusted or corrected to a comparable basis. Correction factors have been developed for each breed to adjust for the length of lactation, the number of milkings per day, age and month of calving, and the fat content of milk. These four adjustments are important for comparing the milk and fat of cows managed under different environmental conditions. Although correction factors are usually

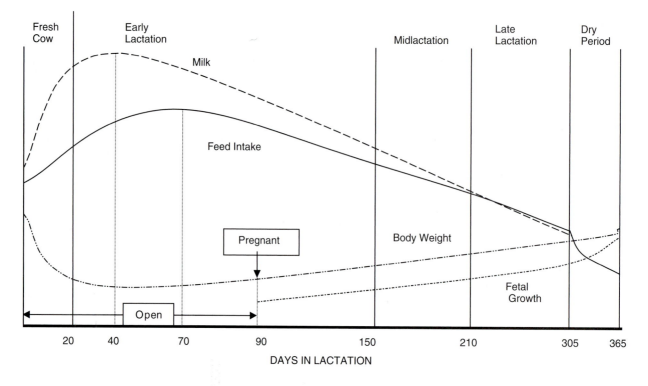

Figure 7.9   Typical Lactation Cycle and Reproductive Cycle of High Producing Dairy Cows.

necessary to reduce two or more records to a common basis, records that are comparable without factors are more reliable.

Length of Lactation

The most generally accepted standard length of lactation records is 305 days. When a cow is milked longer than 305 days, her yield for the first 305 days is used as the standard lactation yield. Partial lactations (those terminated in less than 305 days because of environmental influences having no relation to the cow's genetic ability to complete normal length lactations) are considered legitimate measures of the cow's performance up to the time they were terminated and are corrected to 305 days. The factors commonly used for this projection are listed below:

Days Milked	Factor
95	2.82
125	2.16
155	1.77
185	1.51
215	1.32
245	1.18
275	1.08

Total lactation records, or 365-day records, are often quoted verbally and in promotional literature, with or without an adequate definition of the lactation length. Care should be taken to clarify the length of the lactation when comparing or evaluating production records.

Number of Daily Milkings

Most cows are milked twice daily (usually referred to as 2X). Hence, for most lactations, no adjustment is necessary. To convert three-times-a-day milking to a two-times-a-day basis, multiply by 83 percent (0.83).

Age and Month of Calving

The age of a cow is always based on her age when she calved, which is when her record officially begins. At two years of age a cow produces approximately 70 to 80 percent of her mature production; at three years, 80 to 90 percent; at four years, 90 to 95 percent; at five years, 96 to 100 percent; and at six years, she reaches her mature record. Age-adjustment factors have been developed to standardize 305-day lactation records to a mature equivalent basis and to minimize environmental variation due to the month of the year in which the record began.

Fat-Collected Milk (FCM)

For comparative purposes, the fat content of milk is usually based on calculating the milk and fat production to 4 percent fat (4 percent FCM), but it may be calculated to any desired fat basis. The formula for 4 percent FCM is:

$$4.0\% \text{ FCM} = (0.4 \times \text{milk weight}) + (15 \times \text{fat weight})$$

LIMITATIONS OF RECORDS

Records are useful only if they are used to assist in making decisions that improve the profitability of a dairy operation. They are the first source of information used (if available) in troubleshooting management problems. To use records effectively, however, their limitations must be fully understood.

Records simply provide a snapshot in time of a herd or individual cow parameter. Even sequential records or rolling records are simply a group of snapshots in time. The usefulness of the data depends on the accuracy of that snapshot. For example, the monthly production test data collected for a cow that is unhealthy or in estrus on the day the data is compiled is not an accurate reflection of her performance for that month. The data must be evaluated in the context under which it was collected. Similarly, the herd with a somatic cell count of 750,000 cells/ml this month might be in serious trouble if their typical values have been under 250,000 cells/ml for the last six months. For the herd that has been struggling to bring its somatic cell counts under 1,000,000 for the last six months, however, the producer would look at the same value in a different light.

An additional consideration is that herd records often include only those animals currently in the herd; cows that have been culled recently may not be included. For example, a sudden improvement in reproductive performance or lactational performance may reflect an increase in animals culled for that reason rather than an improvement in management. Removal of a problem through increased culling intensity does not resolve the source of the problem. Superficial examination of records can lead to a false sense of security; it is important to evaluate fully all the information available in the records prior to forming a conclusion.

Finally, evaluation of records should be only the initial step in any troubleshooting protocol. Records are of great value in helping to solve a problem, but they rarely provide a conclusive answer without supplemental investigation.

SUMMARY

- Genetic progress depends on several factors: accurate identification of animals, the ability to identify superior animals accurately, and the ability to utilize those animals most effectively.
- Methods used to identify animals include ear tags, hide brands, neck chains or straps, or tattoos.
- The primary purpose of dairy records is to give the producer detailed information on individual cows. This information can be use to make daily management decisions, allot concentrates; breed, dry, and cull cows; provide treatments; evaluate herd status; determine the strengths, weaknesses, and profitability of the operation; and plan for the future.
- Unofficial record plans include: owner sampler (OS), weight-a-day-a-month (WADAM), milk only record (MOR), and alternate AM-PM.
- A lifetime record for each cow should include complete identification, lactation summaries, breeding records, calving records, and health and veterinary records.
- Good culling decisions depend on the ability to compare accurately the performance of individuals or groups of animals.

QUESTIONS

1. What is the most commonly used means of identifying dairy animals?
2. What are the three most common methods of applying hide brands?
3. What is the foundation of any management system?
4. How does DHIR differ from DHIA?

5. What is a mature equivalent record?

6. What are four adjustment factors important in comparing the milk and fat of cows managed under different environmental conditions?

7. In the month of May, what are the milk and fat adjustment factors for a Guernsey that is 72 months of age?

ADDITIONAL RESOURCES

Articles

Dahl, G. E., R. L. Wallance, R. D. Shanks, and D. Lucking. "Hot Topic: Effects of Frequent Milking in Early Lactation on Milk Yield and Udder Health." *Journal of Dairy Science* 87 (April 2004): 882–885.

Lay, D. C., Jr., T. H. Friend, C. L. Bowers, K. K. Grissom, and O. C. Jenkins. "A Comparative Physiological and Behavioral Study of Freeze and Hot-Iron Branding Using Dairy Cows." *Journal of Animal Science* 70 (April 1992): 1121–1125.

Internet

Florida Cow-Calf Management, 2nd Edition—Practicing Good Management: http://edis.ifas.ufl.edu/BODY_AN121.

National Dairy Herd Improvement Association—http://www.dhia.org.

Records and Record Keeping—The Backbone of Good Management: http://www.wvu.edu/~exten/infores/pubs/livepoul/dirm4.pdf.

Tattooing of Cattle and Goats: http://www.uaex.edu/Other_Areas/publications/HTML/FSA-4015.asp.

8

Fundamentals of Dairy Genetics

OBJECTIVES

- To describe the differences between qualitative and quantitative traits and between dominant and recessive genes.
- To describe terms such as *incomplete dominance* and *lethal traits.*
- To describe the role of chromosomes in sex control and heritability of traits.

W hen the sperm and egg unite, the new cell formed contains thirty pairs of chromosomes, or a total of sixty chromosomes, half of which came from the sperm (male) and half from the egg (female). Dairy animals have thirty pairs of chromosomes in each cell. The number of genes per chromosome is not precisely known; estimates approximate between 50,000 and 100,000 genes for dairy cattle. These genes are responsible for how the animal looks and how well she performs.

QUALITATIVE TRAITS

For traits exhibiting the simplest type of inheritance, known as qualitative traits, only one pair of genes is involved. Examples of qualitative traits in dairy cattle are hair color, horned versus polled, some inherited abnormalities, and blood antigens. Coat color is still important from the standpoint of breed requirements. In recent years, however, most of the breed registry associations have relaxed color or color-pattern requirements. A pair of genes is responsible for hair color in cattle (Figure 8.1). A Milking Shorthorn having two genes for red (RR) is actually red in color, while an animal having two genes for white (rr) is white in color. This situation is illustrated in Figure 8.2. A Milking Shorthorn that has one gene for red (R) and one for white (r) is neither red nor white but roan (Rr), which is a mix-

Figure 8.1 The rare albino trait exhibited by these twin heifers is inherited in the same manner as other coat color traits. *(Courtesy of Iowa State University)*

ture of red and white. Thus, red and white matings in Milking Shorthorn cattle usually produce roan offspring. Likewise, white and white matings generally produce white offspring, even though white in Milking Shorthorns is seldom pure because the face bristles, eyelashes, and ears usually carry red hairs. Roans, having one gene for red and one for white on the paired chromosomes, never breed true and when mated together produce calves in the proportion of one red, two roans, and one white. The most certain way to produce roan Milking Shorthorns is to mate red cows with a white bull, or vice versa; this produces all roan calves. If a roan animal is bred to a red one, one-half of the offspring will be red, whereas the other half will be roan. Likewise, when a roan animal is bred to a white one, approximately an equal number of roan and white calves will be produced.

DOMINANT AND RECESSIVE GENES

In the example of Milking Shorthorn colors, each gene of the pair (R and r) produced a visible effect, whether paired as identical genes (two red or two

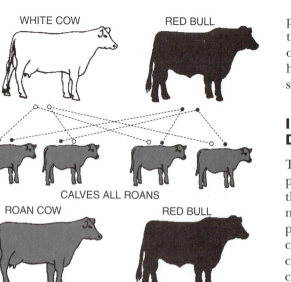

WHITE COW   RED BULL

CALVES ALL ROANS

ROAN COW   RED BULL

ONE HALF CALVES ARE ROANS AND ONE HALF ARE RED

Figure 8.2 Patterns of color inheritance in the Milking Shorthorn. *(Courtesy of M. E. Ensminger)*

white) or as two different genes (red and white). This is not true of all genes. Some of them have the ability to prevent or mask the expression of others, with the result that the genetic makeup of such animals cannot be recognized phenotypically with perfect accuracy. This ability to cover up or mask the presence of one member of a set of genes is called dominance. The dominant gene masks the traits coded for by the recessive gene.

In cattle, the polled trait is dominant over the horned trait. Thus, if a pure polled bull is mated with horned cows (or vice versa), the resulting progeny are not midway between the two parents but are polled (Figure 8.3). Likewise, not every hornless animal is homozygous for the polled trait; many of them carry a recessive gene for horns. A simple breeding test can be used to determine whether a polled bull is homozygous or heterozygous, but it is impossible to determine such purity or impurity through phenotypic inspection. The breeding test consists of mating the polled sire with a number of horned females. If the bull is homozygous for the polled trait, all of the calves will be polled. Heterozygous sires produce half polled offspring, on the average, while half have horns, like the horned parents.

Dominance often makes the task of identifying and discarding all animals carrying an undesirable recessive factor a difficult one. Recessive genes can be passed on from generation to generation, ap-

pearing only when two animals, both of which carry the recessive factor, are mated. Even then, only one out of four offspring produced, on average, will be homozygous for the recessive gene and demonstrate that trait phenotypically.

INCOMPLETE OR PARTIAL DOMINANCE

There are varying degrees of dominance, from complete dominance to an entire lack of dominance. In the vast majority of cases, however, dominance is neither complete nor absent, but incomplete or partial. The results of crossing a trait with horned cattle are clear-cut because the polled character is completely dominant over its allele (horned). If a cross is made between a red and a white Milking Shorthorn, however, the result is a roan (mixture of red and white hairs) color pattern. In the latter cross, the action of a gene is such that it does not cover the allele (incomplete dominance); the roan color is the result of the combined expression of a pair of genes, neither of which is dominant.

Dominance is not always simply the result of single-factor pairs. The degree of dominance depends on the animal's whole genetic makeup together with the environment to which it is exposed and the various interactions between the genotype and the environment. Environment has little effect on hair color except for extreme circumstances, such as molybdenum toxicity, copper deficiency, long exposure to tropical sun, or freeze branding.

LETHALS AND OTHER ABNORMALITIES

Most genetically transmitted lethal traits cause death of the calves prior to or shortly after birth. Documented cases of such abnormalities are numerous. Most lethal traits are recessive and may remain hidden for many generations. The total removal of lethal genes requires test matings and rigid selection.

Ideally, purebred animals should be free of both undesirable and lethal genes (Figures 8.4 through 8.6). This can be accomplished only by eliminating those sires and dams that carry the undesirable recessive character. In addition, producers must DNA-test both phenotypically abnormal and normal offspring produced by these sires and dams because approximately half of the normal animals carry the undesirable character in the recessive condition.

QUANTITATIVE TRAITS (MULTIPLE GENE INHERITANCE)

Relatively few traits of economic importance in farm animals are inherited in as simple a manner as the coat color or the polled conditions described.

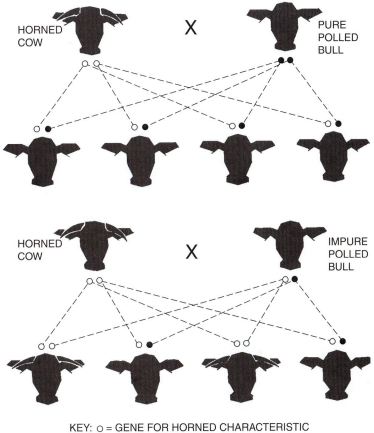

KEY: ○ = GENE FOR HORNED CHARACTERISTIC
      ● = GENE FOR POLLED CHARACTERISTIC

Figure 8.3   Patterns of inheritance for the horned trait in cattle. *(Courtesy of M. E. Ensminger)*

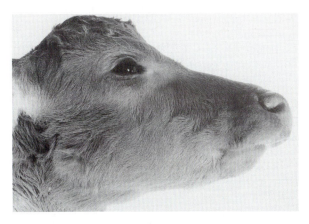

Figure 8.4   Aganthia is a genetic defect that results in the absence of the mandible and other malformations of the head. *(Courtesy of Kansas State University)*

Most traits of economic importance, such as milk yield and composition, conformation, feed efficiency, and disease resistance, are controlled by multiple genes. Because such traits may be expressed in all possible gradations, from high to low performance, they are known as quantitative traits. Estimates of the number of pairs of genes affecting each economically important characteristic vary greatly, but the majority of geneticists agree that ten or more pairs of genes are involved for most such traits. In addition to being influenced by many pairs of genes, quantitative traits differ from qualitative traits because they are frequently strongly influenced by the environment.

SEX CONTROL

On average and when considering a large population, approximately equal numbers of males and females are born in all common species of animals, although the proportion may vary dramatically from year to year in individual herds. In cattle, the female has a pair of similar chromosomes (called X chromosomes), whereas the male has a pair of unlike sex chromosomes (called X and Y chromosomes). The pairs of sex chromosomes separate when the germ cells are formed. Thus, the ovum or egg produced by the cow contains X chromosome, while the sperm of the bull are of two types, one-half containing the X chromosome and the other half the Y chromosome. Because the egg and sperm are assumed to unite at random, half of the progeny will be females (XX) and the other half males (XY).

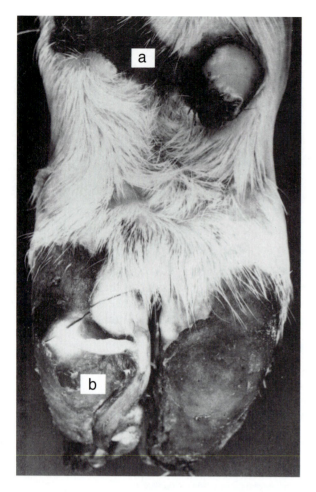

Figure 8.5 Epitheliogenesis imperfecta is a rare genetic defect characterized by areas of abnormal and missing skin. *(Courtesy of Kansas State University)*

Figure 8.6 Limber leg is a recessive genetic disorder in Jerseys. *(Courtesy of Utah State University)*

TABLE 8.1   Heritabilities of Various Production and Type Traits

Trait	$h2$
Milk yield	0.3
Fat percentage	0.5
Reproduction	0.07
Stature	0.42
Strength	0.31
Body depth	0.37
Dairy form	0.29
Rump angle	0.33
Thurl width	0.26
Rear legs—side view	0.21
Rear legs—rear view	0.11
Foot angle	0.15
Fat yield	0.25
Protein percentage	0.5
Milking rate	0.3
Feet and legs score	0.17
Fore attachment	0.29
Rear udder height	0.28
Rear udder width	0.23
Udder cleft	0.24
Udder depth	0.28
Front teat placement	0.26
Teat length	0.26
Final score	0.29

National statistics of births in the United States show that of each 100 dairy calves born, on average, forty-nine are heifer calves and fifty-one are bull calves. Obviously, some method of controlling the sex of offspring would have tremendous economic importance in the dairy field. Control of sex is currently possible; however, it is not yet economically practical in most cases. Predetermination of the sex of six- to twelve-day-old embryos is a reality, and progress is being made in the separation of sperm cells containing X chromosomes from those containing Y chromosomes using flow cytometry.

HERITABILITY OF TRAITS

The phenotypic expression of a trait, such as milk production, depends on two factors: inheritance (the genetic potential for expression of a particular trait) and environment (the opportunity to express the inherited trait). Heritability is 100 percent when the expression of the trait varies solely because of inheritance. A trait that varies solely because of environment has a heritability of zero.

Variation in the expression of most traits is neither totally environmental nor completely hereditary. The heritability of some important dairy cattle traits is given in Table 8.1. Traits with higher heritabilities allow more rapid genetic progress. In addition, traits with high heritabilities increase the value of the animal's own phenotype as an estimator of genotype. Even with the most highly heritable production traits, however, at least 50 percent of the population variation is attributable to environment or heredity-environment interactions. Although some environmental variables are not easily managed (weather or climate, season of calving), many are directly management influenced (feeding, housing, health care, reproductive handling). Most traits of economic importance are primarily controlled by a combination of both management skills and genetics.

The following example shows how heritability can be computed: assume that a herd averages 30,000 pounds of milk on a mature basis. A sire used in that herd is capable of transmitting inheri-

tance for 36,000 pounds of milk production. He is mated to select cows in the herd with production records averaging 36,000 pounds of milk in a normal 305-day lactation period. Because heritability is 30 percent, we expect only three-tenths of the apparent superiority of the parents to be expressed in the offspring. The selected parents averaged 6,000 pounds of milk higher than the herd. Three-tenths of the 6,000 equals 1,800 pounds of milk. Thus, the offspring are expected to average 31,800 pounds of milk in this herd when given the same opportunity (environmental effects) as the parents.

SUMMARY

- The ability to cover up or mask the presence of one member of a set of genes is called dominance. The dominant gene masks the traits coded by the recessive gene.
- Most quantitative traits of economic importance, such as milk yield and composition, conformation, feed efficiency, and disease resistance, are controlled by multiple genes.
- A genetic female usually has a pair of X chromosomes, whereas genetic males have a X and a Y chromosome.
- The phenotypic expression of a trait, such as milk production, depends on two factors: inheritance and environment.
- A trait that varies due to environment alone has a heritability of zero. Even the most highly heritable production traits vary with at least 50 percent of the population variation attributable to environment.

QUESTIONS

1. How many total chromosomes are there in a dairy cow?
2. If a red cow is bred to a roan bull, what percentage of the calves will be white? Red? Roan?
3. Are horns a recessive or dominant trait?
4. If a polled sire is bred to horned females and one-half of the offspring has horns, what does this indicate about the genes of the bull?
5. What type of dominance is illustrated by color pattern in the Milking Shorthorn?
6. The ovum produces which chromosome?
7. What is the heritability of the feet and legs score?

ADDITIONAL RESOURCES

Article

Smith, L. A., B. G. Cassell, and R. E. Pearson. "The Effects of Inbreeding on the Lifetime Performance of Dairy Cattle." *Journal of Dairy Science* 81 (October 1998): 2729–2737.

Internet

Basic Genetics Concepts: http://babcock.cals.wisc.edu/downloads/de/14.en.pdf.

Fundamentals of Inheritance: http://muextension.missouri.edu/explore/agguides/dairy/g03000.htm.

The Genetics and Management of Sound Feet and Legs: http://ianrpubs.unl.edu/dairy/g1197.htm.

Using Heritability for Genetic Improvement: http://www.ext.vt.edu/pubs/dairy/404-084/404-084.html.

9

Selecting Herd Sires

OBJECTIVES

- To evaluate sires based on their standardized transmitting abilities and predicted transmitting abilities.
- To understand the terminology associated with sire summaries.
- To evaluate the advantages and disadvantages of using young sires compared to proven sires obtained through artificial insemination organizations.

Because there is no direct measure of a sire's individual performance, his evaluation is based on the performance of his daughters. The animal model currently used for such evaluations includes records from all identified relatives in the evaluation. The selection of a sire is extremely important because he becomes the parent of many more offspring than does an individual cow. A superior sire may be responsible for 90 percent or more of the genetic improvement in a herd.

A dairy producer has three sources of herd sires: artificial insemination (AI) service, purchase of herd sires, or raising herd sires. The producer must also decide between using proven sires or young sires. A young sire is a sire without recorded daughters, where as a proven sire has records of his progeny. The challenge is to select herd sires that will ensure genetic improvement in the herd. The most reliable source of superior genetics for the breeding program of a herd is bulls that have been accurately evaluated for a large number of traits, including yield, conformation, and calving ease. This approach strongly favors bulls available through bull studs (Table 9.1). Over three-fourths of first-lactation heifers from high Predicted Transmitting Ability bulls outproduce their herdmates, while only one-fourth of these heifers outproduce their herdmates when sired by non-AI bulls. The primary reason for producers not using AI is convenience. Often, heat detection for heifers must be separate from heat detection times for cows. Facilities for breeding animals in estrus are also often inconvenient. These obstacles must be overcome, however, if genetic progress is to be maximized.

If artificial insemination is utilized, the ability to distinguish a superior sire from an inferior one is critical. Twice annually, the U.S. Department of Agriculture (USDA) publishes a USDA-DHIA sire summary. These genetic evaluations are based on information on the bulls' daughters in herds participating in official production testing programs (Dairy Herd Improvement Association (DHIA) and Dairy Herd Improvement Registry (DHIR)).

SIRE SUMMARIES

The USDA sire summaries are publicized and interpreted by the Cooperative Extension Service, state associations, dairy breed registries, AI organizations, dairy magazines, and other channels (Figure 9.1). The bulls are described by the following codes:

Name of bull: Bull's registered name.

Registration number: Bull's registration number.

NAAB code: A three-part code assigned by the National Association of Animal Breeders (NAAB). The number before the letter indicates the stud from which the bull's semen can be purchased. The letter indicates the breed. The number following the letter is an individual bull identification number assigned by the bull stud.

Predicted transmitting ability (PTA): This term applies to the genetic values that rank bulls for production traits. Predicted transmitting ability is the best estimate of expected extra production per daughter per year, when compared to a zero PTA bull (Table 9.2).

TABLE 9.1   Trend in Milk-Breeding Values for the Holstein, Calculated May 2004

Birth Year	Cows	Milk (lbs)	Breeding Value	Relative	Sire BV
2002	43417	25587	1542	0.40	2473
2001	536647	25393	1384	0.43	2277
2000	616218	25361	1182	0.47	2059
1999	614251	25286	972	0.49	1863
1998	614608	24958	759	0.50	1699
1997	624347	24856	520	0.51	1460
1996	615236	24136	267	0.51	1190
1995	638778	23381	0	0.51	886
1994	655420	22261	−246	0.51	657
1993	651909	21701	−485	0.52	405
1992	679597	21679	−733	0.52	143
1991	702779	21235	−993	0.52	−145
1990	702036	20852	−1250	0.52	−434
1989	692020	20573	−1471	0.52	−641
1988	681935	20096	−1733	0.52	−932
1987	664072	19643	−1971	0.52	−1177
1986	637777	19386	−2176	0.52	−1370
1985	615184	18994	−2385	0.52	−1591
1984	594303	18779	−2592	0.52	−1785
1983	609418	18452	−2780	0.51	−1958
1982	595188	18015	−2994	0.51	−2174
1981	579078	17665	−3201	0.51	−2361
1980	539917	17644	−3414	0.51	−2588
1979	486615	17510	−3605	0.52	−2803
1978	431909	17445	−3814	0.52	−3050
1977	396020	17260	−3993	0.52	−3275
1976	354023	16970	−4199	0.52	−3531
1975	315524	16811	−4393	0.52	−3796
1974	295025	16607	−4545	0.52	−3957
1973	273383	16010	−4703	0.52	−4118
1972	265809	15544	−4859	0.52	−4323
1971	255875	15221	−5021	0.52	−4545
1970	246897	15189	−5168	0.52	−4776
1969	234118	15317	−5290	0.51	−4960
1968	224097	15169	−5420	0.51	−5147
1967	212115	14988	−5500	0.51	−5228
1966	198215	14805	−5573	0.51	−5303
1965	188206	14617	−5668	0.51	−5445
1964	178887	14535	−5767	0.51	−5573
1963	170433	14371	−5843	0.50	−5686
1962	165112	14283	−5940	0.50	−5815
1961	154359	14115	−5987	0.49	−5830
1960	137234	13783	−6024	0.49	−5869
1959	112152	13404	−6070	0.48	−5927
1958	95096	13096	−6105	0.47	−5976
1957	31252	12942	−6086	0.46	−5950

Source: From USDA-AIPL: http://www.aipl.arsusda.gov/dynamic/trend/current/trndx.html.

PTAs for pounds protein (LB P), percentage protein (%P), pounds milk (LB M), pounds fat (LB F), and percentage fat (%F): PTAs for pounds protein, percentage protein, pounds milk, pounds fat, and percentage fat indicate how much more (or less) performance to expect from an average daughter of a bull with a PTA of zero for the same trait. The genetic base is selected so that half the cow population will have positive PTAs. It was established by setting to zero the weighted average PTAs of all cows born in the same year, and it is readjusted every five years.

Figure 9.1 Online access to the most recent sire summary information is available. *(Courtesy of California State University, Fresno)*

TABLE 9.2 Animal Improvement Programs Laboratory Means (lbs) for Calculating PTA% for the May 2004 Run

Breed	*Milk*	*Fat*	*Protein*	*Protein Milk**
Ayrshire	16,832	649.2	524.0	16,832
Brown Swiss	19,356	775.9	637.9	19,349
Guernsey	15,427	683.0	506.8	15,438
Holstein	23,382	846.5	691.8	23,378
Jersey	16,053	738.1	568.6	16,055
Milking Shorthorn	16,012	567.9	489.6	16,012

* Protein milk = milk for cows that had protein.
Source: Modified from USDA-AIPL: http://www.aipl.arsusda .gov/dynamic/summary/current/yld_mean.htm.

Fluid milk dollars (FMD): This code weights the PTA milk and fat to reflect the gross income per lactation that future mature daughters of bulls will earn in excess of herdmates sired by bulls having a FM$ equal to zero. It is based on prices of $12.50 per hundred of 3.5 percent fat and 14.8 cents per point fat differential. It assumes that no income is derived from protein. This was the U.S. average milk price for 1989 minus the average hauling and assessments for promotion.

Cheese merit dollars (CMD): Predicted transmitting ability cheese merit dollars reflects the income per lactation future mature daughters of the bull will earn if their milk is priced according to its value in cheddar cheese. It assumes a high differential for protein.

Predicted transmitting ability-type (PTAT): This code is the expected difference in final score between daughters of the bull and the breed average.

Type-production index (TPI) (Holsteins): The type-production index is a value that is determined by placing an emphasis of 2 for PTA protein, 2 for PTA fat, 1 for PTA type, and 1 for udder composite traits.

PTI (all breeds except Holsteins): The Production Type Index is a value which is similar to the TPI for Holsteins. The traits included in the formula and the emphasis placed on each trait varies, and is determined by each breed association.

Calving ease (%DBH): This is the estimate of the percentage of difficult births in first-calf heifers. Producers use this information when choosing bulls to breed heifers. The range of this value is typically 5 to 20 percent, with a median of 9 percent DBH.

Reliability: The PTA values for an individual sire often change as more daughters are added in more herds. Reliability is a percentage figure that indicates the degree of confidence a breeder can place on the PTA. Reliability values range from about 16 percent to 99 percent and increase with the number of daughters, number of herds with daughters, and number of records per daughter. The closer the reliability is to 100, the more reliable the predicted transmitting abilities (PTAs).

Predicted producing ability (PPA): This is the best estimate of a cow's ability to produce either above or below the average of other cows. PPA is standardized for the number of lactations of the cow.

Pedigree index (PI): The pedigree index is an estimate of the animal's genetic transmitting ability based on pedigree information.

Parent average (PA): This is another estimate of the breeding value calculated by using sire and dam information rather than sire and maternal grandsire, PA = (sire's PTA + dam's PTA)/2.

Productive life (PL): This value estimates predicted herd life for cows remaining in the herd. A cow is credited with ten months productive life for each 305-day lactation. No credit is given for dry periods or lactations beyond 305 days. This index also includes correlated information from type, yield, and somatic cell score evaluations. The final index reflects the resistance to culling and is highly correlated with production and udder traits, although additional variation is present beyond that provided by those traits.

Somatic cell score (SCS): This estimates the transmitting ability for somatic cell score. This trait has a relatively low heritable value (around 10 percent heritable).

Net merit index (NM$): This value uses actual income and expenses to estimate the expected lifetime profit that daughters will provide based on production, udder health, longevity, and body size.

TABLE 9.3   Average Mature Daughter Measurement Corresponding to Linear Type Standard Transmitting Ability of Sire When Mated to Breed Average Cows

		STA		
Linear Type Trait	*Measurement*	*−3*	*0*	*3*
Stature	Inches—height at hip.	55.6	56.6	57.6
Rump angle	Inches—slope from hips to pins.	0.6	1.3	2.0
Thurl width	Inches—between the pins.	4.6	5.0	5.4
Foot angle	Degree of the angle that the front of the toes makes with the ground.	41	43	45
Rear udder height	Inches—between the bottom of the vulva and the top of the milk-secreting tissue.	10.6	10.1	9.6
Rear udder width	Inches—width of the rear udder where the udder attaches to the body.	5.5	5.8	6.2
Udder cleft	Inches—depth of the cleft between the rear quarters at the bottom of udder.	1.2	1.4	1.6
Udder depth	Inches—between the lowest point of the udder floor and the point of the hock.	0.5	1.2	1.9
Teat length	Inches—length of the longest teat.	2.2	2.4	2.6

Source: Holstein Association.

Type traits expressed in sire summaries are based on a linear scoring system. This system differs from an official classification score because traits thought to have biological or economic importance are scored over a range of 50 points, from one extreme to another. The scoring system puts no value judgment on the score; there is no optimum or best score. All contemporaries, both registered and grade, are evaluated, and cows are scored without knowledge of the sire's identity or of the previous score. This system provides a better opportunity for individual producers to determine which score is optimum or economically valuable in their herd.

In sire summaries, type traits are expressed as standardized transmitting abilities (STAs), which expresses the predicted transmitting ability on a standardized scale from −3 to +3. The conversion of the STA to an actual measurement can be used to calculate the actual change in various type traits per generation by selecting one sire (or group of sires) relative to another (Table 9.3).

YOUNG SIRES

Theoretically, the most rapid rate of genetic progress is achieved by using young sires. Proven sires are generally seven to eight years old and have a remaining life expectancy of only two to three years. Young sires are typically one to two years old when their semen is being distributed. Thus, they represent a higher level of genetic potential as a group than those bulls entering the program five to seven years previously (Figure 9.2).

Making this theoretical advantage a reality depends on the accuracy of selection of these sires (Figure 9.3). Generally, young bulls are highly selected on pedigree and should, on average, have high PTAs when proven. Pedigree indexes (PIs), although not very reliable from individual to individual, are fairly accurate for predicting future proofs for a group of

Figure 9.2   Curt Van Tassell loads a high-capacity DNA sequencer to find more genetic markers for screening dairy bulls. *(Courtesy of USDA-ARS)*

young sires. For this reason, groups of young sires with high pedigree indexes should be used rather than a single young sire. The same is true for promising bulls with high proofs but low reliability. The average for the group varies only slightly, although any individual bull may change dramatically. By selecting a group of promising young bulls to use, the potential damage to overall herd genetic merit is minimized.

Young sires are priced lower and can be useful to the dairy farmer if they are used properly. It is recommended that a few cows be bred to each of several young bulls and that not more than 25 percent of all matings be to young sires. Sampling bias should be minimized in any breeding plan for young sires because it greatly improves the accuracy of the eventual proof.

One disadvantage with young sires is that their calving ease is an unknown factor. Therefore, for heifers, the group of proven AI sires that have been selected for the herd should be subdivided to use only those sires that will produce offspring with the least expected calving difficulty. For older cows, young sires can be used with some reservations.

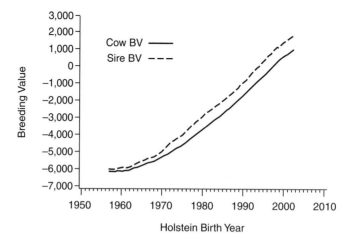

Figure 9.3 Historic trends for breeding values illustrate the speed of genetic progress and the value of young sires.

SUMMARY

- Sire performance is based on the performance of his daughters.
- A dairy producer has three sources of herd sires: artificial insemination, purchase of herd sires, or raising herd sires.
- Theoretically, the most rapid rate of genetic progress is achieved by using young sires.
- For older cows, young sires can be used with some reservations, whereas proven AI sires should be used with heifers.

QUESTIONS

1. What is the primary reason for producers choosing not to use AI?
2. Define predicted transmitting ability (PTA).
3. Define net merit index (NM$).
4. Which type of sire represents a higher level of genetic potential as a group?

ADDITIONAL RESOURCES

Articles

Clay, J. S., B. T. McDaniel, and C. H. Brown. "Variances of and Correlations Among Progeny Tests for Reproductive Traits of Cows Sired by AI Bulls." *Journal of Dairy Science* 87 (July 2004): 2307–2313.

Misztal, I., and T. J. Lawlor. "Supply of Genetic Information—Amount, Format, and Frequency" (review). *Journal of Dairy Science* 82 (May 1999): 1052–1060.

Internet

How To Set Goals For Your Breeding Program: http://ianrpubs.unl.edu/Dairy/g755.htm.

Principles of Selection: http://babcock.cals.wisc.edu/downloads/de/15.en.pdf.

Selection Goals: http://babcock.cals.wisc.edu/downloads/de/17.en.pdf.

Using Net Merit to Select Dairy Bulls: http://www.ansc.purdue.edu/dairy/genetics/nmerit.htm.

10

Herd Strategies for Genetic Improvement

- To describe sire selection strategies.
- To describe systems of selection and breeding for developing and following a breeding program.
- To illustrate how to select cows and replacement heifers.

The ultimate goal for the dairy herd breeding program is to produce replacement heifers that will ultimately calve normally and secrete large quantities of milk with high protein and fat content, breed regularly and without problems, have a minimum of health problems, be structurally sound, and have a long and profitable productive life. Each generation should represent genetic progress over the previous generation. The determinants of genetic progress include:

A: accuracy of selection (predicted versus actual genetic value).

I: intensity of selection (superiority of selected parents).

G: genetic variation.

Combining these determinants into a formula, we have:

$$A \times I \times G = \text{genetic progress per generation}$$

Obviously, the true rate of genetic progress in the industry is also affected by the generation interval (GI). The GI is the average age of the parents when their offspring are born (approximately five years in the dairy industry).

$$\text{Genetic gain per year} = \frac{A \times I \times G}{GI}$$

Genetic variation is beyond the control of the individual producer, while accuracy of selection is primarily determined by the breeding studs because the studs select the parents of the young sires to be tested. The producer affects genetic change by controlling the intensity of selection and controls the rate of this change by controlling generation interval.

In most commercial herds, milk production is the primary source of income and therefore should be the primary selection trait. Fat, protein, or other total solids are also highly important. Some attention to conformation is necessary to avoid serious problems, especially in the udder traits. The best approach for maximizing the intensity of selection is to select sires from among the group that is ranked highest for net merit dollars and to manage matings to avoid inbreeding and inbreeding-related problems (Figure 10.1).

Profit is sacrificed each time a sire ranked below acceptable levels for net merit dollars is used in an

Figure 10.1 Marcus Kehrli tests a calf that has bovine leukocyte adhesion deficiency (BLAD). *(Courtesy of USDA-ARS)*

attempt to improve some other trait. If the possible improvement in that other trait is worth more than the potential loss in profit, the sacrifice should be made. If too many sacrifices are made, genetic improvement for those traits emphasized by net merit dollars may stop.

All dairy herds, regardless of current level of production, can improve their genetic potential to produce milk. A common fallacy from producers in the field is that certain herds don't need to be concerned with sire selection because their management limits their production below their current genetic potential, but this is true of all herds. Management always impinges to some extent on an animal's ability to achieve its genetic potential for production; however, genetic improvement in these herds still results in increases in milk production. The management impairment is best expressed as a percentage of genetic potential, rather than an absolute limit. Thus, all herds can improve through genetic gain, and careful sire selection is equally important to a poorly managed herd as it is to a well-managed herd.

Cow evaluations are available through the national Cooperative Dairy Herd Improvement Program in every state. Accurate and useful sire evaluations are compiled by the U.S. Department of Agriculture (USDA) and distributed through the extension dairy specialists, artificial insemination (AI) associations, and breed associations. Breed associations provide additional programs to improve breeding and management of dairy herds, especially through their type classification programs. Dairy producers must plan, make decisions, and use cow records and sire evaluations for the improvement of their herds. The success of any breeding program depends primarily on the ability of the breeder to select intensively the animals that are to be the parents of the next generation.

SYSTEMS OF SELECTION

Many different approaches toward making genetic progress are available. Single-trait selection selects exclusively for a single trait. This approach allows the fastest progress in any single trait. Tandem selection selects first for one trait, then another, then back to the first trait, and so on. Progress for either trait is minimal unless the traits are highly correlated. The most rapid progress is made in a breeding program by selecting for one trait only. However, when two traits are inherited in a positive relationship, considerable progress may be achieved in both with simultaneous selection strategies. This is true, for example, for total milk production and total production of fat or protein. On the other hand, body type and high produc-tion of milk are negatively associated, with the result that simultaneous selection for both may result in relatively slow progress in either trait. Under these circumstances, herd owners must decide which of the traits in the herd is most crucial to emphasize and make their decision on the trait to be emphasized in selection. If good body type will bring more monetary return than increased milk production, it should be emphasized. Conversely, if higher production will increase income to a greater degree than improved body type, then it should be given greater importance in selection. It is possible to select bulls that allow simultaneous progress in both traits; however, limiting sire selection strictly to bulls that meet both criteria would limit progress compared to selecting sires strictly for one trait.

As a general rule, the total volume of milk and amount of milk components (especially fat and protein) are the traits of greatest economic importance in the breeding of commercial dairy cattle, but other traits should also be considered if they have demonstrated economic value. In many purebred herds, both type and production are important, and selection for both should be made.

Many producers select sires for a primary trait and then apply individual (independent) culling levels for other traits to remove sires based on minimum acceptable levels for those secondary traits they feel are important for their herd. This system does not allow sires with exceptional economic potential to remain in the pool if even one secondary trait is below acceptable standards. Therefore, the pool of sires selected may not provide the best overall genetic progress for the traits important to the producer.

Selection indexes allow producers to weight multiple traits to be selected based on their perceived importance to their individual herds, and they can make their sire selection accordingly. The price of the semen may also be included in these indexes. These indexes are provided by breed associations, or they can be individually designed on a computer. The use of a selection index recommends the use of a sire with exceptional strengths in several important traits, despite weaknesses in other traits that would cause the sire to be removed using the independent culling method.

Some traits, such as productive life, include numerous factors in their value and may be considered as a composite index. The heritability for productive life is low (less than 10 percent), so production-based indexes such as net merit dollars may be more useful.

Traits to consider in selection indexes (other than production) include udder traits. Udder depth is the most heritable of the udder traits, is inversely related to the incidence of mastitis and somatic cell counts, and is highly correlated with productive life. Teat placement is also relatively heritable and inversely related to mastitis. A strong center support (median suspensory ligament), although important, has low heritability. Other udder traits, such as fore and rear udder attachments, rear udder height, and teat length, are less important for inclusion in selection indexes.

In general, traits associated with feet and legs have low heritability. Thus, despite their importance for longevity, they should not receive as much attention. The fastest improvement in these traits is accomplished through improved management and selective culling. Hooves and foot angle are important; cows with short hooves and a steep foot angle have less lameness, improved reproduction, improvements in milk yield from one lactation to the next, and improved productive life. Legs need to be straight as viewed from the rear (not sickle-hocked) with wide hocks. They should also be relatively straight when viewed from the side, without being posty.

The milking rate of dairy cattle is moderately heritable and positively correlated with milk yield and susceptibility to mastitis. Because milking rate impacts group-milked cows more than individually-milked cows, this trait is more important for herds milked in parallel, herringbone, or rotary parlor systems than in side opening parlors, stanchion barns, or tie-stall barns.

Reproductive success has high economic importance because it is necessary for initiation of lactation; however, female reproduction rate has little genetic variation and is lowly heritable. Male fertility, measured as the bull's conception rate, may be of considerable importance and is measurable. Calving ease, measured as the percentage of difficult births in first-calf heifers, is also economically important. It directly affects calf mortality as well as later reproductive problems and production level of the dam.

Resistance to mastitis is also economically important and has some genetic variation, so therefore it is heritable. Currently, somatic cell score data is available for bulls, although it may be truly a composite of improved immune function of daughters in combination with improved udder traits. Longevity, another economically important trait, is measured as productive life. It combines production traits, structural integrity, and health and reproductive success into a single value that may provide a weighted index reflecting what is important to most producers.

Although individual matings appear very important, the merit of the sires selected is more important than the system of mating to determine which cows should be bred to each sire.

SYSTEMS OF BREEDING

In dairy cattle breeding, one of the following general systems of breeding are possible: inbreeding, which utilizes close breeding or line breeding; outcrossing; or crossbreeding. Inbreeding (the mating of animals that are closely related) is usually avoided by dairy breeders, although it will theoretically produce individuals that transmit fairly uniformly. Inbreeding presents a high risk for exposure to undesirable or even lethal recessive traits. Line breeding is practiced much more extensively by dairy cattle breeders and is a far safer breeding program.

Outcrossing (the mating of unrelated animals) is the most widely used system of breeding by dairy producers. It offers considerable opportunity to introduce new genes into the herd simply because a wide choice of animals is available. It often results in producing animals that are highly desirable, but the greatest disadvantage of this system is that the animals may not transmit uniformly. However, this system does not carry the dangers that often go with inbreeding, such as possible reduction in size and scale, lack of vigor, exposure of possible recessive factors that may be undesirable, and a greater concentration of undesirable traits.

Crossbreeding cannot be practiced in registered herds because the offspring cannot be registered. For those milk producers who have no preference as to color and general appearance, crossbreeding may be followed with good results provided good proven sires are used (Figure 10.2). The aims of this system of breeding are to use the best sires available regardless of breed and to gain hybrid vigor in the offspring (reduced calfhood mortality), increased efficiencies of gain and reproduction, and possibly more economical milk production. Crossbred cows produce more than the average of the two breeds incorporated in the cross, but production will not reach the amount expected from the higher-producing parent. Limited data

Figure 10.2 Holstein and Jersey crossbred cows graze in south central Pennsylvania. *(Courtesy of USDA-ARS)*

shows no difference in the lifetime economic value from milk and milk components; this is the best outcome that can be expected. The improved profit from crossbreeding is expected to be derived from decreased health and feed costs and improved reproductive function. The most favorable response would be expected under adverse conditions. Crossbreeding has not historically been a common practice in the dairy industry, so the most profitable crosses are not known. Current interest in crossbreeding is growing, although selection of breeds to cross is based more on anecdotal evidence than on science thus far.

DEVELOP AND FOLLOW A BREEDING PROGRAM

Where replacement animals are raised, in either a purebred or a grade herd, a breeding program must be developed and followed if herd progress and breed progress are to be made. The following steps are pertinent to such a program:

1. Choose a suitable breed of cattle, with consideration given to breed preference, the market for milk, and the value of surplus animals.
2. Select or purchase the best cows available, based primarily on their genetic potential for milk production but with due consideration given to type.
3. Decide on the breeding system to be followed.
4. Evaluate the strong points and the weak points of the cows in the herd (herd analysis).
5. Use sires that offer the greatest promise of improvement in traits offering the greatest potential economic return in the future, based on predicted transmitting ability and reliability of the sire's proof, but with consideration given to the price of the semen.
6. Enroll in the particular testing program (DHIA or DHIR) that best meets the breeding and management programs that will be followed and that achieves your goals.
7. Follow the type evaluation that best meets the needs of the breeding program you are attempting.
8. Follow a feeding and management program that permits the animals in the herd to most fully express their genetic potential.

SELECTING COWS

When milk is the major source of income, selection is simplified. Cows in production are ranked from high to low on the basis of milk production, and the most profitable milk cows are retained and the least profitable ones are sold.

The USDA annually compiles and publishes estimates of the genetic-transmitting ability of cows identified (through the Dairy Herd Improvement Testing Program) as having demonstrated superior genetic merit for milk production. These cow index values are listed in the USDA-DHIA cow performance index and are based on production records of the cow (modified contemporary deviation), her paternal half-siblings (sire's PTA), and her dam's cow index.

A purebred breeder usually finds it desirable and profitable to select animals for type as well as production. Where grade cows are involved and replacement heifers are not being raised, cows can be selected or culled primarily on the basis of production of milk components and on productive life. Where replacement heifers are being selected for retention in the herd or for sale purposes, cows should be selected on the basis of their milk production, pedigree, and progeny, provided all three are available.

SELECTING REPLACEMENT HEIFERS

The number of heifer replacements needed each year to maintain herd size depends on the number of cows eliminated from the herd because of disease, injury, low production, or poor type. Normal turnover in DHIA herds is over 30 percent each year. To meet this need and to allow some opportunity for culling undesirable first-calf heifers, it is necessary to raise approximately one-third as many heifer calves each year as there are milking animals in the herd. Producers who raise their replacements, are in a better position to evaluate the animals genetically than are operators who buy replacements simply because their dams and close relatives are available in the same herd under similar feed and management conditions.

SELECTING PUREBRED CATTLE

Technically, a purebred animal is one that can meet ancestry requirements in one of the breed registry associations, whereas a registered animal is a purebred that has been recorded in one of the registry books. Grade animals are those that are not registered or eligible for registry; however, such animals can frequently approach purebred status as a result of several generations of breeding up by using sires of one breed. Thus, if a registered sire is used successively for seven generations, the final offspring will consist of 99 percent registered parentage. Therefore, in terms of genetic potential for production, the differences between registered and grade animals are often very small.

SUMMARY

- The ultimate goal for the dairy herd breeding program is to produce replacement heifers that secrete large quantities of milk, regularly reproduce, have minimum health problems, are structurally sound, and have a long and profitable productive life.
- Determinants of genetic progress include accuracy and intensity of selection and genetic variation.
- The best approach for maximizing the intensity of selection is choosing sires with the highest ranked net merit dollars.
- Success depends on the ability of the breeder to select animals to be parents of the next generation.
- The most rapid progress is made in a breeding program when selecting for one trait only.
- The merit of sires selected is more important than the system of mating to determine which cows should be bred to each sire.
- The number of heifer replacements needed each year to maintain herd size depends on the number of cows eliminated, and the turnover in DHIA herds is 30 percent each year.

QUESTIONS

1. What is the average age of the parents when their offspring are born?
2. What are the traits of greatest economic importance in breeding commercial dairy cattle?
3. Which economically important traits have low heritability?
4. If a producer has a milking herd of ninety-nine cows, how many heifers should be raised to attain normal turnover?
5. What is the difference between purebred and registered animals?
6. How many generations does it take for a grade animal to consist of 99 percent registered parentage?

ADDITIONAL RESOURCES

Articles

Smith, L. A., B. G. Cassell, and R. E. Pearson. "The Effects of Inbreeding on Lifetime Performance of Dairy Cattle. 81 (October 1998): 2729–2737.

Wiggans, G. R., P. M. VanRaden, and J. Zuurbier. "Calculation and Use of Inbreeding Coefficients for Genetic Evaluation of United States Dairy Cattle." *Journal of Dairy Science* 78 (July 1995): 1584–1590.

Young, C. W., and A. J. Seykora. "Estimates of Inbreeding and Relationships Among Registered Holstein Females in the United States." *Journal of Dairy Science* 79 (March 1996): 502–505.

Internet

Genetic Improvement of Dairy Cattle: http://edis.ifas.ufl.edu/DS094.

Genetic Improvement of Milk and Its Components: http://www.wisc.edu/dysci/uwex/genetics/pubs/05_Genetic_Improvement_of_Milk_and_its_Components.pdf.

Inbreeding: http://www.ext.vt.edu/pubs/dairy/404-080/404-080.html.

National Genetic Improvement Programs for Dairy Cattle: http://www.wisc.edu/dysci/uwex/genetics/pubs/08_National_Genetic_Improvement_Programs_for_Dairy_Cattle.pdf.

Raising Dairy Replacement Heifers: http://edis.ifas.ufl.edu/DS150.

Rate of Crossbreeding in Dairy Cattle Improvement: http://www.wisc.edu/dysci/uwex/genetics/pubs/15_Role_of_Crossbreeding_in_Dairy_Cattle_Improvement.pdf.

3

Concepts in Dairy Nutrition

11

The Ruminant Digestive System

OBJECTIVES

- To outline clearly the anatomy and physiology of the ruminant digestive system.
- To describe the process of digestion.

The digestive system of dairy cattle differs from that of nonruminant animals (e.g., horse, pig, and human) in several important ways. It is important to understand these differences to feed for optimal productivity. Starting at the beginning of the digestive system (the mouth), there are clear differences between ruminants (for example, dairy cattle) and nonruminants. Ruminants have no upper incisor or canine teeth. Thus, they depend on the upper dental pad and lower incisors, along with the lips and tongue, for prehension of feed (Figure 11.1).

Ruminants possess four stomach compartments: rumen, reticulum, omasum, and abomasum (true stomach) (Table 11.1), whereas monogastrics have one. Thus, they have the necessary space for process-ing large quantities of bulky forages to provide their nutrients. The cow, when compared to a human on a proportion-to-weight basis, for example, has about nine times the digestive tract capacity. In addition, the rumen provides a highly desirable environment for an enormous population of microorganisms. Typical counts of rumen bacteria range from 25 to 50 billion/milliliter, and typical counts of protozoa range from 200,000 to 500,000/milliliter. The number of rumen bacteria varies according to the nature of the diet, feeding regimen, time of sampling after feeding, season, and the presence or absence of ciliate protozoa.

Rumen microorganisms serve two important functions; they make it possible for ruminants to utilize roughage and to digest fiber. They break down cellulose and pentosans in feeds into usable organic acids, chiefly acetic, propionic, and butyric acid, commonly referred to as the volatile fatty acids (VFAs). These VFAs are largely absorbed through the rumen wall and provide 60 to 80 percent of the energy needs of the ruminant. Microbial digestion is of great practical importance in the nutrition of ruminants; it is the fundamental reason why they can be maintained almost entirely on a roughage diet.

In a true type of symbiotic relationship, rumen microbes also synthesize nutrients for their host. Rumen microbes synthesize, or manufacture, all the B complex vitamins and all the essential amino acids. Amino acids can even be made from nonprotein nitrogen compounds (NPNs), such as urea or ammoniated products, or from poor quality feed proteins. Finally, the microorganisms that exit the rumen are digested further along in the gastrointestinal tract.

A placid cow lying under a tree slowly chewing her cud conveys a special sense of contentment, symbolic of the tranquility of the countryside. But this activity, or phenomenon, which is peculiar to ruminants, is of great practical significance. During rumination, the animal regurgitates and rechews a

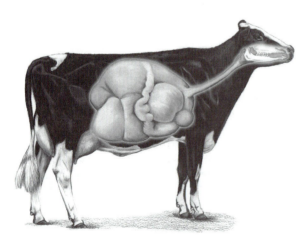

Figure 11.1 Position of the four stomach compartments in relation to the oral cavity in a dairy cow. *(Courtesy of Dana Boeck)*

TABLE 11.1   Percentage of Bovine Stomach Tissue Contributed by Each Compartment

| | Age in Weeks | | | | | | |
Compartment	*0*	*4*	*8*	*12*	*16*	*10–16*	*34–38*
Reticulo-rumen	38%	52%	60%	64%	67%	64%	64%
Omasum	13%	12%	13%	14%	18%	22%	25%
Abomasum	43%	36%	27%	22%	15%	14%	11%

Source: Ruminant Digestive System, p. 2: http://www.afns.ualberta.ca/drtc/dp472-5a.htm.

soft mass of coarse feed particles, called a bolus. Each bolus is chewed for about a minute, then swallowed again. Ruminants may spend eight hours or more per day ruminating, the actual time varying according to the nature of the diet. Coarse, fibrous diets result in more time ruminating. Rechewing serves two important functions: reducing particle size of the forage and stimulating secretion of saliva, an important buffer for rumen pH. Thus, rumination has an important bearing on the amount of feed the animal can eat and utilize. Feed particle size must be reduced to allow passage of the material from the rumen. Because high-quality forages are lower in fiber than low-quality forages, they require much less rechewing and pass out of the rumen at a faster rate; hence, although they allow the cow to eat more, they also decrease saliva flow to the rumen and increase the risk for rumen acidosis.

Substantially more gas is produced following microbial digestion by ruminants than by digestion in simple-stomached animals. The microbial fermentation in the rumen results in the production of gases (primarily carbon dioxide and methane) that must be eliminated; otherwise, bloat results. Normally, these gases are expelled quite freely by eructation (belching) and, to a lesser extent, by absorption into the blood draining from the rumen, from which they are eliminated through exhaled air from the lungs.

PROCESS OF DIGESTION

Taken in a narrow sense of the word, digestion can be defined as the process whereby proteins, fats, and complex carbohydrates are broken down into units that are of small enough size to be absorbed. This process is accomplished primarily through the action of digestive enzymes. Enzymes are organic catalysts that speed biochemical reactions at ordinary body temperatures without being used up in the process. Enzymatic activity is responsible for most of the chemical changes occurring in feeds as they move through the digestive tract. Digestion itself occurs through the entire digestive tract; the

first digestive processes quite literally occur in the mouth of the animal.

Three physical processes occur in the oral region of cattle: prehension, mastication, and the initiation of deglutition. Prehension can be defined as the act of bringing food into the mouth. The cow relies on structures of the mouth, such as the tongue, lips, and teeth. Mastication is the act of chewing food. It involves the physical grinding and tearing of the food in addition to the admixture of saliva that lubricates the food as well as initiates a limited amount of enzymatic digestion. Food that has been masticated and formed into a small compact ball for passage down the digestive tract is called a bolus.

Deglutition is the act of swallowing. This process involves both voluntary and involuntary reflexes. After completion of mastication, the bolus is lifted by the tongue and moved to the back of the mouth. The bolus passes through the pharynx, causing a temporary inhibition of respiration by the reflex closure of the larynx, and finally passes down the esophagus to the gastric region.

The teeth serve primarily as a mechanical aid for mastication. By tearing and grinding the food, they increase the surface area of feed, thereby increasing exposure to the digestive fluids of the tract. Cattle, being entirely herbivorous (plant eaters), do not require many teeth for the tearing of food; hence, they have no canine teeth, and incisors are found only on the lower jaw, where they are used for shearing forages in prehension.

In cattle, the tongue is the primary structure for prehension. The tongue is elongated and covered with rough papillae, making it adapted to wrapping around grass and other forages. The cow then brings the forage into the oral cavity, where it is sheared by the movement of the incisors against the dental pad. Throughout the process of mastication, the tongue serves a threefold purpose. First, movement of the tongue transports the feed to the various areas of the mouth to be torn and ground. While doing this, the tongue is also mixing the feed with the various secretions of the mouth, ultimately forming a bolus. Second, the presence of taste buds on the tongue

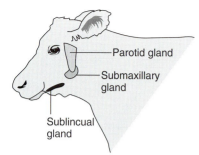

Figure 11.2 The location of the main salivary glands. *(Courtesy of M. E. Ensminger)*

provides neurological control for feed selection and intake. If the feed is bitter or unpalatable, impulses from the taste buds signal the animal to stop eating. Conversely, desirable tastes stimulate appetite. In the cow, the tip of the tongue contains a large number of taste buds, while the middle portion has very few. The back portion of the tongue contains the highest density of taste buds. Finally, the tongue initiates the process of deglutition. When the bolus has been adequately prepared, the tongue moves it to the back of the mouth where neural receptors are stimulated and swallowing commences.

The salivary glands represent a network of accessory structures that are essential to digestion. Three pairs of salivary glands are of primary importance: parotid, submaxillary, and sublingual (Figure 11.2). Salivary secretions act as aids in mastication, the formation of the bolus, and swallowing. Without this moisture, swallowing would be extremely difficult. Saliva provides a means for recycling nutrients back to the rumen; it contains considerable amounts of urea, mucin, phosphorus, magnesium, and chloride, all of which can be readily utilized by the bacteria and protozoa in the rumen. Gas can accumulate in the rumen and cause serious bloating if the eructation process is impaired. Saliva, acting as a surfactant, helps to prevent this problem. Saliva also solubilizes several of the chemicals in the feed, which can be detected by the taste buds once they are in solution. The membranes within the mouth must be kept moist to remain viable. Saliva provides one means by which this is accomplished. Perhaps most important for the lactating cow, a large quantity of sodium and other cations are secreted in saliva, thus serving as a buffer in the ingesta. The buffering capacity of saliva is critical; over 45 gallons of saliva can be secreted daily into the rumen.

The pharynx is the structure that controls the passage of air and feed. In this organ, the openings of the mouth, esophagus, posterior nares, eustachian tubes, and larynx come together. During the act of swallowing, the arytenoid cartilages close

the opening into the larynx, and the epiglottis is passively folded over the opening of the larynx. This forces food into the esophagus, thus preventing any feed from passing into the respiratory tract.

The esophagus is a muscular tube extending from the pharynx to the cardia of the rumen. The musculature and innervation of the esophagus are such that peristaltic waves move the bolus. Peristalsis is the coordinated contraction and relaxation of smooth muscles, which create a unidirectional movement pushing the bolus through the digestive tract.

The primary differences in digestion among domestic livestock can be traced to the specialized development of the gastric region. The ruminant has been described as having four stomachs. In reality, the ruminant possesses a complex stomach consisting of four morphologically distinct compartments. These compartments are rumen, reticulum, omasum, and abomasum.

The rumen and reticulum are closely related in physiological function and are often discussed together (Figure 11.3). The esophagus empties into the atrium ventriculi, a convex area formed by both the rumen and reticulum. The rumen is an extremely large compartment in the adult ruminant and is lined with a large number of papillae that increase the surface area for the absorption of the end products of microbial digestion. The reticulum, a structure with an interior that resembles a honeycomb, acts as a collection compartment for foreign objects as well as an organ for digestion (Figure 11.4). Ingested feed passes into these two compartments and is digested thoroughly through the action of various microorganisms (bacteria and protozoa) present in the rumen (Figure 11.5). The rumen, in effect, is a large physiological fermentation vat.

The microbes of the rumen digest carbohydrates to produce carbon dioxide and volatile fatty acids. Although several volatile fatty acids are produced, the primary end products are acetate, propionate, and butyrate. These products are then absorbed from the rumen and supply much of the energy required by the animal. Quite often, when high-concentrate rations are used, large quantities of lactic acid are produced, and the pH of the rumen falls. Since most of the bacteria in the rumen are pH sensitive, any dramatic shift in pH both decreases the total numbers of microbes and alters the proportions of the various types of microorganisms. When the ruminal pH drops too low, the animal goes off feed, a symptom of acute digestive problems.

Rumen microbes convert ingested lipids to fatty acids and glycerol. Glycerol is then primarily converted to propionate, with the long-chain fatty acids passing down to the small intestine for ab-

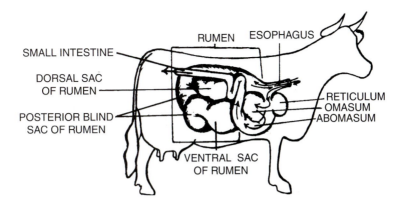

Figure 11.3 The ruminant digestive system. *(Courtesy of M. E. Ensminger)*

Figure 11.4 An inside view of the ruminant stomach compartments. Note the honeycomb appearance of the reticulum. *(Courtesy of Leo Timms)*

Figure 11.5 Hans Jung examines alfalfa sections before and after digestion by rumen bacteria. *(Courtesy of USDA-ARS)*

sorption. Very few dietary proteins escape the degradation process of the rumen. The degree to which dietary protein is degraded depends partially on its solubility. Most dietary proteins are metabolized by the microorganisms and are incorporated as microbial protein. The microbes are then passed down the tract and digested to provide a source of high-quality protein for the animal. With the degradation of the various dietary proteins, ammonia is produced in the rumen that can then either be absorbed through the rumen wall or provide nitrogenous precursors for the synthesis of bacterial protein. If the ration is high in sugars and starches, ammonia concentration is depressed.

Throughout the fermentation process in the rumen, various gases are produced and expelled through eructation (belching). Methane and car-

bon dioxide are the two gases produced most abundantly in the rumen; methane constitutes 30 to 40 percent of the total rumen gas volume and carbon dioxide, approximately 65 percent.

The omasum, or manyplies, is the next compartment for digestion. It contains numerous laminae (tissue leaves) that help grind ingesta. The exact physiological function of this compartment has not been fully elucidated, but it serves to absorb water and volatile fatty acids in addition to its function of grinding ingesta.

The abomasum is analogous to the stomach of the nonruminant. Digestive processes of this compartment are very similar to those of the stomach in the nonruminant.

The pancreatic region involves the pancreas and the pancreatic duct, a duct leading from the pancreas to the small intestine. The pancreas, an accessory organ of digestion, is a glandular structure that plays an essential role in the digestive physiology of animals. The pancreas, being both an endocrine and exocrine gland, serves two physiologically distinct functions. The endocrine function is that of the secretion of the hormones insulin and glucagon. The exocrine function deals with the production and secretion of fluids that are necessary for digestion within the small intestine.

In addition to the pancreas and salivary glands, the liver is an indispensable accessory organ of the gastrointestinal tract. From the stomach and small intestine, most of the absorbed nutrients travel through the portal vein to the liver, the largest gland in the body. The liver not only plays an important part in nutrient metabolism and storage, but also forms bile, a fluid essential for lipid absorption in the small intestine. The primary role of the liver in digestion and absorption is the production of bile. Bile facilitates the solubilization and absorption of dietary fats and also aids in the excretion of certain waste products such as cholesterol and the by-products of hemoglobin breakdown. The greenish color of bile is due to the end products of red blood cell destruction: biliverdin and bilirubin. Bile contains a number of salts resulting from the combination of sodium and potassium with bile acids. These salts combine with lipids in the small intestine to form micelles. Micelles are colloidal complexes of monoglycerides and insoluble fatty acids that have been emulsified and solubilized for absorption. When the micelle has been formed, the lipid can be digested and the resulting products (fatty acids and glycerol) can cross the mucosal barrier of the small intestine and enter the lymphatic system. Bile salts do not travel with the lipid, however; rather, they are recycled into the enterohepatic circulation.

The volume of bile production is highly variable. An animal that has been starved produces little bile. Conversely, an animal that is fed a high-fat ration produces substantial quantities of bile to keep up with absorptive requirements. Generally, the volume of bile depends on blood flow, the nutritive state of the animal, the type of ration being fed, and the enterohepatic bile salt circulation.

The small intestine is divided anatomically into three sections: duodenum, jejunum, and ileum. The first segment, the duodenum, originates at the pyloric sphincter of the stomach and is closely attached to the body wall by a short mesentery. Both bile and pancreatic fluids are emptied into this segment. The next section is the jejunum. There is no clear demarcation between the jejunum and the ileum, but it is arbitrarily defined as the free border of the ileocecal fold.

Throughout the luminal surface of the small intestine lies an extensive network of fingerlike projections called villi. Each villus contains a lymph vessel called a lacteal and a series of capillary vessels. On the surface of the villi are a great number of microvilli that provide further surface area for absorption. The small intestine terminates at the ileocecal valve, a sphincter that controls the flow of ingesta from the small intestine into the cecum and large intestine. This structure prevents the backflow of ingesta into the small intestine.

The cecum and colon in cattle are composed of several layers of muscle. A circular layer of muscle forms the basic tube of the colon and facilitates movement. In addition to this layer of muscle, three strips of longitudinal muscle form the taenia coli. These strips form a series of pouches or sacculations throughout the colon and are called haustrae. Ingesta is held in these saclike structures to facilitate the removal of water. Numerous mucous-secreting goblet cells can be found in the colon, but villi, such as the type found in the small intestine, are absent.

At the proximal end of the colon is a blind sac called the cecum. The cecum of the cow is not very well developed and plays a rather insignificant role in digestion. Some absorption of volatile fatty acids occurs in the cecum, however; and considerable amounts of water and electrolytes are absorbed in the colon.

SUMMARY

- Ruminants possess four stomach compartments: rumen, reticulum, omasum, and abomasum (true stomach), whereas monogastrics have one.
- Rumen microorganisms serve two important functions: they utilize roughage and digest fiber allowing ruminant animals to effectively utilize roughage.
- Rechewing serves two important functions: it reduces particle size and stimulates secretion of saliva.
- Digestion is the process whereby proteins, fats, and complex carbohydrates are broken down so they can be absorbed.

- Enzymes are organic catalysts that speed biochemical reactions at ordinary body temperatures without being used up in the process.
- Three physical processes occur in the oral region of cattle: prehension, mastication, and the initiation of deglutition.
- Three pairs of salivary glands are of primary importance: parotid, submaxillary, and sublingual.
- Saliva aids in mastication, the formation of the bolus, swallowing, and recycling nutrients. It acts as a surfactant and buffer, solubilizes several chemicals, and keeps mouth membranes moist.
- The microbes of the rumen digest carbohydrates to produce carbon dioxide and VFAs, primarily acetate, propionate, and butyrate.
- The pancreas is both an endocrine and exocrine gland that secretes hormones such as insulin and glucagon and produces and secretes fluids necessary for digestion within the small intestine.
- The liver is important in nutrient metabolism, nutrient-storage, and formation of bile.
- The small intestine is divided anatomically into three sections: duodenum, jejunem, and ileum.
- Circular and longitudinal muscles facilitate digesta movement through the gastrointestinal tract.

QUESTIONS

1. What teeth are missing in ruminants?
2. What causes the number of rumen bacteria to vary?
3. What compounds provide 60 to 80 percent of the energy needs of the ruminant?
4. What is the importance of microbial digestion?
5. What is the percentage of bovine stomach tissue in a nine-month-old heifer?
6. Why is it important for the ruminant to eliminate gases such as carbon dioxide and methane?
7. The process of digestion is accomplished through what action?
8. Where does the first digestive process occur?
9. Define mastication, deglutition, and name the primary structure for prehension.
10. How much saliva can be secreted daily into the rumen?
11. Define peristalsis.
12. Which component looks like a honeycomb? Which is covered with papillae? With luminae?
13. What would happen if a dairy cow received a very high concentrate ration?
14. What is the primary role of the liver?
15. What are the functions of bile?
16. What structure prevents backflow of ingesta into the small intestine?

ADDITIONAL RESOURCES

Books

Johnson, D. E., G. M. Ward, and J. J. Ramsey. "Livestock Methane: Current Emissions and Mitigation Potential," pp. 219–233 in *Nutrient Management of Food Animals to Enhance and Protect the Environment*, E. T. Kornegay, ed. Boca Raton, FL: CRC Press, Inc., 1996.

Satter, L. D., and Z. Wu. "New Strategies in Ruminant Nutrition: Getting Ready for the Next Millenium," pp. 1–24 in *Southwest Nutrition and Management, Conference Proceedings*. Phoenix, AZ: University of Arizona, Tucson.

Internet

Dairy Cattle Nutrition and the Environment: http://www.nap.edu/html/dairy_cattle/ch12.pdf.

Digestive Anatomy in Ruminants: http://www.vivo.colostate.edu/hbooks/pathphys/digestion/herbivores/rumen_anat.html.

Rumensin for Lactating Dairy Cows: http://www.westerndairyscience.com/html/U%20of%20A%20articles/html/Rumensin.html.

Rumen Physiology and Rumination: http://www.vivo.colostate.edu/hbooks/pathphys/digestion/herbivores/rumination.html.

12

Fundamentals of Dairy Nutrition

- To list the body composition of cattle.
- To describe nutrient use in dairy cattle.

Dairy animals inherit certain genetic possibilities, but how well this potential is achieved depends on the environment to which they are subjected. The most important influence in the environment is nutrition. In turn, all feed comes directly or indirectly from plants that have their tops in the sun and their roots in the soil. Hence, we have the nutrition cycle as a whole, from the sun and soil, through the plant, to the animal, and back to the soil again.

Thus, nutrition is the science of the interaction of a nutrient with some part of a living organism. It begins with the fertility of the soil and the composition of plants, and it includes the ingestion of feed; the liberation of energy; the elimination of wastes; and all the syntheses essential for maintenance, growth, reproduction, and lactation. A solid understanding of nutrition is important because dairy animals depend on nutrients for the processes of life.

The primary purpose of raising dairy cattle is to transform feeds into milk. But this conversion must be done efficiently and economically. The principles of nutrition must be applied and complemented by superior breeding, good health, and competent management. Like other sciences, nutrition does not stand alone. It draws heavily on the basic findings of biochemistry; physics; microbiology; physiology; genetics; mathematics; endocrinology; and, most recently, animal behavior, ecology, and biotechnology. In turn, it also contributes richly to each of these fields of scientific investigation.

BODY COMPOSITION OF CATTLE

The body composition of cattle varies widely according to age and nutritional state. The bovine body is comprised of several basic components:

1. **Water.** On a percentage basis, the water content shows a marked decrease with advancing age, maturity, and fatness. In cattle, the water content from conception to production changes as follows: embryo soon after conception, 95 percent; newborn calf, 74 percent, 450-pound heifer, 69 percent; mature cow, 60 percent; and choice grade steer, 53.5 percent.
2. **Fat.** The percentage of fat increases with growth and fattening. As the percentage of fat increases, the percentage of water decreases.
3. **Protein.** The percentage of protein remains rather constant during growth but decreases as the animal fattens. On the average, there are 3 to 4 pounds of water per pound of protein in the body.
4. **Ash.** The percentage of ash shows the least change.
5. **Composition of gain.** Gain in weight does not provide an accurate measure of the actual gain in energy of the animal because it tells nothing about the composition of gain, nor does it indicate the relative importance of the components. For example, a tiny amount of carbohydrates (mostly glucose and glycogen) is present in the bodies of animals and found principally in the liver, muscles, and blood. Although these carbohydrates are very important in animal nutrition, they account for less than 1 percent of the body composition.

NUTRIENT INTAKE

The intake of nutrients by cattle is determined by the amount of feed they voluntarily consume (called voluntary dry matter intake) and the quality (or nutrient content) of that feed. The accurate prediction of voluntary dry matter intake and the accurate determination of feed composition are the cornerstones of any dairy nutrition program. Without an accurate estimate of feed intake, it is not possible to balance a ration to meet the nutrient needs of the animal. Therefore, it is important to understand the factors that control voluntary dry matter intake.

Several factors appear to be important in regulating feed intake in dairy cattle. One such factor is simply the physical capacity of the rumen. When the rumen becomes filled with feedstuffs, the stretching of the rumen wall triggers neural impulses that send a message to the appetite centers in the brain, and voluntary intake is reduced until the stretch receptors are no longer stimulated. This fill effect is stimulated by both the weight and volume of the feed that has been ingested. In practical feeding situations, it is usually associated with feeds having a high neutral detergent fiber (NDF) fraction. As the neutral detergent fiber fraction of forages increases, the digestibility of the forage decreases, and the forage spends more time in the rumen before microbial fermentation is completed. So feedstuffs with a slow rate of passage and/or low digestibility are more likely to reduce the dry matter intake of the cow because they are more likely to fill the rumen.

Anything that decreases feed digestibility slows the rate of passage of feed through the digestive tract and reduces dry matter intake. If the appropriate microbial populations are reduced in numbers, as occurs in cases of rumen acidosis, even highly digestible feeds can have a slower digestion pattern. This reduces the rate of passage of these feeds; the rumen becomes filled to capacity, thus activating the fill effect; and dry matter intake is reduced. This concept is important in dairy nutrition. Anything that disrupts the rumen environment and reduces the efficiency of microbial fermentation can decrease dry matter intake of cows in the same way. Therefore, one of the most crucial factors in feeding dairy cows, especially in terms of maximizing dry matter intake, is maintaining a stable and ideal rumen environment. This stable environment encourages a more rapid and complete digestion of forages, a faster rate of passage of feed through the rumen, and therefore an increased ability to consume more feed.

Many factors influence dry matter intake. It has long been theorized that cattle eat to meet their energy needs, and when these needs have been met, there is a metabolic feedback that inhibits further feed intake. In reality, the feedback mechanism appears to be more complex and is affected both by protein intake and energy intake as well as the ratio

Figure 12.1   Important research using dairy cattle is performed at the Henry A. Wallace Beltsville Agricultural Research Center (circa 1994). *(Courtesy of USDA-ARS)*

of protein to energy. There are also sensory and psychological constraints on intake; palatability of the feed, timing of meals, and the presence or absence of adequate lighting can all affect the dry matter intake of dairy cattle. Research in these areas is crucial to achieving the continued production increases producers strive to attain (Figure 12.1).

The science of nutrition goes well beyond providing the proper amounts of nutrients in a ration that meets the daily nutrient requirements of the animal. The best producers employ feeding practices that promote feed intake and enhance feed digestibility to obtain the most profit from each pound of feed provided to the dairy cow.

NUTRIENT USES

Dairy cattle do not utilize feeds as such. Rather, they use those portions of feeds called nutrients that are released by digestion and then absorbed into the body fluids and tissues. Nutrients are those substances, usually obtained from feeds, that can be used by the animal when they are made available in a suitable form for its cells, organs, and tissues. Nutrients include carbohydrates, fats, proteins, minerals, vitamins, and water. More specifically, the term *nutrient* refers to the more than forty nutrient chemicals, including amino acids, minerals, and vitamins. Energy is frequently listed with nutrients because it results from the metabolism of carbohydrates, proteins, and fats in the body.

Of the feed consumed, a portion is digested and absorbed for use by the animal. The remaining undigested portion is excreted and constitutes the major portion of the feces. Nutrients from the digested feed are used for several different body processes, the exact usage varying with the species, class, age, and productivity of the animal. All animals use a portion of their absorbed nutrients to carry on essential functions, such as body metabolism and maintaining body temperature and the replacement and repair of body cells and tissues. These uses of nutrients are referred to as the maintenance requirement. That portion of digested feed used for growth or the production of milk is known as the production requirement. Another portion of the nutrients is used for the development of the fetus and is referred to as the reproduction requirement.

Maintenance

Unlike machines, cattle are never idle. They use nutrients to keep their bodies functioning every hour of every day, even when they are not being used for production. Maintenance requirements may be defined as the combination of nutrients that are needed by the animal to keep its body functioning without any gain or loss in body weight or any productive activity. Although these requirements are relatively simple, they are essential for life itself. Cattle must generate heat to maintain body temperature, sufficient energy to keep vital body processes functional, energy for minimal movement, and the necessary nutrients to repair damaged cells and tissues and to replace those that have become nonfunctional. Thus, energy is the primary nutritive need for maintenance. Even though the quantity of other nutrients required for maintenance is relatively small, it is necessary to have a balance of the essential proteins, minerals, and vitamins. No matter how quietly cattle may be lying in a stall or in a pasture, they require a certain amount of fuel and other nutrients. The least amount on which an animal can exist is called its basal maintenance requirement.

There are only a few times in the normal life of cattle when only the maintenance requirement needs to be met. Such a status is closely approached by mature males not in service and by mature, dry, nonpregnant females. Nevertheless, maintenance is the standard benchmark or reference point for evaluating nutritional needs. One-third to one-half of the feed consumed by cattle as a whole is used to meet the maintenance requirement. Of course, on an individual basis, the higher the production of a lactating cow, the smaller the proportion of nutrients needed for maintenance.

Even though the maintenance requirement might be considered as an expression of the non-production needs of cattle, many factors affect the amount of nutrients necessary for this vital function, including exercise, weather, stress, health,

Figure 12.2 Vic Wilkerson prepares a cow for a feeding test inside a large calorimeter. *(USDA-ARS)*

body size, temperament, and individual variation. For example, the colder or hotter it gets from the most comfortable (optimum) temperature, the greater will be the maintenance requirement (Figure 12.2).

Growth

Growth may be defined as the increase in size of bones, muscles, internal organs, and other parts of the body. Growth is influenced primarily by nutrient intake. Growth is the very foundation of animal production. Heifers may have their reproductive ability seriously impaired if they grow improperly.

Reproduction

Being born and born alive are the first and most important requisites of dairy production. If cows fail to reproduce, the dairy producer is soon out of business. Approximately 70 to 90 percent of all inseminations do not result in viable pregnancies, and on average 25 percent of all cows culled from dairy herds are removed because of reproductive inefficiency. Certainly, there are many causes of reproductive failure, but nutritional inadequacies play a major role.

Overfeeding accompanied by extremely high body condition or underfeeding accompanied by emaciated and rundown body condition may result in temporary sterility. Overconditioned females often experience birth difficulties. Excessive thinness results in low birthweights and weak offspring. Inadequate energy intake during the last trimester of pregnancy and immediately following parturition has a marked adverse effect on rebreeding; fewer of these cows will come in heat and fewer will conceive.

Lactation

The lactation requirements for moderate to heavy milk production are much more rigorous than are the maintenance or pregnancy requirements. Fortunately, heavy milking cows can store up body reserves of certain nutrients before calving to be utilized following parturition. When lactational demands are greater than can be obtained from the feed, the cow draws from the stored body reserves. Thus, both calcium and phosphorus can be stored in the bones and then withdrawn during early lactation when milk production is at its peak.

One of the most dramatic changes in the life cycle of cows occurs at freshening, when a female suddenly makes the transition from nonlactating to lactating. The nutrient needs for lactation depend on the amount and composition of milk secreted. Although high-producing cows require more total feed than do low producers, they utilize proportionately more nutrients for milk production, and generally they return more net income over feed cost.

SUMMARY

- The most important environmental influence on dairy cattle is nutrition.
- Body composition of cattle includes water, fat, protein, and ash.
- Nutrient uses include maintenance, growth, reproduction, and lactation.
- Maintenance requirement may be defined as the combination of nutrients that are needed by the animal to keep its body functioning without any gain or loss in body weight or any productive activity.
- Extremes of overfeeding or underfeeding can result in reproductive failure.
- Nutrient needs for lactation depend on the amount and composition of milk secreted.

QUESTIONS

1. What is the primary purpose of keeping dairy cattle?
2. What percentage of a newborn is water?
3. What is the primary nutritive need for maintenance?
4. Define basal maintenance requirement.
5. List some factors that affect the amount of nutrients necessary for maintenance.

ADDITIONAL RESOURCES

Articles

Rukkwamsuk, T., T. A. Kruip, and T. Wensing. "Relationship Between Overfeeding and Overconditioning in the Dry Period and the Problems of High Producing Dairy Cows During the Postparturient Period." *Veterinary Quarterly* 21 (June 1999): 71–77.

Weiss, W. P., and D. J. Wyatt. "Digestible Energy Values of Diets with Different Fat Supplements When Fed to Lactating Dairy Cows." *Journal of Dairy Science* 87 (May 2004): 1446–1454.

Internet

http://pubs.caes.uga.eud/caespubs/pubcd/b1111-w.html.

13

Protein and Energy Requirements

OBJECTIVES

- To list the protein and energy requirements of dairy cattle including the need for carbohydrates, fats, and proteins.
- To describe ruminal digestion of carbohydrates, proteins, and lipids.

Cows require the same basic nutrients, carbohydrates, fats, proteins, minerals, and vitamins, to function as do all other classes of mammals. The amount of each nutrient needed is determined by the age of the cow and the productive demands.

Lack of energy is the most common deficiency of dairy rations. In young animals, an insufficient supply of energy results in retarded growth and a delay in the onset of puberty. In lactating cows, it results in a decline in milk yields and a loss in body weight; severe and prolonged energy deficiency depresses reproductive performance. Most of the energy requirement is met through intake of carbohydrates, although fats and protein are also used for energy.

The National Research Council energy requirements for dairy cattle are expressed as total digestible nutrients (TDN), digestible energy (DE), metabolizable energy (ME), net energy for maintenance (NE_M), net energy for body gain (NE_G), and net energy for lactation (NE_L). Separate energy values for maintenance (NE_M) and body gain (NE_G) are given because animals use energy for maintenance more efficiently than they do for growth. However, the efficiency of energy use by lactating cows for maintenance, pregnancy, and milk production is similar; so only one energy value, net energy for lactation (NE_L), is used for these functions.

The energy value of a feed may be separated into the losses that occur in digestion and metabolism and the net energy (NE) that is available to the animal for maintenance and production (Figure 13.1). The total energy in feed, which is determined by complete oxidation (burning) of the feedstuff and measurement of the heat produced, is known as gross energy and is expressed as calories. Common feedstuffs are similar in gross energy content, but they differ in feeding value because of variations in digestibility. About 60 percent of the total energy in grain and 80 percent of the total energy in roughage is lost in feces, urine, gases, and heat. Energy is also expressed as total digestible nutrients (TDN). TDN is comparable to digestible energy. It has been in use longer than the net energy system, and more values are available for feedstuffs. TDN is computed as follows:

$$TDN = \text{digestible carbohydrate}[1] + \text{digestible crude fiber} + \text{digestible protein} + (\text{digestible fat}[2] \times 2.25)$$

[1]Calculated as nitrogen-free extract
[2]Calculated as ether extract

CARBOHYDRATES

Carbohydrates are the major source of energy for dairy cattle. They constitute 50 to 80 percent of the dry matter of forages and grains. Three major categories of carbohydrates exist in feeds: simple sugars (glucose or sucrose), stored carbohydrates (starch), and structural carbohydrate or fiber (cellulose and hemicellulose). Sugars are found in the cells of growing plants and in feeds such as molasses. Starch is the main component of grain. Cellulose and hemicellulose, which are classified as fiber, are made up of sugar molecules, as is starch, but they are bound together differently. Adult ruminants can digest fiber because the microbial population in the rumen breaks it down into usable products (Figure 13.2). However, lignin, which is not a true carbohydrate, is almost indigestible.

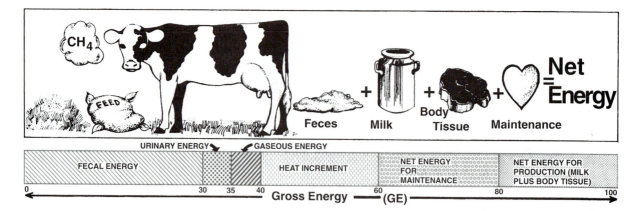

Figure 13.1   The partitioning of gross energy by a lactating cow.

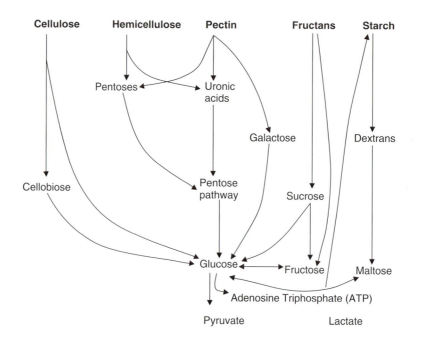

Figure 13.2   Carbohydrate diagram.

Nonstructural carbohydrates (NSC) are often estimated rather than directly determined chemically by using the following formula:

$$100 - (NDF + crude\ protein + fat + ash)$$

This estimate includes the cumulative errors of the analyses of all the included nutrients; it does *not* include any indication of rumen solubility or degradability. It is a fairly accurate estimate of the actual NSC content of grains but typically underestimates the true NSC content of forages.

FAT

Fat is mainly used in the rations of young calves, but it may also be added to the ration of lactating cows to increase energy density and reduce feed dusti-

ness. Fat contains over twice the energy density of carbohydrates or proteins, so it is typically used to increase the energy density of rations in animals where dry matter intake is limited. In addition to the fat present in natural feedstuffs (generally less than 2 to 3 percent), dairy cows can utilize 1 to 1.5 pounds of additional fat per day. Fat intakes that exceed this limit have direct detrimental effects on microbial function and especially on fiber digestion. Microbes do not use fat as an energy source; it is important to remember that although the ration may meet the needs of the animal, it may not meet the needs of the rumen microorganisms. Fat sources high in unprotected polyunsaturated fatty acids (soybean oil or corn oil) exert a greater negative effect on rumen microbes and depress fiber digestibility more than do saturated fatty acid sources (such as tallow). Whole oil seeds (soybeans, sun-

PROTEIN DIGESTION IN RUMINANTS

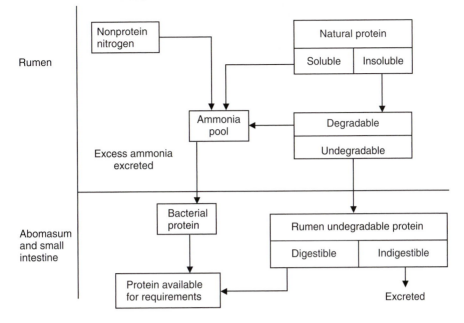

Figure 13.3 Protein digestion in ruminants.

flowers, or cottonseed) are good fat sources because they are slowly digested in the rumen, with the result that the oil is slowly released. Also, feeding rumen-protected fats that are resistant to microbial action in the rumen may allow feeding of higher levels of fat.

PROTEIN

Protein is essential for maintenance, growth, milk production, and the development of the fetus in dairy cattle. Also, it is required for the formulation of enzymes and certain hormones that control or regulate chemical reactions in the body. The protein requirement of the cow is really a requirement for amino acids.

Proteins are complex chemical structures that are made up of amino acids linked together in many different ways. Amino acids contain carbon; hydrogen; oxygen; nitrogen; and, in some cases, sulfur. There are twenty-two naturally occurring amino acids. Amino acids are supplied to the ruminant following digestion of microbial protein or feed protein that escapes microbial breakdown in the rumen (Figure 13.3). As milk production increases, a substantial amount of additional dietary protein from protein supplements must escape rumen fermentation to meet the cow's requirement for protein (Figure 13.4).

The protein composition of feeds and the protein requirements of dairy cattle may be expressed as crude protein, digestible protein, rumen degradable protein (RDP), rumen undegradable protein (RUP) and/or nonprotein nitrogen (NPN):

1. **Crude protein.** Chemically, most proteins contain 16 percent nitrogen; so crude protein is determined by finding the nitrogen content, then multiplying the result by 6.25 ($100 \div 16 = 6.25$). It is called crude protein because not all of the nitrogen in feeds is in the form of protein; rather, it is a combination of true protein and nonprotein nitrogen. Some feeds, particularly green roughage, contain one-third or more of their nitrogen as nonprotein nitrogenous substances such as amides, ammonium salts, amino acids, alkaloids, and other nitrogenous compounds. However, ruminal microorganisms make use of the various nitrogen sources for synthesis of microbial proteins that, in turn, are used by the cow (Figure 13.4). Consequently, for dairy cows and other ruminants, the amount of crude protein is about as good a measure for protein allowances as is the amount of true protein. The amount of protein needed in the total ration of lactating cows is determined primarily by the amount of milk produced. Milk is a rich source of high-quality

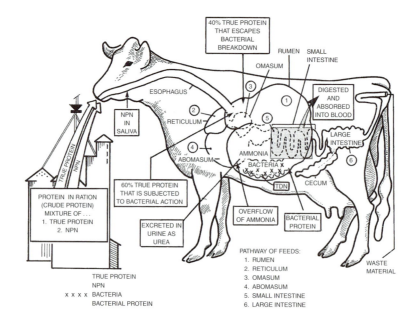

Figure 13.4   Protein utilization in a cow.

protein; as milk production increases, a substantial amount of dietary protein is necessary. When more protein is fed than needed, the excess is used as a source of energy. Because protein feeds are generally more expensive than carbohydrate feeds, it usually is more economical to feed only the amount needed. Besides, a large excess of dietary protein may decrease the energy supply because excess protein must be deaminated to ammonia and, for the most part, transformed back into urea for excretion.

2. **Digestible protein.** This is the amount of crude protein consumed less the crude protein excreted in the feces. The term *apparent digestibility* is more accurate, however, because it is recognized that a portion of the fecal nitrogen is derived from the animal and is not a feed residue.

3. **Rumen degradable protein (RDP).** This refers to the intake crude protein that is broken down (degraded) by microorganisms in the rumen. Approximately 60 percent of the crude protein in the typical dairy cow ration is broken down (degraded) to ammonia by microbial digestion. The rumen microbes must convert the ammonia to microbial protein in their own cells if the dairy animal is to receive any benefit. Fermentable energy must be available for the microorganisms to grow and synthesize the necessary amino acids. If rumen ammonia levels are excessively high, the ammonia is absorbed into the blood and either recycled or excreted in the urine as urea.

4. **Rumen undegradable protein (RUP).** This is the crude protein that is not broken down in the rumen; instead, it is swept out of the rumen to the abomasum and small intestine for breakdown there and absorption as peptides and amino acids. All feed protein sources are not degraded in the rumen to the same extent. The optimal ration meets both the nitrogen requirement of rumen microorganisms for maximum synthesis of microorganism protein and allows for maximum escape or bypass of high-quality feed protein for digestion in the small intestine. Protein synthesis by rumen microbes depends on feed intake, organic matter digestibility, feed type, protein level, and feeding system. Because 3.5 pounds of microbial protein synthesis per day is near the maximum, the remainder of the protein must be derived from nondegraded protein (RUP) sources. Young, fast-growing heifers and high-producing cows generally require additional RUP sources beyond their normal ration to meet total protein requirements. The more rapid the growth and the higher the milk production, the greater the quantities of RUP needed. Brewers' grain, distillers' grain, corn gluten meal, fish meal, meat meal, and heat-treated soybeans are examples of feed with reduced rumen degradability that may be substituted in rations in which excess rumen ammonia exists and less than optimal amounts of quality protein (undegraded) pass into the small intestine.

5. **Nonprotein nitrogen (NPN).** Feedstuffs that contain nitrogen in a form other than proteins or peptides are called nonprotein nitrogen (NPN). Nonprotein nitrogen compounds, such as urea and ammonium salts, have a crude protein value, but they do not supply any amino acids directly. The billions of microorganisms in the rumen convert nitrogen from NPN sources into amino acids for their growth and use. Then the microbes pass into the small intestine, where they are digested and release amino acids for absorption and utilization, the same as amino acids released from the digestion of true proteins (composed of amino acids) in feeds. Urea is a nonprotein nitrogen (NPN) compound (Figures 13.5 and 13.6), containing about 45 percent nitrogen, with a protein equivalent of 281 percent (45 percent N × 6.25).

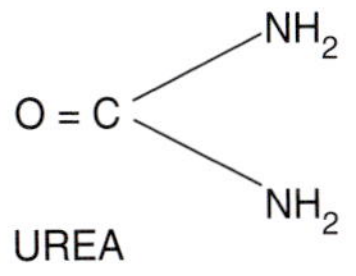

Figure 13.5 The chemical structure of urea reveals why it is such a potent source of nitrogen for microbial protein synthesis. *(Courtesy of M. E. Ensminger)*

$$\begin{array}{c} NH_2 \\ \diagdown \\ \diagup \\ NH_2 \end{array} C = O + H_2O \rightarrow 2NH_3 + CO_2$$

Figure 13.6 Urea is hydrolized by the urease activity of the rumen microorganisms with the production of ammonia. *(Courtesy of M. E. Ensminger)*

Urea use in the ration is similar to using degradable intake protein. It and other nonprotein nitrogen (NPN) compounds, such as ammonium salts, can be used to replace part of the protein required in dairy cattle rations after rumen function has become established. Urea is not palatable; therefore, it should be mixed thoroughly with the grain mix or silage. Molasses can improve acceptability. If cattle have not been fed urea previously, a seven- to ten-day adjustment period, in which the urea is gradually increased, helps to maintain feed intake and production. High levels of urea can be toxic; so excessive intakes should be avoided. Urea should not be top-dressed; it should be mixed in the feed.

Several ammoniated products or ammonium salts are used successfully as sources of nitrogen. Monoammonium phosphate, which contains about 11 percent nitrogen (crude protein equivalent of 68.25 percent), is also used for phosphorus supplementation. Ammonia (cooled to form a liquid or in a water solution) may be added to corn silage at the rate of 7 pounds (5 pounds of nitrogen) per ton. Urea should not be fed in the concentrate when ammonia or other NPN has been added to the corn silage. Because NPN products do not provide any energy, minerals, or vitamins, these nutrients must be provided through other sources.

SUMMARY

- Cows require carbohydrates, fats, proteins, minerals, and vitamins to function.
- Three major categories of carbohydrates exist in feeds: simple sugars, stored carbohydrates, and structural carbohydrates or fiber.
- Fat is added to the diet of young calves and to lactating cows to increase the energy density of the ration and reduce feed dustiness.
- Protein is essential for maintenance, growth, milk production, and the development of the fetus in dairy cattle.
- The protein composition of feeds may be expressed as crude protein, digestible protein, rumen degradable protein (RDP), rumen undegradable protein (RUP), and/or nonprotein nitrogen (NPN).
- Protein synthesis by rumen microbes depends on feed intake, organic matter digestibility, feed type, protein level, and feeding system.

QUESTIONS

1. What is the most commonly found deficiency in dairy rations?
2. Where are sugars found? Where is starch found?
3. How many naturally occurring amino acids are there?
4. Most proteins contain what percentage of nitrogen?

5. What is the protein percentage of soybean meal?
6. How is the amount of protein needed in the total ration determined?
7. What is digestible protein? Rumen degradable protein?
8. What converts nitrogen from NPN sources into amino acids?
9. What pathway does pectin and hemicellulose use to form glucose?
10. What three volatile fatty acids are formed from glucose?

ADDITIONAL RESOURCES

Book

Butler, W. R., D. J. R. Cherney, and C. C. Elrod. "Milk Urea Nitrogen (MUN) Analysis: Field Trial Results on Conception Rates and Dietary Inputs," p. 89.

Articles

Butler, W. R., J. J. Calaman, and S. W. Beam. "Plasma and Milk Urea Nitrogen in Relation to Pregnancy Rate in Lactating Dairy Cattle." *Journal of Animal Science* 74 (April 1996): 858.

Faust, M. A., and L. H. Kilmer. "Determining Variability of Milk Urea Nitrogen Reported by Commercial Testing Laboratories." *1996 Dairy Report.* Iowa State University, Ames, IA.

Harris, B. Jr. "MUN and BUN Values Can Be Valuable Management Tools." *Feedstuffs.* (October 9, 1995): 14.

Jonker, J. S., R. A. Kohn, and R. A. Erdman. "Milk Urea Nitrogen Target Concentrations for Lactating Dairy Cows Fed According to National Research Council Recommendations." *Journal of Dairy Science* 82 (June 1999): 1261–1273.

Roseler, D. K., J. D. Ferguson, C. J. Sniffen, and J. Herrema. "Dietary Protein Degradability Effects on Plasma and Milk Urea Nitrogen and Milk Nonprotein Nitrogen in Holstein Cows." *Journal of Dairy Science* 76 (March 1993): 525.

Internet

Amount and Degradability of Protein for Lactating Cows: http://www.traill.uiuc.edu//dairynet/paperDisplay.cfm?ContentID=277.

Effect of Breed and Concentrations of Dietary Crude Protein and Fiber on Milk Urea: http://ohioline.osu.edu/sc169/sc169_11.html.

Microbial Protein Synthesis in the Rumen: http://www.calfnotes.com/pdffiles/CN031.pdf.

Urea and NPN for Cattle and Sheep: http://www.ext.colostate.edu/pubs/livestk/01608.pdf.

14

Requirements for Minerals, Vitamins, and Water

OBJECTIVES

- To identify the macrominerals and microminerals.
- To distinguish between the fat-soluble and water-soluble vitamins.
- To describe the requirements for minerals, vitamins, and water.

Minerals are inorganic elements, frequently found as salts with either inorganic elements or organic compounds. Dairy cattle require at least fifteen mineral elements. They are needed for both structural and regulatory functions. Minerals are also needed for bone and teeth formation and to maintain acid-base balance, water balance, and enzyme and hormone systems. They are components of certain substances within the body, such as iron in hemoglobin. A lactating cow needs minerals for the developing fetus and for milk production. Milk contains about 0.7 percent minerals; thus, a cow producing 30,000 pounds of milk during lactation secretes 210 pounds of minerals per year in her milk.

Mineral excesses should be avoided because of interaction with other minerals and possible toxicities and undesirable interactions. The maximum tolerable level for a mineral element has been defined as the dietary level that, when consumed for a limited period, does not impair animal performance and should not produce unsafe residues in human food derived from the animal. When dairy cattle are fed mixed feed, in part or totally, the needed minerals are usually incorporated in the ration to meet their requirements. When animals are fed an unmixed ration or are on pasture, ad libitum supplementation of minerals is commonly practiced.

MACROMINERALS

The macrominerals of importance in dairy cattle nutrition are salt (sodium chloride), calcium, phosphorus, magnesium, potassium, and sulfur.

Salt (sodium chloride [NaCl]). Sodium and chlorine are usually provided in the form of common salt (NaCl). However, potassium chloride may be used as a source of chlorine also. Excessive levels of chlorine without sodium or potassium can contribute to acidosis in dairy cattle.

Calcium (Ca). Whole milk contains 0.12 percent calcium. A deficiency of calcium may cause slow growth and poor bone development, easily fractured bones (Figures 14.1 and 14.2), reduced milk yield, and increased incidence of milk fever (Figure 14.3). Feeding calcium (Figure 14.4) at more than 0.95 to 1.0 percent dry matter (DM basis) in mixed rations may reduce dry matter intake and lower performance.

Phosphorus (P). Whole milk contains 0.09 percent phosphorus. A deficiency of phosphorus may result in fragile bones, stiff joints, poor growth, low blood P (less than 4.6 mg/100 ml), depraved appetite (chewing wood, hair, and bones) (Figure 14.5), and poor reproductive performance. Excessive phosphorus intakes may cause bone resorption, elevated plasma phosphorus levels, and urinary calculi.

Magnesium (Mg). Milk contains a substantial amount of magnesium (about 0.015 percent). Thus, when expressed as a percentage of the ration, the magnesium requirement increases with the cow's level of milk production. Under practical conditions, magnesium deficiencies may occur when dairy cattle, especially older and lactating cows, are grazing lush, rapidly growing pastures that have been highly fertilized with nitrogen or potassium, or both, during cool seasons. Under conditions conducive to grass tetany and for high-producing cows in early lactation, the

Figure 14.1   A calcium-deficient Jersey cow. *(Courtesy of the University of Florida)*

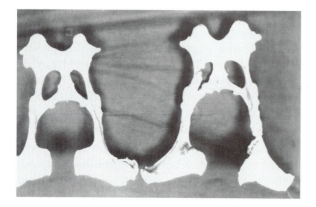

Figure 14.2   Depletion of bone is evident in these pelvic bones from calcium-deficient cows. *(Courtesy of the University of Florida)*

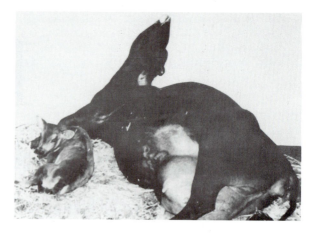

Figure 14.3   A Jersey cow with milk fever. *(Courtesy of Washington State University)*

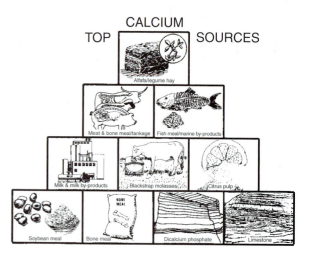

Figure 14.4   Sources of Calcium. *(Courtesy of M. E. Ensminger)*

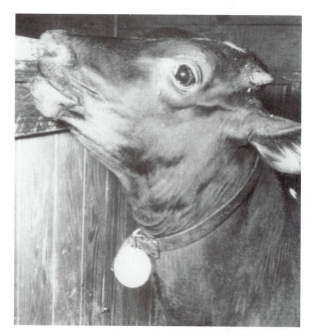

Figure 14.5   A cow with phosphorus deficiency exhibiting depraved appetite, or pica. *(Courtesy of Cornell University)*

suggested requirement is 0.25 to 0.3 percent dietary magnesium, with the supplemental magnesium provided in a readily available form such as magnesium oxide. Magnesium toxicity is not known to be a practical problem in dairy cattle.

Potassium (K). Milk contains about 0.15 percent potassium. Stress, especially heat stress, appears to increase the need for potassium perhaps because of greater loss of potassium through sweat. The signs of relatively severe potassium deficiencies in lactating cows include a marked decrease in feed intake, loss in weight, decreased milk yield, pica, loss of hair glossiness, decreased pliability of the hide, lower plasma and milk potassium, and higher hematocrit readings. Generally, forages contain considerably more potassium than is required by dairy cattle. High levels of potassium (3 percent or above) in very lush forages grown on high potassium soils in cool weather are considered to be a factor in causing both grass tetany and milk fever of lactating cows.

Sulfur (S). Milk contains 0.03 percent sulfur, much of which is in the form of the amino acids methionine and cystine. Sulfur is needed for microbial protein synthesis, especially when nonprotein nitrogen is fed to cattle.

MICROMINERALS

The trace minerals, or microminerals, of importance in dairy cattle nutrition are cobalt, copper, iodine, iron, manganese, molybdenum, selenium, and zinc.

Cobalt (Co). Normal cow's milk averages 0.38 to 1.04 micrograms of cobalt per quart. Colostrum contains four to ten times more cobalt than milk. Because cobalt is a component of vitamin B_{12}, ruminal microorganisms can synthesize this vitamin only when adequate cobalt is in the ration of the cow. Supplements of 30 to 45 grams of cobalt sulfate or 20 to 25 grams of cobalt carbonate with 100 pounds of salt have prevented any cobalt deficiency problems (Figure 14.6).

Copper (Cu). Colostrum contains more copper than does milk. The amount of copper in milk decreases with the length of lactation. Copper is needed for hemoglobin formation, although it is not actually contained in it. A deficiency of copper results in anemia and bleaching of the hair (Figure 14.7). Black hair turns gray and red hair becomes yellow. Higher levels may be required for cattle grazing pastures or consuming feedstuffs that contain

high levels of molybdenum or other interfering substances. Copper supplementation may be advisable under certain conditions (Figure 14.8), but it should be done with discretion. Excess copper is toxic (Figure 14.9) and is a primary cause of oxidized flavor in milk.

Figure 14.7   A dairy calf exhibiting signs of copper deficiency. *(Courtesy of the University of Florida)*

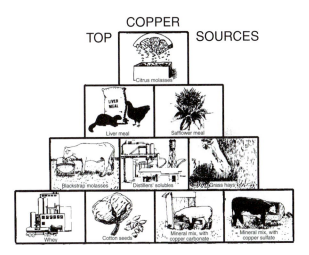

Figure 14.8   Sources of copper. *(Courtesy of M. E. Ensminger)*

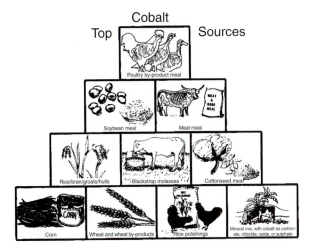

Figure 14.6   Sources of cobalt. *(Courtesy of M. E. Ensminger)*

Figure 14.9   Copper toxicity causes a darkening of kidney tissue and urine. *(Courtesy of University of Tennessee)*

Iodine (I). About 10 percent of the iodine intake of lactating cows is normally excreted in milk. Iodine deficiency can be detected by analyzing milk or blood serum. Iodine concentration of less than 9.5 to 19 micrograms per quart of milk or 37.8 micrograms per quart of serum indicate iodine deficiency. Goiter (an enlargement of the thyroid gland) occurs in newborn calves if their mothers are fed iodine-deficient rations; the necks of the calves are swollen. They are weak at birth, or they are born dead. Much of the small amount of iodine in the body is contained in the thyroid gland as thyroxin and diiodotyrosine, both of which are contained in the protein thyroglobulin, part of the thyroid hormone. The principal function of the thyroid gland is to regulate the metabolic rate. Many protein supplements (including soybean meal and cottonseed) are mildly goitrogenic because they reduce the availability of dietary iodine, and Brassica forages (cabbage, kale, rape) are highly goitrogenic. When stabilized iodine is used, a level of 0.0076 percent in salt is adequate. The Northwest and Great Lakes regions are the most iodine-deficient areas of the United States. Lactating cows should not receive excessive dietary iodine because the resulting high iodine content of the milk is considered undesirable for humans. The use of iodine disinfectants as teat dips or udder washes can increase the iodine content of milk, but the main cause of high iodine levels in milk is dietary iodine (Figure 14.10).

Iron (Fe). Iron is essential because it is a constituent in hemoglobin, the oxygen carrier in the blood. Cow's milk is low in iron, about 10 ppm. The iron requirements of a young calf are higher than those of a mature cow; the iron reserves of a newborn calf, which are primarily in the liver, are generally adequate to prevent serious anemia if calves are fed dry feeds at a few weeks of age. When calves are fed a milk diet exclusively for several weeks, however, they may develop iron deficiency anemia.

Manganese (Mn). Manganese deficiency in dairy cattle is seldom a problem (Figure 14.11). In general, forages contain higher levels of manganese than do grains. The manganese requirement for cattle is higher for reproduction than for growth. Little experimental work has been done on the manganese requirements of dairy cattle. Trace mineralized salt and commercial mineral supplements usually contain manganese. Manganese toxicity in cattle is unlikely.

Molybdenum (Mo). Molybdenum is an indispensable component of the enzyme xanthine oxidase, which is found in milk and distributed widely in animal tissue. Yet a deficiency of molybdenum has never been developed or observed in cattle. Molybdenum is known largely for its toxic characteristics; molybdenum toxicosis is a practical problem in grazing cattle in several areas of the world. There is an antagonistic relationship between molybdenum and copper. Elevated dietary molybdenum increases both the animal's requirements for copper and the amount of copper that causes toxicosis; increased dietary copper can reduce the toxic effect of molybdenum. Thus, the relative amounts of copper and molybdenum in the diet are important in determining the occurrence of molybdenum toxicosis. If the level of copper in the body is low, a lesser amount of molybdenum is toxic; as dietary copper increases, so does tolerance to molybdenum. High levels of both molybdenum and sulfur interfere with copper absorption. Molybdenum and sulfur also influence the metabolism of copper; added dietary molybdenum decreases the metabolism of copper, whereas added dietary sulfur en-

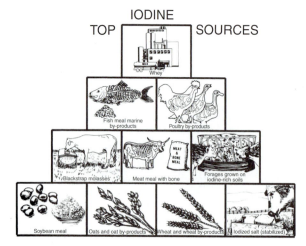

Figure 14.10   Sources of iodine. *(Courtesy of M. E. Ensminger)*

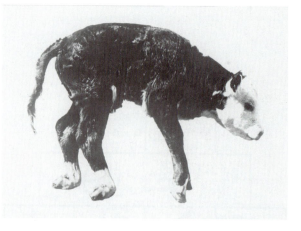

Figure 14.11   Manganese-deficient calves have weak legs, enlarged joints, stiffness, and twisted legs. *(Courtesy of Washington State University)*

hances it. The signs of molybdenosis have appeared in cattle that were fed about 6 ppm of molybdenum for several months.

Selenium (Se). Selenium, like molybdenum, was known for its toxic characteristics long before it was discovered to be an essential nutrient. However, research has firmly established that selenium is essential for ruminants; it is needed in trace amounts to prevent retarded growth; reproductive problems; retained placenta; white muscle disease, a condition that occurs in calves and lambs in selenium-deficient areas; and some mastitis problems. Also, it is closely associated with vitamin E. Both selenium and vitamin E protect cells from the detrimental effects of peroxidation, but each takes a different approach. Vitamin E is present in the membrane components of the cell and prevents free-radical formation, whereas selenium functions throughout the cytoplasm to destroy peroxides. This explains why selenium corrects some deficiency symptoms of vitamin E, but not others. Deficient or toxic selenium areas are widely scattered throughout the United States and the world.

Zinc (Zn). Milk generally contains about 4 ppm of zinc, but the concentration can double by increasing zinc intake. Zinc is involved in several enzyme systems and is affected adversely when excess quantities of calcium are present. Moderate excesses of zinc are not toxic to dairy cattle. Galvanized pipes and galvanized buckets, which are commonly used to provide water to cattle, contribute zinc along the way. Thus, it is unlikely that a zinc deficiency (Figure 14.12) would occur under normal circumstances; so zinc supplementation of dairy cattle may be considered as precaution only.

VITAMINS

Vitamins are complex organic compounds that are required in minute amounts by one or more animal species for normal growth, production, reproduction, and/or health. Dairy cattle, like other animals, require vitamins for optimum performance and health. Vitamins are classified as fat-soluble or water-soluble. The fat-soluble vitamins include vitamins A, D, E, and K; the water-soluble vitamins include the B vitamins and vitamin C.

Fat-Soluble Vitamins

Dairy cattle require fat-soluble vitamins A, D, E, and K. Generally, all classes of dairy cattle require a dietary source of vitamins A and E. Vitamin D must either be synthesized in the skin by action of ultraviolet radiation or be included in the ration. Rumen microbes synthesize adequate amounts of

Figure 14.12 Severe zinc deficiency results in hair loss, unhealthy appearance, and stiffness of joints. *(Courtesy of University of Georgia)*

vitamin K to meet the needs of most dairy cattle, except the young calf, whose rumen is not yet fully functional.

Under normal conditions, natural feeds furnish most fat-soluble vitamins or their precursors in adequate amounts. High-quality forages contain large amounts of vitamin A precursors, and vitamin E is abundant in most feeds. Vitamin D is found in large quantities in sun-cured forages. Cattle can store adequate reserves of the fat-soluble vitamins to meet their needs for several months. When dairy producers feed limited or low-quality forage, use high levels of ensiled forage, expose cattle to little sunlight, or use milk replacers for young calves, additional vitamins will probably be needed for optimum health and high performance.

Vitamin A. Vitamin A supplementation may be desirable when poor quality or limited amounts of forage are used, when using forage that has been stored for a long period and has lost its carotene through oxidation, or when high levels of corn silage and low-carotene concentrates are used. A deficiency of vitamin A causes many problems. Some or all of the following symptoms may occur, depending on the length and severity of the deficiency: night-blindness, a condition that is readily detected when animals are driven among obstacles in dim light; watery eyes; lack of coordination; convulsive seizures; complete blindness; stratified keratinized epithelium; increased susceptibility to infection; loss of appetite; rough hair coat; scaly skin; abortion; shortened gestation period; birth of dead, weak, or blind calves; and retained placenta (Figure 14.13). Several indicators of vitamin A deficiency may be used before clinical signs of deficiency become evident, one of the most sensitive of which in growing

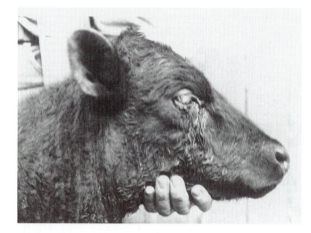

Figure 14.13   Blindness is a symptom of severe vitamin A deficiency. *(Courtesy of University of California)*

calves is the elevation of cerebrospinal fluid pressure. The vitamin A requirements of cattle can be met by carotene in feeds, supplements of vitamin A in a stabilized form, or a combination of both. For cattle, 1 milligram of carotene is considered to be equivalent to 400 international units (IU) of vitamin A.

Vitamin D. Cows that are fed sun-cured forage or that are exposed to sunlight do not need supplemental vitamin D. Even green forage, barn-cured hay, and silage have some vitamin D activity due to the irradiation of dead tissue of the stems and leaves of growing plants. When animals are exposed to sunlight, vitamin D is synthesized by the skin in sufficient amounts for maintenance, growth, reproduction, and lactation. However, calves housed indoors need vitamin D supplementation due to lack of exposure to sunlight. Animal sources of vitamin D (called D_3) and plant sources (called D_2) are biologically equivalent in dairy cattle. A vitamin D deficiency leads to a failure of the bones to calcify normally, resulting in rickets in calves (Figure 14.14) and osteomalacia in adults (Figure 14.15). Vitamin D deficiencies in calves kept indoors do occur, but deficiencies in mature cattle under normal conditions are extremely unlikely because exposure to sunlight provides adequate vitamin D. Some of the first signs of rickets caused by vitamin D deficiency are decreases in the blood plasma concentrations of calcium or inorganic phosphorus, or both, and increases in serum phosphates.

Vitamin E. Vitamin E is an antioxidant associated with selenium. It stimulates the immune system and reduces the incidence of oxidized flavor when consumed at high levels (400 to 1,000 mg/cow/day); it may aid in protection against white muscle disease, caused by a deficiency of selenium. White muscle disease is characterized by a weakening of the leg muscles, resulting in calves walking with a typical crossing of the hind legs; relaxation of the pasterns

Figure 14.14   A dairy calf with severe rickets—a vitamin D deficiency disorder. *(Courtesy of Michigan State University)*

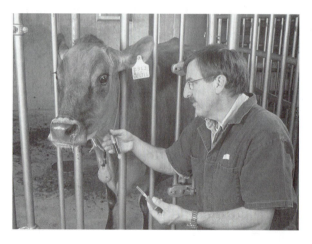

Figure 14.15   Ronald Horst collects blood via jugular venipuncture to assess the vitamin D and calcium status of a Jersey cow. *(Courtesy of USDA-ARS)*

and splaying of the toes; impaired ability to suckle because the musculature of the tongue is affected; and in advanced cases, the calf may be unable to hold up its head and to stand. All green feeds are good sources of vitamin E. Cows on pasture or being fed green chop receive adequate vitamin E. However, the vitamin E content of dry feedstuffs decreases during storage. One IU of vitamin E is defined as 1 milligram of dl-alpha-tocopherol acetate. Vitamin E and selenium play a synergistic role in the nutrition of calves. Some deficiency signs, such as white muscle disease, may respond to either vitamin E or selenium; some deficiencies may require both. The requirements for these nutrients may also be influenced by the type of liquid in the diet.

Vitamin K. Vitamin K functions as a stimulant to blood coagulation. Either vitamin K_1 (phylloquinone) or vitamin K_2 (menaquinone) meets the needs of cattle. Green, leafy materials of any kind, both fresh and dry, are good sources of vitamin K_1. Normally, vitamin K_2 is synthesized in large amounts in the rumen; so dietary supplementation is not recommended. When cows consume moldy sweet-clover hay, which is high in dicoumarol, blood coagulation may be impaired, followed by generalized hemorrhaging. This syndrome, commonly called sweet-clover disease or sweet-clover poisoning, responds to treatment with vitamin K.

Water-Soluble Vitamins

The water-soluble vitamins include biotin, choline, folacin (folic acid), inositol, niacin (nicotinic acid, nicotinamide), pantothenic acid (vitamin B_3), para-aminobenzoic acid (PABA), riboflavin (vitamin B_2), thiamin (vitamin B_1), vitamin B_6 (pyridoxine, pyridoxal, pyridoxamine), vitamin B_{12} (cobalamins), and vitamin C (ascorbic acid, dehydroascorbic acid). However, a physiological need in cattle of all of these vitamins has not been demonstrated.

Until recently, it was assumed that dairy cattle with a functional rumen did not require supplemental B vitamins. The rumen microflora were believed to synthesize adequate amounts of these nutrients for the host's requirements. In addition, the B vitamins are relatively abundant in dairy feeds. But recent evidence suggests a need for supplemental niacin under certain conditions and possibly supplemental choline and thiamin in the case of mature cattle, for which microbial synthesis and quantities in feeds may be inadequate, especially during diseased conditions or periods of stress. Dairy cattle of all ages have a physiological need for most of the B vitamins, especially biotin, choline, niacin, pantothenic acid, riboflavin, thiamin, vitamin B_6, and vitamin B_{12}. In young calves, deficiency signs have been demonstrated when there is inadequate intake of these vitamins, but even without a functioning rumen, their needs for these B vitamins appear to be met when they are fed whole milk. When young calves are fed milk replacers, however, it is advisable to ascertain the adequacy of vitamin intakes until their rumens are functional.

Biotin. A biotin deficiency in calves is characterized by paralysis of the hindquarters. Signs of deficiency did not develop when synthetic milk was supple-

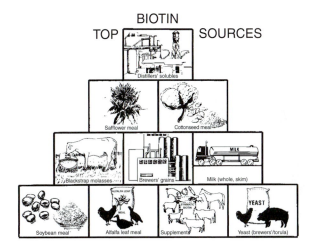

Figure 14.16 Sources of biotin. *(Courtesy of M. E. Ensminger)*

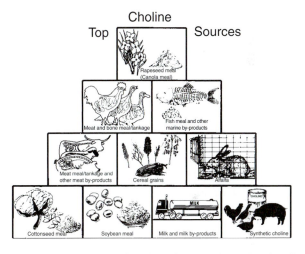

Figure 14.17 Sources of choline. *(Courtesy of M. E. Ensminger)*

mented with 4.5 micrograms of biotin/pound of feed and fed at 10 percent of live weight (Figure 14.16).

Choline. Researchers have produced choline deficiency in calves by feeding a synthetic ration containing 15 percent casein. Within seven days, the calves developed extreme weakness and labored breathing and were unable to stand. Supplementation of the ration with 236 milligrams of choline per quart of synthetic milk prevented the development of these signs. Adding choline to the ration may increase the percentage of milk fat in lactating cows (Figure 14.17)

Niacin (nicotinic acid, nicotinamide). Niacin is required by the young preruminant calf. In addition, rumen microorganisms may not synthesize adequate amounts of niacin to meet the needs of high-producing cows in early lactation. The major reason for improvement in milk production that occurs with added niacin may be related to the role of niacin in carbohydrate and lipid metabolism and the resultant decrease in ketosis. Niacin may also influence rumen fermentation, as evidenced by greater microbial protein synthesis and increased levels of rumen propionate with niacin supplementation. When cows are fed heated soybean meal, rumen response to niacin is greater than it is when cows are fed unheated soybean meal.

Pantothenic acid (vitamin B$_5$). Pantothenic acid deficiency in the calf is characterized by a scaly dermatitis around the eyes and muzzle, loss of appetite, diarrhea, weakness (unable to stand), and convulsions. Pantothenic acid deficiency in animals with functioning rumens is unlikely due to microbial production of pantothenic acid.

Riboflavin (vitamin B$_2$). Riboflavin deficiency in the calf is characterized by hyperemia of (presence of blood in) the mucosa of the mouth; lesions in the corners of the mouth and along the edges of the lips; loss of hair, especially on the belly; and excess salivation. A riboflavin deficiency in lactating cattle is unlikely because of the amounts of riboflavin that are present in feedstuffs and synthesized in the rumen.

Thiamin (vitamin B$_1$). Thiamin deficiency in the calf may cause polioencephalomalacia, characterized by listlessness; lack of muscular coordination; progressive blindness; convulsions; and sudden death, which may be preceded by diarrhea and dehydration. The condition is found primarily in cattle fed high-concentrate rations, and it has been linked to increased microbial thiaminase activity and the production of thiamin analogs in the rumen. Treatment consists of the intravenous or intramuscular administration of thiamin at a rate of 1 mg/pound of live weight.

Vitamin B$_6$ (pyridoxine, pyridoxal, pyridoxamine). Vitamin B$_6$ deficiency has been produced in calves fed a synthetic diet. It is characterized by loss of appetite; cessation of growth; and epileptic seizures in some, but not all, calves after about three months. Calves respond to vitamin B$_6$ therapy if it is initiated in the early stages of the disease.

Vitamin B$_{12}$ (cobalamin). Vitamin B$_{12}$ deficiency has been produced in calves under six weeks of age by feeding them a diet containing no animal protein. Deficiency signs include poor appetite and growth, muscular weakness, and poor general condition. It has been suggested that the vitamin B$_{12}$ requirement for dairy cattle is between 0.15 and 0.3 grams-pound of live weight. Vitamin B$_{12}$ is of special interest in the mature ruminant because of

Figure 14.18   Lactating cows may require over 50 gallons of fresh water daily to maintain high levels of milk production. *(Courtesy of Iowa State University)*

its role in propionate metabolism and because of the incidence of B$_{12}$ deficiency as a secondary result of cobalt deficiency. Certain soils have insufficient cobalt to produce levels of the element in plants that are adequate to support optimum vitamin B$_{12}$ synthesis in the rumen.

WATER

Large amounts of water are essential if a cow is to produce to her maximum capacity (Figure 14.18). Water is necessary for maintaining body fluids and proper ion balance; digesting, absorbing, and metabolizing nutrients; eliminating waste material and excess heat from the body; providing a fluid environment for the fetus; and transporting nutrients to and from body tissues.

The water that dairy cattle need is supplied by drinking, by water in the feed that they consume, and by metabolic water produced by the oxidation of organic nutrients. Cows drink an average of 100 to 200 pounds of water per day, with heavy producers drinking over 400 pounds per day (1 gallon of water = 8.33 pounds). Cows need 4 to 5 pounds of water for each pound of milk produced. The amount of water a cow drinks depends on her size and milk yield, quantity of dry matter consumed, temperature and relative humidity of the air, temperature of the water, quality of the water, and amount of moisture in her feed (Table 14.1).

Fresh, clean water is of the utmost importance for a dairy cattle feeding program to be successful. This factor is often neglected by many producers. Water

TABLE 14.1   Water Intake Guideline (Gallons per Day) for Dairy Cattle

		Temperature (F)		
Weight (lb)	Milk (lb)	40 and below	60	80
Heifers				
200	0	2.0	2.5	3.3
400	0	3.7	4.6	6.1
800	0	6.3	7.9	10.6
1,200*	0	8.7	10.8	14.5
Dry Cows				
1,400	0	9.7	12.0	16.2
1,600	0	10.4	12.8	17.3
Lactating cows[†]				
1,400	20	12.0	14.5	17.9
	60	22.0	26.1	24.7
	80	27.0	31.9	38.7
	100	32.0	37.7	45.7

* Maintenance and pregnancy.
† Maintenance and milk production.
Source: Linn, J. G., M. F. Hutjens, W. T. Howard, L. H. Kilmer, and D. E. Otterby. Feeding the Dairy Herd. 1998, p. 10. http:www.inform.umd.edu/EdRes/To...eeding/ FEEDING_THE_DAIRY_HERD.html.

Figure 14.19 Water quality is as crucial as is water availability. *(Courtesy of Howard Tyler)*

troughs should be cleaned routinely to ensure that the water is free from dirt and pathogenic bacteria (Figure 14.19). A common drinking trough provides an excellent means of spreading parasites and disease.

Dairy cattle lose water from the body in saliva, urine, feces, and milk; through sweating; and by evaporation from body surfaces and the respiratory tract. The amount of water lost from the body of cat-tle is influenced by the activity of the animal, ambient temperature, humidity, respiratory rate, water intake, feed consumption, and other factors. Several factors including age, body weight, production, weather (heat and humidity), and type of ration can affect the amount of water a particular animal consumes. The intensity of production dramatically affects the water requirement.

Cattle can survive for a longer period without feed than they can without water. Water is one of the largest constituents in the animal body, ranging from 40 percent in very fat, mature cattle to 80 percent in newborn calves. Deficits or excesses of more than a few percentage points of the total body water are incompatible with health, and large deficits of about 20 percent of body weight lead to death.

SUMMARY

- Dairy cattle require at least fifteen mineral elements for structural and regulatory functions, bone and teeth formation, acid-base balance, water balance, enzyme and hormone systems, a developing fetus, and milk production.
- Macrominerals include salt, calcium, phosphorus, magnesium, potassium, and sulfur.
- Microminerals include cobalt, copper, iodine, iron, manganese, molybdenum, selenium, and zinc.
- Fat-soluble vitamins include vitamins A, D, E, and K.
- Water-soluble vitamins include the B vitamins and vitamin C.
- Water is necessary for a cow to produce to her maximum capacity; maintain body fluids and proper ion balance; digest, absorb, and metabolize nutrients; eliminate waste material and excess heat; provide a fluid environment for the fetus; and transport nutrients to and from body tissues.

QUESTIONS

1. What are the differences between calcium deficiency and phosphorus deficiency?
2. A producer notices some of the black-colored cattle are turning gray. What mineral deficiency would cause this condition?
3. Define goiter.
4. Which regions are the most iodine-deficient areas in the United States?
5. Why is selenium essential for a dairy cow?
6. What symptoms may occur from vitamin A deficiency?
7. Which vitamin is needed for the prevention of rickets and osteomalacia?
8. Which vitamin stimulates blood coagulation?
9. What is the difference in water consumption per day of an average-producing cow versus a high-producing one?
10. For each pound of milk produced, how much water does the cow need to drink?

ADDITIONAL RESOURCES

Articles

Kincaid, R. L., L. E. Lefebvre, J. D. Cronrath, M. T. Socha, and A. B. Johnson. "Effect of Dietary Cobalt Supplementation on Cobalt Metabolism and Performance of Dairy Cattle." *Journal of Dairy Science* 86 (April 2003): 1405–1414.

Weiss, W. P. "Macromineral Digestion by Lactating Dairy Cows: Factors Affecting Digestibility of Magnesium." *Journal of Dairy Science* 87 (July 2004): 2167–2171.

Internet

An Update on Vitamin Levels: http://www.traill.uiuc.edu//dairynet/paperDisplay.cfm?ContentID=561.

Measuring Dry Matter: http://www.tocal.nsw.edu.au/reader/2868.

Mineral and Vitamin Nutrition of Dairy Cattle: http://ianrpubs.unl.edu/dairy/g1111.htm.

Water for Dairy Cattle: http://cahe.nmsu.edu/pubs/_d/D-107.pdf.

Water Intake and Quality for Dairy Cattle: http://www.das.psu.edu/dcn/catnut/PDF/Water.PDF.

Water Quality and Requirements for Dairy Cattle: http://ianrpubs.unl.edu/dairy/g1138.htm.

15

Ration Formulation Strategies and Feeding Systems

OBJECTIVES

- To describe ration formulation strategies.
- To analyze rations according to available feeds.
- To identify methods of chemical analysis.

A successful nutrition program for a dairy operation requires far more than a computer printout. It requires a thorough understanding of the digestive anatomy and physiology of the animal being fed; knowledge of the nutrient composition of the available feeds, the interactions of those feeds, the nutrient requirements of the animal, and the environmental effects on these requirements; and, perhaps most important, a feeding system that allows the animal to utilize effectively the benefits inherent in the ration.

Feeding requirements indicate the amounts of one or more nutrients required by different species of animals for specific productive functions, such as growth and lactation. Most feeding requirements are expressed as either quantities of nutrients required per day or as a concentration of those same nutrients in the ration. Quantities are more descriptive of the actual need of the animal and are used when animals are provided a given amount of a feed during a twenty-four hour period. Nutrient concentrations are more commonly provided when animals are provided a ration without limitation on the time in which it is consumed (Table 15.1).

The most widely used feeding standards in the United States are those published by the National Research Council (NRC) of the National Academy of Sciences. The current recommended nutritive requirements for dairy cattle are contained in *Nutrient Requirements of Dairy Cattle,* 7th edition (revised) 2001, prepared by the Subcommittee on Dairy Nutrition, National Research Council, and published by the National Academy Press, Washing-

TABLE 15.1 Nutrition Guidelines for High-Producing Herds

Dry matter (DM) intake	4–5% of body weight
Neutral detergent fiber (NDF)	26–30% of DM
Forage NDF	20–22% of DM
Nonstructural carbohydrates	35–40% of DM
Fat	5–7% of DM
Crude protein (CP)	17–19% of DM
Degradable protein	60–65% of CP
Undegradable protein	35–40% of CP

Source: Adapted from Chase, L. E. "Feeding Programs to Achieve 13,600 kg of Milk," in Advances in Dairy Technology 1998, 5, 13–20.

ton, D.C. Although these feeding standards are excellent and needed guides, there are still many situations where nutrient needs cannot be specified with great accuracy for animals. Also, in practical feeding operations, the economics of the ration must be considered. Dairy producers are interested in obtaining a level of milk production that will earn the largest net return in light of current feed costs and the market price of milk. Although higher-producing cows are usually more profitable than lower-producing cows, optimal profit is rarely obtained through maximal production.

In addition, feeding requirements tell nothing about the palatability or physical nature of a ration. They do not account fully for individual animal differences, management differences, and the effects of stresses such as disease, parasites, and surgery (Figure 15.1). Thus, many variables alter the nutrient needs and utilization of animals, variables that are difficult to include quantitatively in feeding requirements, even when feed quality is well known.

Ration formulation consists of combining feeds that will be eaten in the amount needed to supply the daily nutrient requirements of the animal. This may be accomplished by several methods. No matter which method is employed, however, the resulting

Figure 15.1   Poor growth and performance is not always related to feeding problems. *(Courtesy of Monsanto)*

ration is simply the first step in a feeding program. In computing rations, more than simple arithmetic should be considered; no set of figures can substitute for experience and animal intuition. Formulating rations is both an art and a science; the art is a result of experience and continuing observations, while the science is largely founded on mathematics, biochemistry, physiology, and bacteriology. A combination of both approaches is essential for success.

The first step in ration formulation is to determine which feeds can be considered for inclusion in the ration. It is important to consider availability, quality, and cost of the different feed ingredients. The quantities of high-moisture feeds must be limited in rations where intake is limiting, such as early lactation cows and young calves. Similarly, the palatability of the ingredients can have a dramatic effect on intake in these animals.

Rations should be formulated to nourish the billions of bacteria in the rumen, which maximizes digestion of forages and permits utilization of lower quality, cheaper proteins and other nitrogenous products. For example, it is possible to use urea to constitute up to one-third of the total protein of the ration of many classes of ruminants, provided care is taken to supply enough readily available carbohydrates.

All rations should be balanced as the least-cost feeding alternative that provides optimal performance. The ideal ration is one that maximizes production at the lowest cost. A costly ration may yield phenomenal production in cows, but the cost per pound of milk may make the ration impractical. Likewise, the cheapest ration is not always the best because it may adversely affect the productivity and performance of cows. Therefore, the cost per unit of production is the ultimate determinant of what constitutes the best ration. Awareness of this fact is sometimes all that separates a successful producer from the marginal or unsuccessful ones.

Typically, this balance requires the maximum use of feeds available in the area or on the farm. It also requires that forage quality be carefully assessed because forage quality typically dictates the economy of the ration. Ration balancing can be seen as matching the nutrient composition of the feeds to the nutrient needs of the animal; therefore, the composition of the feeds to be used must be accurately identified.

FEED ANALYSIS

Useful chemical analysis of a feed depends on the accuracy of the sample that is analyzed. If the sample is not representative of the entire batch, the evaluation is useless, no matter how extensively it is analyzed. This point bears emphasis because some feeds are highly variable in composition. Feeds are routinely analyzed through highly sophisticated chemical procedures. Many agricultural experiment stations, as well as most large feed companies, have facilities to analyze feeds for both the prevention and diagnosis of nutritional problems.

A chemical analysis gives a solid foundation on which to start in the evaluation of feeds. Thus, feed composition tables serve as a basis for ration formulation and for feed purchasing and merchandising. Commercially prepared feeds are required by state law to be labeled with a list of ingredients and a guaranteed analysis. Although state laws vary slightly, most of them require that the feed label (tag) show the percentage of the minimum crude protein and fat and the maximum crude fiber and ash. Some feed labels also include maximum salt, minimum TDN, and/or minimum calcium and phosphorus. These figures are the buyer's assurance that the feed contains the minimal amounts of the higher-cost items (protein and fat) and not more than the stipulated amounts of the lower-cost items (the crude fiber and ash).

The most important feeds to analyze chemically are the forages. They comprise the backbone of any dairy ration, and they are more variable in content than grains.

Chemical Analysis

For more than 100 years, feeds were analyzed by a method called proximate analysis. Feeds were broken down into six components: moisture, ash, crude protein, ether extract, crude fiber, and nitrogen-free extract. The inadequacies of proximate analysis gave rise to the detergent system of feed analysis for estimating energy content of forages. This system separates fibrous feeds into two fractions: a neutral detergent fibrous fraction and an acid detergent fibrous fraction. Further chemical analysis and calculations allow us several useful methods to evaluate forage quality.

Neutral Detergent Fiber (NDF).

Neutral detergents separate the feed into two fractions: (1) neutral detergent solubles, representing the highly digestible portion of the feed and consisting of proteins, fats, and carbohydrates (along with nonprotein nitrogen, pectin, and soluble materials) and (2) neutral detergent fiber (NDF), representing the less digestible portion of the feed, consisting of plant cell walls, including lignin, cellulose, and hemicellulose. The NDF content of a feedstuff is closely related to feed intake because it contains all the fiber components that occupy space in the rumen and are slowly digested. Thus, the lower the NDF percentage, the more the animal will eat; it is inversely related to voluntary feed consumption. Hence, a low percentage of NDF is desirable. Milk production of lactating cows is more highly correlated with the NDF portion of the ration than with the ADF portion of the ration.

Acid Detergent Fiber (ADF).

Acid detergent solutions are used to separate the feed into two fractions: (1) acid detergent solubles, containing the more readily digestible hemicellulose and (2) acid detergent fiber (ADF), representing the less digestible portion of the feed and consisting of lignin (indigestible) and cellulose (digestible). The ADF is an indicator of forage digestibility because it contains a high proportion of lignin, which is the indigestible fiber fraction. NDF will always be a higher number than ADF because ADF does not contain hemicellulose. The lower the ADF, the more feed an animal can digest. Hence, a low ADF percentage is desirable.

Acid detergent lignin (ADL). Sulfuric acid may be used to separate the ADF further into cellulose, which is digestible, and lignin, which is indigestible.

Digestible dry matter (DDM). The accepted equation for predicting DDM of legumes, grasses, and legume-grass mixtures from ADF is:

$$DDM \% = 88.9 - (0.779 \times ADF \%)$$

Dry matter intake (DMI) estimates. The amount of DM an animal can consume is affected by how fast forages are digested and pass through the digestive tract. The fiber fraction that appears to be most clearly related to the DMI of forages is neutral detergent fiber (NDF). However, the exact NDF level in rations necessary to achieve optimum performance is uncertain. Wisconsin research indicates maximum feed intake in alfalfa-based dairy rations occurs when NDF is 1.2 lb per 100 lb of body weight:

$$DMI \, (\% \text{ of body weight}) = \frac{120}{\text{forage NDF } (\% \text{ of DM})}$$

Relative feed value (RFV). Relative feed value is an index that combines important nutritional factors (potential intake and digestibility) into one number for a quick, easy, and effective method of evaluating feeding value or quality. The formula for calculating RFV is the estimated digestibility and potential intake of a forage calculated from ADF and NDF fractions, respectively. The calculation of RFV is done by multiplying DDM by DMI, then dividing by 1.29. The number derived from the RFV calculation has no units and is used only as an index for evaluating the quality of hay or haylage made from legumes, grass, or legume-grass mixtures. The RFV concept should be used to evaluate quality only for those forages listed previously. The RFV does not include protein estimates because they are influenced by factors unrelated to those affecting RFV. Protein should be considered, however, in pricing forages. Various studies show that both visual appraisal and a forage test are necessary to assess quality properly. For example, forage should be inspected for absence of mold and for the presence of good green color, as well as for foreign matter. However, descriptions of forage quality for marketing should include RFV. In Wisconsin, prices in quality-tested hay auctions have averaged about $1 per ton higher for each 1 percent increase in relative feed value (RFV). This means that a lot of hay with an RFV of 125 would be worth about $5 more per ton than would hay with RFV of only 120. Furthermore, surveys of hay buyers indicate that four out of five buyers had confidence in using RFV for pricing hay.

Near Infrared Reflectance Spectroscopy (NIRS)

Near infrared reflectance spectroscopy (NIRS) is a nonconsumptive instrumental method for fast, accurate, and precise evaluation of the chemical composition and associated feeding value attributes of forages and other feedstuffs. The instrument, known as a near infrared analyzer, produces infrared radiation over a given range of wavelengths, and this radiation is focused onto the sample being tested. Because of the chemical structure of the sample material, certain combinations of infrared wavelengths are reflected, and certain combinations are absorbed for each chemical characteristic tested, for example, energy values, crude protein, digestibility, minerals, NDF, and ADF. By using a system of filters and detectors, the instrument senses these reflected wavelengths and passes this information to the computer. The computer sorts the appropriate wavelength combinations and their relative magnitudes for each chemical characteristic and transforms these data into percentages. The near infrared reflectance measures feed quality by comparing the energy reflected back from a hay sample with computerized standards established by conventional laboratory analysis of a large number of reference samples.

The NIRS method of analysis has four main advantages: speed, simplicity of sample preparation, multiplicity of analysis with one operation, and non-consumption of the sample. (It can be analyzed again by the same or another procedure.) With the NIRS method of analysis, it is possible to take a sample from a truckload of hay and provide, in less than three minutes, an analysis for crude protein, NDF, ADF, dry matter, lignin, and in vitro dry matter digestibility. The chief disadvantages of the NIRS method are instrumentation requirements and costs, dependence on calibration procedures, complexity in the choice of data treatment, and lack of sensitivity for minor constituents.

Calorimetry

When compounds are burned completely in the presence of oxygen, the resulting heat is referred to as gross energy or the heat of combustion. The bomb calorimeter is used to determine the gross energy of feed, waste products from feed (for example, feces and urine), and tissues. The calorie is defined as the amount of heat required to raise the temperature of 1 gram of water 1°C (precisely from 14.5° to 15.5°C).

Computer-Formulated Rations

Once feeds to be considered for inclusion in the ration are determined and nutrient content of those feeds are definitively known, then a least-cost ration can be created for any class of dairy animal using various software programs designed for this purpose. Until the late 1970s, only those dairy producers with access to a large mainframe computer could formulate a ration using the computer. Usually this number was limited to those producers working through a university (extension personnel) or subscribing to a time-sharing system. Many rations were formulated by using a pencil, an eraser, paper, and a calculator. Balancing rations was time-consuming, and options were limited. The advent of microcomputers changed the future of ration balancing. Currently, over 60 percent of individual dairy producers own a home computer. Rations can be adjusted to match changes in availability, price, composition, and moisture content of ingredients. Many ration-balancing programs are available from software companies and universities that convert the tabular NRC requirements into equations that compute animal requirements. In addition, the NRC publication *Nutrient Requirements of Dairy Cattle* publishes predic-

tion equations to facilitate ration balancing and the software programs to use these equations.

Computers formulate rations objectively from the information that is fed into them. The ration that is generated is therefore the best solution to the mathematical problem, but it may not be practical or realistic. For example, radical changes in ration composition cannot be made without causing digestive disorders, especially in ruminant animals. Microbes in the rumens of dairy animals need time to adapt to changes in rations, a fact that the computer does not consider.

One of the major costs involved in computer formulations is the continual review and revision of the information that must go into the program. The user must be constantly updating costs to maximize the use of the computer program.

When the computer formulates rations, it uses average values for the nutrient composition of the various feeds unless it is programmed otherwise. Feeds can often vary in their nutrient composition; so there is a good possibility that the chemical analysis of the formulated feed will not be the same as the formulated analysis. In most cases, however, programs allow producers to plug in their own values based on the analyzed feed samples.

Finally, the results obtained from the computer depend on the person who feeds the information into the machine. If the data given to the computer are outdated or wrong, the ration that is formulated will be of little value.

FEEDING SYSTEMS

Traditional individual feeding of lactating cows in stanchioned barns or milking parlors is giving way to feeding systems that allow for considerable savings in labor and facilities. The system of feeding is often dictated, or at least limited, by the facility design. In the design of a new facility or the expansion of an existing facility, the feeding system should be an important consideration.

Individual Feeding Systems

Individual cow feeding is still utilized as the primary system for feeding lactating cows in many types of facilities. The primary advantage of feeding cows in this manner is the ability to feed each cow according to her individual needs. The labor costs associated with individual feeding and the requirements for specialized housing or specialized feeding equipment have increased the popularity of group feeding systems in recent years. However, individualized feeding can still be a profitable way to feed cows if the system is managed properly.

Component-Fed Herds

Many herds housed in tie stalls or stanchions are individually fed both their roughage and concentrate feeds. This allows total control over the ration for each individual animal, but it has the highest associated labor costs. Typically, roughage (hay or silage, or both) is provided and the concentrate is top-dressed based on the production, age, and body condition requirements of the animal (Figure 15.2).

Many dairy producers use modified versions of the component feeding regimen. Typically, the concentrate is fed to cows at milking time, either in the parlor (via drop feeders) or at the tie stall or stanchion. The roughage portion of the diet is fed at a feed bunk (group feeding).

The primary challenge of component feeding is to manage the feeding order. It is crucial to provide forage prior to concentrate, thus allowing the production of saliva (the primary rumen buffer) during cud chewing prior to the acid load associated with grain feeding. This can be a challenge in many management systems. The frequency of feeding of the concentrate portion of the ration is also crucial in these systems; more frequent feedings reduce the amount of acid produced in the rumen at any individual feeding and also improves digestibility of the feeds (Figure 15.3).

Automatic Grain Feeders

The advent of computerized grain feeders was heralded as a way in which producers could manage cows in loose housing but feed them individually. If not properly managed, however, such feeders may result in overfeeding grain, accompanied by health problems and lower profits. Additionally, some types of feeds do not handle well in these types of feeders.

The following general types of mechanized grain-feeding systems are available:

1. **Free-choice, electronic grain feeders.** These units allow cows equipped with an identification unit (magnet, key, or chain) access to a feeding station. This system does not restrict access time or the amount of grain consumed per feeder visit. Careful management is necessary to avoid digestive problems. The major advantages of this system are low initial investment and a simple design.

2. **Preset or computerized grain feeders.** This system controls the maximum amount of grain that individual cows receive during a set period of time. The feedings can be split into small feedings over an extended period, thus normalizing rumen function and allowing component feeding with a lower risk of acidosis (Figure 15.4). The initial cost of this system varies widely depending on herd size and complexity of the system.

Figure 15.2 Concentrate is top-dressed on forage for this component-fed herd so that cows have the potential to consume ration components selectively. *(Courtesy of Iowa State University)*

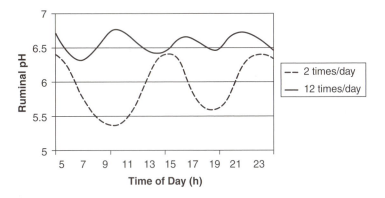

Figure 15.3 Effect of feeding frequency on ruminal pH: two times daily and twelve times daily feeding.

Figure 15.4 A computerized feeding station regulates the timing and quantity of concentrate intake for individual cows. *(Courtesy of Iowa State University)*

Figure 15.5 A feed-mixing truck dumps feed along a fence-line feeder. *(Courtesy of Iowa State University)*

Regardless of which feeder is selected, successful adoption requires superior management. The entire daily grain allocation for individual cows and/or the herd can be fed through computer-operated feeders. Computerized feeders generally accommodate twenty to twenty-five cows, although this number should be decreased for high-producing cows in early lactation and/or cows in negative energy balance (NEB cows). Stall length, protection of the unit, lighting, and location of the unit relative to cow traffic patterns can affect the success of this system of feeding.

Group Feeding Systems

Individual feeding of lactating cows has largely given way to mechanized group feeding. Feeding generally takes place in bunks along the fenceline of pens (Figure 15.5). Group feeding was developed for convenience and labor savings rather than for improved animal well-being or feed efficiency. To design a nutritional program for herds with hundreds or thousands of cows that can be adapted to the specific needs of the cows, the cows are separated into groups according to nutritional needs (whether for lactation, reproduction, or body reserves). When producers decide to utilize group feeding strategies, they must decide on the number of groups into which to divide the herd. Consideration should be given to total herd size, types and costs of available feeds, current type of housing (Figure 15.6), feeding and milking systems, and overall economic integration of the operation.

Group feeding facilitates the use of total mixed rations (TMRs), where the concentrates, roughages, and supplements are mixed into one feed rather than being fed separately. Some producers who use TMRs prefer to feed dried roughages, especially long-stemmed hay, separately to enhance stimula-

tion of the rumen and to facilitate mixing. In addition, long hay is often fed separately because it does not always lend itself to mixing in a mixer.

There are several advantages to feeding TMRs. They permit precise definition of the ration that is consumed. They facilitate mechanized feeding; hence, less labor is required. Feeding TMRs minimizes problems associated with preferential consumption of a certain feedstuff, or feed-sorting behavior. Thus, cows have fewer digestive disturbances, such as displaced abomasums. Every mouthful of the ration is the same, stabilizing the rumen environment. The feeding of TMRs increases intake in cows and facilitates masking of certain unpalatable feeds, such as urea.

There are also disadvantages to feeding TMRs. This type of feeding necessitates specialized blending equipment to ensure thorough mixing (Figures 15.7 and 15.8). It is typically not feasible to divide small herds into production-based groups to feed separate rations. Inadequate numbers of groups or mismanagement can quite easily result in overfeeding many cows in a group, leading to fat cow syndrome and related health problems, such as calving difficulties, poor reproduction, low production, low dry matter consumption, and metabolic disorders. In many cases, these problems do not become immediately obvious; rather, they may take many months to develop.

Regardless of herd size, a maximium of 100 cows per group is recommended because of potential feeding problems and social considerations. One of the problems inherent with group feeding concerns the behavioral adaptation of a cow newly introduced to a group. Group acceptance and establishment of a social hierarchy can pose occasional problems with a new cow, but the magnitude of the problem is usually not very great. One means

Figure 15.6   Cows access feed from either side of this drive-through feeding alley in an open barn in southern California. *(Courtesy of Iowa State University)*

Figure 15.7   Portable mixer wagons are used to mix ration ingredients into a total mixed ration. *(Courtesy of Mark Kirkpatrick)*

Figure 15.8   Inside a mixer wagon, augers are used to mix feedstuffs thoroughly prior to feeding. *(Courtesy of Mark Kirkpatrick)*

Figure 15.9   Heifers in this corral are fed at a fenceline feed bunk. *(Courtesy of South Dakota State University)*

of reducing this effect is to move several cows into a new group at the same time and just before feeding, rather than moving them individually.

When group feeding, first-calf heifers should be placed in a separate group and fed for both milk production and growth (Figure 15.9). Their nutrient requirements for milk production are similar to the requirements of their older counterparts producing milk at the same level, but because of their growth, they should receive about 20 percent more nutrients than are required for maintenance.

SUMMARY

- A successful nutrition program requires a thorough understanding of digestive anatomy and physiology; knowledge of the nutrient composition of the available feeds, nutrient requirements, and environmental effects; and an effective feeding system.
- Ration formulation consists of combining feeds that will be eaten in the amount needed to supply the daily nutrient requirements of the animal.
- A useful chemical analysis depends on the accuracy of the sample that is analyzed.
- Proximate analysis breaks feeds into six components: moisture, ash, crude protein, ether extract, crude fiber, and nitrogen-free extract.
- The detergent system separates fibrous feeds into two fractions: a neutral detergent fibrous fraction and an acid detergent fibrous fraction.
- The lower the NDF percentage, the more the animal will eat.
- The NIRS method has four main advantages: speed, simplicity of sample preparation, multiplicity of analysis with one operation, and nonconsumption of the sample.
- The amount of dry matter that an animal consumes is affected by how fast forages are digested and pass through the intestinal tract.
- Radical changes in ration cannot be made without causing digestive disorders, especially in ruminant animals.
- Two types of mechanized grain-feeding systems are (1) free-choice, electronic and (2) preset or computerized grain feeders.

QUESTIONS

1. What is the first step in ration formulation?
2. What are the most important types of feeds to analyze chemically?
3. What are the two fractions of neutral detergents?
4. Define calorie.
5. What are the advantages and disadvantages of feeding total mixed rations (TMRs)?
6. What is the maximum number of cows per group in a group feeding system?
7. Why should first-calf heifers be handled in a separate feeding group?

ADDITIONAL RESOURCES

Book

Reeves, III, J. B., and J. L. Weihrauch. *Composition of Foods: Fats and Oils, Raw, Processed, and Prepared.* Agriculture Handbook No. 8-4. Washington D.C., U.S. Department of Agriculture, 1979.

Articles

Baldwin, R. L., and B. W. Jesse. "Propionate Modulation of Ruminal Ketogensis." *Journal of Animal Science* 74 (July 1996): 1694–1700.

Whitt, J., G. Huntington, E. Zetina, E. Casse, K. Taniguchi, and W. Potts. "Plasma Nutrient Flow and Net Nutrient Flux Across Gut and Liver of Cattle Fed Twice Daily." *Journal of Animal Science* 74 (October 1996): 2450–2461.

Internet

Feeding the Dairy Herd: http://www.ces.uga.edu/caespubs/pubcd/B816-w.htm.

Nutrient Conservation and Metabolism Laboratory Recent Literature: http://www.lpsi.barc.usda.gov/ncml/bibligr.htm.

Nutrient Requirements for Dairy Cattle: http://www.calfnotes.com/pdffiles/CN069.pdf.

Understanding Dry Matter Consumption by Dairy Cows: http://edis.ifas.ufl.edu/DS079.

Using Forage Analysis to Optimize Dairy Rations: http://www.ansc.purdue.edu/dairy/forage/forqual.htm.

Using NDF and ADF to Balance Diets: http://muextension.missouri.edu/explore/agguides/dairy/g03161.htm.

4

Concepts in Dairy Feeds

16

Fundamentals of Hay Quality

OBJECTIVES

- To identify the types of hay and hay quality.
- To outline the process of harvesting.

Hay is the most important harvested forage for U.S. dairy cattle. The proportion of concentrates to roughage is largely determined by economics. When buying forage, this means it is important to consider the cost of energy and protein units per dollar. However, forages will always remain the foundation of the dairy ration.

Drying, or making hay, is the most common method of preserving forage for storage primarily because it is relatively easy to handle. It can be stored or transported long, chopped, pelleted, cubed, or baled into various types and sizes of bales. Modern equipment, including conditioners, hasten drying time; automated systems facilitate handling.

TYPES OF HAYS

Hays are made from legumes, grasses, and cereal crops (Table 16.1). In terms of total tonnage produced annually, alfalfa accounts for well over half of the U.S. hay crop (Figure 16.1). Many different kinds of hay make up the remainder of the nation's hay supply: among them, the clovers, lespedeza, soybeans, and cowpeas; the cereal hays made from oats, barley, wheat, or rye; and grass hays made from Bermuda grass, prairie grass, redtop, johnsongrass, orchard grass, fescue, and timothy.

Whenever feasible, it is recommended that legumes be grown for hay. In comparison with grasses, legumes are higher in protein, vitamins, and minerals; they provide higher yields; and they are nitrogen-fixing when inoculated: the bacteria (rhizobia) on their roots take free atmospheric nitrogen from the air. However, a mixture of grasses and legumes is often preferred for reasons of palatability, ease in curing, erosion control, and lessening the risk of bloat.

Legume hays are higher in protein, calcium, and carotene than grass hays, and they are usually more palatable. Alfalfa yields the highest tonnage per acre and has the highest protein content of the legume hays. Of course, poor-quality legume hays, those cut at a late stage of maturity (Figure 16.2) and exposed to weathering, are not as good as high-quality grass hays. Lespedeza is an excellent hay for dairy cattle, provided it is cured without weather damage, it is fine stemmed, and free of foreign material.

Grass hays include prairie grass, redtop, johnsongrass, orchard grass, and timothy. Grass hays grow under a wider range of conditions than alfalfa does, but they yield less dry matter per acre. When cut at the usual stage of maturity, grass hays are less palatable than legume hays and lower in protein and mineral content. However, when grass hays are heavily fertilized and cut at an early stage of maturity, they are very palatable and about equal to alfalfa in protein content.

Clover (Figure 16.3) is usually grown with timothy, as a grass-legume mixture. In comparison with alfalfa, clover-timothy mixed hays are lower in protein and not as high in quality. The lower quality is due to the fact that at cutting time the timothy is at the right stage of maturity whereas the clover is overly mature.

Soybeans, cowpeas, and vetch are often made into hays and fed to dairy cattle. They are not as valuable as alfalfa; generally, they are stemmy and difficult to cure. If they are cut at the proper stage and cured without loss of leaves, however, they make good feed.

Cereal hays are made from oats, barley, wheat, and rye. When cut sufficiently early in the flower stage and before the milk stage, they retain much of their feeding value. They are low in protein; hence, they must be fed with legume hay, grass silage, or a protein supplement. If cereal hays are cut too early, yield is reduced; if cut too late, they become fibrous and are of low feeding value.

TABLE 16.1   Nutrient Content of Several Common Forages, as a Percentage of Dry Matter

Forage	Maturity	CP	NDF	ADF	Ca	P	Mg	K
Alfalfa	Bud storage	21	40	30	1.4	0.30	0.34	2.5
	Early bloom	19	44	34	1.2	0.28	0.32	2.4
Bromegrass	Vegetative	19	51	31	0.6	0.30	0.26	2.0
	Early head	15	56	38	0.5	0.26	0.25	2.0
Corn Silage	Dough	8	50	27	0.3	0.20	0.20	1.0
	Black layer	8	46	26	0.3	0.20	0.20	1.0
Small Grain Hay	Heading	11	60	40	0.5	0.25	0.23	1.0
	Dough	10	65	43	0.5	0.25	0.23	1.0

Source: Adapted from Undersander, Howard and Shaver. U.W. Extension Bulletin No. A3325.

Figure 16.1   An alfalfa plant in full bloom. *(Courtesy of USDA-ARS)*

Figure 16.2   Cutting at the optimal stage of plant development dramatically improves the feeding value of forage. *(Courtesy of Iowa State University)*

Figure 16.3   A field of red clover. *(Courtesy of USDA-ARS)*

HAY QUALITY

The quality of hay greatly affects its consumption. High-quality forage is more digestible and passes through the digestive tract more rapidly than does low-quality forage; hence, animals will consume more of it (Figure 16.4).

Fortunately, hay quality and value can be estimated by certain characteristics. Hay of high feeding value is made from plants cut at an early stage of maturity, thus ensuring the maximum content of protein, minerals, and vitamins and the highest di-

gestibility. It is leafy, thus giving assurance of high protein content. It is bright green in color, thus indicating proper curing, a high carotene or provitamin A content, and palatability. It is free from foreign material, such as weeds, and it is free from must or mold or dust. It is fine stemmed and pliable, not coarse, stiff, and woody, and has a pleasing, fragrant aroma; it smells good enough to eat. During the curing process, the quality and feeding value of hay is decreased rapidly by rain, sun, bleaching, raking, handling when too dry, and storing with too much moisture.

Figure 16.4   At some stages of production, high-quality forage can meet most of the nutrient needs of the cow, thus minimizing feed costs. *(Courtesy of Iowa State University)*

Figure 16.5   Visual appraisal is the initial approach to determining hay quality. *(Courtesy of Iowa State University)*

Visual estimates of hay quality are valuable and should be used (Figure 16.5), but the most precise way to determine the nutrient value of hay is through chemical analysis. Analyses are not infallible, however; a Pennsylvania study revealed errors of as much as 5 percent in crude protein and 9 percent in Total Digestible Nutrients (TDN) (energy) content of forages, with evaluations made by trained individuals. Fortunately, there is a high relationship between the chemical composition, especially the protein and fiber, of hay and its feeding value for animals. As hay matures, protein decreases and fiber increases. Likewise, weathering lowers the protein and raises the fiber content because soluble nutrients are washed out by rain and leaves are lost during harvest.

By using detergents, hay samples are chemically separated into two fibrous fractions: a neutral detergent fibrous (NDF) fraction and an acid detergent fibrous (ADF) fraction. In comparison with traditional proximate analysis, NDF provides a better estimate of dry matter intake (consumption) by animals, and ADF provides a better estimate of the in vivo (inside the animal) dry matter digestibility.

No forage test is any better than the sample taken. Thus, the most important single step in determining the chemical composition of hay is sampling. No matter how accurate the chemical analysis, a poor sampling technique can easily invalidate the results and lead to an erroneous conclusion. It is difficult to obtain a representative, meaningful sample of forage because of its bulky nature and variability within a given lot of hay as compared to most other crops. For the sample to be representative of a given lot of hay, it should have been produced under the same cultural conditions, be from the same cutting, and be at the same stage of maturity, and all of it should have been baled within a forty-eight hour period using only one harvesting method. With conventional, rectangular bales, at least twenty bales should be sampled at random, by probing every third bale, for example. The probe, or core sampler, should be at least ⅜ inch in diameter. The center of either end of a rectangular bale may be probed by inserting the probe at a right angle to the face of the bale and to a depth of 12 to 18 inches. Hay samples should be placed in a plastic bag or freezer carton; otherwise, the moisture content will not be meaningful.

Certainly, poor-quality hay can be fed, and under certain circumstances, it may even be economical. When buying poor hay, however, the purchase price should be lowered accordingly, and the feed analysis should also be used as a basis of balancing the ration. By the same token, it is usually good business to pay a premium for high-quality hay. Some dairy producers very wisely apply an escalator principle to hay purchases. They may pay a premium per ton for each 1 percent of protein above an agreed-upon figure, or they may dock the price by a corresponding amount if the content is lower. For example, if a vendor guarantees to deliver alfalfa with 15 percent crude protein and it is agreed that a $1.50 per ton premium will be paid for each 1 percent protein in excess of this figure, a $4.50 per ton premium would be added for alfalfa containing 18 percent crude protein.

HARVESTING

Whether the crop is a grass or a legume or a combination of the two, the stage of maturity of the plants at the time of harvest affects digestibility, yield, and feeding value. Young, immature plants are high in protein and low in fiber or lignin. As hay crops mature, feeding value goes down, and fiber content increases. Digestibility of the forage (TDN) declines about 0.5 percent each day that cutting is delayed beyond the early bloom stage, and the intake of forage decreases during this same period at more than 0.5 percent each day. Thus, in total, the feeding value of forage drops more than 1 percent for each day's delay after early bloom. With increasing maturity of the hay, the crude protein, digestible dry matter decreases, as do the dry matter intake and milk production of the cows fed the hay. Maturity also increases the NDF and ADF fractions. The increase in NDF and ADF with maturity is expected because NDF is inversely correlated with intake, whereas ADF is highly correlated with digestibility.

Forage dry weight yields increase until midbloom to late-bloom stages. Timothy and bromegrass fully headed and red clover and alfalfa at full bloom give maximum yield of dry matter. However, the maximum feeding value of first cutting forage is reached at least ten days before the time of maximum dry weight yield; this usually corresponds to less than 10 percent bloom for alfalfa hay.

Stage of maturity also affects the vitamin content of hay. Levels of carotene (precursor of vitamin A) and the B vitamins decrease as plants mature. Vitamin D content is the one exception: it increases as the forage is sun-cured. Everything considered, there is a loss of about 1 percent in nutrient value for each day that the hay harvest is delayed beyond the late vegetative stage of growth.

Cutting (Figure 16.6), followed by curing in the swath or windrow (Figures 16.7 and 16.8), are the first two steps in haymaking, regardless of the subsequent method or type of equipment employed. Any one of several types of mowers may be used because all of them are designed to get the hay down. The most important thing is that the hay be cut at the proper stage of maturity.

After the hay has wilted sufficiently in the swath but while it is still tough and the leaves will not shatter, it should be windrowed. For this assignment, the side-delivery rake is preferred to the dump rake. The side-delivery rake rolls hay into fluffy, cylindrical windrows, which allow for good circulation of air. Dump rakes, on the other hand, produce large windrows that are apt to remain damp underneath and bleach excessively on top. Where the hay crop is exceedingly heavy, windrow size can be kept small by limiting the width of each windrow.

If considerable shattering appears probable, it may be desirable to do the raking early in the morning, when the dew makes the hay a bit tougher. When windrowed hay is rained on, wait until the top half dries out and then turn it upside down with the side-delivery rake. The use of the tedder for windrowing again is not recommended because of excessive leaf shattering.

Figure 16.6   Freshly cut hay is typically 75 to 80 percent water and must be field-dried prior to harvest. *(Courtesy of Iowa State University)*

Figure 16.7   Hay is raked into windrows by a wheel rake. *(Courtesy of Iowa State University)*

Figure 16.8   Windrows of alfalfa in an irrigated California hay field. *(Courtesy of USDA-ARS)*

Proper curing ensures that the hay can be stored safely without overheating or becoming moldy, and the maximum leafiness, green color, aroma, nutrient value, and palatability will be retained. Freshly cut forage contains 75 to 80 percent moisture, whereas the maximum moisture content for safe hay storage is as follows:

For loose hay, 25 percent moisture.

For baled hay, 20 to 22 percent moisture (the lower figure is for larger bales).

For chopped hay, 18 to 20 percent moisture.

For cubes, 16 to 17 percent moisture.

Hay of a higher moisture content than indicated should not be stored because its value may be greatly lowered due to mold or to nutrient losses accompanying fermentation and because of the danger of spontaneous combustion and a costly fire.

Legume forages contain a larger proportion of leaves than do grasses, but the fine, thin legume leaves dry out more rapidly than do the coarse stems to which they are attached. This results in considerable shattering losses unless great care is taken. In alfalfa, for example, 50 percent of the total weight of the plant is contained in the leaves, but the leaves contain 70 percent of the protein and 90 percent of the carotene content of the entire plant. In field-curing hay, losses from leaf shattering range from 2 to 5 percent for grass hay and 3 to 39 percent for legume hays, with as much as 15 to 20 percent for legume hays field-cured under the most favorable conditions (Figure 16.9).

In general, the carotene or provitamin A content of freshly cured hay is proportional to the greenness. With severe bleaching, more than 90 percent of the vitamin A potency may be destroyed. Even under the best of conditions, unavoidable loss, especially losses in sugars, starch, and carotene, occurs through fermentation. With good weather and proper curing methods, however, these losses are not excessive. The leaching losses from rain are less severe soon after mowing, but they increase in severity as curing progresses. Also, repeated showers are more damaging than is one heavy rain. Damaging rains may lower the feeding value of hay by one-fourth to one-third or even more with severe exposure. Losses from weather damage may be reduced by using haymaking equipment that reduces the field drying time and by using proven chemical conditioning and preserving agents.

Chemical hay-drying agents and preservatives assist haymakers in decreasing haymaking losses and improving hay quality. These products speed up the haymaking process and reduce exposure to weather

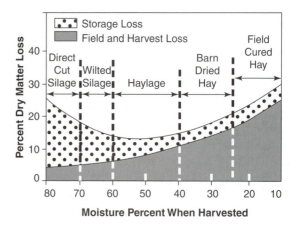

Figure 16.9 Estimated forage losses for hay and silage harvested at various moisture stages. *(Courtesy of M. E. Ensminger)*

damage. In comparison with no treatment, the use of a desiccant, or drying agent, along with mechanical conditioning can reduce the moisture content by an additional 2 to 10 percent during a twenty-four hour period. Adding a preservative to hay that is in the 25 to 35 percent moisture range allows it to be baled and stored without undue heating.

Chemical drying agents, which are sprayed on the crop at mowing time, break down the waxy cutin layer on the wall of the stem and allow moisture to escape, thereby promoting faster drying time. The drying rate of the stems approaches that of the leaves. Several chemicals are used for conditioning, including potassium carbonate, sodium carbonate, and sodium silicate. Also, methyl esters of fats, vegetable oils, or animal fats have been mixed with potassium carbonate in an attempt to increase the effectiveness of chemical conditioning.

Chemical conditioners are effective on legumes such as alfalfa, birdsfoot trefoil, and red clover, but generally they are not effective on grasses. Although they reduce drying time on all cuttings of legumes, they are most effective on second and third cuttings and are least effective on first and late autumn cuttings. This situation is attributed to the fact that conditioners work best when drying conditions are best (in the summer) and that first cutting has heavier yields and heavier swaths than later cuttings, conditions that hamper drying because moisture movement inside the swath is inhibited. Drying agents are more effective as an addition to and not as a substitute for mechanical conditioners. The chemical of choice is applied at the time of cutting by either a spray boom mounted ahead of the reel or spray nozzles mounted behind the reel but in front of the conditioning rollers so that the rollers help distribute the spray.

Under normal conditions and for safe baling, moisture content of 20 percent or less is a must. If properly treated with an adequate amount of the right preservative, however, alfalfa hay can be baled at 25 to 30 percent moisture, thereby speeding harvesting and decreasing losses significantly. Preservatives act as fungicides and inhibit the growth and reproduction of microorganisms that cause heating and molding in wet hay.

Propionic acid is the organic acid of choice. It is sometimes mixed with acetic acid, inorganic acids, formaldehyde, water, flavoring ingredients, and/or antioxidants. But to be most effective, organic acid formulations should have at least 60 percent propionic acid; should be applied at the proper rate, depending on the moisture content of the hay; and must be uniformly distributed throughout the hay mass.

When properly applied, anhydrous ammonia stops bacteria and mold growth; when applied to poor-quality hay, it has the added advantage of increasing protein and digestibility. As a preservative, however, it is not as effective as propionic acid. Also, unless large round bales are covered and/or contain less than 28 percent moisture, too much ammonia escapes. For high-quality alfalfa hay, which is already high in protein and digestibility, it is doubtful that the added expense of ammonia use can be justified.

Some claim that most bacterial inoculants on the market produce lactic acid, which acts as a fungicide and inhibits mold growth. More experimental work is needed to substantiate the effectiveness of bacterial inoculants as hay preservatives.

In many areas, there is a decided preference among livestock producers in favor of a certain cutting of alfalfa hay. Generally, first-cut alfalfa hay is coarser stemmed and less leafy than later cuttings and therefore of somewhat lower feeding value when the different cuttings are equally well cured. Also, the weather is often less favorable for curing the first cutting. On average, each successive cutting is lower in crude fiber and higher in crude protein.

SUMMARY

- Hay is the most important harvested forage for U.S. dairy cattle.
- It is recommended that legumes be grown for hay because they are higher in protein, calcium, carotene, vitamins, and minerals; they provide higher yields; they are more palatable; and they are nitrogen-fixing.
- High-quality forage is more digestible and passes through the digestive tract more rapidly than does low-quality forage.
- Hay of high feeding value is made from plants cut at an early stage of maturity, which ensures maximum content of protein, minerals, and vitamins and the highest digestibility.
- As hay matures, protein, DDM, DMI, and milk production decrease, while fiber, NDF, and ADF fractions increase.
- The stage of maturity of the plants at the time of harvest affects digestibility, yield, and feeding value.
- Chemical hay-drying agents and preservatives assist haymakers in decreasing haymaking losses and improving hay quality.
- Each successive cutting of alfalfa is lower in crude fiber and higher in crude protein.

QUESTIONS

1. What forages and cereals are hays made from?
2. Give some examples of grass hays and cereal hays.
3. When using detergents to analyze hay samples chemically, which fraction provides a better estimate of dry-matter intake by animals?
4. How deep should a probe be inserted into the end of a rectangular bale when taking a representative sample?
5. Which vitamin increases as the forage is sun-cured?
6. What is the first step in haymaking?
7. What is the advantage of using a side-delivery rake compared to a dump rake?
8. Why should a producer avoid storing loose hay at a moisture greater than 25 percent?
9. What is the purpose of adding a preservative?

ADDITIONAL RESOURCES

Articles

Golding, E. J., M. F. Carter, and J. E. Moore. "Modification of the Neutral Detergent Fiber Procedure for Hay." *Journal of Dairy Science* 68 (November 1985): 2732–2736.

Mansfield, H. R., M. D. Stern, and D. E. Otterby. "Effects of Beet Pulp and Animal By-Products on Milk Yield and In Vitro Fermentation by Rumen Microorganisms." *Journal of Dairy Science* 77 (January 1994): 205–216.

Internet

Know Your Forages: http://www.ext.nodak.edu/extpubs/ansci/dairy/as991w.htm.

The Importance of Fiber in Feeding Dairy Cattle: http://edis.ifas.ufl.edu/DS064.

17

Haymaking Systems and Marketing Hay

OBJECTIVES

- To compare and contrast long loose hay, chopped hay, packaged hay, stacks, cubes, and pellets.
- To understand storing, buying, and selling hay.

In haymaking, the term *system* refers to a team of processes and machines that does the work from field through feeding, saves crop nutrients, reduces labor requirements, and eliminates drudgery. When each step is mechanized, it must be coordinated; otherwise, workers and machines end up waiting. Automation has had a great impact on haymaking.

HAY HANDLING SYSTEMS

Some haymaking systems are completely mechanized from field to feeding. There is no single best haymaking system for all conditions. Nevertheless, all good systems are fast, make handling easy, save labor and nutrients, and increase profits. Baling is the most popular hay-handling system in North America; however, there are many methods to handle hay that are still utilized.

Long, Loose Hay

The acreage harvested as long, loose hay has declined sharply in recent years, especially in the humid areas, because of high labor cost and because long hay is too bulky for mechanized feeding. Nevertheless, long, loose hay is still popular in many western areas where specialized handling equipment is used. Some of the newer systems of handling and self-feeding loose hay show promise.

The two common methods of handling long hay are:

1. **Loading with a hay-loader directly from windrows.** In this method, cured hay is loaded on a truck or wagon directly from the windrow by means of a hay-loader. Usually the hay is then transported to a barn or stack, where it is unloaded by fork or sling and moved away by hand. Sometimes it is chopped into the barn or other storage area.

2. **Stacks.** These are loaf-shaped (one system makes a circular stack), mechanically pressed haystacks. Long, loose hay is blown into a wagon and pressed down by a hydraulically operated canopy roof. Stacks range in size from 7 to 10 feet wide, 8 to 22 feet long, and 8 to 11 feet high and weigh from 1 to 6 tons. In the West, much of the hay is cured in windrows or cocks and then transported by buck rakes, sweep rakes, or sleds to field stacks, where it is stacked by hay stackers or other large mechanical devices (Figure 17.1). Then after going through a sweat in the stack, the hay is fed out as loose hay, or it is baled if intended for market. Without doubt, this method results in the production of the highest percentage of good-quality hay of any known method primarily because more latitude is permissible in the moisture content when stacking than when baling or

Figure 17.1 Loaflike stacked hay is released from the stacking wagon. *(Courtesy of Iowa State University)*

Figure 17.2 Hay from the windrow is field-chopped and blown into dump wagons for transport to the storage facility. *(Courtesy of Iowa State University)*

chopping. The practice predominates in areas that normally have good haying weather, and the method is prevalent on farms and ranches where haymaking is frequently a major enterprise.

Chopped Hay

Chopped dry hay fits into some feeding systems, particularly in the West. For safe storage, the moisture of chopped hay should not exceed 18 to 20 percent. Two common methods of chopping freshly cured hay are:

1. **Field chopping cured hay directly from the windrow.** In this method, a field chopper gathers the cured hay from the windrow, chops it, and blows it into a truck or trailer (Figure 17.2). The chopped hay is then blown into the barn, stack, or other storage area.
2. **Chopping into the barn or other storage area.** In this method, the cured hay is generally hauled from the windrow or cock to the barn or other storage area, where it is chopped by a hay chopper or silage cutter and blown directly into the storage area. This method is slower and requires more labor than chopping field-cured hay directly from the windrow, but less expensive equipment is necessary.

Packaged Hay

Great strides have been made in hay packaging in the last fifty years with the advent of round bales, large rectangular bales, and cubes. Although large round bales and compressed stacks are better adapted to outside storage than are small round bales, unrestricted access at feeding results in excess

Figure 17.3 A small rectangular bale of alfalfa hay is thrown into a hay wagon by a baler with a thrower attachment. *(Courtesy of National Alfalfa Alliance)*

waste. Large rectangular bales are suitable for commercial marketing and long-distance shipping. Compressing hay into cubes and pellets has many advantages, including completely mechanized handling, high density, more economical transportation and storage, easier self-feeding and higher intake by animals, and lower feeding losses.

The following choices of bales are available:

1. **Rectangular bales.** Conventional, small, rectangular packages (square bales), weighing 60 to 140 pounds (Figure 17.3) are still popular on farms and ranches that produce hay for their own use. Today, hay is also packaged in large rectangular bales, weighing from 1,000 to 2,000 pounds (Figure 17.4), that are designed for custom operators or hay growers with large volumes of hay or straw.
2. **Large round bales.** Many makes and models of large round balers are on the market. All of

Figure 17.4   Handling large square bales requires less labor but more equipment. *(Courtesy of National Alfalfa Alliance)*

Figure 17.5   A round baler rolls long-stem alfalfa hay into large round bales. *(Courtesy of National Alfalfa Alliance)*

them can be classified by the method of rolling the hay into a bale. One method is to pick up the hay from the windrow and roll it in a chamber between a series of belts or chains (Figure 17.5). The other method is to roll the windrow on the ground, similar to rolling up a carpet. Both methods produce a rounded shape that gives weather protection. Large round bales range in weight from 850 to 2,000 pounds, depending on the make of the equipment.

Cubes

Field cubers are machines that move across hayfields; pick up windrows of forage; and produce dense, high-quality forage cubes or wafers. Stationary cubers are used to produce similar cubes from loose haystacks or bales. Hay cubes are of special interest to dairy producers (Figure 17.6) because they

have the advantages of pellets, without their disadvantages. Like finely ground forage that is pelleted, cube-feeding can be readily automated. In comparison with long hay, there is less transportation and storage cost.

Pellets

Pelleted forages are finely ground and then condensed. The biggest disadvantage to increased pelleting at the present time is the difficulty of processing chopped forage coarse enough so that it does not cause digestive disturbances or lower the fat content of milk. A minimum of a ¼-inch chop is recommended.

Storing

Good hay should never be improperly stored. Naturally, the type of storage varies from area to area. In the more arid sections where little rainfall comes during the fall and early winter, a good stack of loose or baled hay may be entirely satisfactorily stored (Figure 17.7). On the other hand, in areas with high rainfall amounts, more expensive waterproof storage should be provided.

Brown hay results when hay wilted to about 50 percent moisture content due to inclement weather is stocked or placed in storage. The damp mass soon ferments extensively and heats. As a result of this action, the hay darkens in color, from dark brown to nearly black. Also, it becomes sweet, aromatic, and palatable. But as a result of the fermentation and heating to which it is subjected, the forage is lower in digestible protein, total digestible nutrients, vitamin content, and feeding value.

Figure 17.6   A hay cuber in action. *(Courtesy of Iowa State University)*

Figure 17.7   In dry climates, large hay bales can be simply stacked outdoors. *(Courtesy of Howard Tyler)*

Figure 17.8   Purchased hay is delivered to an Iowa dairy farm. *(Courtesy of Mark Kirkpatrick)*

Wet hay ferments and generates heat, which can result in spontaneous combustion and fire, usually about one month to six weeks after storing. The warning signals are hay that feels hot to the hands, strong burning odor, and visible vapor. Hot spots may be located by probing the hay with a steel rod. Then the temperature of the hot spots may be tested with a thermometer (a dairy thermometer or other type) attached to a wire and dropped down a pipe. If the hay is over 140°F, it should be checked periodically during the day. If the hay is 160°F, it should be checked hourly. If the hay is 180°F, there are apt to be fire pockets, and it should be removed from a barn. Either dry ice or liquid carbon dioxide can be fed through pipes into the hot areas of hay to help cool it down.

Buying and Selling Hay

Historically, most hay has been fed on the farms where it was produced, but this practice is changing. Today, about 25 percent of U.S. production, with a cash value of over $2 billion, is sold off the farm. As dairy producers become larger and more specialized, they prefer to care for animals and rely on other specialists to grow hay (Figure 17.8).

New methods of marketing hay have made selling easier, but visual inspection is still the most common method of assessing quality and price when either selling or buying. However, the development of near infrared reflectance spectroscopy (NIRS) as a rapid, accurate, and precise method of measuring hay quality is having a major impact on the marketing of hay.

Hay dealers or brokers are important suppliers of hay, as evidenced by the continued growth of the National Hay Association. In California, a high percentage of the hay is marketed through this channel. For the nation as a whole, dealers and brokers market 5 to 10 percent of the hay produced. They purchase hay from growers and sell it to consumers.

Auctions are the most rapidly growing method of marketing hay. Originally, auctions simply brought together producers with loads of hay, an auctioneer, and an assembly of prospective buyers. Hay was sold by visual inspection, and the price was determined by supply and demand. This approach is still widely used in many areas. Individual contracts are agreements between a hay grower and a livestock producer to supply hay of a specified quality at an agreed-upon price. It ensures both the buyer and the seller an orderly market. Such eventualities as weather-damaged hay should be covered in the contract.

SUMMARY

- Two common methods of handling long hay are loading with a hay-loader directly from windrows and hauling cured hay from windrows or cocks with buck rakes, sweep rakes, or sleds.
- Two common methods of chopping freshly cured hay are field chopping cured hay directly from the windrow and chopping into the barn or other storage area.
- Two choices of bales are available: rectangular bales and large round bales.
- Field cubers are machines that move across hayfields to produce high-quality forage cubes or wafers, whereas stationary cubers are used to produce similar cubes from loose haystacks or bales.
- The development of near infrared reflectance spectroscopy (NIRS) as a rapid, accurate, and precise method of measuring hay quality had a major impact on the marketing of hay.

QUESTIONS

1. Which haymaking method results in the highest percentage of good-quality hay?
2. Which type of bale is better adapted to outside storage?
3. What is the biggest deterrent of pelleted forage?
4. At what temperature should hay with hot spots be checked hourly?

ADDITIONAL RESOURCES

Articles

Leonardi, C., and L. E. Armentano. "Effect of Quantity, Quality, and Length of Alfalfa Hay on Selective Consumption by Dairy Cows." *Journal of Dairy Science* 86 (February 2003): 557–564.

Plaizier, J. C. "Replacing Chopped Alfalfa with Alfalfa Silage in Barley Grain and Alfalfa-Based Total Mixed Rations for Lactating Dairy Cows." *Journal of Dairy Science* 87 (August 2004): 2495–2505.

Internet

Buying and Selling Hay and Straw: http://www.uwex.edu/ces/ag/haybuying.html.

Management Tips for Round Bale Hay Harvesting, Moving, and Storage: http://www.ext.vt.edu/pubs/ageng/442-454/442-454.html.

Round Bale Hay Storage: http://server.age.psu.edu/extension/Factsheets/i/I112.pdf.

18

Grazing Systems and Pasture Management

OBJECTIVES

- To identify the advantages and disadvantages of various types of grazing-based operations.
- To identify the factors affecting the value of pasture.
- To describe pasture management.

Approximately 50 percent of the total land area of the United States is devoted to pasture and grazing lands; much of this land is unsuitable for cultivation. It is estimated that over half of the nation's milk is produced from forages. Top-quality pasture alone can provide cows with sufficient nutrients for body maintenance and for the production of 20 pounds to more than 40 pounds of milk daily (Figure 18.1).

When larger numbers of lactating cows are concentrated on smaller acreages and when milk production per cow increases, dairy producers utilize less pasture and depend more on other feeds. A major reason for this development is the inability of cows with the genetic potential for high milk production to consume enough feed to supply their energy requirements when pasture is their main feed source. The physical form and volume of pasture fill the rumen to capacity before the nutrient needs of high producers are fulfilled. Pastures continue to be practical in some smaller herds, however, especially for heifers and dry cows.

ADVANTAGES OF GRAZING-BASED OPERATIONS

Grazing-based operations have several advantages over confinement production:

1. **They decrease feed costs.** Grass is cheaper than harvested and stored hay or silage. Cows consume from 100 to 200 pounds (as fed) of pasture per head daily (Figure 18.2). Pasture normally contains 70 to 85 percent moisture, pature fed cows consume 15 to 60 pounds of dry matter per day. For cows with a high genetic potential for milk production, grass alone cannot provide enough nutrients for the cows to reach their genetic potential.
2. **They decrease health problems.** Animals on pasture come in contact with each other less than do animals in confinement, thus reducing the risks of communicable disease. In addition, hoof health is typically better than it is in confinement herds (Figure 18.3).

Figure 18.1   Cows grazing on an open pasture. *(Courtesy of USDA)*

Figure 18.2 Calculating milk produced per acre provides a benchmark for the efficiency of pasture utilization. *(Courtesy of Iowa State University)*

Figure 18.3 Jerseys on pasture have fewer health problems, especially hoof health problems, than do those in confinement. *(Courtesy of Iowa State University)*

Figure 18.4 Land that is unsuitable for growing row crops can be an excellent source of forage for grazing cows. *(Courtesy of Iowa State University)*

3. **They decrease costs for buildings and equipment.** Lower-cost buildings and equipment can be used in a pasture system than can be used in confinement, with the result that a grazing-based operation requires less capital investment on a per-cow basis.
4. **Labor requirements are decreased.** It requires more skilled labor to operate a fully automated, highly mechanized, confinement complex than a pasture operation.
5. **They maximize noncrop areas.** Pastures permit the maximum utilization of areas not suited to crop production (Figure 18.4).

DISADVANTAGES OF GRAZING-BASED OPERATIONS

Several disadvantages of pasturing dairy cattle have caused and will continue to cause a shift away from its use to confinement production. The disadvantages sometimes attributed to pastures follow:

1. **They mitigate against enlarging dairy production without enlarging the farm.** With high-priced land, this fact must be weighed if one wants to increase the size of the dairy operation (Figure 18.5).

Figure 18.5 In larger grazing operations, cows may have to travel some distance to and from the milking parlor twice daily. *(Courtesy of John Smith)*

2. **They may prevent more profitable uses of land.** On many dairy farms, operators can make more money from growing corn, soybeans, and other crops than they can from allowing cows to graze pastures.

3. **They do not facilitate manure handling.** Although less manure has to be handled in a pasture system than in confinement, it is more difficult to automate and handle manure when animals are scattered over a large area.

4. **They decrease yields.** Pasturing results in considerable waste of the crop through trampling, with lower yield of nutrients per acre than from harvested crops (Figure 18.6).

5. **They result in cows expending energy in grazing.** The activity of cows on pasture (Figure 18.7) increases the energy required for maintenance in proportion to the amount of energy expended in grazing; up to a 40 percent increase in the maintenance requirement may occur.

6. **They rarely provide sufficient nutrients.** Lush pasture may be so high in water content that comparatively few nutrients are ingested, even though cows are full.

7. **They do not maintain uniform growth and quality.** Because of variable weather conditions, it is difficult to maintain uniform growth and quality in pastures, with the result that ration balancing is extraordinarily complex and milk production fluctuates with pasture conditions.

8. **They reflect soil composition.** The nutritive value of pasture is directly related to the composition of the soil. Hence, a soil that is low in certain minerals produces pasture that is low in those minerals. Conversely, toxicities can result in animals grazing on soil with a high concentration of certain trace minerals, for example, selenium. Soils that produce the best forages for grazing are probably more profitably utilized for other purposes.

Forage plant choices for a grazing-based dairy operation are critical to the profitability of that operation. The choice must be adapted to local soil and climatic conditions. Although this is a prime requisite, producers cannot afford to disregard the grazing qualities of available forages. The pasture must be highly palatable and provide a nutrient balance that meets closely the class of animal being grazed. It must also be able to withstand tramping and grazing and fit satisfactorily into the crop rotation. It must be free of disease or parasites and not induce toxicities or disorders, such as bloat, at the feeding rate anticipated. Over most of the United States, one or more adapted grass-legume mixtures possess most of these qualities.

Figure 18.6   The amount of usable forage per acre is reduced for grazing because of trampling losses and inefficiencies in harvest. *(Courtesy of Iowa State University)*

Figure 18.7   The energy expended during grazing increases the maintenance requirements of pasture-fed cattle. *(Courtesy of Iowa State University)*

FACTORS AFFECTING THE VALUE OF PASTURE

Many factors affect the value of pasture, including the type of soil and level of fertilizer application, the plant species selected and the stage of maturity of that plant, and the intensity of grazing. The chief benefit from applying fertilizer to pasture is an increase in yield. In grass-legume pastures, proper fertilizing can influence the proportion of legumes; in turn, this increases the protein, calcium, phosphorus, and vitamin content of the mixture. Generally, legume-grass pastures with about 50 percent legumes do not require nitrogen (N) fertilization. However, it is important to maintain adequate levels of lime, phosphorus (P), and potassium (K) in the soils of these pastures. Properly fertilized pasture plants begin growth earlier in the spring and continue growth later in the fall, thus extending the grazing season.

The protein content of young, immature nonlegume pasture is increased appreciably by nitrogenous fertilization, unless there is already plenty of nitrogen in the soil. This increase may be sufficient to add substantially to the palatability and the feeding value of grass pasture. Calcium-deficient soils affect pastures in two ways: the percentage of calcium in nonlegume crops is considerably reduced, and the legume crop, if present, will not thrive.

On phosphorus-deficient soils, the phosphorus content of grasses may drop so low that a phosphorus deficiency in animals results unless a supplement is provided. The phosphorus content of legumes is less affected by a deficiency of soil phosphorus than that of nonlegumes. However, most legumes do not thrive or produce high yields on phosphorus-deficient soils.

Plant species affect the feeding value of pasture. Generally, legumes contain a higher percentage of protein and calcium than do nonlegumes. Also, marked differences exist between different kinds of pasture plants as growth advances. For example, bromegrass retains its palatability and nutritive value over a longer period compared to most grasses. By contrast, reed canarygrass is readily eaten when young, but it becomes woody, high in alkaloids, and unpalatable with maturity. Most pasture legumes retain their palatability and nutritive value better as they mature compared to most grasses; an exception is lespedeza sericea, which becomes bitter and distasteful with maturity due to the accumulation of tannin in the plants. However, plant breeders have developed sericea that is low in tannin, thereby overcoming this problem to some degree.

Great differences in nutritive value also exist between young, immature pasture and the same plants when they are mature or even at the usual hay stage. Young plants are much richer in protein, on a dry basis, than are the same plants at maturity. Young, actively growing grasses may contain over 20 percent protein; young legumes, such as alfalfa and clover, may contain 25 percent protein or more. The protein content of grasses decreases greatly as they head out. Very mature, dormant grasses may contain as little as 2 percent protein. The protein content of legumes decreases with maturity, but to a lesser extent than it does in grasses.

The percentage of fiber and lignin in plants influences both palatability and digestibility. Although inherent differences exist among plant species as well as among plants of the same species, young plants and regrowth are always lower in fiber and lignin than are mature plants. As a result, they are more tender and digestible. On a dry-matter basis, new growth in a grass legume pasture ranges up to 68 percent total digestible nutrients in comparison with 51 percent at the normal hay stage. In humid areas, the nutrient content of some grasses is leached rapidly following maturity, and the digestibility and nutrient value is greatly reduced. To a considerable degree, the decrease in digestibility as plants mature is due to an increase in lignification, which lowers digestibility.

The calcium content of plants decreases with maturity. However, the percentage change in calcium is much less than that occurring in phosphorus. Young grasses or legumes grown on phosphorus-rich soils usually contain 0.25 percent or more phosphorus. Although phosphorus in pasture decreases with maturity, there is generally plenty to meet the requirements of dairy cattle unless it is produced on phosphorus-deficient soils or it is left to cure on the stalk, which allows for weathering and bleaching. Early spring grass may contain five times more phosphorus than leached and weathered grass in the winter.

The actively growing, green parts of plants are high in carotene. Likewise, such plants are usually rich in most of the B complex vitamins, vitamin E, and ascorbic acid. But the content of vitamins, especially of carotene, decreases as plants mature.

Rapidly growing grass is usually rich in protein and in other nutrients. Therefore, it is important that pasture plants be managed so that they keep growing and that they be prevented from heading out. This can be accomplished by mowing, but it is more practical when done with animals through intensive grazing programs. Grass is usually higher in protein and other nutrients early in the spring than later in the season. If plant growth is sharply checked in the summer due to drought, hot weather, and/or lack of available plant food, the protein content and the digestibility will be lower than that found earlier in the season. If pasture resumes growth after the fall rains come, it may be nearly as high in protein and other nutrients as spring growth.

When pastures are grazed closely throughout the season, the total yield of dry matter is usually 30 to 50 percent less than when they are allowed to grow to the normal hay stage. This is due to the smaller leaf surface and lowered photosynthesis. Rotational and strip grazing or feeding green chop yields more nutrients per acre than does close continuous grazing. Overgrazing reduces the yield of tall-growing plants such as timothy, orchard grass, alfalfa, and the erect clovers to a greater extent than that of low-growing spreading plants, such as bluegrass, Bermuda grass, and white clover. Animals allowed to graze selectively in an extensive grazing program pick and choose the leaves and finer parts of stems, which are more digestible and more nutritious, and reject the courser, stemmy parts (Figure 18.8). Thus, the portion consumed under such circumstances always differs appreciably from the chemical composition of the entire plant.

MANAGEMENT OF DAIRY PASTURES

Many good dairy pastures have been carefully established only to be lost in succeeding years through poor management. Efficient and profitable pasture management requires controlled grazing. Nothing contributes more to good pasture management than controlled (proper) grazing. First-year seedings

Figure 18.8 Grazing cows consume the best parts of the plants first. *(Courtesy of Iowa State University)*

Figure 18.9 This two-strand electric fence separates paddocks in a rotational grazing scheme. *(Courtesy of Iowa State University)*

should be grazed lightly or not at all to allow them to become established. Where practical, instead of grazing, it is best to mow a new seeding about 3 inches above the ground and to utilize it as hay or silage, provided there is sufficient growth.

When portable salt containers are used, more uniform grazing may be obtained simply by the practice of shifting the location of the salt to less grazed areas of the pasture. When possible and practical, water locations should also be well distributed. Development of more and smaller pastures and employing rotational grazing programs permits greater control of when and where cattle graze. Many more intensive systems are designed with stationary salt and water locations, and the area being grazed may change every three to four days or even several times daily in some cases. Various forms of electric fencing are used in several different pasture arrangements, all designed to facilitate grazing greater numbers of animals on an area for shorter periods of time (Figure 18.9).

When possible, allow 6 to 8 inches of growth before turning out to pasture in the spring, which allows roots to become firmly established. This is not always practical on year-round pasture programs. When two or more pastures are available, however, early spring grazing can be rotated to the benefit of plants in all pastures. Lush pasture presents other problems. Grass tetany is a metabolic condition that affects cows (especially lactating cows) grazing lush pasture high in nitrogen, which results in low absorption of magnesium and is most common in the spring. Afflicted animals develop tetany, walk with a stiff gait, go into convulsions, and may die. During the danger period, cows grazing lush pasture should be supplemented with magnesium oxide.

Figure 18.10 Pearl millet should be grazed down to about 4 inches. It should not be regrazed until regrowth is back to at least 16 inches. *(Courtesy of Iowa State University)*

Pastures that are closely grazed late in the fall start growing late in the spring. Plants should be allowed to replenish root reserves prior to going dormant. With most pastures, 3 to 5 inches of growth should be left for winter cover. An exception to this close-grazing rule should be made where winter annual clovers are to be seeded or are expected to volunteer, especially on Bermuda grass or Bahia grass pastures. Under such circumstances, close grazing or mowing of the Bermuda grass or Bahia grass is recommended.

Pastures should not be grazed more closely than 2 to 3 inches during the pasture season (Figure 18.10). Continued close grazing reduces the yield, weakens the plants, encourages weeds to invade, and increases runoff and soil erosion. The use of temporary and sup-

Figure 18.11   Grazing cattle on open, continuous pastures results in some areas of undergrazing and some areas of overgrazing. *(Courtesy of Iowa State University)*

plemental pastures may save regular pastures through seasons of drought and other pasture shortages and thus alleviate overgrazing.

Conversely, undergrazing seeded pastures should be avoided because rank growth is less palatable and less nutritious; tall-growing grasses may drive out desirable low-growing plants such as white clover due to shading; and weeds, brush, and coarse grasses are more apt to gain a foothold when the pasture is grazed insufficiently (Figure 18.11). If the stocking rate is inadequate, pastures should be clipped as necessary to control competing weeds and to get rid of unpalatable course growth left after incomplete grazing. Good grazing management reduces the amount of clipping needed. Pastures that are grazed continuously may be clipped at or just preceding the usual haymaking time; rotated pastures may be clipped at the close of the grazing period. Mowing is expensive and should be implemented only when results are clearly beneficial. Clipping solely for cosmetic reasons should be critically evaluated.

Droppings should be scattered at the end of each grazing season to prevent animals from leaving ungrazed clumps and to guarantee distribution over a larger area. This can best be done by the use of a brush harrow or a chain harrow. When animals are concentrated under intensive systems such as strip grazing, droppings tend to be broken up by hoof action, and the need for harrowing is reduced.

Fence management cannot be overlooked when managing pastures. Nails, bits of fencing material, and other foreign objects must be carefully removed from pastures following fence construction or repair procedures. Ruminants are sometimes indiscriminate in their selection of feed, and they often consume these foreign objects. Due to the motility patterns of the gastric region, these objects tend to accumulate in the reticulum; the presence of these sharp objects can pose serious problems, especially if the reticulum should be punctured and the for-

Figure 18.12   Large open pastures are still commonly used on many dairy farms. *(Courtesy of Monsanto)*

eign object allowed to migrate to the pericardial region. The condition in which the collection of foreign objects irritates or punctures the reticulum is called reticulitis, or hardware disease. Placing magnets in the reticulum of grazing animals can prevent these foreign objects from doing great damage following ingestion and is recommended for all grazing animals.

Several systems of grazing management have been applied successfully to pastures. Generally, more intensive management systems provide both higher yields of forage and of milk. The basic types of systems are continuous grazing and rotational grazing systems.

Continuous Grazing

Continuous grazing is the uninterrupted grazing of a specific pasture by animals throughout the year or grazing season (Figure 18.12). It requires a moderate stocking rate, with periodic adjustment in animal numbers to reduce the severity of under- or

Figure 18.13 Rotationally grazed cattle in Ohio are restrained by an electrified temporary fence. *(Courtesy of Ohio State University)*

overgrazing. The advantages of continuous grazing compared to rotational grazing include lower costs for fencing and watering facilities, fewer management decisions, and decreased labor requirements.

Continuous grazing operations have extreme limitations, however. Animal numbers are not as flexible as in other systems; in addition, pastures must be stocked lighter than desired when forage growth is maximal to avoid overgrazing during periods of minimal forage growth. Animals typically graze some species in preference to others and return to graze the regrowth of the same plants, thus selectively reducing plant vigor. Finally, animals often show a preference for grazing certain portions of pastures, resulting in uneven fertilization.

Rotational Grazing

Rotational grazing systems have two or more pastures that are grazed and rested in a planned sequence (Figure 18.13). This system allows major forage species to be harvested and then provided a period of rest, enabling the plants to remain thrifty and vigorous. Rotational grazing involves the concept of time as a management variable for either the grazing period or the regrowth in each pasture. Duration of grazing and rest generally are governed by herbage growth rate, which depends on the time of year, moisture, fertility, and species. The number of animals grazing each system may be fixed or variable.

Rotational grazing permits the dairy farmer to match grazing more adequately to the growth habit of the forage species, condition of the pasture, and animal needs than does continuous grazing. It improves stand persistence and production. Plants are given recovery periods during the growing season for more or less unhampered development of tillers and leaves, which is essential to replenish root reserves. This system of grazing enables the tall-growing legumes and grasses to survive. It increases carrying capacity. Greater amounts of feed nutri-

ents can be removed in the form of herbage, with reduced losses due to trampling, fouling, and herbage death and decay.

Rotational systems also encourage equalization of grazing; they help prevent overgrazing and undergrazing, and maintain a better balance of the legumes and grasses. Both the palatable and the inferior species are grazed at more nearly the same rate. It often provides more nutritious herbage because the herbage is at the most ideal pasture stage. It helps prevent the grasses from heading out, which is accomplished by concentrating grazing animals or by mowing when animals are shifted to new pastures. This allows new growth to come back uniformly and keeps it more palatable, making it more convenient to harvest surplus forages as hay or silage. These systems also help control animal parasites, especially intestinal worms. The life cycles of worms can be broken by proper planning of grazing and rest periods.

The limitations of rotational grazing include a higher input of capital and management compared to continuous grazing. A continuous day-to-day decline occurs in the quality of the available forage, especially on the more intensive systems. At first turn-on, animals have access to leafy, high-quality forage, but the quality of the forage gets poorer and poorer during the grazing period.

Rotational grazing systems come in several variations, including strip grazing and other intensive grazing systems.

Strip Grazing

In this system, cows access a strip of pasture that may be large enough for several days of grazing or small enough for less than one day of grazing. Heavy stocking rates of up to fifty animal units per acre are accomplished by fencing each strip with movable electric fences both in front and behind the grazing animals. Strip grazing provides increased utilization of herbage, with waste reduced to 10 to 20 percent relative to continuous systems. Increased milk yields per acre result. In addition, stability of milk yield is improved because the nutritive value of the pasturage consumed is quite constant. More animal units are maintained on a given area, even when individual animal productivity is not increased. Less herbage is soiled by feces, urine, and tramping in this system. When strip grazing, animals are quieter and settle down quickly for steady grazing.

Intensive Grazing

Several ingenious intensive grazing systems have evolved. All of them are designed to provide and harvest the maximum amount of high-quality forage, utilize the highest quality pastures for the highest producing animals, and increase profits. Many designs are utilized for intensive short-duration grazing systems. Generally, dairy producers favor the conventional (rectangular) system, which usually consists of small pastures of about equal size or production capacity, fenced in a grid arrangement. Water and salt may be located in each of the pastures, or a single source may be used, with cattle gaining access by a lane or alley.

Some short-duration grazing systems involve two groups of animals: first grazers and second grazers. This system uses the best quality pastures for lactating cows. High-producing lactating cows, with the highest energy requirements, are first grazers; they are allowed to graze the higher-quality (leafy) portion of the first pasture, and then they are moved to a second, fresh pasture. Dry cows or replacement heifers, which have a low energy requirement, are second grazers; they are turned into pastures immediately following the removal of the high producers. This progression is continued through all the pastures, and then the cycle is repeated.

The chief advantage of the system of first and second grazers is the enhanced productivity of the first grazers. The main limitations are the necessity of maintaining two balanced groups of animals of different productivity levels and maintaining balanced stocking rates and pasture sizes.

SUPPLEMENTAL FEEDING FOR GRAZING CATTLE

Pasture alone cannot meet all the nutritional requirements of high-producing dairy cattle; therefore, supplemental feeding systems are beneficial. Graziers that supplement pasture are more profitable than those that do not provide supplemental feed. Greater challenges are inherent in balancing rations for grazing cattle than for cattle fed in confinement systems. Pasture quality varies throughout the growing season; this influences the nutrient density, digestibility, and intake of the forage. Estimates of pasture intake and pasture quality need to be made frequently to adjust the supplemental feeding program accordingly. In addition, many graziers provide supplemental feeds during milking, when cows are conveniently grouped closer to feeds storage areas. However, feeding concentrates as two or three large meals per day is more likely to disrupt rumen function and reduce feed digestibility, especially forage digestibility. Therefore, it is better for the cows to receive supplemental feeds while they are on pasture; however, this is much more labor-intensive and requires bunk availability in each pasture. Many unique feeding systems are being tested for supplementing grazing cattle, including portable computerized feeders. Whatever system is ultimately implemented, profitability must be the primary factor in the decision-making process.

SUMMARY

- Grazing-based operations have several advantages over confinement production, including a decrease in feed costs, health problems, costs for buildings and equipment, and labor requirements. They also maximize noncrop areas.
- Disadvantages of pasturing dairy cattle include the need for more land, prevention of profitable uses of land, more difficult manure handling, decreased yields, cows expending more energy in grazing, nutrient deficiency, lack of uniform growth and quality, and the effects of soil composition.
- The type of soil, level of fertilizer application, plant species, stage of maturity of the plant, and intensity of grazing affect the value of pasture.
- Efficient and profitable pasture management requires controlled grazing.
- The more intensive system of management, the higher the yield of forage and of milk.
- Basic types of systems are continuous grazing, rotation grazing, intensive grazing, and strip grazing.
- Continuous grazing is the uninterrupted grazing of a specific pasture by animals throughout the year or grazing season.
- Rotational grazing involves the use of two or more pastures that are grazed and rested in planned sequence.
- Intensive grazing is designed to provide and harvest the maximum of high-quality forage, utilize the highest quality pastures for the highest producing animals, and increase profits.
- Strip grazing is a system where cows access a strip of pasture that may be large enough for several days of grazing or small enough for less than one day.

QUESTIONS

1. Which grasses and legumes are found in all areas of the United States?
2. What is the chief benefit from applying fertilizer to pasture?
3. Which grass becomes woody and high in alkaloids with maturity?
4. What happens to calcium and phosphorus content as plants mature?
5. How much growth is necessary before turning cattle out to pasture in the spring?
6. Define grass tetany.
7. Why should pastures not be grazed closer than 2 to 3 inches during the pasture season?
8. Define reticulitis or hardware disease. What is the treatment?
9. What are some of the advantages of continuous grazing?
10. What kind of growth does rotational grazing yield?
11. What is the difference between a first grazer and a second grazer?
12. List some of the advantages of strip grazing.

ADDITIONAL RESOURCES

Articles

Cros, M. J., M. Duru, F. Garcia, and R. Martin-Clouaire. "Simulating Rotational Grazing Management." *Environment International* 27 (September 2001): 139–145.

Fike, J. H., C. R. Staples, L. E. Solleberger, B. Macoon, and J. E. Moore. "Pasture Forages, Supplementation Rate, and Stocking Rate Effects on Dairy Cow Performance." *Journal of Dairy Science* 86 (April 2003): 1268–1281.

Kolver, E. S. "Nutritional Limitations to Increased Production on Pasture-Based Systems". (review). *Proceedings of the Nutrition Society* 62 (May 2003): 291–300.

Parsons, R. L., A. E. Luloff, and G. D. Hanson. "Can We Identify Key Characteristics Associated with Grazing-Management Dairy Systems from Survey Data?" *Journal of Dairy Science* 87 (August 2004): 2748–2760.

Internet

Comparing Various Grazing Management Systems: http://www.uaex.edu/Other_Areas/publications/HTML/FSA-2129.asp.

Controlled Grazing: http://www.mda.state.mn.us/crp/GRAZING.htm.

Developing a Pasture Management Program: http://ohioline.osu.edu/agf-fact/0017.html.

Efficient Pasture Systems: http://www.ca.uky.edu/agc/pubs/agr/agr85/agr85.htm.

Grazing Systems: http://www.caf.wvu.edu/~forage/5712.htm.

Permanent Pasture Management: http://www.caf.wvu.edu/~forage/5728.htm.

Rotational Grazing: http://www.ca.uky.edu/agc/pubs/id/id143/id143.htm.

19

Ensiled Feeds

OBJECTIVES

- To understand the ensiling process.
- To describe various kinds of silage.

Ensiling is an old method of preserving feed by fermenting forage plants. Columbus found that the Indians used pits or trenches in which to store their grain, and centuries earlier, in the Old World, silos were used as a means of preserving both grain and green forage. Until about 1910, silage was generally thought of as feed for dairy cows. Even today, most silage is used on dairy farms.

The surplus forage produced during the growing season may be preserved for feeding during the winter months and other periods of pasture scarcity by haymaking, which is the most efficient method during dry weather next to grazing. But weather conditions are not always favorable for haymaking. Ensiling, on the other hand, can be done in inclement weather. Also, it has the advantage of preserving a higher proportion of the nutrients of the plant than can be accomplished in haymaking, although a slightly greater cost may be involved than in normal field-curing of hay.

The importance of silage in this country is attested to by the fact that about 120,000 tons are made annually. It is especially important in all dairy regions of the United States that have humid climates and cold winters. A wide variety of silo types continue to appear, and the use of preservatives has grown enormously since 1980. Most of the silage is fed on the farms on which it is produced, rather than being bought and sold. It is important, therefore, that dairy producers know how to produce good silage, as well as how to feed it, because most of them determine their own destiny from the standpoint of quality.

THE ENSILING PROCESS

The ensiling process refers to the changes that take place when forage or feed with sufficient moisture to allow microbial fermentation is stored in a silo in the absence of air. The basic strategy of silage preservation is to exclude oxygen and to acidify the forage through bacterial fermentation. An understanding of these changes is likely to lead to the production of more high-quality silage.

The entire ensiling process requires two to three weeks. Initially, the living plant cells of the forage continue to respire, or breathe, consuming the oxygen of the silage-entrapped air, producing carbon dioxide and water and thus in turn releasing energy or heat. Simultaneously, aerobic yeasts and molds thrive and multiply. In good ensiling conditions and with proper management, the aerobic phase is very short, and silage temperatures seldom rise above 100°F. However, slow filling, inadequate packing of the forage, or leaky sealing around the silage lengthens the aerobic phase, increases losses, and causes excessive heating.

When the available oxygen of the entrapped air has been consumed, anaerobic bacteria multiply at a prodigious rate. Simultaneously, the molds and the yeasts die. Certain plant enzymes continue to function. The nonstructural carbohydrates, especially sugars, are converted to lactic acid, some acetic acid, a small amount of other acids, alcohol, and carbon dioxide. The plant proteins are broken down into peptides, ammonia, amino acids, amines, and amides. These nonprotein nitrogen compounds are utilized less effectively by the animal than is true protein. If the acidity becomes too great, the bacteria die, and the silage stabilizes. This occurs when the silage pH reaches a low of 3.5 to 4.5 for corn and cereals or 4.0 to 5.0 for grasses and legumes.

Proper fermentation requires sugars, oxygen-free conditions, moisture content in the correct range, and bacteria. The optimum moisture content for corn or cereal silages is 60 to 70 percent. Grass or legume silages should be in the range of 50 to 70 percent. With the right storage conditions and careful management, however, good silage can be made outside these ranges.

In grasses, corn, and cereals, there are normally enough sugars for fermentation. Legumes with over 70 percent moisture are frequently low in sugars, which may limit fermentation and prevent the attainment of a stable pH. If the forage is harvested without rain damage and the silo is filled and sealed quickly, there is less respiration and more sugar is available for fermentation.

Grass and legume forages lose valuable nutrients during drying in the field. Rapid wilting and the avoidance of rain damage are two ways to increase the quality of the silage. Minimizing the time from cutting to ensiling also conserves valuable crop nutrients. Good packing of the forage is essential for keeping out air and allowing fermentation. This may require chopping the forage into sufficiently small particles and applying pressure to the top surface of the silage. Maintaining an airtight seal around the surfaces of the silage keeps out oxygen, promotes good fermentation, conserves nutrients, and reduces heating. Finally, careful management is needed after the silo is opened to minimize exposure to air before the silage is fed to the animals.

Several silage preservatives are currently available, including inoculates that speed fermentation by supplying either sugars to supplement the available plant sugars for fermentation or enzymes that break down complex carbohydrates such as starch, cellulose, and hemicellulose to fermentable sugars. In addition, some preservatives supply acids that reduce the pH without fermentation and inactivate the enzymes that break down protein or aerobic inhibitors such as propionates and antioxidants that restrict the yeasts, molds, bacteria, and enzymes that in turn degrade the forage in air. The use of additives can increase the preservation of crop nutrients and maintain a higher quality of silage. But additives increase the cost of silage making, may not be reliable, and in some cases may be dangerous to handle.

The decision to use a silage additive should be based on a clear need to increase nutrient retention and on scientific evidence that the additive is effective. Application rate is of critical importance in terms of the active ingredient supplied to each unit of forage. Proper application at the most appropriate point in the harvesting/storing process is necessary to achieve good mixing of the additive with the forage without slowing down the harvesting operation.

KINDS OF SILAGE

A great variety of crops can be and are made into silage. Most silage in the United States is made from either corn or sorghum; over fifteen times as much corn silage as sorghum silage is made. About 80 million tons of corn silage and 5 million tons of sorghum silage are produced in the United States. About 75 percent of the nation's silage is made from corn and sorghum and 25 percent from grasses, legumes, and other feeds. In addition to the kinds of silage already mentioned, silage is made from sunflowers, the small grains, sugar beet tops, crop residues, wastes from food processing (sweet corn, green beans, green peas), root crops, and various vegetable residues.

Corn and Sorghum Silage

For the United States as a whole, corn ranks first in importance as a silage crop (Figure 19.1). Generally, more total digestible nutrients can be obtained from an acre of corn as silage, which yields from 5 to 25 tons of forage per acre with an average of over 14 tons, than can be obtained from an acre of any other crop. Also, corn ensiles easily without the aid of a preservative, keeps almost indefinitely in a good silo, is highly palatable, is well adapted to mechanized feeding, and may be fed with little waste. There are four kinds of corn silage:

1. **The whole corn plant.** When at the peak of its nutritive value and right for ensiling, the whole corn plant contains 1.5 times the nutrients of the ripened grain that the plant would have yielded. Also, in corn silage made from the whole crop, more than 90 percent of the nutrients produced are saved (Figure 19.2).

Figure 19.1   Corn is the most commonly ensiled crop, and corn silage is an extremely palatable high-energy feedstuff for dairy cattle. *(Courtesy of Iowa State University)*

Figure 19.2 Harvesting corn silage from an irrigated field in California. *(Courtesy of Howard Tyler)*

Figure 19.3 Sorghum is typically harvested for silage when the heads are at the soft- to medium-dough stage. *(Courtesy of Iowa State University)*

2. **Ear corn silage.** The ensiled ears contain up to 68 percent of the nutrients of the entire corn plant.
3. **Corn stover silage.** The forage remaining after harvesting a grain crop accounts for about one-third of the total nutritive value of the crop.
4. **Shelled-corn silage.** This consists of the kernels only. At 70 percent dry matter (30 percent moisture), shelled-corn silage contains 61 to 66 percent of the nutrients in the whole crop.

The sorghums are more dependable and higher yielding than corn in certain areas, particularly in unirrigated, relatively dry areas of the western and southwestern United States (Figure 19.3). Sorghum for silage is harvested with the same equipment as that used for corn silage. It should not be harvested for silage until the heads are soft- to medium-dough stage. Harvesting at this stage provides the highest yields of total feed material, enhances preservation, and makes palatable silage.

On a dry-matter basis, corn silage contains an average of 8.3 percent crude protein, 68 percent total digestible nutrients, 0.31 percent calcium (Ca), and 0.27 percent phosphorus (P). Grain sorghum silage contains less protein and total digestible nutrients than does corn silage. Grass/legume silages contain more protein and less TDN than does corn silage. The carotene content of corn silage is variable, but it is on the low side.

Corn and Sorghum Residue Silage

Corn and sorghum residues, the forages that remain after harvesting a grain crop of corn or sorghum, may be used as cattle feed in three ways: grazed, harvested (stacked or baled) and fed dry, or ensiled and fed as silage. When ensiled, cornstalks (stover) produce a product known as corn stover silage or cornstalk silage. When stalks are processed as silage, the use of a forage harvester equipped with a screen or a recutter-blower at the silo is necessary to chop the material finely. Fine chopping ensures good packing and improves consumption by avoiding selectivity.

When making corn stover silage, the residue should be harvested as soon as possible after the grain is taken off, before the residue loses any moisture. At that time, the grain moisture is generally under 30 percent, and the refuse moisture is over 48 percent. In an airtight silo, 40 to 45 percent moisture is very satisfactory for ensiling. In an unsealed or bunker silo, the moisture content should be 48 to 55 percent for proper lactic acid formation. Water may be added at the silo if necessary. As a precaution, the addition of 56 pounds of corn meal (or other finely ground grain) per ton of corn stover silage provides readily fermentable carbohydrates for better microbial fermentation. The acids provided act as a preservative. With husklage, the latter precaution is not necessary because sufficient grain remains in the husk and cob.

The biggest deterrent to harvesting stalklage, in either dry or ensiled form, is the cost, primarily for equipment. Rather than own such expensive equipment, which is used for a short period only, custom harvesting of stalklage is likely cheaper for most operators.

Husklage, the forage discharged from the rear of a corn combine and consisting of the husks, cobs, and any grain carried through the combine, may also be ensiled. Ensiling husklage, following recutting and adding water, results in increased cow consumption and less rejection of cobs.

Like corn, sorghum stover may be grazed or harvested and stored as dry feed or silage. Because the sorghum plant stays green late in the fall, good sorghum stover silage can be made without additional water.

Grass/Legume (Hay Crop) Silage

Grass/legume (hay crop) silage refers to silage made from any of the green crops that might otherwise be grazed or dried and made into hay. This includes grasses (such as timothy or fescues), legumes (such as alfalfa or clovers), grass-legume mixtures, and cereal grains (such as oats). In practical operations, any adapted grasses and/or legumes may be used three ways: for grazing, hay, or silage. Grass/legume silage can be produced in areas where the climate is too cool and the growing season too short for corn or sorghum silage (Figure 19.4). Although grass and legume crops have been ensiled in Europe for hundreds of years, the practice did not become widely used in the United States until the 1930s.

Ensiling minimizes field, harvest, and storage losses of grass/legume forages. It minimizes dependence on favorable weather to harvest the crop. Like all silages, however, grass/legume silage is heavy, bulky, and costly to transport. Thus, its off-farm market value is limited. Existing upright silos may not be in sufficiently good condition to store grass/legume silage.

Once grass/legume silage is removed from the silo, it must be fed within twelve to twenty-four hours to alleviate spoilage. Once feeding begins during warm weather, it is necessary to feed a minimum of 4 inches off the exposed surface daily to prevent spoilage in the silo, unless you are using an oxygen-limiting silo.

Direct-cut grass/legume silage is forage that is harvested and stored without field-drying and usually contains more than 70 percent moisture. Although direct-cut ensiling is the standard practice with mature corn and sorghum, it is not recommended for grass/legume crops because of the difficulty in getting good preservation (due to the high moisture content) and the increased nutrient losses (due to seepage). The higher the moisture content of grass/legume forage, the more critical the need for a low pH to obtain good preservation. Direct-cut silage is subject to excessive deterioration of protein, undesirable odor, excess loss of nutrients due to seepage, and deterioration of concrete stave silos. Also, when fed as the major forage, direct-cut silage usually results in lower consumption of dry matter and consequently lower animal production than with wilted or low-moisture silage.

When making direct-cut silage, harvest at the proper stage of maturity; avoid cutting when the forage is wet with dew or rain. Drainage should be provided for seepage and an additive or preservative should be added. (Coarsely ground cereal grain absorbs some of the excess moisture in addition to providing sugars for fermentation.) Finally, the silage must be distributed evenly in the silo, packed thoroughly, and covered with plastic or other suitable material.

A high percentage of grass/legume forage is dried to some degree prior to ensiling. Wilting gets rid of some water in the field, so less weight is handled. Also, in comparison with direct-cut silage, odor and seepage problems are reduced, and additives or preservatives are usually not needed. Forages are allowed to wilt in the swath and/or windrow until the moisture reaches about 65 percent, which may take from 1 to 4 hours or longer, depending on the weather, but which may be expedited by the use of a forage conditioner. A short cut (about ⅜ inch) is used on the forage harvester. No additive or preservative is needed with properly wilted silage, although it may be added if desired.

Livestock producers frequently have the option of producing corn silage or grass/legume silage. Each has its advantages and disadvantages. Corn or sorghum generally produces a greater tonnage of feed per acre than does grass/legume silage. High-quality corn or sorghum silage can be made more consistently and with greater ease than can high-quality grass/legume silage. Corn or sorghum silage may be more palatable than grass/legume silage, even when the latter is carefully preserved. However, grass/legume silage is generally higher in protein and carotene but lower in total digestible nutrients and vitamin D than corn or sorghum

Figure 19.4 A row of wilted, chopped green alfalfa is collected into a wagon before it is taken to the silo. *(Courtesy of USDA-ARS)*

silage. Generally, grass/legume silage contains about 90 percent as much TDN as corn silage, but it will equal corn silage in TDN when 150 pounds of grain per ton has been added. Thus, grass/legume silage generally requires the addition of less protein supplement to the ration but more total concentrates than does corn or sorghum silage. In addition, grass/legume silage can be produced in areas where the climate is too cool and the growing season too short for corn or sorghum silage. The production of grass/legume silage results in less soil washing than the production of corn or sorghum silage on lands subject to erosion.

OTHER SILAGE CROPS

In the Northwest and North Central states, where the weather is cool and the growing season is short, sunflowers are sometimes grown for silage. Although they produce high yields and ensile well, sunflower silage is neither as palatable nor as nutritious as corn, sorghum, or grass silage. Pound for pound, sunflower silage is about 80 to 85 percent as valuable as corn silage.

Throughout the United States, many by-product feeds are ensiled, especially in the less expensive and temporary types of silos. Among such by-products are grain chaff, pea and bean vines, beet tops and pulp, sunflower hulls and chaff, potatoes, cannery refuse, cull and surplus fruits and vegetables, pulp and trimming wastes from market vegetables and fruits, wet brewers' and distillers' grains, almond hulls, and poultry litter. Sometimes even Russian thistles and other weeds are ensiled.

When potatoes, which contain about 80 percent moisture, are ensiled for cattle, it is recommended either that 20 to 25 pounds of dry hay, straw, or chaff be run through the ensilage cutter with each 100 pounds of potatoes or that 1 ton of corn or sorghum silage be chopped with each 500 pounds of potatoes. Frozen and sprouted potatoes should not be ensiled. Potato-processing wastes (cull potatoes, off-flavor french fries and chips, etc.) can be ensiled in the same manner as unprocessed potatoes. Either of the methods recommended for ensiling potatoes for cattle is equally adaptable for the preservation of other high-moisture crops, such as apples, beets, pears, tomatoes, cauliflower, broccoli, kale, and trimming wastes from market vegetables, provided the added forage is in proportion to their respective moisture contents. Cabbage, rape, and turnips should not be ensiled because they make unsatisfactory, watery, foul-smelling silage.

To lower the moisture content, alleviate the necessity of a preservative, and ensure better quality silage, forages of high-sugar content are sometimes combined with forages of low-sugar content. Thus,

excellent silage can be made by mixing 1 ton of sorghum forage with each 3 tons of grass/legume silage material or a ton of corn forage with each ton of grass/legume forage material. (Less sorghum forage is necessary than corn forage because of the higher sugar content of the former.) At times, such combination silage crops are even grown together, for example, corn and soybeans; millet or Sudan grass; and soybeans, oats, and peas. A major difficulty in combining ensiling crops is that it is almost impossible to synchronize the stage of maturity of different crops so that they reach maximum yield and nutrient level at the same time.

Corn, sorghum, sunflowers, small grains, beans, and other crops, which may or may not have been intended for silage, are sometimes frosted before they reach the silage cutting stage. Corn that has been frosted before reaching maturity is commonly known as soft corn. Such frosted crops may be salvaged as silage. They should be cut at recommended moisture contents and ensiled according to directions. If they are too dry, water should be added. Frosted crops, especially frosted sorghum, may be high in cyanide (HCN).

Corn or sorghum, or other crops, are sometimes drought stricken to the extent that little or no grain is produced. Such crops may be harvested for silage and used as an energy source for ruminants. They should be cut and ensiled like any other silage crop. If they are too dry, water should be added. Drought-stricken crop silage may be used in the same manner as any other low-energy source.

The danger of cyanide toxicity is much greater from sorghum than from corn. Drought-stricken plants can accumulate cyanogenetic glycoside that hydrolyzes to form free cyanide (HCN). The danger is increased when crops are grown on heavily nitrogen-fertilized soils or if any of the following have occurred: frosting, wilting, trampling, or hail. Any combination of these conditions can lead to a dangerous buildup or release of cyanide.

HAYLAGE (LOW-MOISTURE SILAGE)

Low-moisture grass/legume silage (or haylage), containing 40 to 60 percent moisture, is made with limited bacterial growth and fermentation (Figure 19.5). The term *oatlage* is sometimes used specifically to indicate low-moisture silage made from oats. Fermentation is of minor concern in making low-moisture silage because little acid is produced and pH is not a useful criterion of quality. The most important factor is the establishment and maintenance of air-free conditions through fine chopping, rapid filling, and the use of a good silo. Because of the difficulty of maintaining air-free conditions, stacks, bunkers, and trench silos are seldom used for storing low-moisture

Figure 19.5 Haylage is a highly palatable feed source for dairy cattle. *(Courtesy of Iowa State University)*

Figure 19.7 Long-stem hay can be ensiled as a low-moisture round bale by bagging in plastic, which provides a highly palatable and flexible feeding option. *(Courtesy of Iowa State University)*

Figure 19.6 Round bales of hay are ensiled by baling at a low-moisture content and sealing in plastic. *(Courtesy of Iowa State University)*

silage. Infiltration of air into the silage mass results in growth of yeasts and molds and an increase in temperature. Temperatures about 95°F for a few days cause certain proteins to combine chemically with carbohydrates to form a product, termed *bound protein*, that is indigestible. When more than 12 percent of the total protein is in the bound form, the silage has undergone excessive heating.

Baleage refers to round bales of hay that are rolled from the field while still at a dry matter of 40 to 50 percent (Figure 19.6). Increasing dry matter above this range increases the risk of heating and molding, while wetter bales increases the risk of spoilage and poisoning from mycotoxins. Bales should be stacked and wrapped or individually wrapped in white plastic to reflect heat and minimize levels of entrapped oxygen. Storage conditions should not be in direct sunlight. The same risks apply as with haylage.

Properly made and stored low-moisture haylage or baleage has a pleasant aroma and is a palatable, high-quality feed (Figure 19.7). Animals usually receive more dry matter and net feed value from low-moisture silage than they do from wilted silage made from the same cut. Low-moisture silage is increasing in popularity, especially as a dairy feed. It may be fed like wilted silage, with adjustment for the difference in moisture content.

HIGH-MOISTURE GRAIN

High-moisture grain refers to grain that is harvested at a moisture level of 22 to 40 percent and stored without drying. Interest in storing and feeding high-moisture grain was prompted by the shift toward more field shelling of corn, instead of picking ear corn, because much shelled corn must be dried for safe storage, whereas ear corn can be safely stored at moisture contents up to 24 percent without drying. Interest in high-moisture grain increased with the high energy cost for drying. It takes approximately 2 gallons of propane fuel or 1 kilowatt hour of electricity using conventional high temperature drying to reduce the moisture content of ten to twelve bushels of wet corn ten percentage units (i.e., to dry from 25 percent down to 15 percent moisture). High-moisture grain therefore alleviates the high energy cost for drying grain, decreases field losses at harvest time, and permits harvesting earlier at higher moisture content and usually during more desirable weather. As a result, it releases land for fall plowing in the North and for fall seeding of a second crop in the South. It makes it practical to use later-maturing, higher-yielding varieties of corn and sorghum in the northern areas, which often have early frost. Finally, it improves the feeding value of grain for dairy cattle.

Conversely, high-moisture grain limits market flexibility for the grain because it must be fed to livestock. It may result in higher storage losses than for dry grain if proper ensiling or acid treatment is not followed. Also, it may freeze in the bunk in the winter, and flies may be a problem in the summer. Considerable quantities of high-moisture ear corn, shelled corn, sorghum (milo), and small grains (wheat, barley, and oats) containing about 30 percent moisture are stored and fed to cattle. Some high-moisture grains are planned for and intended. Others are the result of crops planted late, early frost damage, or harvesting when wet.

The feeding value of high-moisture grain is equal or slightly superior to that of dry grain, with some variation according to class and productivity of livestock. The feeding value of properly ensiled or acid-treated, high-moisture corn is equal to that of dry corn for lactating dairy cows. When high-moisture shelled corn supplies more than 50 percent of the total ration dry matter, however, depressed milk-fat percentages may occur, with inadequate fiber the probable cause. Some form of processing (e.g., rolling or grinding) of high-moisture corn improves utilization. Whole kernels appearing in feces indicate incomplete digestion. Processed corn is higher in digestible, metabolizable, and net energy for dairy cows than rations containing whole shelled corn.

SUMMARY

- Ensiling is an old method of preserving feed by fermenting forage plants.
- The ensiling process refers to the changes that take place when forage or feed, with sufficient moisture to allow microbial fermentation, is stored in a silo in the absence of air.
- Most silage in the United States is made from either corn or sorghum (75 percent) and from grasses, legumes, and other feeds (25 percent).
- There are four kinds of corn silage: whole corn plant, ear corn, corn stover, and shelled-corn silage.
- Corn and sorghum residues can be used as cattle feed in three ways: grazed, harvested and fed dry, or ensiled and fed as silage.
- The higher the moisture content of grass/legume forage, the more critical the need for a low pH to obtain preservation.
- Corn or sorghum silage generally produces a greater tonnage of feed per acre, can be made more consistently and with greater ease, and may be more palatable than grass/legume silage.
- Grass/legume silage is generally higher in protein and carotene and can be produced in areas where the climate is too cool and the growing season too short for corn or sorghum silage.
- Low-moisture grass/legume silage is made with limited bacterial growth and fermentation.
- High-moisture grain refers to grain that is harvested at a moisture level of 22 to 40 percent and stored without drying.

QUESTIONS

1. How much time does the entire ensiling process require?
2. What are two ways of increasing the quality of silage?
3. What is the biggest deterrent to harvesting stalklage?
4. Why should cabbage, rape, and turnips not be ensiled?
5. What is the most important factor in producing haylage?
6. What are some advantages to high-moisture grain?

ADDITIONAL RESOURCES

Articles

Broderick, G. A., R. G. Koegel, R. P. Walgenbach, and T. J. Kraus. "Ryegrass or Alfalfa Silage as the Dietary Forage for Lactating Dairy Cows." *Journal of Dairy Science* 85 (July 2002): 1894–1901.

Chandler, P. T., C. N. Miller, and E. Jahn. "Feeding Value and Nutrient Preservation of High Moisture Corn Ensiled in Conventional Silos for Lactating Dairy Cows." *Journal of Dairy Science* 58 (May 1975): 682–688.

Einarson, M. S., J. C. Plaizier, and K. M. Wittenberg. "Effects of Barley Silage Chop Length on Productivity and Rumen Conditions of Lactating Dairy Cows Fed a Total Mixed Ration." *Journal of Dairy Science* 87 (September 2004): 2987–2996.

Knowlton, K. F., B. P. Glenn, and R. A. Erdman. "Performance, Ruminal Fermentation, and Site of Starch Digestion in Early Lactation Cows Fed Corn Grain Harvested and Processed Differently." *Journal of Dairy Science* 81 (July 1998): 1972–1984.

Internet

Haylage: http://www.ces.ncsu.edu/onslow/AG/hay/haylage.html.

Phases of the Ensiling Process: http://cvm.msu.edu/courses/Lcs643/forage1/tsld034.htm.

Silage Basics: http://www.oznet.ksu.edu/pr_silage/basic_principles.htm.

Silage Fermentation and Preservation: http://www.ext.nodak.edu/extpubs/ansci/range/as1254.pdf.

20

Silage Quality

OBJECTIVES

- To understand the factors affecting silage quality.
- To describe the silage gases, molds, and storage losses.
- To describe how to put up silage.

Harvesting silage crops at the proper stage of maturity ensures the maximum yield and nutrient content. Corn should be harvested after the early dent stage and prior to the black layer forming. Corn harvested for silage in early dent stage is highly digestible, but total yield per acre and nutrient content of the silage will be lower than if the corn is harvested after this stage. On the other hand, when the grain reaches full physiological maturity, several layers of cells near the tip of the kernel turn black, forming the black layer (Figure 20.1). This layer can be detected by removing several kernels from the middle of the ear and splitting them lengthwise or cutting off the tip. If the black layer is present, usually the grains are dented and glazed, the lower four to six leaves of the corn plant are brown, and the plant contains 60 to 67 percent moisture. If the corn is harvested for silage at this stage, yield per acre is highest, but silage digestibility is lower than if the corn is harvested prior to this stage. Optimal harvest stage occurs at somewhere around one-half milkline stage, and cow productivity is highest when they are fed silage harvested at this stage.

Sorghum should be cut for silage when the seeds are hard. Grass silage forages (grasses, legumes, and cereal crops) should be cut at the same stage at which they would make the best hay. The length of the cut sections affects the packing and hence the quality of the silage (Table 20.1). Also, the proper length of cut varies with the crop and the moisture content. Thus, for corn and sorghum crops, forage harvesters should be set to make a theoretical cut of ⅜ inch. If the knives are sharp and set up to the cutter bar, the result is about 15 percent of the particles being 1.5 inches and over, 25 percent of the particles being ¾ to 1½ inch, and 60 percent being ⅛ to ¾ inch in length. Grass silages should be more finely chopped than corn or sorghum silage to improve packing. Also, wilted and dry forage and forage with hollow stems should be chopped more finely than forage of high-moisture content, thus permitting more thorough packing and eliminating most air pockets.

Moisture content is one of the most important factors in determining the quality of silage. Moisture content between 60 and 67 percent is best

TABLE 20.1 Effect of Chop Length on Whole Kernels in Silage and Passage into Manure

Chop (Inches)	Silage Dry Matter (%)	Whole Kernels in Silage (% of Dry Matter)	Whole Kernels in Manure (% of Whole Kernels Fed)
⅛	28	0.02	0
¼	26	0.1	93
⅛	34	0.31	68
¼	32	0.53	66
⅛	48	1.4	63
¼	50	2.1	66

Source: Modified from J. G. Linn, D. E. Otterby, and N. P. Martin, Feeding Corn Silage to Dairy Cattle and Agricultural Extension Service, University of Minnesota.

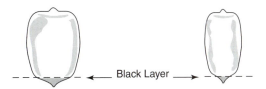

Figure 20.1 The black layer test is useful for determining readiness for harvest of corn silage. *(Courtesy of M. E. Ensminger)*

(Figure 20.2). However, low-moisture silage of 40 to 60 percent moisture can be preserved successfully in oxygen-limiting silos or tall, conventional silos that are properly topped off with heavy, wet forage or sealed with a plastic cover. Forage containing more than 60 to 67 percent moisture is heavier and more costly to handle; is apt to produce slimy, putrid silage due to the presence of butyric and other undesirable acids; produces excessive seepage (Figure 20.3) and loss of nutrients; results in excessive deterioration in the silo walls due to the high acidity; and exerts greater pressure on the silo walls (the greater the moisture content, the greater the pressure.) Cows also have reduced intakes when consuming these excessively high-moisture silages in part because of the different end products of fermentation produced during the ensiling process. If corn and sorghum are harvested at the stage recommended, their moisture content is right. However, freshly cut grass and/or legume forage contains 75 to 80 percent moisture, which means that for proper ensiling its moisture content must be lowered by 10 to 15 percent. The moisture content of silage material may be lowered by conditioning and/or wilting; adding dry hay or straw; combining with corn or sorghum silage; or adding a dry preservative of grain, dried molasses, or dried by-products of citrus or beets.

Conditioning and/or wilting of grass silage increases the percentage of sugar in the forage, decreases seepage losses from the silo, lessens the pressure on the silo walls, and decreases the destructive action of the acids on the silo walls. The needed 10 to 15 percent reduction in the moisture content of grass silage material can be accomplished by wilting for about 2 hours on a good drying day and up to 1 day or longer in slow drying weather. The combination of conditioning and wilting is the method most commonly followed. Excellent equipment is available for conditioning. Excess drying should be avoided because it results in the forage becoming too dry for proper ensiling.

The moisture content of any wet silage material can be lowered effectively by mixing dry hay or straw with it at the time of filling. Thus, during poor wilting weather, the moisture content of grass forage can be brought within the desired range by adding 5 to 20 percent hay or straw. This is the standard method of lowering the moisture content of high-moisture products such as potatoes. Conditioning and wilting is the preferred method of lowering the moisture content of grass silage, rather than adding dry hay or straw which reduces digestibility and energy content. Dry preservatives such as ground grain; corn-and-cob meal; dried molasses; and dried citrus meal, citrus pulp, or beet pulp reduce the moisture content of freshly cut and unwilted forage and, in turn, decrease seepage from the silo.

If the crop is overripe and too dry when cut or if it becomes overwilted, water should be added to the silo. A preferred alternative is to harvest the crop as hay rather than ensiling it. Drier material may be used for silage by cutting shorter and packing more thoroughly. If necessary, water should be added or the dry material should be mixed with very green, freshly cut material by alternating loads.

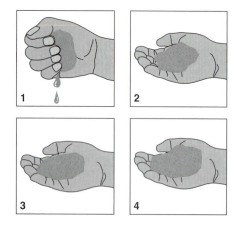

Figure 20.2   Quick determination of moisture content of silage: 1, 75 to 85 percent; 2, 70 to 75 percent; 3, 60 to 65 percent; 4, less than 60 percent. *(Courtesy of M. E. Ensminger)*

Figure 20.3   Seepage stains are apparent on the side of this upright silo. *(Courtesy of Mark Kirkpatrick)*

SILAGE ADDITIVES AND PRESERVATIVES

Silage additives provide supplemental nutrients that enhance the feeding value of silage. Silage preservatives enhance the keeping qualities of silage. High-quality silage can be made without the use of additives or preservatives if you start with good material and all proven good practices are followed. But there are times when the ensiled material is either too wet or too dry, does not contain sufficient fermentable carbohydrates, is deficient in certain nutrients, and lacks palatability, and/or the proven good practices cannot be followed. Under such circumstances, silage additives or preservatives may reduce silage losses and/or improve the feeding value of the silage. Normally, additives or preservatives are not needed for corn or sorghum silages. But additives or preservatives may be very helpful if ensiling a grass/legume forage with over 70 percent moisture. The decision about using any silage additive or preservative is based on how much it improves animal performance and net profit, not whether it merely makes silage look and smell better.

Silage made from legumes or grasses may be improved under certain conditions by the addition of ground grain (corn, wheat, or barley), ground ear corn, beet pulp, citrus meal, citrus pulp, or other appropriate feed ingredient. The ground material provides a readily available source of carbohydrates (sugar and starch) for bacterial fermentation and the production of acids, increases the feeding value of the silage because 75 to 85 percent of the feeding value of the grain is retained if the silo excludes air properly, and likely improves palatability. If the primary concern is to reduce the moisture content of the silage, cheaper materials, such as ground corn cobs, cottonseed hulls, oat hulls, or chopped straw or hay, may be more appropriate than ground grain. Dry, finely ground corn cobs absorb nearly 200 pounds of water per 100 pounds of the cob material; dried beet pulp absorbs even more. These materials may be added by feeding them into the blower from a properly adjusted hopper attachment. When green forage is ensiled at a proper moisture content, there is usually no advantage to adding grain for the purpose of preservation or palatability.

Some green forages, such as legumes and certain grasses, are rather low in sugar content. Hence, adding molasses, which is high in sugar, may increase lactic and acetic acid production and improve silage quality and preservation. Also, molasses improves the palatability of silage and increases its nutritive value. For legumes, about 80 pounds of molasses is added per ton of silage; for grasses, about 40 pounds is added per ton (molasses weighs 12 pounds/gallon). Addition of much less than these amounts, as an ingredient in mixed preservatives, is of little value. Much of the feeding value of the molasses is retained in the silage under good storage conditions and when there is no seepage loss. Molasses may be added in either liquid or dehydrated form as the forage enters the blower. When a grass and/or legume is wilted to 50 to 60 percent moisture content and adequately

protected from air, an excellent feed with a good aroma and keeping quality can be obtained without the addition of molasses.

Adding dried whey to alfalfa silage or haylage slightly improves the quality and digestibility. There is also indication that adding dried whey to urea-treated corn silage may help reduce nitrogen losses and improve the feeding value of the silage.

Urea, ammonia, and other nonprotein nitrogen products can be added to corn or sorghum silage at the time of ensiling as a source of nonprotein nitrogen. Urea increases the crude protein content of the silage and the amount of lactic and acetic acids produced. The addition of 10 pounds of urea per ton of ensiled corn material makes the following approximate increases on a dry-matter basis: crude protein, from 8.3 to 12.3 percent; lactic acid, from 4.2 to 5.4 percent; and acetic acid, from 0.9 to 1.2 percent. Since the amount of nonprotein nitrogen that can be converted to microbial protein by the organisms in the rumen is limited, no more than 10 pounds of urea should be added to a ton of ensiled corn material. The urea can be added by spreading it over the top of each load of chopped corn, or it can be added to the chopped corn through the blower by commercially manufactured metering equipment. Ammonia-containing materials, including ammonia-water solutions, ammonia-mineral solutions, ammonia-mineral-molasses solutions, anhydrous ammonia gas, and cold-flow ammonia, can also be added to corn silage as a source of nonprotein nitrogen. For dairy cattle, 5 pounds of actual nitrogen (about 6 pounds of ammonia) may be added per ton of wet silage. Ammonia-treated corn silage has been found to contain increased concentrations of true protein, lactic acid, and acetic acid. Also, it may have a higher pH and be more stable than untreated silage when exposed to air. Special equipment is required to add ammonia or ammonia-containing materials.

When corn silage spiked with urea or ammonia is the major source of protein fed to ruminants, there may be inadequate sulfur for the rumen organisms to manufacture their own protein. In such cases, the addition of sulfur to achieve a nitrogen-sulfur ratio of less than 15:1 improves both growth and the milk production of cattle. The most practical way to provide additional sulfur is to add gypsum (calcium sulfate) at the rate of 1.8 pounds per ton of silage. Even under ideal conditions, about 10 percent of the urea may be lost; under average conditions, up to 30 percent of the urea may seep away or be lost as ammonia gas. Ammonia gas is dangerous, causes irritation and can accumulate in the silo.

Limestone (calcium carbonate) may be added at a level of 0.5 to 1 percent to corn silage to increase acid production. It neutralizes some of the initial acids as they are formed, allowing the lactic acid bacteria to perform longer and to produce more desirable acids. The addition of limestone at ensiling time raises the naturally low calcium content of corn silage, a fact that should be considered when balancing rations.

Both inorganic (mineral) and organic acids may be used as additives. Mineral acids lower the pH immediately, while organic acids have a limited effect on lowering pH. Both mineral and organic acids limit microbial growth and help to stabilize silage. Inorganic acids (hydrochloric acid, sulfuric acid, phosphoric acid) have been used as silage preservatives, almost entirely in Europe, in connection with the ensiling of high-moisture material. These acids substitute for the acids produced by bacterial action; however, they are very corrosive, creating application difficulties including corrosion of silo walls and silage-handling equipment. Of the three acids, phosphoric acid is preferred because it is less corrosive than sulfuric acid or hydrochloric acid, it may enhance the phosphorus content of the silage, and it increases the residual manure value from the silage. But phosphoric acid may introduce a problem of proper calcium-phosphorus ratio. When mineral acid preservatives are used, it is recommended that ground limestone or some other form of calcium or sodium carbonate be fed to animals at the rate of approximately 1 ounce for each 10 pounds of silage to neutralize the acid.

In general, the use of mineral acid preservatives is not considered as desirable as the use of molasses or grain because they produce a more sour and less palatable silage; they may damage clothing, machinery, and/or masonry silo walls due to their corrosiveness; and they do not add to the nutrient value of the silage except by enhancing the preservation of carotene. In general, the use of mineral acids has more disadvantages than advantages.

Propionic, acetic, lactic, citric, and formic acids are used in a manner similar to inorganic acids, but they are much less corrosive and not so difficult to handle, although precautions must be taken. Organic acids enhance the preservation of forage without the loss of palatability. Also, they serve as mold inhibitors. Even so, like all additives, they cost money; hence, the economics of using them when making silage must be considered. Organic acids are of the greatest benefit in the preservation of high-moisture grain.

Fermentation aids include bacterial cultures, yeast cultures, and enzyme supplements. Controlled experiments support the claims made for some of these products, but not all of them; they should be purchased only from reputable sources that have valid research data to support the claims made for them. Silage additives containing cultures of acid-forming bacteria (Lactobacillus) provide an inoculum to increase the numbers of these bacteria and ensure rapid fermentation. Advocates of these products claim that they increase the dry matter, energy, and protein of the silage. The number of bacteria provided through such additives is insignificant compared to the numbers already present on the ensiled material, and inclusion of these products must be justified on some other basis.

Yeast cultures have also been included in certain silage additives. However, yeasts sometimes grow in silage without an inoculum being added. When this happens, the silage has a yeasty odor and taste, which is considered undesirable. Yeast does have nutritional value, but because of the small quantity involved in additives, the contribution is minimal.

Cultures of molds or of molds with other microorganisms are sometimes added to silage to provide a source of enzymes. Some claim that these enzymes improve the nutritive value of the silage by increasing its digestibility or digestible nutrient content. Although the enzyme activity of these preparations has not been measured experimentally, the quantity of enzymes added is insignificant compared to those already present in the silage.

Preservatives include antibiotics, salt, and sterilants. These products preserve silage by inhibiting microbial action or undesirable fermentations. All of them are of questionable value if air is properly excluded from the silage. If air is not excluded, they must be added at very high levels to be effective. Theoretically, antibiotics can preserve silage by selective action: by inhibiting undesirable microbial activity while allowing the desirable organisms to develop. So far, the results have been inconsistent. Also, at an appropriate level, salt inhibits certain microorganisms without preventing the action of bacteria that produce the desirable acids.

Sterilants include sulfur dioxide, sodium diacetate, sodium metabisulfite (sodium sulfite), sodium benzoate, and sodium nitrate. Sodium propionate and other organic acids have also been used as preservatives because of their mold-inhibiting properties. Each of these products appears to reduce carotene losses, improve the odor of silage, and/or lessen the production of toxic gases. But their effect on palatability is variable. The cost and inconvenience of application of these products may not justify their advantages.

Additives or preservatives are not essential to good silage formation when conditions of moisture and storage are right. Under special circumstances, however, they can be recommended for use. For example, molasses, grain, or grain by-products might be a wise addition to silage when conditions do not allow for proper wilting prior to ensiling or when an all-in-one silage is being made. Urea may be an appropriate addition when increasing the protein content of the silage simplifies its feeding. It is doubtful that there is any justification for adding limestone unless this is a convenient method of calcium supplementation. The economy of most nutritive additives of this type depends largely on how well their nutrients are retained in the silage and the use made of them in balancing the rations.

When forages are stored at the proper moisture content and when air is properly excluded, nutrient losses are low, and a good-quality silage forms. Additives such as lactic acid bacteria, mold inhibitors, antibiotics, salt, enzymes, yeast cultures, and mineral acids can therefore do little, if anything, to improve the preservation of the silage or its feeding value. When high-moisture material is ensiled, grain is superior to any of these additives. When air is not properly excluded, none of these additives correct the large fermentation and spoilage losses. There is no substitute for good management of forage crops for silage, with proper control of factors such as stage of maturity at harvest, harvesting methods, moisture content, fineness of chopping, distribution and packing, and exclusion of air.

PUTTING UP SILAGE

Once silo filling is started, it should be rapid to avoid spoilage before the silo is filled and sealed (Figure 20.4). Typically, a silo should be filled in two days or less. To avoid the presence of air pockets and spoilage, it is essential that any kind of chopped forage be distributed uniformly in the silo and that it be packed well. Proper silage distribution is obtained by keeping the material nearly level or slightly higher at the center. Silage-distributing equipment is available for keeping the material in an upright silo level. These devices are very helpful, especially in silos of 14 feet or larger in diameter.

When corn, sorghum, and sunflower forage is harvested at a green, immature stage and cut into short lengths, tramping in an upright silo is not necessary, but uniform distribution is very important. The only filling precaution under these conditions is to ensure that the top is carefully leveled, well-packed, and covered whenever filling is completed. Grass silage (especially when wilted), hollow-stemmed forages, and forages that have matured or dried beyond the best silage stage should always be tramped well, especially near the wall. Packing in a trench silo should be ensured by use of a large tractor.

Figure 20.4 Molds form in areas where silage is exposed to air. *(Courtesy of Mark Kirkpatrick)*

Sealing or topping off is necessary to avoid excess spoilage, especially with grass silage, which tends to dry out on the surface and to shrink away from the silo walls. This may be accomplished by leveling off the top and thoroughly tramping the last few feet, especially near the walls; topping off the silo with two to three loads of wetter material; or covering the top with plastic cut to fit the silo diameter and turned up against the silo wall a distance of 5 to 8 inches.

Seepage losses vary with the moisture content, depth of silage, distribution of the silage, and the amount of nutrients in the seepage. Seepage losses may be as high as 14 percent of the dry matter stored. The nutrient losses vary, but generally they are in proportion to the runoff. The nutrients lost in seepage from a 100-ton silo may equal the nutrients in ¾ ton or more of hay.

SILAGE GASES AND MOLDS

Two types of toxic gases may be formed when making silage: carbon dioxide (CO_2) and/or nitrogen dioxide (NO_2). Carbon dioxide forms soon after filling begins and continues until fermentation stops. It is a colorless, suffocating gas that is heavier than air and tends to collect in low places. Under drought conditions, corn, sorghum, and other grass species may accumulate higher than normal levels of nitrates. When ensiled, nitrates are converted to nitrites, and then nitrites are converted to nitrogen oxide by bacteria and plant cells. As the nitrogen oxide comes in contact with air, it is oxidized to form nitrogen dioxide, a reddish brown gas that is heavier than air. This gas is highly toxic to both humans and farm animals. Precautions against hazards caused by silage gases include operating the blower for a 15-minute period if it is still connected or using proper life-support equipment when entering an oxygen-limiting or sealed silo. Also, adequate provision for ventilation of the silo through the roof is essential. A victim of silo gas should be moved into fresh air immediately, and artificial respiration should be applied. A physician should be called immediately.

Mycotoxicosis (silo filler's disease or farmer's lung disease) is a reaction to molds on silage. It occurs most often when opening a silo that has been previously sealed. It is a flu-like illness; the best cure is to avoid the problem. Wearing a protective mask when opening a silo is a useful precaution.

SILAGE STORAGE LOSSES

Tight structures, good distribution and packing, and the proper use of plastic covers minimize silage storage losses (Figure 20.5). Losses from different types of silo storage vary widely based primarily on length of time and season of feedout; this is especially critical for silages continuously exposed to air. Losses in trench and open stack silos are also influenced by depth; less surface is exposed in deeper silos.

Losses in the silo include surface or top spoilage, seepage, gas production, and heating losses (browning reaction and spontaneous combustion). Surface or top spoilage losses of 20 percent or more may occur in stack silos and in any uncovered bunk, trench, or pit silo (Figure 20.6). These losses can be reduced with the use of suitable protection, such as a plastic cover.

Seepage losses can be high in high-moisture silage stored in upright silos. The higher the silo, the greater the pressure and the higher the losses through seepage. The seepage carries soluble feed nutrients with it. Horizontal silos have less seepage loss than do upright (tower) silos because of lower vertical pressure. Seepage losses are reduced by wilting forages to less than 65 percent moisture before ensiling.

Gas production is unavoidable as long as the plant material respires and there is subsequent fermentation. These losses can be minimized, however, by keeping air out of the silo, having the pH decline rapidly, and encouraging favorable fermentations. Lowering the moisture without excluding the air may lead to heat damage, known as the browning reaction or Maillard reaction. Spontaneous ignitions sometimes occur in low-moisture silage (haylage). For such losses to occur, there must be a buildup of temperature to the combustion point in the silo mass, combined with a low transfer of heat. These fires are difficult, often impossible, to extinguish. The addition of water may build up pressure and lead to an explosion. Most silo fires should be allowed to burn.

Figure 20.5   This extended pile of silage is covered with plastic that is held in place with half-tires until it is ready to be fed to the cattle. *(Courtesy of Howard Tyler)*

Figure 20.6   Under adverse conditions, plastic coverings protecting silage can be damaged or lost, which can result in spoiled or moldy silage. *(Courtesy of Howard Tyler)*

SUMMARY

- Harvesting feed at the proper stage of maturity ensures the maximum yield and nutrient content.
- The length of the cut sections affects the packing and quality of silage.
- Moisture content is one of the most important factors in determining the quality of silage.
- Silage additives provide supplemental nutrients while enhancing the feeding value of silage.
- Urea, ammonia, and other NPN products can be added to corn or sorghum silage at the time of ensiling as a source of nonprotein nitrogen.
- Inorganic acids have been used as silage preservatives.
- The higher the silo, the greater the pressure and the higher the losses through seepage.
- Two types of toxic gases may form when making silage: carbon dioxide and/or nitrogen dioxide.
- Tight structures, good distribution and packing, and the proper use of plastic covers all help minimize silage storage losses.

QUESTIONS

1. What test can be applied quickly and easily to determine when to harvest corn for maximum yield and nutrient quality?
2. What are some disadvantages to silage with a moisture content greater than 60 to 67 percent?
3. What is mycotoxicosis (silo filler's disease or farmer's lung disease)?
4. What are the advantages of conditioning and/or wilting grass silage?
5. Why would a dairy farmer add limestone to the corn silage?
6. What are the three inorganic acids used as silage preservatives?
7. What is the greatest benefit in the preservation of high-moisture grain?
8. What is the typical time to fill a silo?

ADDITIONAL RESOURCES

Book

McDonald, P. *The Biochemistry of Silage*. Chichester, UK: Wiley, 1981.

Articles

Dewar, W. A., P. McDonald, and R. Whittenbury. "The Hydrolysis of Grass Hemicelluloses During Ensilage." *Journal of the Science of Food Agriculture* 14 (1963): 411–417.

Muck, R. E. (Trans.). "Effect of Inoculation Level on Alfalfa Silage Quality." Transactions of the American Society for Agricultural Engineers 32 (April 1989): 1153–1158.

Nagel, S. A., and G. A. Broderick. "Effect of Formic Acid or Formaldehyde Treatment of Alfalfa Silage on Nutrient Utilization by Dairy Cows." *Journal of Dairy Science* 75 (January 1992): 140–154.

Pitt, R. E., and R. E. Muck. (Trans.). "Enumeration of Lactic Acid Bacteria on Harvested Alfalfa at Long and Short Wilting Times." Transactions of the American Society for Agricultural Engineers ASAE 38 (June 1995): 1633–1639.

Weinberg, Z. G., A. Ashbell, and A. Azrieli. "The Effect of Applying Lactic Acid Bacteria at Ensilage on the Chemical and Microbiological Composition of Vetch, Wheat and Alfalfa Silages." *Journal of Applied Bacteriology* 64 (January 1988): 1–7.

Internet

Bacterial Inoculants and Enzyme Additives: http://edis.ifas.ufl.edu/DS161.

Evaluating Silage Quality: http://www1.agric.gov.ab.ca/$department/deptdocs .nsf/all/for4909.

Harvesting Silage Safely: http://edis.ifas.ufl.edu/DS078.

Harvesting, Storing, and Feeding Silage to Dairy Cattle: http://edis.ifas.ufl.edu/ DS166.

Inoculation of Silage and Its Effects on Silage Quality: http://www.dfrc.wisc.edu/ Research_Summaries/ind_meet/dfrc7.pdf.

Silage Preservation: http://cecommerce.uwex.edu/pdfs/a3544.pdf.

Silo Gases—The Hidden Danger: http://server.age.psu.edu/../extension/factsheets/ e/E16%20.pdf.

21

Silage Storage Systems

OBJECTIVES

- To describe silage storage systems.
- To list the classifications and sizes of silos.
- To provide estimates of silage inventory.

Silage may be stored in almost any kind of container. The proper size for a silo is determined by the number and kind of animals to be fed daily, length of the feeding period, and amount of forage available for ensiling. It must exclude air from the stored material, including entrance of air around the doors of tower silos. The sidewalls must be straight and smooth to prevent the formation of air pockets and to allow for unimpeded packing.

Silos must also be properly reinforced. This point is especially important when direct cut grass silage is made because it exerts up to 2.5 times as much pressure on the walls as does corn silage. Thus, tower silos originally built for corn or sorghum silage are not usually suitable for wet grass silage or high-moisture corn. They must be either reinforced with extra bands placed around the lower part to strengthen the walls, if an inspection reveals that the existing strength is not adequate, or filled to no more than half capacity. Adequate provision should be made for the escape of seepage by either a drain or a gravel bottom, along with containment to avoid surface pollution.

Silos may be classified according to the five basic methods used for processing forages. Each method is associated with the shape and material of the structure, which also influences the efficiency of preserving the silage. The different shapes of the structures are also adapted to different methods of filling and unloading. Within each classification are many variations of each type, depending on the manufacturer.

The kind of silo to use and the choice of construction material should be determined primarily by the cost and by the suitability to the particular needs of the farm. Silos may be classified as follows (Figure 21.1):

1. Conventional upright (tower) silos
   a. Concrete stave
   b. Galvanized steel
   c. Wood stave
   d. Monolithic concrete (poured in place)
   e. Tile block
   f. Brick
2. Gas-tight (oxygen-limiting) silos
   a. Glass-lined structures
   b. Concrete stave
   c. Galvanized steel
   d. Monolithic concrete
3. Pit silos
4. Horizontal silos
   a. Trench silos (below ground level)
   b. Bunker (above ground level)
5. Temporary silos
   a. Enclosed stacks
   b. Open stacks
   c. Modified trench-stack silos
   d. Plastic or polyethylene bag silos
   e. Round bale bagged or wrapped silage

CONVENTIONAL UPRIGHT (TOWER) SILOS

The upright or tower silo is a cylinder built above ground (Figure 21.2). The round shape withstands pressure well and is adapted to good packing. The tower silo is a permanent farm structure and, as such, should be constructed to withstand long usage. Although tower silos are convenient, they are, in their initial cost, generally the most expensive of all types. However, they provide durability, minimum top and side spoilage, and convenience for feeding during periods of inclement weather, and

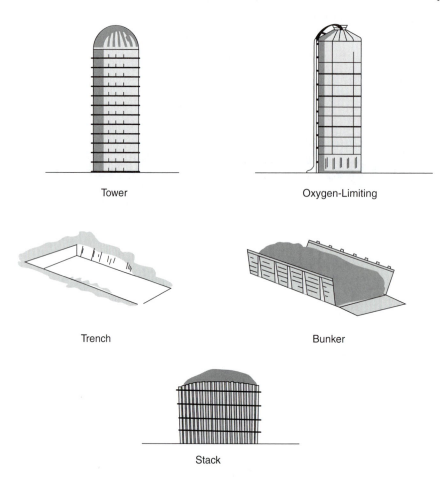

Figure 21.1 Types of silos. *(Courtesy of M. E. Ensminger)*

Figure 21.2 Upright tower silos made of concrete stave. *(Courtesy of Iowa State University)*

they are well adapted to automation (loading and unloading machinery). Unloading mechanisms in tower silos include bottom unloaders, with elimination of doors; center-core unloaders, with elimination of most of the doors; and top-unloaders. These features are also available in gas-tight (oxygen-limiting) silos.

GAS-TIGHT (OXYGEN-LIMITING) SILOS

Gas-tight silos resemble conventional tower silos, but they are more expensive because of their construction (Figure 21.3). Sealed silos are designed for storage of wilted or even overwilted forage with as little as 40 to 55 percent moisture content or for the storage of high-moisture grain containing 22 to 30 percent moisture. These structures may be partly filled on widely separated dates provided they are sealed between fillings. Packing and tramping of forage is not necessary, although distribution is desirable.

Practically all outside air is kept out of the oxygen-limiting silo. Carbon dioxide formed during fermentation is kept in a plastic breather bag, and a pressure-relief valve located in the top of some of these structures compensates for differ-

Figure 21.3 Upright oxygen-limiting silos. *(Courtesy of USDA)*

Figure 21.4 Trench silos are simple to construct and can withstand tremendous pressures with minimal investment in concrete. *(Courtesy of Iowa State University)*

ences in inside and outside pressures without allowing outside air to contact the forage. Before each filling, the plastic breather bag should be checked for holes and flaws, pressure-relief valves should be inspected, structure should be checked for leaks, door seals should be inspected, and unloaders should be checked for wear to prevent malfunctioning during unloading.

There is never sufficient oxygen inside a gastight (oxygen-limiting) silo to sustain life. Producers should never enter a filled sealed silo without proper life-support equipment.

PIT SILOS

The pit silo is shaped like the tower silo, but it is inverted into the ground. It resembles a well or cistern. The walls of a pit silo may or may not be lined. If the water table is low enough that the silo will not fill with water, such as in semiarid areas, the pit silo is very satisfactory.

In comparison with tower silos, pit silos are not damaged by storms or by fire, and they require less reinforcing, minimize silage losses because they don't have doors, and avoid frozen silage. But they are dangerous due to the frequent presence of suffocating carbon dioxide gas. Before entering a pit silo, it is recommended that a lighted candle or lantern be lowered into the silo. If the flame goes out, assume that the pit is dangerous to enter and replenish it with fresh air before entering. Considerable work is involved in removing the silage from a pit silo, despite the development of several hoist devices.

HORIZONTAL SILOS

Trench Silos

The trench silo is a horizontal, trenchlike structure that can be built quickly and at low cost (Figure 21.4). It is most popular in areas where the weather is not too severe and where there is good drainage. The walls of a trench silo may or may not be lined, but for

making good silage, they should always be smooth. There may or may not be a floor. A trench silo should be wider at the top than at the bottom, and the bottom should slope away from one end so that excess seepage drains off.

Trench silos have the following advantages: low initial cost, inexpensive filling machinery (a blower is not necessary), relative freedom from freezing, and ease of construction. The chief disadvantages of trench silos in comparison with tower silos are the larger area to seal, higher spoilage losses, and inconvenience in feeding during inclement weather. Because of shallowness, the forage should be packed very thoroughly in a trench silo by driving a tractor back and forth over it. When filling is complete, the top should be carefully sealed by 3 to 6 inches of limestone or by polyethylene, plastic, aluminum, or other materials.

Bunker Silos

Aboveground horizontal silos are usually constructed with concrete floors and sidewalls of wood, concrete, or other materials (Figure 21.5). Bunker silos were originally intended for animals to self-feed directly, but skid loaders or front-end loaders on tractors are now the common methods of unloading. In warm periods, the exposed face of the silo may deteriorate rapidly due to the growth of yeasts and molds. Silo size and dimensions should be matched to the animal feed needs to obtain an unloading rate of at

Figure 21.5 A skid loader is used to move silage from this bunker silo. *(Courtesy of Howard Tyler)*

Figure 21.6 Long piles of silage covered by plastic in central Kansas. *(Courtesy of Howard Tyler)*

least 6 to 8 inches per day in warm weather. Packing is critical for success; silage density must be at least 15 pounds of silage per cubic foot, but well-packed silage generally exceeds 18 pounds per cubic foot. To accomplish this density, a large tractor must be used. An easy method to determine tractor weight is to multiply the tons of wet feed harvested per hour by 80. For example, if you bring 100 tons of wet feed to the bunker per hour, 80,000 pounds of tractor weight is required for adequate packing.

TEMPORARY SILOS

Several kinds of aboveground temporary silos are used. Generally, this kind of storage is used to meet emergencies, supplement permanent silos, or ensile by-product feeds such as cannery refuse, pea vines, and beet tops or pulp. Aboveground temporary silos are low in cost, can be erected on short notice, require no special foundation, and can be set up on almost any level site convenient for filling and feeding.

The amount of spoilage in aboveground temporary silos can be kept to a minimum with straight sides; considerable height; proper packing; and protection with fiber-reinforced paper, plastic, or other suitable material. Also, the use of propionic acid, formic acid, or other effective organic acids

applied to harvested materials at the time of chopping is very effective in reducing spoilage in temporary silos. It is important that stacked material treated with organic acid be covered with plastic to prevent dilution by rain and snow.

The spoilage on the sides of temporary silos varies from 4 to 20 inches, with greater spoilage in grass silage than in easier-packing corn and sorghum silage. Most aboveground temporary silos can be classified as one of the following four kinds:

1. **Enclosed stacks.** These stacks are built entirely aboveground, without trenches or holes. They are upright; generally circular; and enclosed by snow or picket fences, poles, wooden staves, heavy woven wire, or other materials. Most of them are lined with tar paper, plastic, or tough fiber-reinforced paper made especially for the purpose. Because of the relatively weak walls of such silos, their height should not be greater than twice their diameter unless poles are set at four to six points around their circumference and tied together at the top.

2. **Open stacks or piles.** These are similar to enclosed stack silos except that no supports or walls are used (Figure 21.6). As expected,

greater spoilage is encountered in the open stack than in the enclosed stack because of the greater evaporation and spoilage that accompanies the exposed sides. Less spoilage, as a percentage, occurs in stacks of considerable size (those that contain 500 to 1,000 tons or more of silage.) The key to success, as in most ensiling systems, is proper packing. Stacks should be round rather than oblong because the shoulders of oblong piles become difficult to pack and have high spoilage rates.

3. **Modified trench-stack silo.** This silo, which is a cross between a trench and a stack silo, is adapted to areas where the groundwater level is high. It is constructed by excavating a shallow trench 12 to 18 inches deep and piling the excavated earth on either side of the trench to support the silage and to keep out surface water. The silage is packed thoroughly in and over the trench to a height of 10 to 15 feet and is covered with any one of the materials recommended for covering the trench silo. The modified trench-stack silo is designed to give greater protection and less spoilage than can be accomplished by open or closed stacks. Also, it is easier to feed from this type of silo than from a trench silo.

4. **Plastic silos.** Plastic (polyethylene) is available for use as temporary silos; for use as covers for trench, bunker, and tower silos; and as silo liners. If not punctured, it is nearly airtight. Plastic thicknesses range from 4 to 9 millimeters. The thicker grades have better resistance to tears and punctures and low permeability by both air and moisture; however, they cost more and are difficult to tie tightly. Thinner-grade plastics are less costly, more pliable, and easier to seal. The two common types of plastic silos are enclosed plastic bag or tube silos (Figure 21.7) and round bale, plastic-covered silage. Enclosed plastic bag or tube silos are made of heavy plastic in the form of a tube. Forage is forced into the tube by a special machine (much like stuffing sausage). The machine needed to pack the tube is generally rented or owned cooperatively. The filled structure is 8 feet in diameter and up to 100 feet long. Preservation of silage is excellent provided the ends are kept sealed and the plastic is not torn or damaged by rodents or other animals. To remove or self-feed the silage, the plastic is cut and folded back at one end to expose as much silage as needed each day. The plastic cannot be reused.

Figure 21.7 Silage bags are an inexpensive and flexible method of storing ensiled feeds. *(Courtesy of Mark Kirkpatrick)*

The most common methods for using plastic material to produce round bale silage are:

1. **Individual bags.** Bags come in various lengths, diameters, and thicknesses. A tractor-mounted spear device is needed to lift the bale while wrapping the bag. Then the bale is placed in storage position before it is tied off. If possible, the bales should be stacked in cordwood fashion to reduce exposed surface area. A plastic cover over the entire stack may reduce storage damage.

2. **Plastic tubes.** Several round bales are stuffed by a machine into a long plastic tube that is then sealed at both ends. The filled plastic tube consists of a row of round bales covered with plastic. Plastic tubes can be effective and time-saving, but storing multiple bales in one package tends increase the loss if the bag is torn, punctured, or opened for feeding. However, the tube can be easily tied off into one-bale (or more) segments for feeding.

Several round bales can also be stacked under two sheets of plastic with the plastic ends on the ground covered with soil, sand, or other effective sealing procedure. The hazard with this type of storage is that air leaks are more likely to develop, which may result in a large number of bales being spoiled.

Round bale silage may serve as a supplement to, rather than a replacement for, other stored forages on most livestock farms. When silo capacity is lacking because of a surplus of forage, round bale silage can offer an effective method of storing excess forages. It saves about one-third of the harvesting energy plus the fuel required for chopping silage, and it can be self-fed if properly presented, thereby saving both labor and fuel by not requiring daily silage feeding.

But round bale silage has disadvantages. Conditions associated with round bale silage are not optimum for fermentation; thus, extreme care must be taken to eliminate air leaks. In addition, machines for lifting and moving heavy, high-moisture bales must be available. The plastic is easily damaged, which can result in forage losses greater than in conventional silos.

For the most part, users of baled silage have liked the method, but not the machinery for bagging it and not the occasionally poor fermentation. Automatic equipment is available that individually shrink-wraps big bales with tough plastic and provides an airtight seal. Additives and preservatives have been added successfully to large round bales.

SILO SIZE

The size of the silo to build should be determined by needs. With tower type and pit silos, the diameter should be determined by the quantity of silage to be fed daily, and the height (depth in a pit silo) should be determined by the length of the silage feeding period. Similar consideration should be accorded trench silos.

Tower Silo

If the diameter of the silo is too great, the silage will be exposed too long before it is fed; unless a quantity is thrown away each day, spoiled silage will be fed to the animals. The minimum recommended rate of removal of silage varies with the temperature. In most sections of the United States, a minimum of 2 inches of silage should be removed from tower silos daily during the winter feeding period, with the quantity increased to a minimum of 3 inches during summer feeding. Of course, the total daily silage consumption on any given farm or ranch is determined by the class and size of animals, number of animals, and rate of silage feeding.

Silo height should be determined primarily by the length of the intended feeding period. In general, however, the height should not be less than twice or more than 3.5 times the diameter. The greater the depth, the greater the unit capacity. Extreme height should be avoided because of the power required to elevate the cut silage material and the cost of the heavier construction material required. Figure 21.8 shows the capacities of tower silos of different heights and diameters, based on well-eared corn silage harvested in the early dent stage, cut in 0.25-inch lengths, and well-packed when filled and the silo refilled after settling for a day.

Trench Silo

As in an upright silo, the cross-sectional area of a trench silo should be determined by the quantity of silage to be fed daily. The length is determined by the number of days in the silage feeding period. The only difference is that a greater allowance for spoilage is made in the case of trench silos. Under most conditions, it is recommended that a minimum of 4 inches of silage be fed daily from the face (from the top to the bottom of the trench) of a trench silo during the winter months, with a somewhat thicker slice preferable during the summer months.

To calculate the capacity of a trench silo, we typically assume that silage weighs 35 pounds per cubic foot, which is an average figure for corn or sorghum silage. Thus, a trench silo 8 feet deep by 6 feet wide at the bottom and 10 feet wide at the top has a cross-sectional area of 64 square feet. This size silo will hold 747 pounds of silage for each 4-inch slice, or 2,240 pounds of silage for each 1-foot slice, or 112 tons in a trench 100 feet long.

ESTIMATING SILAGE INVENTORY

Sometimes, either for inventory purposes or for purposes of buying or selling, dairy producers need to estimate the amount of silage remaining in a silo after part of it has been fed out. For a tower silo, this may be done by referring to Figure 21.8. The depth indicated in the left-hand column is the actual depth of the settled silage and not the height of the silo. As noted, silage is more compact and heavier as the depth increases.

To estimate the amount of corn or sorghum silage in a trench silo (whether it is full or partly empty), multiply the average width in feet by the depth in feet to get the cross-sectional area. Then multiply by the length to get the volume. Finally, multiply the volume by 35 (the average weight of 1 cubic foot of corn silage in a trench silo) to obtain the pounds of silage. For example, the amount of silage in a trench silo 8 feet wide at the bottom, 12 feet wide at the top, 8 feet deep, and 40 feet long is computed as follows:

1. $8 + 12 = 20$ divided by $2 = 10$ feet; average width
2. $10 \times 8 = 80$ square feet cross-sectional area
3. $80 \times 40 = 3,200$ cubic feet; volume of the silo
4. $3,200 \times 35 = 112,000$ pounds, or 56 tons capacity

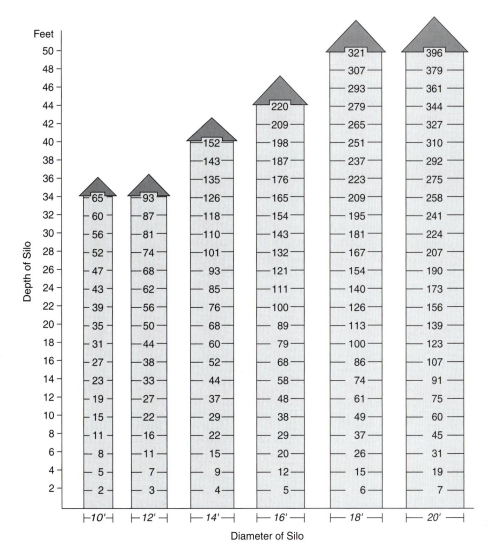

Figure 21.8   The capacity in tons of settled corn silage in upright silos of varying sizes. *(Courtesy of M. E. Ensminger)*

SUMMARY

- The proper size for a silo is determined by the number of animals to be fed daily, length of the feeding period, and amount of forage available for ensiling.
- Silos may be classified as conventional upright, gas-tight, pit, horizontal, and temporary.
- Conventional silos are permanent farm structures and should be constructed to withstand long usage.
- Sealed silos are designed for storage of wilted or overwilted forage or for the storage of high-moisture grain.
- Pit silos are inverted into the ground, like a well or cistern.
- Trench silos are horizontal, with smooth walls that are wider at the top than at the bottom.

- Bunker silos are aboveground horizontal silos with concrete floors and sidewalls of wood, concrete, or other materials.
- Temporary silos are aboveground silos used to meet emergencies, to supplement permanent silos, or to ensile by-product feed such as cannery refuse.
- The size of the silo should be determined on needs and by the length of the intended feeding period.

QUESTIONS

1. Describe the unloading mechanisms in tower silos.
2. What does a producer need before entering a filled, sealed silo?
3. How does a producer know if it is safe to enter a pit silo?
4. What are some advantages and disadvantages of trench silos?
5. What are the four kinds of temporary silos?
6. What are the two types of plastic silos?
7. A minimum of how many inches of silage should be removed daily from a tower silo versus a trench silo?

ADDITIONAL RESOURCES

Articles

Douglas, W. W., N. G. Hepper, and T. V. Colby. "Silo-Filler's Disease." *Mayo Clinic Procedures* 64 (March 1989): 291–304.

Tixier, M., O. Pitois, and P. Mills. "Experimental Impact of the History of Packing on the Mean Pressure in Silos." *European Physics Journal of Experimental Soft Matter* 14 (July 2004): 241–247.

Internet

Assessing and Reducing the Risk of Groundwater Contamination from Silage Storage: http://www.uky.edu/Agriculture/AnimalSciences/extension/pubpdfs/ip47.pdf.

Horizontal Silos: http://server.age.psu.edu/extension/Factsheets/h/H76.pdf.

Reducing the Risk of Groundwater Contamination by Improving Silage Storage: http://www.cahe.nmsu.edu/farmasyst/pdfs/9fact.pdf.

Siloge Storage Structures: http://edis.ifas.ufl.edu/DS083.

Temporary Grain Storage: http://www.bae.umn.edu/extens/postharvest/TempStorage/sld001.htm.

22

Feed Sources for Concentrates

OBJECTIVES

- To list the various feed sources for concentrates.
- To list high-moisture and high-energy feeds.
- To identify protein feeds.

Grains are seeds from cereal plants, members of the grass family, *Gramineae*. They provide an excellent source of highly digestible energy for cattle that are either on high levels of production or unable to utilize forages (e.g., young calves). Grains are more costly than forage on a weight basis. However, the energy content, digestibility, and other nutrients must be considered on a per-unit feed basis. Thus, a relatively expensive grain containing large amounts of highly digestible energy may in reality be a better buy than low-cost, low-quality roughage. Grains are extremely deficient in calcium; most contain less than 0.1 percent calcium. Adequate amounts of phosphorus are generally present in grain, but the calcium to phosphorus ratio is highly unbalanced. Grains are also deficient in certain vitamins; for example, vitamin A is low in all grains except fresh yellow corn.

The easily recognizable characteristics of good grains and other concentrates follow:

1. Seeds are not split or cracked.
2. Seeds are of low-moisture content, generally containing about 88 percent dry matter.
3. Seeds have a good color, one that is characteristic of the species.
4. Concentrates and seeds are free from mold.
5. Concentrates and seeds are free from rodent and insect damage.
6. Concentrates and seeds are free from foreign material, such as iron filings.
7. Concentrates and seeds are free from rancid odor.

Figure 22.1 Ear corn. *(Courtesy of USDA)*

CORN

Corn is the leading US dairy feed (Figure 22.1). It is palatable, nutritious, and rich in energy-producing carbohydrate and fat, but it has definite limitations. Protein quality is low (especially low in the amino acids lysine and tryptophan), and the quantity of proteins is also low (it averages about 9 percent). It is deficient in minerals, particularly calcium. Corn is higher in fat than barley or wheat (4 percent versus less than 2 percent). Fat not only contributes to the high-energy content of corn, but it also improves its palatability. Corn may be fed to dairy cattle shelled (to young stock that masticate well), cracked, as corn-and-cob meal, or flaked.

BARLEY

Barley is fed mostly in the western United States (Figure 22.2). Compared with corn, it contains more protein (barley, 13 percent; corn, 9 percent) and fiber (due to the hulls) and less carbohydrate and fat. Like oats, the feeding value of barley is

Figure 22.2   Barley plants. *(Courtesy of USDA-ARS)*

Figure 22.3   Over 144 million bushels of oats were harvested in the United States in 2003. *(Courtesy of USDA-ARS)*

Figure 22.4   Horticulturist Eric Brennan harvests a bundle of a late-summer rye cover crop from a square-meter quadrant. *(Courtesy of USDA-ARS)*

quite variable because of the high variability in test weight per bushel. For dairy cattle, barley is about 90 percent as valuable as corn. When fed to dairy cattle, barley is typically steam rolled or ground to improve feeding value.

OATS

Oats rank third in acreage among the cereal grains in the United States (Figure 22.3). About three-fourths of the acreage of oats is in the North Central states. Oats normally weigh 32 pounds per bushel, but the best oats are heavier. The feeding value varies according to the hull content and test weight per bushel. On the average, oats contain about 30 percent hulls.

RYE

Rye grows on poor, sandy soils (Figure 22.4). In such areas, it is sometimes marketed for animals, although it generally sells at a premium for bread-making or brewing. Rye is similar to wheat in chemical composition, but it is less palatable. For the latter reason, it should be limited to no more than one-third of the ration. When fed in limited amounts, the feeding value of rye is equal to corn. Because of its small, hard kernel, rye should always be processed.

SORGHUM

The grain sorghums, which include several varieties of earless plants bearing heads of seeds, are assuming an increasingly important role in American agriculture (Figure 22.5). New and higher-yielding

varieties have been developed and have become popular. As a result, more and more grain sorghums are being fed to dairy cattle.

Kafirs and milos are among the more important sorghums grown for grain. Other less widely produced types include sorgo, feterita, durra, hegari, kaoliang, and shallu. These grains are generally grown in regions where climatic and soil conditions

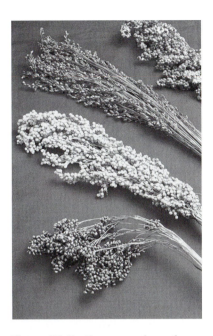

Figure 22.5 Representatives of several sorghum varieties from the USDA sorghum collection. *(Courtesy of USDA-ARS)*

Figure 22.6 Wheat. *(Courtesy of USDA-ARS)*

are unfavorable for corn. Like corn, the grain sorghums are low in carotene (vitamin A value). They are also deficient in other vitamins and in proteins and minerals. Feeding trials show that the grain sorghums are worth about 90 to 95 percent as much as corn for dairy cattle or about the same feeding value as barley. Sorghum should be ground or rolled for dairy cattle.

WHEAT

The total annual U.S. wheat tonnage is second only to corn (Figure 22.6). Because it is produced mainly for the manufacture of flour and other human foods, however, it is generally too high in price to feed to dairy cattle. When the price is favorable or when the grain has been damaged, it may be more profitable to market it for animals than for human consumption.

Compared with corn, wheat is higher in protein and carbohydrates, lower in fat, slightly higher in total digestible nutrients, and more palatable. Wheat, like white corn, is deficient in carotene. Pound for pound, it is equal to or up to 5 percent more valuable than corn for dairy cattle. Because the kernels are small and hard, wheat should be ground coarsely or rolled for the most economical utilization by cattle.

HIGH-MOISTURE GRAIN

Feeding high-moisture corn or other wet grains to dairy cattle allows grain to be harvested two to three weeks earlier than normal, thereby reducing field losses and harvest problems associated with adverse weather. In addition, storage and handling losses are reduced. High-moisture grain eliminates the expense of drying grain, increases palatability, and decreases the labor associated with processing and grinding grain.

High-moisture shelled corn should be stored at a moisture content of 25 to 30 percent. Ground ear corn should contain 28 to 32 percent moisture for proper preservation. High-moisture shelled corn and ear corn should be ground before storing in conventional silos. Silos made for corn silage often require extra banding to support the extra weight associated with high-moisture corn. In airtight silos, shelled corn can be stored whole and then rolled after removal from the silo. Propionic acid can be used effectively to treat and preserve high-moisture corn or barley for dairy cattle.

Moisture is important when buying feeds. When buying grains, dairy producers should never lose sight of how much water they may be purchasing. Moisture is also important in formulating rations. Careful feeders must constantly watch the moisture content of the feeds they buy and the effect of moisture on their nutritional quality control. Feeders must readjust feeding formulas whenever moisture in a leading ingredient changes by over 1 percent.

OTHER HIGH-ENERGY FEEDS

Although feed grains and their milling by-products comprise the vast majority of the energy feeds, numerous other feeds are routinely used to supply energy to dairy cattle. Seeds from plants other than *Gramineae* can be used effectively (for example, beans). Fats and oils provide an extremely concentrated source of energy. Molasses is a liquid energy feed that is highly palatable and digestible.

Seeds from plants other than cereal grains are used in dairy feeds when they are readily available and when the price is right. Legume seed, such as soybeans and peanuts, and whole cottonseed can be used for their energy content in addition to their protein content. Many types of seeds are by-products of cash-crop enterprises, representing culls of processing or marketing. On occasion, a surplus of a certain seed generally used for human consumption may reduce the cost to a level where it becomes economically feasible to incorporate it in dairy feeds.

Fats and Oils

Fat serves the following functions when added to dairy rations: it increases the caloric density of the ration without lowering the forage (fiber) content; controls dust; decreases wear and tear on feed-mixing equipment; facilitates pelleting of feeds; and helps to homogenize and stabilize certain feed additives, especially those of a very fine particle size. Most forages and grains are low in lipids; they contain less than 2 to 3 percent fat. Dairy cows should be able to utilize 1 to 1.5 pounds of fat per day in addition to the fat present in natural feedstuffs.

The type of fat (saturated or unsaturated) added to the ration greatly influences the animal's nutrient utilization, milk production, and feeding behavior; ration acceptability; the amount of fat that can be fed; and milk composition. Unsaturated fats are less desirable for dairy cows because of their inhibitory effects on rumen fermentation and digestion. Animal fats (which are more saturated) and blended animal-vegetable fats have generally produced the most positive responses in animal performance.

Because vegetable oils are high in unsaturated fat, they are less satisfactory than saturated fats as ration supplements. Whole seeds, such as cottonseed, soybeans, and sunflower seeds, have been used successfully, but the added fat derived from them should not exceed 1 pound per cow per day. Unsaturated fats that contain high levels of oleic acid apparently exceed the hydrogenation ability of rumen microorganisms and result in the greatest milk fat depression. An increase in long-chain fatty acids in the ration inhibits the synthesis of short- and medium-chain fatty acids in mammary tissue.

Added ration fat, including feeding whole cottonseed and soybeans, decreases the protein content of milk by about 0.1 percent, primarily by decreasing casein content. Whole cottonseed also increases the proportion of long-chain fatty acids in milk. In some cases, feeding rumen-protected fats that are resistant to microbial action in the rumen has increased milk fat percentage and the effi-

ciency of milk production. However, feeding polyunsaturated fats may result in milk with an increased proportion of polyunsaturated fatty acids and an increased susceptibility to oxidative rancidity. Limited data indicate that calcium salts of fatty acids (which are 82 percent fat) and prilled forms of saturated fats are effective sources of protected fat for dairy cows and that their use may allow the total fat in the ration to reach 5 to 7 percent of the dietary dry matter of the ration. Until the rumen becomes functional, young dairy calves require some fat in the diet. A level of 10 percent fat in milk replacers appears to be sufficient to supply essential fatty acids and to carry fat-soluble vitamins, but it may be insufficient to supply adequate energy for maximal gains.

Molasses

Molasses (including cane or blackstrap, beet, citrus, and wood molasses) is used extensively as a livestock feed: 1,605,000 metric tons are used for animal feeds in the United States annually. Cane and beet molasses are by-products of the manufacture of sugar from sugarcane and sugar beets, respectively. Citrus molasses is produced from citrus wastes. Wood molasses is a by-product of the manufacture of paper, fiberboard, and pure cellulose from wood; it is an extract from the more soluble carbohydrates and minerals of the wood material. Cane or blackstrap molasses is by far the most extensively used type for cattle feed.

When used at levels of 10 to 15 percent of the ration, molasses has about three-fourths the energy value of corn. However, molasses has added value as an appetizer, a way to reduce the dustiness of a ration, and a binder for pelleting. Also, cane molasses is a good source of certain minerals. In hot, humid areas, molasses should be limited to 5 percent of the ration; otherwise, mold may develop. Where mustiness is a problem, it may be controlled by adding calcium propionate to the feed, according to the manufacturer's directions.

PROTEIN FEEDS

Protein supplements are feedstuffs that contain more than 20 percent protein or protein equivalent. The amount of protein needed in the total ration of lactating cows is determined primarily by the amount of milk produced. Milk is a rich source of high-quality protein; as milk production increases, a substantial amount of dietary protein is necessary.

The amount of protein needed in the concentrate mix depends on the kind and quality of the forage that is fed. As the amount of legume increases, the percentage of protein in the concentrate can be lowered. When more protein is fed than needed, the excess is used as a source of energy. Because high-protein feeds are generally more expensive than high-energy feeds, it usually is more economical to feed only the amount needed. Besides, a large excess of dietary protein may decrease the energy supply because excess protein must be deaminated to ammonia and transformed back into urea for excretion. High-protein feeds are usually named and classified according to their origin and method of processing. On the basis of origin, they are usually grouped into two general categories: plant proteins and animal proteins.

Plant Protein Sources

This group, which supplies the bulk of protein supplements for dairy cattle, includes the common oilseed by-products: soybean meal, cottonseed meal, linseed meal, peanut meal, safflower meal, sunflower meal, rapeseed meal (canola meal), and coconut (or copra) meal. They vary in protein content and feeding value, depending on the seed from which they are produced, amount of hull and/or seed coat included, and method of oil extraction used. Protein quality is less important with ruminants because of microbial synthesis. The unprocessed seed is sometimes used to provide both a source of protein and a concentrated source of energy. The oil-bearing seeds are especially high in energy because of the oil that they contain.

Additional plant proteins are obtained as by-products from grain milling, brewing and distilling, and starch production. Most of these industries use the starch in grains and seeds and then dispose of the residue, which contains a large portion of the protein of the original plant seed.

Several rich oil-bearing seeds are produced for vegetable oils for human food (oleomargarine, shortenings, and salad oil) and for paints and other industrial purposes. In processing these seeds, protein-rich products of great value as dairy feeds are obtained; among them are soybean meal, coconut meal, cottonseed meal, linseed meal, peanut meal, rapeseed meal, safflower meal, sesame meal, and sunflower seed meal. Oil is extracted from these seeds by one of the following basic processes or modifications thereof: solvent extraction, hydraulic extraction, or expeller extraction.

Soybean Meal

Soybean meal has the highest nutritive value of any plant protein source. It is now the most widely used protein supplement in the United States. Soybean meal is the ground residue (soybean oil cake or soybean oil chips) remaining after the removal of most of the oil from soybeans (Figure 22.7).

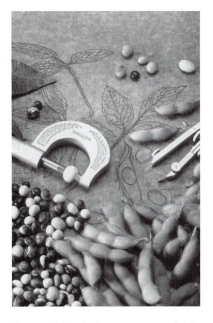

Figure 22.7 Soybeans are second only to corn in importance to US farmers. *(Courtesy of USDA-ARS)*

Almost all soybeans are solvent extracted. Soybean meal normally contains 41, 44, 48, or 50 percent protein, depending on the amount of hull removed. Because of its well-balanced amino acid profile, the protein of soybean meal is of better quality than other protein-rich supplements of plant origin. However, it is low in calcium, phosphorus, carotene, and vitamin D.

Coconut Meal

Coconut meal is the by-product from the production of oil from the dried meats of coconuts (Figure 22.8). The oil is generally extracted by either the hydraulic process or the expeller process. Coconut meal averages about 21 percent protein content.

The lipid component of copra meal is very low in unsaturated fatty acids. Hence, dairy producers use copra meal to produce a pleasant-flavored, rather hard (highly saturated) butterfat. The maximum level of coconut meal in dairy rations is 3.3 to 6.5 pounds per day; higher amounts tend to produce tallowy butter.

Cottonseed Meal

The U.S. cotton crop ranks fourth in value (Figure 22.9), exceeded only by soybeans, corn, and wheat. Among the oilseed meals, cottonseed meal ranks second in tonnage to soybean meal. The oil is extracted by mechanical means or solvent, or it may be partially mechanically extracted

Figure 22.8 A Manila dwarf coconut palm on the grounds of the Tropical Agriculture Research Station in Mayaguez, Puerto Rico. *(Courtesy of USDA-ARS)*

Figure 22.9 Close-up of a cotton boll, the source of fiber and cottonseed. *(Courtesy of USDA-ARS)*

and then solvent extracted. The protein content of cottonseed meal varies from about 22 percent in meal made from undecorticated seed to 95 percent in flour made from seed from which the hulls have been removed completely. Thus, by screening out the residual hulls, which are low in protein and high in fiber, the processor can make a cottonseed meal of the protein content desired, usually 36, 41, 44, and 48 percent. The protein content of cottonseed meal varies with the geographical location in which it is grown. Meals manufactured from cottonseed produced on the West Coast generally contain higher protein levels than those produced throughout the rest of the United States.

Figure 22.10   Flax seed. *(Courtesy of USDA)*

Linseed Meal

Linseed meal is a by-product of flax (Figure 22.10). In this country, most of the flax is produced as a cash crop for oil from the seed and the resulting by-product of linseed meal. Most of the nation's flax is produced in North Dakota, South Dakota, and Minnesota. The oil is extracted from the seed by either the mechanical process or the solvent process.

Linseed meal is the finely ground residue (known as cake, chips, or flakes) remaining after the oil extraction. It averages about 35 percent protein content (33 to 37 percent). Linseed meal is lacking in carotene and vitamin D and is only a moderate source of calcium and the B vitamins.

Peanut Meal

Peanut meal is ground peanut cake, the product that remains after the extraction of part of the oil of peanuts by pressure or solvents. It is a palatable, high-quality vegetable protein supplement used extensively in livestock and poultry feeds. Peanut meal ranges from 41 to 50 percent protein and from 4.5 to 8 percent fat. It is low in methionine, lysine, and tryptophan and in calcium, carotene, and vitamin D.

Peanut meal and hulls is ground peanut meal with added hulls or the ground by-product remaining after extraction of part of the oil from whole or unshelled peanuts (Figure 22.11). Because about one-fourth of peanut meal and hulls consists of peanut hulls, it is high in fiber, averaging about 22.5 percent. Peanut meal tends to become rancid when held too long, especially in warm, moist climates; it should not be stored longer than six weeks in the summer or two to three months in the winter.

Figure 22.11 A healthy peanut crop prior to harvest. *(Courtesy of USDA)*

Figure 22.12 A field of safflower. *(Courtesy of USDA-ARS)*

Rapeseed Meal (Canola Meal)

Rapeseed is grown extensively in Canada, where it is called canola. Canola was created from specially selected rapeseed by Canadian plant scientists in the 1970s. The old rapeseed was high in glucosinolate compounds that, when fed to animals at high levels, made for palatability problems and lowered performance due to goitrogenic action. Canola changed this predicament. The new canola is low in glucosinolates in the meal and low in erucic acid (a long-chain fatty acid). Canola meal averages about 36 percent crude protein, and its amino acids compare favorably with soybean meal. When the price is favorable, canola meal may be used as a protein supplement for dairy cattle.

Safflower Meal

A large proportion of the safflower seed (Figure 22.12) is composed of hull, about 40 percent. Once the oil is removed from the seeds, the resulting product contains about 60 percent hulls and 18 to 22 percent protein. Various means have been tried to reduce this high hull content. Most meals contain seeds with part of the hull removed, thereby yielding a product of about 15 percent fiber and 40 percent protein. Safflower meal is not very palatable, but it can be used effectively when mixed with other feeds.

Sesame Meal

Little sesame is grown in the United States despite the fact that it is one of the oldest cultivated oilseeds. The oil meal is produced from the entire seed. Sol-

Figure 22.13 Sunflowers. *(Courtesy of USDA-ARS)*

vent extraction yields higher protein (45 percent) but lower fat levels (1 percent) than either the screw press or hydraulic methods, which produce meals containing about 38 percent protein and 5 to 11 percent oil. Excessive levels (greater than 7 pounds per head per day) can produce soft butter.

Sunflower Meal

Sunflower meal (Figure 22.13), a relative newcomer to the oilseed industry in the United States, is rapidly gaining acceptance as a high-quality source of plant protein. The oil meal varies considerably depending on the extraction process and whether the seeds are dehulled. Meal from prepressed solvent extraction of dehulled seeds contains about 44 percent protein compared to 28 percent for whole seeds. Screw-pressed sunflower meal ranges from 28 to 45 percent protein. Sunflower hulls can be used effectively for roughage in dairy feeds.

Animal Protein Sources

Animal proteins are derived from animal products. Milk and milk products, along with a limited amount of feather meal, constitute the bulk of animal proteins that are fed to dairy cattle.

Nonprotein Nitrogen (NPN)

Certain nonprotein nitrogen sources may be substituted for all or much of the supplemental protein required in most ruminant rations, provided such rations are adequate in minerals and readily available carbohydrates. Among such products are urea and ammoniated feed sources. The rumen microorganisms, which are able to use inorganic compounds much like plants utilize chemical fertilizers, build proteins of high quality in their cells from sources of inorganic nitrogen that nonruminants cannot use. These microorganisms pass down the digestive tract and serve as a source of top-quality protein for the animal. In ruminant nutrition, therefore, even nonprotein sources of nitrogen such as urea and ammonia have a protein replacement value. An exception is the very young calf, whose rumen and its ability to synthesize microbial protein are not yet well developed.

Because of the increasing shortage and higher price of oilseed proteins, more and more urea is being fed to dairy cows. Approximately 265,000 metric tons of urea is fed to ruminants annually in the United States as a source of protein. Urea is a white crystalline, odorless, nonprotein nitrogen compound of the formula N_2H_4CO. It is manufactured synthetically using nitrogen from the air. When properly used, it is a safe, low-cost source of protein for dairy cattle; when used improperly, it is potentially toxic. No more than one-third of the protein requirements should be met by urea. Most state laws limit the amount of urea that can be put in commercial feeds. One pound of 45 percent nitrogen (281 percent protein equivalent) urea provides as much protein value as 6.8 pounds of 41 percent soybean meal or cottonseed meal. However, urea does not provide energy, minerals, or vitamins, which must be provided through other sources when urea is substituted for protein meals.

When added to the dairy cattle ration, urea must be mixed thoroughly to ensure even distribution, and it must be used according to directions. It may be toxic when used improperly or when fed in excessive amounts. Normally, dairy producers limit urea to 1.5 to 2.0 percent of the concentrate ration. Higher levels (up to 2.75 percent of the concentrate) are unpalatable and depress appetite. However, the unpalatability may be masked by pelleting the urea with alfalfa or inclusion in a total mixed ration (TMR).

Urea can be mixed in feed either as a powder or as an aqueous solution. Both methods are relatively simple and inexpensive. When urea is added as a powder, it could sift through the grain and be unevenly distributed, thereby increasing the chance of toxicity. If careful mixing procedures are followed, however, this hazard can be minimized. If urea is added to feed in aqueous form, uniform distribution throughout the feed is ensured. However, this method has two disadvantages: special mixing equipment is required, and once in solution, urea may be degraded during prolonged storage.

Liquid supplements combining molasses for energy and urea as a protein precursor are widely used. This type of supplement can also be used as a carrier for micronutrient and nonnutritive additives. Several problems are inherent in feeding such liquid supplements. If they are to be used effectively, the urea must remain in solution or suspension. The supplement must keep its chemical integrity throughout varying environmental temperatures over prolonged periods. Liquid supplements are extremely palatable; hence, there is a danger of overconsumption if intake is not monitored. These supplements also tend to be highly corrosive.

One way of feeding urea to dairy cattle is through the addition of urea to crops that are being ensiled. If chopped, whole-plant corn is ensiled at 35 to 40 percent dry matter, urea can be added at a level of 0.5 percent of wet material. This level should increase the crude protein level of the silage on a dry-matter basis by about five percentage points. Urea levels higher than 0.5 percent can create palatability problems as well as storage problems. When the silage contains little or no grain, the amount of urea to be added should be reduced. Silage tends to be rather variable in moisture, and this variability can affect the benefits of added NPN urea. Hence, one should have a reasonable estimate of the moisture content of the material that is to be ensiled. Likewise, water in silage creates some leaching out of the urea. Also, ammonia is produced during the ensiling process, representing an additional loss of urea.

GRAIN PROCESSING

Grains are often processed prior to their inclusion in dairy rations. Processing alters the feeding characteristics in a predictable way, and the strategic use of processing can enhance the feeding value and the profitability of a ration. For example, starches in high-moisture grains are degraded more rapidly than starches in dry grains, and grinding of dry grains also enhances degradability. Starch digestibility of grains can also be enhanced through steam flaking.

SUMMARY

- Grains are seeds from cereal plants that provide an excellent source of highly digestible energy for cattle.
- Good grains and concentrates are seeds that are not split or cracked; have a low moisture content and good color; and are free from mold, rodent, and insect damage, foreign material, and rancid odor.
- Corn, barley, oats, rye, sorghum, and wheat are examples of feed sources for concentrates.
- Corn has a low protein quality and quantity, is deficient in minerals, and is higher in fat than barley or wheat.
- Barley contains more protein and fiber and less carbohydrate and fat.
- Rye grows on poor, sandy soils and generally sells at a premium for bread-making or brewing.
- The grain sorghums, kafirs and milos, are among the more important sorghums grown for grain.
- Compared to corn, wheat is higher in protein and carbohydrates and lower in fat, slightly higher in total digestible nutrients, and more palatable.
- High-moisture grain eliminates the expense of drying grain, increases palatability, and decreases the labor associated with processing and grinding grain.
- When added to dairy rations, fat increases caloric density, controls dust, decreases wear on feeding equipment, facilitates pelleting of feeds, and helps to homogenize and stabilize certain feed additives.
- The amount of protein, usually grouped into either plant or animal proteins, depends on the kind and quality of forage that is fed.
- Plant protein sources include soybean, coconut, cottonseed, linseed, peanut, rapeseed, safflower, sesame, and sunflower meal.
- Linseed meal is a by-product of flax.
- Certain nonprotein nitrogen sources, such as urea, may be substituted for some of the supplemental protein required in most ruminant rations.

QUESTIONS

1. In what vitamins and minerals are grains deficient?
2. The total annual U.S. wheat tonnage is second to what other concentrate?
3. What acid can be used effectively to treat and preserve high-moisture corn or barley for dairy cattle?
4. How does the type of fat added to a ration influence the rumen?
5. What is the most common type of molasses used for cattle feed?
6. In hot, humid areas, why should molasses be limited to 5 percent of the feed ration?
7. Define protein supplements.
8. List three basic processes for extracting oil from seeds.
9. What plant protein has the highest nutritive value?
10. What proportion of the protein requirement can urea fulfill?

ADDITIONAL RESOURCES

Articles

Adesogan, A. T., M. B. Salawu, S. P. Williams, W. J. Fisher, and R. J. Dewhurst. "Reducing Concentrate Supplementation in Dairy Cow Diets While Maintaining Milk Production with Pea-Wheat Intercrops." *Journal of Dairy Science* 87 (October 2004): 3398–3406.

Keady, T. W., C. S. Mayne, D. A. Fitzpatrick, and M. A. McCoy. "Effect of Concentrate Feed Level in Late Gestation on Subsequent Milk Yield, Milk Composition, and Fertility of Dairy Cows." *Journal of Dairy Science* 84 (June 2001) 1468–1479.

Reist, M., D. Erdin, D. von Euw, K. Tschuemperlin, H. Leuenberger, C. Delavaud, Y. Chilliard, H. M. Hammon, N. Kuenzi, and J. W. Blum. "Concentrate Feeding Strategy in Latating Dairy Cows: Metabolic and Endocrine Changes with Emphasis on Leptin." *Journal of Dairy Science* 86 (May 2003): 1690–1706.

Stockdale, C. R., G. P. Walker, W. J. Wales, D. E. Dalley, A. Birkett, Z. Shen, and P. T. Doyle. "Influence of Pasture and Concentrates in the Diet of Grazing Dairy Cows on the Fatty Acid Composition of Milk." *Journal of Dairy Research* 70 (August 2003): 267–276.

Internet

Alternative Feeds for Dairy Cattle in Northwest Minnesota: An Update: http://www.ansci.umn.edu/dairy/dairyupdates/du 126.htm.

Canola Meal in Cattle Diets: http://www.canola-council.org/pubs/meal7.html.

Citrus Feedstuffs for Dairy Cattle: http://edis.ifas.ufl.edu/DS149.

Concentrate Feeding Guide: http://babcock.cals.wisc.edu/downloads/de/07.en.pdf.

Concentrates for Dairy Cattle: http://www.das.psu.edu/dcn/catnut/PDF/Concentrate.PDF.

Corn Gluten Feed: http://cecommerce.uwex.edu/pdfs/A3518.PDF.

Curds Meal: http://cecommerce.uwex.edu/pdfs/A3514.PDF.

Distillers Grains for Dairy Cattle: http://agbiopubs.sdstate.edu/articles/EXEX4022.pdf.

23

Industry By-Products and Feed Additives

OBJECTIVES

- To understand how industry by-products act as feed sources for dairy cattle.
- To understand the purpose of supplements and additives to a dairy cow ration.

Dairy cattle can utilize a host of by-products, animal wastes, and crop residues. As the world grows more populous, the dairy industry will increasingly provide a practical and economical outlet for human-inedible feeds.

By-product feeds from plant processing are important in dairy rations. The milling, sugar, vegetable oil, and fermentation industries provide by-products of special significance to most commercial dairy rations and for many home-mixed feeds. If dairy producers plan to incorporate a by-product feed in their rations, however, they should first determine the moisture content of the product, its relative feeding value, and the appropriate amount to feed. By-products are often extremely variable in feeding value. For example, almond hulls can range from excellent to poor as a feedstuff. Some by-product feeds may also contain pesticide residues that can be excreted in the milk. For the latter reason, producers should make sure that their feeds are free from environmental contaminants.

Many of the residues and by-products from the vegetable industry can be used effectively in dairy rations. Feeds such as cannery wastes, pea vines and pods, and cottonseed hulls can be used as economic alternatives to the more traditional feeds.

Wheat bran and wheat middlings (or mill run) are the most widely used by-products of the milling industry. Wheat bran has always been considered an excellent dairy feed when the quality is high and the price is competitive. Wheat mill run (middlings) is usually more price competitive than wheat bran. When of high quality, it is an excellent feed, although it is not as high in digestibility or energy as whole grains.

Beet pulp and beet molasses are used extensively in dairy rations, where they add to the texture and palatability and provide good sources of energy. Molasses reduces the dustiness of feeds and is an excellent carrier for minerals and vitamins, as well as other ingredients where uniform mixing is important. It is usually restricted to about 5 percent of the grain-concentrate mix. At higher levels, it may make the feed sticky and cause mechanical mixing and feeding difficulties. Also, excess molasses may be too laxative because of its high mineral and sugar content; hence, it may reduce the utilization of other less digestible ingredients. When available at competitive prices, the dried or wet mash from the brewing and distilling industries, as well as some of the pharmaceutical fermentations, have been used extensively in dairy rations.

The vegetable oil industry extracts the oil from certain seeds, leaving a residue that is high in protein and beneficial in dairy rations. The oil meals are incorporated in dairy rations primarily as a source of protein. Soybeans are a popular oil crop, and the extracted meal is one of the most common protein supplements in dairy rations. In cotton-growing areas, cottonseed meal is widely used in dairy rations. Other oil meals used in dairy rations include peanut meal, copra (coconut) meal, safflower meal, and several others of lesser importance. Some heating is desirable for most oil meals to destroy toxic or inhibitory materials that they contain. However, the heating must be controlled so that the digestibility of the proteins is not reduced to the point that it erases the benefits derived from eliminating the undesirable substances.

Fruit and vegetable by-products may come from three sources: cull, unmarketable or damaged commodities; crop residues left in the field; or canning, juicing, or processing wastes. These products can be used successfully in many feeding programs. But problems involving their continued availability, storage, and handling must be considered because many of them are highly perishable. Most of these by-product feeds are generally restricted to areas where processing and canning operations are located.

Many of the oilseeds are hulled (decorticated) prior to processing. Cottonseed hulls and gin trash are widely used fibrous by-products. Soybeans, peanuts, and sunflowers are three additional oilseeds from which hulls are routinely removed in processing. Hulls from the covered grains provide satisfactory roughage for dry cows. Rice, buckwheat, oats, and barley are commonly dehulled. Hulls from many nuts, such as almonds and walnuts, can be used in dairy feeds when price and availability warrant such use.

Two crops, sugarcane (Figure 23.1) and sugar beets (Figure 23.2), account for the bulk of refined US sugar, with sugar beets being the primary crop. Several by-product feeds are produced in various stages of processing these two crops for sugar. In the processing of sugarcane, the white, crystalline sucrose is separated from the molasses and brown sugars. The molasses product, which is referred to as blackstrap molasses, is incorporated in many dairy rations. After the juice has been extracted from the cane, the remaining by-product is known as bagasse. Bagasse is high in fiber, and the fiber is low in digestibility, only about 25 percent. Additionally, its Total Digestible Nutrients (TDN) is extremely low, ranging from 20 to 25 percent. However, bagasse has been used effectively as a carrier of molasses, the combination of which yields a relatively high-fiber, high-energy feed. The resulting beet pulp from the processing of sugar beets can be fed wet if it is used within a short time, or it can be ensiled or dried when long-term storage is desired. Molasses is often added to the dried beet pulp to increase the energy content. On occasion, beet pulp is ammoniated to provide a source of nonprotein nitrogen. Beet tops and crowns can be fed fresh, dried, or ensiled and are highly palatable.

Considerable quantities of grains are used in the brewing of beers and ales and in the distilling of liquors. After processing, the remaining by-products can be readily adapted to dairy feeding programs. Usually brewing and distilling by-products contain some of the B vitamins and other nutrients that are produced during fermentation.

Wood and by-products from the wood and paper industry offer an abundant source of potential feedstuffs for ruminants. These by-products can be classified as milling and processing wastes, treated wood scraps, pulp scraps, paper products, wood sugar products, and torula yeast. The digestibility

Figure 23.1   Billeted sugarcane with trash (top) and free of trash (bottom). *(Courtesy of USDA-ARS)*

Figure 23.2   Shallow grooves of an SR96 smooth-root sugar beet (left) and a traditional sugar beet. *(Courtesy of USDA-ARS)*

and energy content of many of these products is low; their use must be restricted to those animals with low energy requirements, such as dry cows.

Crop residues are the portions of crops that are normally left in the field following harvest. These include corn stalks and husklage, sorghum stalks, soybean refuse, small grain straws and chaff, and legume and grass seed straws. Of all the crop residues, the residue of corn is produced in greatest abundance in the United States and offers the greatest potential for expansion in animal numbers. Corn usually produces an amount of residue equal to the quantity of the grain produced.

Crop residues may be grazed, processed as dry feed, or made into silage. The important point to remember is that their relatively low value, in comparison with grains, necessitates low-cost harvesting, storing, and feeding. Also, they must be fed to the right class of animals, primarily dry cows, and they must be supplemented properly. The term *supplement* refers to feedstuffs that are used to improve the value of basal feeds. Thus, supplements are products that provide an additional nutrient or nutrients. They can be used in large quantities, such as protein supplements, or in extremely small quantities, such as trace minerals. An *additive* is a substance that is added to a basic feed, usually in small quantities, for the purpose of fortifying it with certain nutrients, stimulants, or medicines. In general, the term *feed additive* refers to a nonnutritive product that affects the utilization of the feed or productive performance of the animal. In a legal sense, however, many nutritional supplements, for example, methionine hydroxy analog (MHA), are considered to be additives.

ADDITIVES

Many additives are used by dairy producers to increase milk production, affect milk composition, and/or improve feed efficiency. New products are constantly evolving, including antibiotics, buffers, and ionophores.

Antibiotics

Antibiotics, which are widely used in the diet of young dairy calves, are especially beneficial for calves exposed to adverse conditions of housing, sanitation, and disease. However, they should not be used as a substitute for good management and a clean, sanitary environment. Antibiotics are mainly effective in increasing feed intake and growth, along with preventing diarrhea. Generally, antibiotics are fed at the following concentrations: in the milk replacer (dry basis) or in an equivalent amount of whole milk, 20 to 40 ppm and in the starter ration, 10 to 20 ppm. For the prevention and control of disease, higher concentrations may be necessary: 50 to 100 ppm in the milk replacer and 25 to 50 ppm in the starter.

Buffers

Buffers are used primarily to improve the feed intake, rumen function, milk production, milk composition, and health of lactating cows. The common buffers are sodium bicarbonate ($NaHCO_3$), magnesium oxide (MgO), sodium bentonite, sodium sesquicarbonate, and calcium carbonate or limestone ($CaCO_3$). Buffers maintain the hydrogen ion concentration in the rumen, intestines, tissues, and body fluids; they increase the rate of passage of liquids from the rumen; or both. Buffers are of greatest benefit to cows under the following circumstances: during early lactation, when large amounts of rapidly fermentable carbohydrates are fed, especially when they are fed at infrequent intervals; when fermented forage, primarily corn silage, is the major or only forage in the ration; and when the particle size of the total ration dry matter has been reduced by chopping, grinding, or pelleting to the extent that it increases the rate of ruminal fermentation and depresses salivary secretion and buffering capacity.

Ionophores

Ionophores are feed additives that change the metabolism within the rumen by altering the rumen microflora to favor propionic acid production. Currently, two ionophores, Bovatec (lasalocid) and Rumensin (monensin), are approved by the Food and Drug Administration (FDA) for replacement heifers. Both are antibiotics. Feeding Bovatec or Rumensin to replacement heifers improves live-weight gains and the efficiency of feed utilization. Rumensin is also approved by the FDA for feeding to lactating cows.

SUMMARY

- Many residues and by-products from the vegetable industry, such as cannery wastes, pea vines and pods, and cottonseed hulls, can be used effectively in dairy rations as an economic alternative to more traditional feeds.
- Beet pulp and beet molasses add texture, palatability, and energy to a ration.
- Supplements refer to feedstuffs that are used to improve the value of basal feeds.
- A feed additive is a nonnutritive product that affects the utilization of the feed or productive performance of the animal.
- Vitamins A, D, and E are of concern for dairy cattle, with vitamin A the most likely to be deficient.
- Cobalt must be supplied in adequate amounts.

- Additives such as antibiotics, buffers, and ionophores are used to increase milk production, affect milk composition, and/or improve feed efficiency.
- Buffers function to maintain hydrogen ion concentration in the rumen, intestines, tissues, and body fluids; increase the rate of passage of liquids from the rumen; or both.
- Common buffers include sodium bicarbonate, magnesium oxide, sodium bentonite, sodium sesquicarbonate, and calcium carbonate or limestone.
- Ionophores are feed additives that change the metabolism within the rumen by altering rumen microflora to favor propionic acid production and improving liveweight gains and efficiency of feed utilization.

QUESTIONS

1. What are the most widely used by-products of the milling industry?
2. Fruit and vegetable by-products come from what three sources?
3. Crop residues should be fed to dairy cows during which stage of their life cycle?
4. Why are buffers added to feed?
5. When are buffers of greatest benefit to cows?
6. What are two FDA-approved ionophores for replacement heifers?

ADDITIONAL RESOURCES

Articles

Cunningham, K. D., M. J. Cecava, and T. R. Johnson. "Nutrient Digestion, Nitrogen, and Amino Acid Flows in Lactating Cows Fed Soybean Hulls in Place of Forage or Concentrate." *Journal of Dairy Science 76* (December 1993): 3523–3535.

Firkins, J. L., and B. S. Oldick. "Molasses, Fat Blends Are Highly Available Energy Sources." *Feedstuffs 59* (February 1987): 14–17, 27.

Holden, L. A., L. D. Muller, and S. M. Emanuele. "Effect of Replacing Concentrate with Liquid Supplement for Late Lactation Cows Grazing Grass Pasture" (abstract). *Journal of Dairy Science 78* (Supplement 1, 1995): 209.

Huhtanen, P. "The Effect of Barley, Unmolassed Sugar-Beet Pulp and Molasses Supplements on Organic Matter, Nitrogen and Fiber Digestion in the Rumen of Cattle Given a Silage Diet." *Journal of Animal Feed Science Technology* 20 (1988): 259–278.

Maiga, H. A., D. J. Schingoethe, and F. C. Ludens. "Evaluation of Diets Containing Supplemental Fat with Different Sources of Carbohydrates for Lactating Dairy Cows." *Journal of Dairy Science 78* (May 1995): 1122–1130.

Mansfield, H. R., and M. D. Stern. "Effects of Soybean Hulls and Lignosulfonate-Treated Soybean Meal on Ruminal Fermentation in Lactating Dairy Cows." *Journal of Dairy Science 77* (April 1994): 1070–1083.

Nocek, J. E., and J. B. Russell. "Protein and Energy as an Integrated System: Relationship of Rumen Protein and Carbohydrate Availability to Microbial Synthesis and Milk Production." *Journal of Dairy Science 71* (August 1988): 2070–2107.

Nombekela, S. W., and M. R. Murphy. "Sucrose Supplementation and Feed Intake of Dairy Cows in Early Lactation." *Journal of Dairy Science 78* (April 1995): 880–885.

Oldick, B. S., J. Pantoja, and J. L. Firkins. "Efficacy of Fat Sources in Liquid Supplements for Dairy Cows" (abstract). *Journal of Dairy Science 80* (Supplement 1, January 1997): 243.

Internet

By-Products and Regionally Available Alternative Feedstuffs for Dairy Cattle: http://www.ext.nodak.edu/extpubs/ansci/dairy/as1180w.htm.

Feeding Fibrous By-Products to Lactating Dairy Cows: http://www.asrc.agri .missouri.edu/deptrpt99/99dr18-11.pdf.

Importance of Trace Minerals in Dairy Heifer, Dry Cow, and Lactating Cow Rations: http://www.traill.uiuc.edu//dairynet/paperDisplay.cfm?ContentID=551.

Recent Applications of Liquid Feed Supplements in Rations for Lactating Dairy Cows: http://www.wisc.edu/dysci/uwex/nutritn/pubs/LPSrev.pdf.

Recommendations for Feeding Selected By-Product Feeds to Dairy Cattle: http://www.cals.ncsu.edu/an_sci/extension/dairy/205-d.pdf.

Selecting By-Product Feedstuffs for Feeding Dairy Cattle: http://edis.ifas.ufl.edu/ BODY_DS077.

Use of By-Products in Growing Heifer Diets: http://agbiopubs.sdstate.edu/ articles/ExEx4030.pdf.

Vitamin Needs of Dairy Cattle: http://edis.ifas.ufl.edu/DS153.

5

Concepts in Dairy Reproduction

Topic 24
Fundamentals of Female Reproduction

Topic 25
Reproductive Physiology and Management of the Bull

Topic 26
Heat Detection and Estrus Synchronization

Topic 27
Assisted Reproductive Technologies

Topic 28
Pregnancy and Parturition

24

Fundamentals of Female Reproduction

OBJECTIVES

- To describe the reproductive physiology of the cow, including hormone production and pattern, reproductive structures, and the estrous cycle.
- To describe the process of fertilization.
- To identify the causes and prevention of reproductive failures.

Reproduction is the first and most important requisite of successful dairy cattle production. Simply stated, milk production is a by-product of the reproductive process. Reproductive problems are the second most common reason for culling cows from dairy herds, accounting for over one-fourth of cows culled.

REPRODUCTIVE ANATOMY AND PHYSIOLOGY

Understanding reproductive function first requires an understanding of female reproductive anatomy (Figures 24.1 and 24.2). Reproductive function in the dairy cow involves interplay among the ovary, uterus, hypothalamus, and pituitary gland, primarily through reproductive hormones. If any of these tissues fail to communicate effectively with the others, infertility results.

The main functions of the ovary are to produce ova, estrogen, and progesterone (Figures 24.3 through 24.6). The newborn heifer has the potential for over 100,000 ovulations, although this potential is rarely fully utilized. The right ovary is larger than the left in most heifers and cows, and about 60 percent of all ovulations come from the right ovary. The ovary has two regions: the cortex (outer shell) and the medulla (central core). The cortex contains germinal epithelial cells where oogenesis occurs; these ova develop in Graafian follicles.

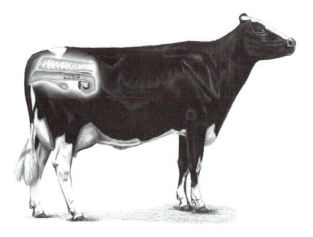

Figure 24.1 The position of the reproductive tract in a dairy cow. *(Courtesy of Dana Boeck)*

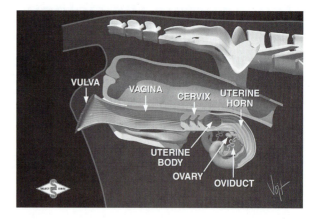

Figure 24.2 Visualization of the reproductive tract of the cow, with the bladder beneath and the rectum and large intestine above. *(Courtesy of Select Sires)*

Puberty is defined as the point at which regular ovarian cycles begin to occur. Although waves of follicular growth and regression are initiated in the prepubertal period, all follicles become atretic and regress. At the time of puberty, the hypothalamus matures to the point that it is no longer sensitive to

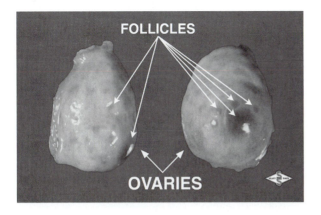

Figure 24.3 Developing follicles form blisterlike structures on the surface of the ovary. *(Courtesy of Select Sires)*

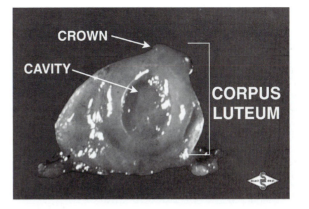

Figure 24.5 The cavity and the crown both represent different stages of luteal development. *(Courtesy of Select Sires)*

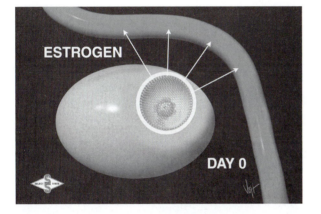

Figure 24.4 Increasing levels of estrogen produced by the dominant follicle enter the bloodstream of the cow, inducing estrus behaviors. *(Courtesy of Select Sires)*

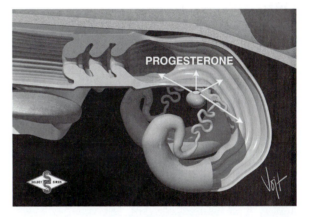

Figure 24.6 Progesterone produced by the corpus luteum has a quiescent effect on the uterus. *(Courtesy of Select Sires)*

the inhibitory effects of the estradiol produced by the follicles. At this point, the hypothalamus produces gonadotropin-releasing hormone (GnRH), which in turn stimulates the secretion of luteinizing hormone (LH). The release of LH primes the dominant follicle to complete the maturation process and ovulate, and the onset of ovarian activity is finally observed (Figure 24.7). Puberty typically occurs when dairy heifers reach about 35 percent of their mature weight (at about seven to nine months of age).

Ovulation is the process whereby the follicle releases the ova into the uterine horn. The coordination of the ovulation process with estrus, or the period of female receptivity to the male, is critical for reproductive success. The coordination of these events is controlled through the endocrine system. The hypothalamus produces GnRH, which stimulates the release of follicle-stimulating hormone (FSH). Increases in circulating FSH induce the growth of the dominant follicle until ovulation and also stimulate ovarian production of estrogen or estradiol (Figure 24.8). Estradiol causes the cow to

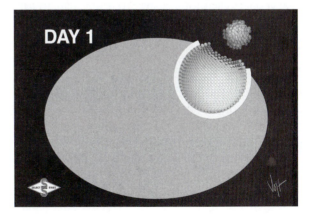

Figure 24.7 Ovulation occurs on day 1 of the estrous cycle. *(Courtesy of Select Sires)*

show behavioral signs of heat and also prepares the uterus for conception. In addition, estradiol suppresses FSH secretion and stimulates LH secretion by the anterior pituitary; LH starts the ovulation process and stimulates luteinization of the follicle

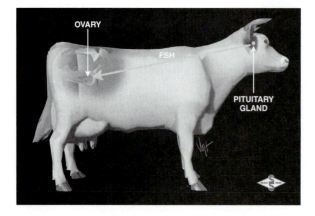

Figure 24.8 Follicle-stimulating hormone (FSH) from the anterior pituitary stimulates follicular development at the ovary. *(Courtesy of Select Sires)*

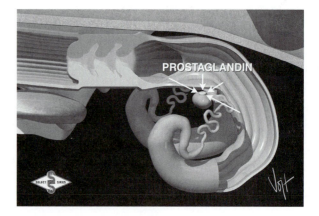

Figure 24.10 Prostaglandins produced by the uterus initiate luteal regression in nonpregnant cows. *(Courtesy of Select Sires)*

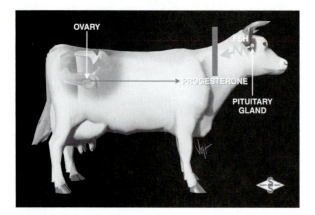

Figure 24.9 Progesterone blocks the hypothalamic production of GnRH, ultimately blocking FSH release from the pituitary. *(Courtesy of Select Sires)*

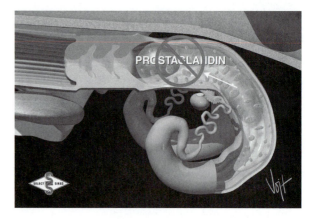

Figure 24.11 The developing embryo blocks the production of prostaglandins and prevents luteal regression. *(Courtesy of Select Sires)*

following ovulation. The corpus luteum (CL) is the luteinized tissue left after the follicle ruptures; it produces predominantly progesterone. Progesterone completes the uterine preparation for pregnancy and feeds back to the anterior pituitary, inhibiting both FSH and LH and preventing ovulation during pregnancy (Figure 24.9).

If an animal does not conceive, then the uterus produces prostaglandin $F_{2\alpha}$ ($PGF_{2\alpha}$) which causes luteal regression at about the sixteenth day of the estrous cycle (Figure 24.10). If the animal stays pregnant (Figure 24.11), then eventually placental production of progesterone takes over from luteal production to maintain pregnancy.

The estrous cycle (Figures 24.12) can be divided into several phases (Figure 24.13):

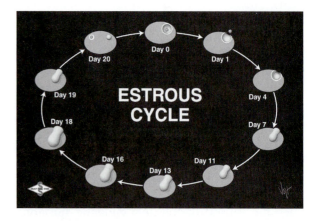

Figure 24.12 The bovine estrous cycle averages twenty-one days, with day 0 representing the period of standing heat, or estrus. *(Courtesy of Select Sires)*

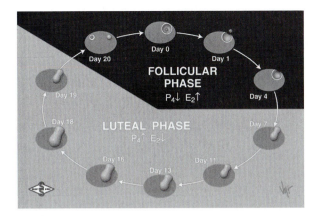

Figure 24.13 The estrous cycle is comprised of a luteal phase and a follicular phase, each of which has distinct hormonal profiles. *(Courtesy of Select Sires)*

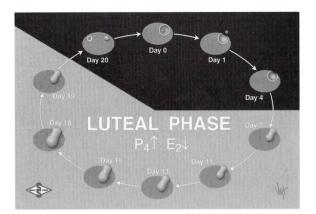

Figure 24.15 During the luteal phase of the estrous cycle, progesterone levels remain high, and estrogen levels are much lower. *(Courtesy of Select Sires)*

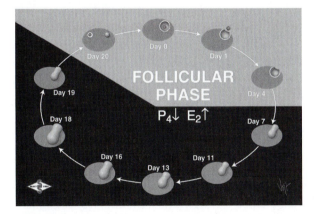

Figure 24.14 During the follicular phase of the estrous cycle, progesterone levels are low, and estrogen levels are high. *(Courtesy of Select Sires)*

1. **Proestrus (two days):** The first period of follicular growth and regression of the previous CL; it starts the follicular phase of the cycle (Figure 24.14).
2. **Estrus (twelve to eighteen hours):** The period when the female is receptive to the male.
3. **Metestrus (two to three days).** Ovulation occurs early in metestrus; the CL is formed from the site of ovulation (corpus hemorrhagicum). Bleeding is often noted in the uterine discharge during the second or third day.
4. **Diestrus (fifteen days).** Often called the luteal phase (Figure 24.15). The CL becomes fully developed by day 7 or 9. By day 16, the CL regresses if there is no pregnancy, and it imbeds in the ovary if the cow is pregnant. In the pregnant animal, the CL may eventually occupy two-thirds to three-fourths of the total ovary size.

FOLLICULAR WAVES

Four different categories of follicles are normally present on the ovary. The first category of follicles are simply the small (100 micrometers in diameter), dormant primordial follicles. During the next stage of development, the oocyte is surrounded by a single layer of granulosa cells and is called a primary follicle. Secondary follicles are identified by the development of numerous cell layers, including thecal cells, surrounding the oocyte. The development of the antral cavity is the defining characteristic of tertiary, or antral, follicles. At this point, the developing follicle averages 0.6 millimeters in diameter. The growth and development of follicles (folliculogenesis) is a complex process. It takes well over one month for the antral follicle to reach preovulatory size.

Groups of follicles are recruited by the action of FSH to begin the final stages of development. Cattle typically have two or three surges of FSH during each estrous cycle. Each surge of FSH stimulates a wave of follicular development (Figures 24.16 and 24.17). Cows or heifers experiencing three follicular waves have longer estrous cycles (extended luteal phases) than those animals experiencing two follicular waves. The high concentrations of progesterone that dominate the luteal phase of estrus prevent these follicles from fully maturing. Luteal regression removes the source of progesterone and allows the LH surge to stimulate the final development of a dominant follicle that grows to ovulation (Figure 24.18).

The selection of the final dominant follicle occurs on about day 16 or 17 of the estrous cycle. The follicle grows to about 20 mm in diameter at the time of ovulation (Figures 24.19 and 24.20). The

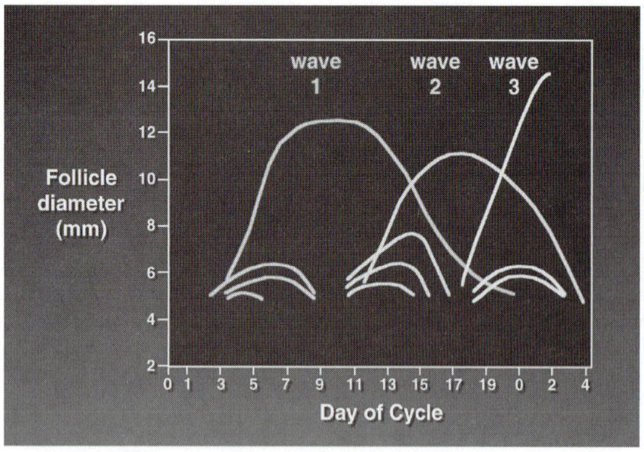

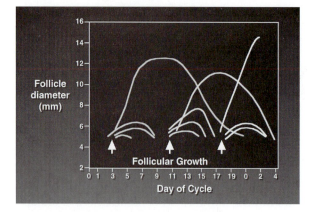

Figure 24.16   Dairy heifers average three waves of follicular development per estrous cycle, while cows more frequently have two-wave cycles. *(Courtesy of Select Sires)*

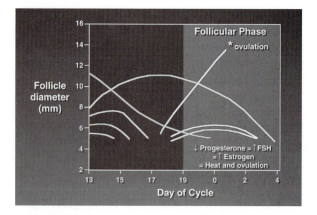

Figure 24.19   The regression of the corpus luteum reduces progesterone levels and allows for the continued development of the ovulatory follicle. *(Courtesy of Select Sires)*

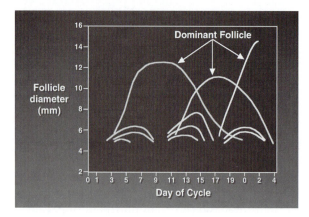

Figure 24.17   A new cohort of follicles is recruited for growth every seven to eight days in three-wave cycles. *(Courtesy of Select Sires)*

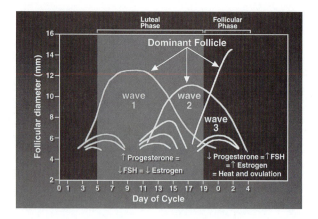

Figure 24.20   The complex interrelationships among hormones of the ovary and the pituitary regulate follicular growth and atresia. *(Courtesy of Select Sires)*

just ovulated. Typically, only a single dominant follicle emerges to ovulate. Twin ovulations do occur at a fairly high rate in dairy cattle, however, especially in high-producing cows. The development of twins is still a relatively rare occurrence.

FERTILIZATION

Fertilization is the union of the sperm and ovum. The sperm are deposited in the vagina during natural service (Figure 24.21) or in the uterus during artificial insemination (Figure 24.22) at the time of service and migrate up the oviducts (Figure 24.23). Under favorable conditions, they meet the egg and one of them fertilizes it in the upper part of the oviduct near the ovary (Figure 24.24).

In cows, fertilization is an all-or-none phenomenon because only one ovum is ordinarily involved. Thus, the breeder's problem is to synchronize ovulation and insemination and to ensure that large numbers of vigorous, fresh sperm are present in the fallopian tubes at the time of ovulation. Ovulation

Figure 24.18   Tpically, a single dominant follicle develops from the initial cohort stimulated during each follicular wave. *(Courtesy of Select Sires)*

anovulatory follicles simply regress and undergo atresia, or cell death. At ovulation, the follicle ruptures, releasing the ova, still surrounded by cumulus cells. This mass of cells is brought into the oviduct by action of the fimbriae adjacent to the ovary that

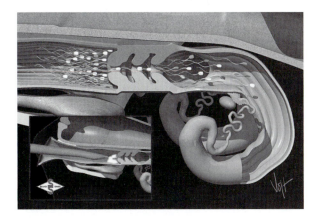

Figure 24.21 Only a small percentage of sperm initially ejaculated by the bull find their way through the cervix to the uterus. *(Courtesy of Select Sires)*

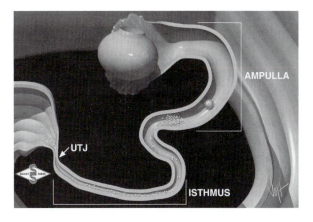

Figure 24.24 Within twelve hours of insemination, sperm are distributed throughout the oviduct and are ready to fertilize the ova. *(Courtesy of Select Sires)*

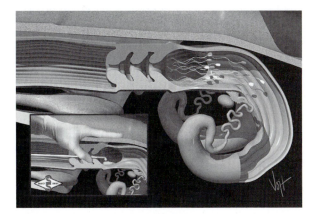

Figure 24.22 During artificial insemination, a smaller number of sperm is deposited directly in the uterus. *(Courtesy of Select Sires)*

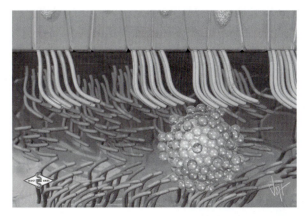

Figure 24.25 Cells in the oviduct are ciliated to facilitate ova transport and prevent oviductal implantation. *(Courtesy of Select Sires)*

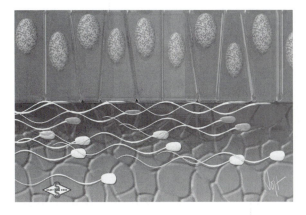

Figure 24.23 During transport, sperm adhere and release from the oviductal epithelium. *(Courtesy of Select Sires)*

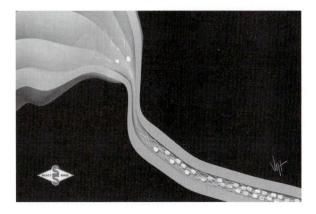

Figure 24.26 The first event in sperm capacitation, hyperactive motility, occurs primarily in the oviduct. *(Courtesy of Select Sires)*

occurs about twenty-eight hours after the initiation of estrus; the egg is optimally functional for only six to eight hours. At the time of ovulation, the follicle ruptures, allowing the ovum (egg) to move into the upper fallopian tube (Figure 24.25). Sperm have a lifespan of less than thirty hours and require about

six hours to migrate to the oviduct (Figure 24.26). Thus, for optimal results from insemination, cows should be bred between eight and twelve hours after the start of standing heat. Although inseminations occurring after this point may result in a pregnancy, the pregnancy rate is markedly lower.

REPRODUCTIVE FAILURE

The potential areas of concern for reproductive failure include errors in heat detection, errors in insemination technique, and lack of cow fertility (including fertilization failure, early embryo loss, and abortions). Inaccuracy in heat detection is easily the most important factor in reproductive failure. Cows are often bred at the wrong stage of the estrous cycle. Either they are bred too late following estrus, or they were not in estrus at all. Heat detection rates average less than 50 percent on most farms, and significant improvements are possible in this area.

Insemination techniques are often overlooked as a source of reproductive failure. Two-thirds of all inseminations fail to place the semen in the body of the uterus, and one-fourth of inseminations never reach the uterus at all. These numbers provide the justification for horn breeding. Improper semen-handling techniques are also common sources of reproductive failure.

Other causes of failures in cow fertility are more difficult to document. The source of failure of cow fertility that is the easiest to document is genital infections (vaginitis, metritis, or cervicitis). These conditions are often accompanied by cloudy discharge and may be cleared up by antibiotic infusions. Fertilization rates on most dairies average less than 50 percent, and significant improvement is also possible in this area.

Progesterone must be produced in adequate amounts by the developing corpus luteum for pregnancy to be maintained. Lack of adequate amounts has been theorized to be a problem in many repeat breeders. Between day 15 and day 17 of the estrous cycle, the embryo must produce a luteotrophic signal of sufficient strength to overcome the luteolytic effects of the prostaglandins normally produced by the uterus. Sufficient concentrations are also important because of the immunosuppressive nature of steroids.

Nutritional inadequacies can also prevent expression of estrus, result in a failure to conceive, or contribute to early embryo losses. Poor management during the transition period results in a longer period of negative energy balance and reduces fertility until a positive energy balance is achieved. Other factors influencing the fertility of cows include the environmental temperature one day after artificial insemination, the age of the cow (see Table 24.1), environmental temperature on the day of breeding, and fertility of the sire.

TABLE 24.1   The Influence of Age on Conception Rate

Location Number	Conception Rates (%)
Heifers	59.3
First	50.8
Second	48.9
Third	48.3
Fourth or more	38.2

Modified from F. C. Gwasdauskas, and J. A. Lineweaver. Should You Breed 1 Time or 2 Times a Day with AI? Advanced Animal Behavior 15 (September 1981): 8.

Ovarian cysts occur most often following failure of a follicle to ovulate. These cysts often become thick-walled (5 mm thick) and large; many are greater than 2 inches in diameter and easily palpable. Milk progesterone assays verify the high progesterone concentrations that distinguish luteal cysts from their follicular counterparts. Luteal cysts can be regressed with prostaglandins. Follicular cysts, as mentioned here, are characterized by low progesterone output. These cysts can also be distinguished by ultrasonography by their relatively thin wall (less than 3 millimeters). Behaviorally, cows with follicular cysts exhibit symptoms of nymphomania, showing behavioral signs of heat every few days. Mucus discharge from the reproductive tract is watery and abundant. They do not respond to prostaglandins; however, treatment with GnRH followed by prostaglandins ten days later is often successful. Alternatively, the use of an intravaginal progesterone-secreting device immediately stops the behavioral signs; cows ovulate three to five days after withdrawal of the device. Regardless of their origin, cysts should never be manually ruptured.

No reproductive program can overcome these problems; each problem must be solved at the source. Many reproductive failures are predestined through management practices that were initiated in the previous lactation. Overfeeding in late lactation can lead to overconditioned cows and increases susceptibility to a host of problems in the transition period. The reproductive difficulties encountered in these cows are not easily overcome. Many management deficiencies ultimately manifest themselves as reproductive failure; the challenge for the producer is to elucidate the source of the problem. Pregnancy rates (the number of cows that are eligible to be bred in any twenty-one day period that actually maintain a pregnancy) average about 15 percent on midwestern dairy farms. This number has the potential to be much higher with appropriate focus on factors affecting reproductive management and the appropriate use of reproductive management tools.

The recommendations for minimizing infertility in a dairy herd are straightforward and simple:

1. Breed at the proper time using highly fertile and properly preserved and handled spermatozoa.

2. Provide optimal transition cow management that allows cows to return rapidly to a positive energy balance.

3. Feed early lactation cows a diet that maximizes milk yield and minimizes body condition score loss.

4. Perform regular reproductive checks on heifers and postpartum cows to detect uterine infections and ovarian abnormalities.

SUMMARY

- Reproductive function in the dairy cow involves interplay among the ovary, uterus, hypothalamus, and pituitary gland.
- The main function of the ovary is to produce ova, estrogen, and progesterone.
- The cortex contains germinal epithelial cells, where oogenesis occurs.
- Ova develop in Graafian follicles.
- If an animal stays pregnant, the placental progesterone eventually takes over from luteal progesterone production.
- Progesterone production by luteal tissue suppresses FSH production by the anterior pituitary.
- The estrous cycle can be divided into four stages: proestrus, estrus, metestrus, and diestrus.
- Cows are in standing heat for about eight hours and should be bred between eight to twelve hours after the start of standing heat.
- The dominant follicle is also called the ovulatory follicle.
- Causes of reproductive failure includes errors in heat detection, errors in insemination technique, lack of cow fertility, inaccuracy in heat detection, improper semen-handling techniques, genital infections, nutritional inadequacies, poor management, environmental temperature one day after AI, age of the cow, environmental temperature on the day of breeding, overfeeding, and fertility of the sire.
- The embryo must produce a luteotropic signal of sufficient strength to overcome the luteolytic effects of the prostaglandins normally produced by the uterus.

QUESTIONS

1. What is the second most common reason for culling cows from dairy herds, accounting for over one-fourth of cows culled?
2. List the two regions of the ovary.
3. Define *ovulation*.
4. What is the term for the period of female receptivity to the male?
5. What does the hypothalamus produce?
6. What causes luteal regression? On which day of the estrous cycle?
7. How many days are in the estrous cycle?
8. In what phase of the estrous cycle is estrogen high? When is it low?
9. Define *fertilization*.
10. How long is the egg optimally functional? How long are sperm cells optimally functional?
11. Where does fertilization normally occur?
12. What are the recommendations for minimizing infertility in a dairy herd?

ADDITIONAL RESOURCES

Articles

Brackett, B. G., Y. K. Oh, J. F. Evans, and W. J. Donawick. "Fertilization and Early Development of Cow Ova." *Biological Reproduction* 23 (January 1980): 189.

Britt, J. H., R. G. Scott, J. D. Armstrong, and M. D. Whitacre. "Determinants of Estrous Behavior in Lactating Holstein Cows." *Journal of Dairy Science* 69 (August 1986): 2195.

Dransfield, M. B. G., R. L. Nebel, R. E. Pearson, and L. D. Warnick. "Timing of Insemination for Dairy Cows Identified in Estrus by a Radiotelemetric Estrus Detection System." *Journal of Dairy Science* 81 (July 1998): 1874.

Gwazdauskas, F. C., J. A. Lineweaver, and M. L. McGilliard. "Environmental and Management Factors Affecting Estrous Activity in Dairy Cattle." *Journal of Dairy Science* 66 (July 1983): 1510.

Nebel, R. L., W. L. Walker, M. L. McGilliard, C. H. Allen, and G. S. Heckman. "Timing of Artificial Insemination of Dairy Cows: Fixed Time Once Daily Versus Morning and Afternoon." *Journal of Dairy Science* 77 (October 1994): 3185.

Senger, P. L. "The Estrus Detection Problem: New Concepts, Technologies, and Possibilities." *Journal of Dairy Science* 77 (September 1994): 2745.

Walker, W. L., R. L. Nebel, and M. L. McGilliard. "Time of Ovulation Relative to Mounting Activity in Dairy Cattle." *Journal of Dairy Science* 79 (September 1996): 1555.

Internet

Managing Reproductive Disorders in Dairy Cows: http://www.wisc.edu/dysci/uwex/nutritn/pubs/NutrAndMgt/MngReproDisorders.pdf.

Physiology and Endocrinology of the Estrous Cycle: http://www.wvu.edu/~exten/infores/pubs/livepoul/dirm2.pdf.

When Should Dairy Cows Be lnseminsted?: http://www.cahe.nmsu.edu/pubs/_b/b-117.html.

25

Reproductive Physiology and Management of the Bull

OBJECTIVES

- To describe the reproductive physiology and reproductive behavior of the bull.
- To outline the process of collecting bull semen.
- To outline the management and proper use of natural-service herd sires.

Although the cow is the primary focus of reproductive management on most dairy farms, the bull is the second part of the reproductive equation. The successful union of sperm and egg requires both a functional egg and a functional sperm. The production of this functional sperm depends on a full understanding of the reproductive physiology and reproductive behavior of the bull.

REPRODUCTIVE ANATOMY AND PHYSIOLOGY

Understanding reproductive function first requires an understanding of male reproductive anatomy. Reproductive function in the dairy bull involves interplay among the testes, hypothalamus, and pituitary gland, primarily through reproductive hormones. If any of these tissues fail to communicate effectively with the others, infertility results.

The testes are about 6 by 3 by 3 inches in size and oval in shape in the mature bull. They weigh approximately ¾ pound and are fairly equal in size. Within the testes are masses of convoluted seminiferous tubules where sperm develop and mature, a process called spermatogenesis. The interstitial tissue surrounding these tubules contains the Leydig cells that produce and secrete testosterone. Spermatogenesis takes about fifty-four days in the bull; during this time, the number of chromosomes is reduced to half the normal amount.

The germs cells that develop from the basement membrane of the seminiferous tubules become either Sertoli (nurse) cells or spermatazoa. The Sertoli cells produce tubule fluids containing many compounds that support or regulate spermatogenesis, including pyruvate, lactate, inhibin, estrogens, and proteins. The spermatogonia that ultimately develop into mature spermatozoa depend on functional Sertoli cells for their development.

Spermatogenesis is stimulated by the function of follicle-stimulating hormone (FSH). Pituitary-derived FSH stimulates pyruvate and lactate secretion from Sertoli cells, thus providing an energy source for the developing sperm cells. FSH also causes the Sertoli cells to maintain a supportive endocrine environment of spermatogenesis via the secretion of androgen-binding proteins and the production of enzymes that convert testosterone to estrogens.

The other reproductive hormone secreted by the anterior pituitary, luteinizing hormone (LH), stimulates testosterone production by the Leydig cells. Testosterone ultimately controls libido, imparts secondary sex characteristics to the bull, and stimulates accessory gland function.

The bull has a fibro-elastic penis with a sigmoid flexure. During arousal, the flexure is straightened and the penis extends from the sheath. There is little increase in actual length of the penis during this process.

THE FUNCTIONAL BULL

Puberty is often defined as that point when bulls begin to exhibit libido, but bulls often mount well before puberty. The onset of puberty is technically defined as the onset of fertility; the ability to produce functional spermatazoa is usually apparent at about nine months of age. As in the female, puberty is both age- and weight-dependent and is influenced by the environment, especially the nutritional status of the bull. Although puberty signals the onset of fertility, full sexual maturity is not achieved until two to three years of age.

Copulatory behaviors in the bull are relatively short-lived activities. The bull typically identifies the cow in estrus by smell. Pheromones are chemicals secreted by the cow that typically stimulate the bull through olfactory mechanisms. The bull locates the estrus female, sniffs and licks the vulva area to confirm her identity, and rests his chin on her rump to assess her willingness to stand for mounting. If the cow responds to the chin-resting activities, the bull mounts, copulates, and ejaculates with a characteristic thrust within a matter of seconds. Failure to thrust indicates a failure to ejaculate. The bull then dismounts rapidly, although he may service the cow several times in rapid succession.

SEMEN COLLECTION

Several methods are available for collecting semen, including aspiration of ejaculate from the vaginal area of a recently serviced cow, rectal massage of the ampullae, electro-ejaculation techniques, and collection of the semen directly into an artificial vagina. The only truly satisfactory method of routinely collecting semen is with an artificial vagina. Bulls in service are trained to mount dummies or steers for collection.

The artificial vagina (AV) is a rigid cylindrical device with a latex rubber liner (Figure 25.1). The area between the rigid cylinder and the latex liner becomes a water jacket and is filled with water that is between 108 and 115°F. The inner surface of the latex is coated with obstetrical jelly. A latex cone and collecting tube are connected to the distal end of the apparatus to collect the ejaculate. When the bull mounts, either on a teaser animal or on a mounting dummy, the penis is deflected into the AV prior to ejaculation (Figure 25.2). The bull then thrusts and ejaculates into the AV, and the semen drains immediately into the collecting tube (Figure 25.3). The semen is sensitive to both light and temperature; precautions must be taken to protect the ejaculate immediately for these potential sources of sperm loss. A typical ejaculate volume is about 6 milliliters, although it can range from 2 to 12 milliliters. The total sperm output in this ejaculate is about 7.5 bil-

lion sperm, and the concentration at this point is about 1.2 billion sperm per milliliter.

Semen is evaluated by several standards. The most easily evaluated aspect of quality is the volume of the ejaculate. The color of the ejaculate can indicate the presence of contamination or infection, and normal ejaculates should be creamy yellow to milky white. Sperm motility and morphology are noted, the percentage of dead sperm is determined, and the concentration determined.

After collection and evaluation, semen can be used fresh or frozen for later use. The vast majority of collected semen is frozen. Diluent is added to the fresh semen as a cryoprotectant, but it also serves several other important functions. It includes a sugar as an energy source for spermatazoa, a buffer to protect sperm from pH fluctuations, and antibiotics to protect against bacterial contamination. Diluent also increases greatly the number of doses of semen available to inseminate eligible cows because the bull produces far more spermatazoa per ejaculate than is necessary for fertilization. Most diluents

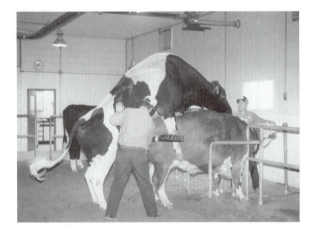

Figure 25.2   Bulls in active service are typically collected two or three times per week, with two or three ejaculates per collection day. *(Courtesy of Genex)*

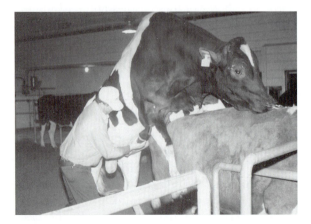

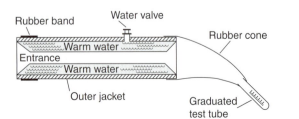

Figure 25.1   A schematic of an artificial vagina for semen collection. *(Courtesy of M. E. Ensminger)*

Figure 25.3   Semen is collected in an artificial vagina. *(Courtesy of Genex)*

are egg-yolk citrate or heated skim milk with egg yolk, fructose, and glycerol. The glycerol acts as a cryoprotectant, and the fructose serves as an energy source. The diluted semen is then placed in plastic straws with a 0.25-milliliter capacity, and the straws are carefully frozen and stored in liquid nitrogen at a temperature of $-320°F$ until they are used.

MANAGING THE NATURAL SERVICE SIRE

Many dairies rely on artificial insemination entirely and do not keep mature bulls. However, over 25 percent of dairy farms maintain one or more mature bulls. There is no debate that the bulls currently available through bull studs are vastly superior to those currently in use as natural-service sires. In addition, most economic analyses that compare costs of artificial insemination (direct costs of semen, equipment, and labor) and natural service (maintenance costs associated with housing and handling a bull) show an advantage to the use of artificial insemination. It is estimated that the costs associated with the use of herd bulls, on a per-breeding basis, are more than the cost of high-quality semen through a bull stud. When the economic benefits associated with the production of daughters with superior genetic merit are included, the efficient use of artificial insemination seems to be an easy choice.

So why are so many herd bulls still in service? The primary reason for using a herd bull is to improve pregnancy rate. The average pregnancy rate for dairy herds is under 15 percent primarily because heat detection rates are quite poor. Natural service does not provide improvements in conception rate, and although good data is scarce, conception rates may actually be lower. Therefore, the improvements are based on the perception that a highly motivated bull will find more cows in heat than an unmotivated farm employee. The impact of reproductive efficiency on cow profitability and cow culling rates and ultimately on herd profitability cannot be overstated. Solving the obstacles to improved pregnancy rates can be a challenge on many farms. In most cases, the problems are strongly related to a reluctance of management or labor to adopt the practices necessary for success. In these cases, the use of natural-service sires may be the only practical solution to maintaining herd profitability.

It is important, therefore, to understand the management and proper use of natural-service herd sires. They present a danger to employees, cows, and facilities. They present a health risk to the entire breeding herd if proper health management protocols are not in place. They can contribute to, rather than solve, reproductive problems in the herd if proper selection and use practices are not observed closely. Breeding soundness exams should

be administered in addition to the usual diagnostic testing and quarantine practices normally employed for new herd additions.

Age and Intensity of Use

Young bulls (less than two years of age) should not be used as heavily as older bulls. They can functionally service only two to four estrus cows per week, while fully mature bulls can service up to twelve estrus cows per week for short periods. Therefore, young bulls should not be grouped with more than fifteen cows or heifers. Mature bulls can be exposed to up to twenty-five cows in a group. However, young bulls are widely preferred in the dairy industry because they are less dangerous than older bulls, they exhibit fewer destructive dominance behaviors when maintained in groups, and they are more likely to be both reproductively and functionally sound than their older counterparts. Bulls less than 1,000 pounds in weight are generally not physically large enough to service mature cows and should be restricted to use with heifers.

Several options are available for using natural service in dairy herds. Each method has significant drawbacks that emphasize the problems inherent in using natural service. The first option is simply to house bulls with the breeding age females. The number of bulls included with each group depends on the number of females to be serviced. There are several drawbacks with this approach. First, in situations where several bulls are housed with a single group of females, the dominant bull may prevent breeding activity by the subordinate bulls to the detriment of the pregnancy rate in the group of females. In addition, bulls housed in this manner can be extraordinarily destructive to fences, bunks, parlors, and other facilities created for the more docile females (Figure 25.4). The second approach solves

Figure 25.4 Herd bulls are a safety hazard for people, and they are also hard on facilities. *(Courtesy of Mark Kirkpatrick)*

these problems by hand-servicing females. Bulls are permanently housed in facilities that are built to handle their aggressive nature. The females that are observed in heat are then brought to the bull facility for breeding. This option, while eliminating the drawbacks associated with housing bulls and cows together, also eliminates the primary reason for maintaining a herd sire: to improve heat detection. If heat detection is adequate enough to allow successful use of hand-breeding cows to natural-service bulls, then artificial insemination should be utilized.

All mature dairy bulls should be considered dangerous and handled as such. Safety pens and stalls are essential, and the bull should always be handled with a nose ring. A satisfactory (but inexpensive) shelter should be provided; preferably in a separate barn, a little distance away from other animals. The stall should be at least 12 square feet. It should open into a strongly fenced paddock to which the bull has access or is turned daily. For safety, the feed and watering arrangement should be such that the bull can be fed and watered without going into the pen.

Venereal Diseases

Natural-service bulls can transmit venereal diseases to the breeding herd. The most important reproductive diseases in this respect are campylobacteriosis and trichomoniasis. Dairy breeders ensure the health of their bulls through regular testing and appropriate vaccination protocols, as well as through the addition of antibiotics to semen that are effective against campylobacteriosis.

Campylobacteriosis is caused by *Campylobacter fetus*. The organism infects the bull's penis and prepuce, but clinical signs are not visible in the bull. The organism is transmitted to the cow or heifer at the time of breeding. The fertilized ova is destroyed by the organism, and subsequent breedings are infertile. Spontaneous recovery of the female commonly occurs at about four months after infection, and there is some level of immunity to reinfection. Some abortions may occur, especially in the second trimester. However, most cows and heifers are culled prior to recovery. This disease in essence is responsible for eliminating infected animals from the herd. Bulls do not recover or develop immunity because they are never truly infected. They are simply carriers for the organism. Testing and identification of the causative organism is therefore best accomplished through infected females rather than through the bull. The best preventive practice, vaccination, is also most effective in females, although it is somewhat effective in bulls.

Trichomoniasis is caused by the protozoan organism, *Trichomonas foetus*. It also presents as infer-

Figure 25.5 Although this bull appears healthy, he can spread trichomoniasis to cows during natural service. *(Courtesy of University of Illinois)*

tility, with occasional abortions, and clinically it may be indistinguishable from campylobacteriosis. The most common carriers are older bulls (Figure 25.5), and again the infection is transmitted to the female at the time of breeding. Most abortions occur at about three to four months of gestation, and there may be some apparent uterine discharge or palpable pus in the uterus. Occasionally, abortions occur in later pregnancy. With this organism, the most effective animal for testing and identification of the infective agent is the bull. Females may be vaccinated, and infected bulls can be effectively treated.

Factors Affecting Bull Fertility

Heat stress is as detrimental to male fertility as it is to female fertility. However, the effects of heat stress on bull fertility do not appear for thirty to sixty days because it affects the development of immature spermatazoa early during spermatogenesis. Likewise, it takes at least one month for the detrimental effects of heat stress to disappear.

Overconditioned bulls are not only more susceptible to heat stress, they are also more likely to have poor semen quality under ideal conditions. Overfeeding young bulls (for example, allowing access to the lactating cow ration) can adversely affect both semen quality and libido. In addition, feeding the lactating cow ration to bulls also provides excessive calcium, leading to bone lesions in the spine and hip areas and eventually to lameness.

Housing can also affect the reproductive function of bulls; the quality of footing is an obvious issue in many facilities. Slick, unstable, or muddy conditions dramatically decreases mounting behavior in bulls. Mounting behavior in general is enhanced with dirt lots in comparison to behavior on concrete flooring. Any factors that might inhibit normal female estrus behaviors also inhibit male copulatory behaviors. In addition, bulls are less active within a totally enclosed facility. Housing conditions that are adequate for cow or heifer comfort may not consider the needs of the bull. Free stalls are generally too small for bulls, and cleanliness is important for reduction of disease transmission.

Reproductive Failure

There are several causes of bull infertility, including lack of libido, impotence, or poor-quality semen. Generally, dairy bulls are more aggressive and exhibit greater libido than do beef bulls. Low libido is still a common problem in dairy bulls, however, and the exact cause of the problem can be challenging to pinpoint. Libido declines with age, although young bulls can have libido problems due to bullying by older bulls or even by older cows in the herd. Noise or other distractions, a change in environment or of the person caring for or handling the bull, or poor footing can all affect the bull's desire to mount. Other common problems that affect libido include boredom, lack of exercise or overconditioning, overuse, and disease or injury. Libido is moderately heritable, and selection for this trait is productive. Treatment of poor libido with gonadotropins or testosterone is not only ineffective, but in some cases can also actually be counterproductive.

Impotence is defined as a lack of fertility despite good service behavior. The libido may be excellent and the semen may be highly fertile, but the cow fails to conceive. There are several potential causes for this problem. Congenital defects or disease conditions can interfere physically with the ability to protrude the penis or achieve erection. The exact cause should be determined and corrected if possible.

Failure of fertilization can also be due to poor-quality semen. Poor-quality semen can be caused by poor nutrition, stress associated with a new environment or a change in handling procedures, high environmental temperatures, congenital defects in the reproductive system leading to abnormal spermatogenesis, or a venereal disease. Again, the exact cause should be pinpointed to determine whether treatment options are available or warranted.

SUMMARY

- FSH causes Sertoli cells to maintain a supportive endocrine environment of spermatogenesis via the secretion of androgen-binding proteins and the production of enzymes that convert testosterone to estrogens.
- The only truly satisfactory method of routinely collecting semen is with an artificial vagina.
- All mature dairy bulls should be considered dangerous and handled as such.
- Failure of fertilization can be due to poor nutrition, stress associated with a new environment or a change in handling procedures, high environmental temperatures, congenital defects in the reproductive system leading to abnormal spermatogenesis, or a venereal disease.

QUESTIONS

1. How long does spermatogenesis take?
2. Which cells regulate spermatogenesis?
3. What are three functions of testosterone?
4. When is full sexual maturity achieved in the bull?
5. What are four methods available for collecting semen?
6. What is the primary reason for using a herd bull?
7. What are two venereal diseases that natural-service bulls can transmit to the breeding herd?
8. Name three factors affecting bull fertility.
9. How many cows in a group can mature bulls be exposed to at a given time?
10. Define *impotence*.

ADDITIONAL RESOURCES

Book

Monke, D. R. "Bull Management: Artificial Insemination Centres." In *Encyclopedia of Dairy Sciences,* edited by H. Roginski, J. W.Fuquay, and P. F. Fox. London: Academic Press, 2003.

Articles

Andersson, M., J. Taponen, E. Koskinen, and M. Dahlbom. "Effect of Insemination with Doses of 2 or 15 Million Frozen-Thawed Spermatozoa and Semen Deposition Site on Pregnancy Rate in Dairy Cows." *Theriogenology* 61 (May 2004): 1583–1588.

Howard, T. H., B. Bean, R. Hillman, and D. R. Monke. "Surveillance for Persistent Bovine Viral Diarrhea Virus Infection in Four Artificial Insemination Centers." *Journal of the American Veterinary Medical Association* 196 (December 1990): 1951–1955.

Hueston, W. D., D. R. Monke, and R. J. Milburn. "Scrotal Circumference Measurements on Young Holstein Bulls." *Journal of the American Veterinary Medical Association* 192 (March 1988): 766–768.

Monke, D. R. "Noninfectivity of Semen from Bulls Infected with Bovine Leukosis Virus." *Journal of the American Veterinary Medical Association* 188 (August 1986): 823–826.

Monke, D. R. "On-Farm Management of Bulls Intended for AI Service." *Proceedings of the Technical Conference on Artificial Insemination and Reproduction* 12 (XXXXX 1988): 68–79.

van Os, J. L., M. J. de Vries, N. H. den Daas, and L. M. Kaal Lansbergen. "Long-Term Trends in Sperm Counts of Dairy Bulls." *Journal of Andrology* 18 (November–December 1997): 725–731.

Internet

Hidden Expenses and Problems with Natural Service Bulls: http://www.wisc.edu/dysci/uwex/rep_phys/pubs/bulls.pdf.

Managing Herd Bulls on Large Dairies: http://www.wdmc.org/2003/Managing%20Herd%20Bulls%20on%20Large%20Dairies.pdf.

Semen Collection from Bulls: http://www.vivo.colostate.edu/hbooks/pathphys/reprod/semeneval/bull.html.

The Modern Dairy Bull: http://www.selectsires.com/news_lifecycle.html.

26

Heat Detection and Estrus Synchronization

OBJECTIVES

- To understand the purpose of, methods for, and devices for heat detection.
- To apply synchronization to estrus.
- To describe hormonal control of estrus and synchronized ovulation regimens.

Obviously, the key component in any reproductive management program is efficiency of heat detection. The regression of the corpus luteum and the development of the dominant follicle signal the hormonal changes that control the behavioral signs of heat. The follicle produces predominantly estradiol; estrogen brings the cow into heat. The keys to successful heat detection are proper animal identification, good records with all observations permanently recorded, and knowledge and understanding of the signs of heat.

Many observable behavioral changes occur in the eight hours prior to standing heat. Cows vocalize or bawl more than normal. The activity level of animals in estrus increases dramatically. Cows coming into estrus attempt to mount or ride other animals. They spend increasing amounts of time sniffing other animals, although they usually will not stand to be ridden at this time (Figure 26.1). The vulva is slightly swollen, moist, and red, and there may be a clear mucus discharge from the vulva. Feed consumption is typically low during this period, and milk production also decreases.

During the period of standing heat, cows stand to be mounted by other cows (Figure 26.2). They attempt to mount other animals, often mounting from the front. They vocalize even more frequently than before, they are nervous and excitable, and their activity level peaks during this period. Cows in

Figure 26.1 Heifers entering estrus often exhibit sniffing behaviors, although they will not stand to be mounted. *(Courtesy of Iowa State University)*

Figure 26.2 Mounting activity is the best behavioral indicator of estrus. *(Courtesy of John Smith)*

179

Figure 26.3   Cows exhibit more intense estrus behaviors when they are housed on dirt rather than on concrete. *(Courtesy of Iowa State University)*

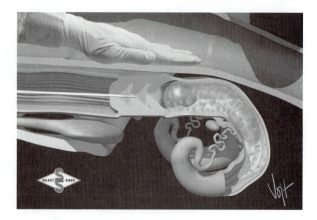

Figure 26.4   Inseminating a pregnant cow breaches the cervical plug, greatly increasing the risk of abortion. *(Courtesy of Select Sires)*

estrus have reduced feed intake and decreased milk production. Their vulva is noticeably swollen, often with an abundant clear mucus discharge.

As cows come out of estrus, they will no longer stand to be ridden; however, they continue to attempt to mount other cows. Feed consumption and milk production return to normal during this period. The mucus produced in the reproductive tract becomes sticky and viscous. On the second or third day after standing heat (metestrus), a slight bloody discharge may occur. This discharge has no relationship to conception, lack of conception, or even to injury, but rather is due to a normal sloughing of some of the uterine lining.

Mounts typically last 4 to 7 seconds, with cows standing to be mounted anywhere from four to fifty times during standing heat (Figure 26.3). Therefore, a typical cow is mounted from a total of less than 1 minute up to 6 minutes during the time she is in heat.

The percentage of heifers that stand to be mounted that are actually in estrus is greater than 90 percent, while mounting behavior is exhibited by less than 80 percent. Thus, standing to be mounted is the cardinal sign of a cow in heat, and breeding practices based on this behavior are far more successful than those based on secondary signs of heat.

Accurate heat detection is a challenge on most dairy farms. Although the true incidence of silent heats (those cows exhibiting no behavioral signs) is very low, on the average farm over 40 percent of cows in estrus are missed. In addition, 15 to 20 percent of all cattle that are bred were not really in heat. The costs associated with these two problems are enormous. Some surveys indicate that less than

half of dairy producers consistently devote specific times twice daily to heat detection, and close to one-third of producers fail to devote any committed time periods to observing heat, instead relying completely on casual observation throughout the work day. These figures indicate that management is often more of a problem in herds experiencing breeding problems than are cow-related factors.

Why is a casual approach to heat detection such a problem? First, up to 70 percent of all estrus-related activity occurs between 6 PM and 6 AM, and up to 90 percent of all activity occurs between 6 PM and 12 noon. Because 25 to 45 percent of cattle show heat for less than eight hours, heat detection programs that operate only during daylight hours are inefficient. To complicate matters further, false heats occur in up to 5 to 10 percent of pregnant heifers and cows, even in late gestation. If these animals have been confirmed pregnant, they should not be bred unless they have been confirmed as not pregnant (Figure 26.4).

Even for dairy producers willing to devote specific times to heat detection, there can be significant challenges to meeting the time commitments involved (Figure 26.5). For example, many farms have multiple groups of animals requiring observation. While two thirty-minute observation periods (at the appropriate times) can detect 80 percent of all heats, with three periods detecting 90 percent of heats, the time required must be multiplied by the number of groups requiring observation. If a producer has three groups of cows containing breeding-eligible animals, this three- to five-hour investment in observation is difficult to accomplish without utilizing some type of heat detection aid.

Figure 26.5   Cows in the follicular phase of the estrous cycle often form sexually active groups with increased mounting activity. *(Courtesy of Iowa State University)*

HEAT DETECTION METHODS AND DEVICES

Under ordinary farm conditions, an estimated 25 to 50 percent of heat periods are missed on dairy farms. On the average, a missed heat period prolongs the calving interval by thirty to forty days and means a loss of more than $60 in a dairy herd. For these reasons, producers are interested in heat detection aids. Heat expectancy calenders allow producers to focus attention on the animals most likely to exhibit estrous behavior. Good recordkeeping is truly the key to a successful reproductive program and can greatly facilitate the success of heat detection programs based on visual observation. However, visual observation is labor intensive and difficult to manage. On many operations, three and sometimes four separate groups contain animals eligible for breeding, with each group requiring several periods of observation daily for optimal efficiency of heat detection. Several recent technological advances have the potential to reduce labor inputs required for this critical management task. These aids can be divided into several categories: those that detect the primary sign of heat (standing to be mounted), those that detect secondary signs of heat (increased activity, changes in mucous secretions, etc.), and those that measure the hormonal changes associated with estrus.

Aids Measuring Mounting Behavior

Tailhead Markers (Paint Sticks and Paints)

There are several options in this category, but all of them are similar in effectiveness. A coat of wax crayon (Figure 26.6) or latex paint is applied over the tailhead region of the cow, often on a daily basis. Mounting activity smears or removes this application. In some systems, two coats of paint are applied, using dif-

Figure 26.6   Marking crayons are an inexpensive method for marking tailheads to detect mounting activity. *(Courtesy of Leo Timms)*

ferent colors for each coat, for greater ability to detect the amount of mounting that occurred. Although costs of such markers are low, these systems are labor intensive. They help draw attention to cows that may be active, but they cannot replace visual observation. Accuracy of these systems is greatly affected by rainy weather and the amount of oil in the haircoat.

Dye-Filled Tailhead Patches

One example of a tailhead patch is a KaMaR patch. This type of heat-mount detector is typically a 2- by 4-inch fabric base to which is attached a white plastic capsule (Figure 26.7). Inside the capsule is a small plastic tube containing dye. The tube is constructed so the dye is released following direct, sustained pressure. When enough dye is released from the tube (after about 4 to 5 seconds of pressure), it spreads over the inner lining of the capsule, typically causing it to turn red, or less often, blue (Figure 26.8). The pressure from the brisket of a mounting animal causes the dye to be released; if the cow does not stand for the mounting animal, there will not be enough pressure

Figure 26.7   Heat mount detectors are glued to the tailhead region of the cow and are activated by pressure. *(Courtesy of M. E. Ensminger)*

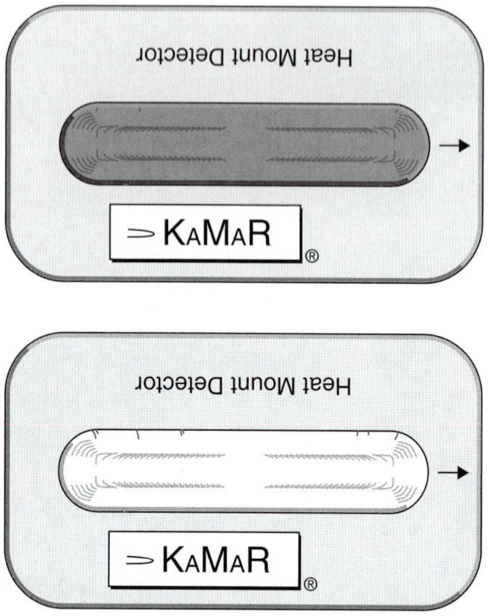

Figure 26.8   Mounting activity ruptures the internal dye capsule in a heat mount detector, changing the detector color to bright red. *(Courtesy of Iowa State University)*

Figure 26.9   Electronic mount detectors are secured to the cow's tailhead with a large fabric patch. *(Courtesy of Mark Kirkpatrick)*

to release the dye. These are not very useful as the sole means of heat detection, but they can be a useful adjunct to visual detection.

Electronic Pressure-Sensitive Sensors

The most recent technology introduced to the market are pressure-sensitive transmitters. One example is HeatWatch. Producers using this system glue the transmitter (in a fabric patch) to the tailhead region of the cow (Figure 26.9). Mounting behavior activates the transmitter and either signals estrus at the patch or sends a signal to a receiver, which is linked to a computer. The computer then stores the cow identification and the time of each mount; this information is available to the producer through a software interface. This technology has the advantage of recording the definitive sign of estrus: standing heat. Research utilizing this system has determined that ovulation typically occurs about eleven hours after the first mount, and optimal pregnancy rates are achieved by breeding from four to sixteen hours after the first mount. This system can simultaneously and continuously monitor all animals eligible for breeding on a dairy operation, effectively eliminating the need for and costs of vi-

sual observation. Disadvantages of the system include a high initial capital investment and a relatively high rate of transmitter loss, which requires management time in locating and reapplying lost fabric patches and transmitters.

Teaser Animals

These animals are typically either testosterone-treated heifers or surgically modified teaser bulls, often called Gomer bulls. One of these animals can accurately heat detect approximately eighty cows. In large pastures and in large herds, it is best to have two such animals. The dangers of having such inherently aggressive animals present cannot be minimized. In addition to the risk to personal safety, these animals are expensive to feed and manage, and they are destructive to fencing, buildings, and equipment. There are less expensive and more accurate options currently available.

Chin-Ball Marker

This device was developed in New Zealand. It is similar to a ball point pen attached to a halter under the chin of a teaser animal. Prior to mounting, bulls commonly place their head over the shoulders, back, and rump of the cow. This action causes a smearing of the colored ink over the hindquarters of the cow. One filling of the stainless steel container is sufficient to mark fifteen to twenty-five cows. This system relies on the presence of a teaser animal; thus, it has all the inherent disadvantages of that system. It cannot replace visual detection because cows not in estrus are also marked. It merely draws the attention of the producer to those cows that need closer observation.

Aids Measuring Secondary Signs of Heat

Pedometers

Pedometers are electronic devices strapped to a rear leg of cows eligible for breeding. A mercury switch effectively measures the activity level of the animal, and a microchip within the device compares this activity to baseline levels previously recorded and stored in memory. Cows entering estrus are more active than cows in other phases of the estrous cycle, and this technology can be useful for animals in certain management schemes. This system has the advantage of providing information regarding the time of onset of increased activity, increasing the accuracy of the timing of insemination. Increases in activity can also be induced by changes in management routine or by changes in weather, and so interpretation of the data provided by the system is somewhat more complicated than it is with other electronic systems.

Vaginal Probes

Vaginal probes measure electrical impedence of vaginal mucus. Decreases in electrical resistance correspond with increases in estrogens occurring at estrus, with the lowest readings corresponding with the luteinizing hormone peak. The hormonal changes increase ion flux across the vaginal epithelium, which reduces electrical resistance as measured by the probe. This technology is more accurate than visual observation for determining the best time to breed cows; however, it requires individual response curves for each animal. Therefore, each animal must be probed daily to establish baseline values. The potential for pathogen introduction into the reproductive tract is also high if strict sanitation guidelines are not followed. Industrywide acceptance of this technology is not likely at this stage of development; however, research into vaginal implant devices based on the same principles provides hope for a future application.

Aids Directly Measuring Hormones

Milk Progesteron Kits

Milk progesterone kits provide qualitative measurements of the amount of progesterone present in milk samples, labeling them as either "high" or "low." They can be used in several ways. First, they can be used every other day to identify the cows with low progesterone. Because progesterone levels are low for several days around the time of estrus, they cannot be used to confirm directly that a cow is in estrus. However, the alternate use for these kits is simply to confirm cows that have shown secondary signs of heat. A reading of "high" is then used as an indication not to breed a cow that was suspected of being in estrus. These kits are fairly expensive and labor intensive.

Summary of the Use of Heat Detection Aids

Properly used, all of these aids improve heat detection. Their usefulness for any individual operation depends on their cost, the value put on labor for visual observation and to what extent this cost will be reduced, their projected accuracy for this operation (which depends on the facilities available and management expertise), semen costs, and costs associated with days open calculated for the operation. They do not solve all the problems associated with reproductive failure; for example, cows unable to conceive or maintain a pregnancy due to poor nutritional status remain open until the nutritional problems are resolved. Breeding, health, and lifetime records are as essential as milk production records. Among other things, they are the only way to diagnose and reduce infertility. About 5 percent of the dairy cows in the United States become infertile each year.

HORMONAL CONTROL OF ESTRUS

Controlled estrus greatly facilitates both artificial insemination and embryo transfer. Synchronization of estrus or ovulation can be accomplished by injection or implant.

Prostaglandins

These are hormonelike substances that play a key role in regulating cellular metabolism. The name *prostaglandin*, which is a misnomer, was given to these substances because initially they were believed to have originated in the male's prostate gland. These highly potent substances have been called local hormones or tissue hormones because they do their work in the immediate area in which they are produced, as distinguished from circulating hormones that may target distant tissues. Prostaglandin F,2-alpha initiates regression of the corpus luteum. The time interval between injection with a single prostaglandin dosage (25 mg intramuscularly) and the onset of estrus and ovulation is very short (in cows, about forty-eight to seventy-two hours after injection) and reasonably predictable. Following single-injection protocols, cows must be bred following observed heat.

Because prostaglandins are effective only in cows with a functional corpus luteum, often a two-shot regimen is provided, with the second shot given eleven to fourteen days after the first injection. The timing of these shots ensures that a vast majority of the cows treated with these shots come into heat following the second shot. Visual observation is still required despite the synchrony induced by this regimen; timed breeding eighty hours after the second injection results in conception in less than 40 percent of cows or heifers. The subsequent cycle for those cows failing to conceive is often longer than normal, although the effects of this extended luteal phase on fertility are not known. The true usefulness of prostaglandins comes from their ability to allow producers to focus heat detection labor over smaller periods of time.

Progestagens

Exogenous sources of either progesterone or synthetic progestagen serve to mimic the function of the corpus luteum (CL), suppressing the normal cyclical activity of the ovary. In the absence of a functional CL, the removal of this exogenous source initiates ovarian function. Intravaginal devices, such as progesterone-releasing intravaginal devices (PRIDs and CIDRs), are available that release progesterone over a period of time.

The PRID is a flat stainless steel coil covered with an inert elastomer imbedded with progesterone along with an estradiol benzoate capsule. The PRID is inserted into the anterior vagina of healthy, nonpregnant heifers or cows and removed twelve days later. Cows can be bred either by timed artificial insemination at forty-eight and seventy-two hours (or a single insemination at fifty-six hours), or they can be bred following the observation of visual signs of estrus. Synchrony is not great but can be improved by injecting prostaglandins on day 7 and removing the PRID on day 8.

The CIDR is a Y-shaped nylon device that is also coated with a progesterone-imbedded elastomer (Figure 26.10). Because no estradiol is associated with the CIDR, its withdrawal is always followed by

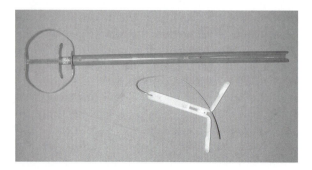

Figure 26.10 A controlled internal drug releasing device (CIDR) is imbedded with progesterone that is slowly released and absorbed in the bloodstream. *(Courtesy of Mark Kirkpatrick)*

prostaglandin. The protocol for timing of insemination is identical to that followed after PRID removal.

Norgestamet is a synthetic progestagen that is more potent than progesterone. It is not cleared for use in lactating cows. It is applied subcutaneously as an ear implant, along with intramuscular injections of norgestamet and estradiol at the time of implantation. It is removed after nine to ten days, and the protocol for insemination is the same as for the PRID. Again, like the PRID, synchrony is improved by injecting prostaglandins twenty-four hours prior to implant removal.

Synchronized Ovulation Regimens

The inclusion of injections of gonadotropin-releasing hormone (GnRH) with the prostaglandin regimen allows producers to synchronize ovulation of mature cows. Several variations on this theme have been developed, but all are designed to minimize the labor of heat detection.

Ovsynch

This protocol uses an injection of gonadotropin-releasing hormone (GnRH) followed by an injection of prostaglandins seven days later and a second injection of GnRH two days after that. Using this protocol, a high percentage of cows are brought into estrus within a sixteen-hour window of time, allowing the potential for timed insemination (without heat detection). Cows are typically inseminated twelve to sixteen hours after the second GnRH injection. Pregnancy rates are high when cows are inseminated at any time within a twenty-four-hour period after the second GnRH injection, and although actual conception rates are slightly lower than in cows bred following visual heat detection, pregnancy rates are higher because all eligible cows are bred. The two primary disadvantages of this protocol are that costs per pregnancy are very high and it is relatively ineffective in synchronizing ovulations in heifers. Conception rates for OvSynch are best when the program is initiated between days 5 and 8 of the estrous cycle.

PreSynch

Providing two injections of prostaglandins prior to initiating the OvSynch protocol allows all cows to initiate the OvSynch program at the same stage of the cycle, and conception rates of cows on this protocol are higher than for cows on OvSynch alone (43 percent versus 29 percent). This protocol is most effective for setting up the first postpartum breeding; however, the length of time required for administering all these injections (thirty-five to thirty-seven days) makes it a poor resynchronization program.

HeatSynch

Substituting estradiol cypionate (ECP) for the second GnRH shot is also a viable synchronization protocol. The use of ECP at this time (low progesterone levels with a follicle present) stimulates endogenous GnRH release. This stimulates LH release and subsequent ovulation. In addition, ECP enhances estrus behavior and uterine tone. However, it appears not to work as well in anovulatory cows. Overall, conception rates for HeatSynch and OvSynch are similar.

SUMMARY

- The key component in any reproductive management program is efficiency of heat detection.
- Keys to successful heat detection are proper animal identification, good records with all observations permanently recorded, and knowledge and understanding of the signs of heat.
- Heat detection aids are divided into several categories: those that detect the primary and secondary signs of heat and those that measure the hormonal changes associated with estrus.
- Prostaglandins are local hormones because they do their work in the immediate area in which they are produced.

QUESTIONS

1. What controls the behavioral signs of heat?
2. What structure predominantly produces estradiol?
3. Compare and contrast the behavior of animals prior to standing heat and during the period of standing heat.
4. What is the cardinal sign of a cow in heat?
5. Between what times of the day does 90 percent of all estrus-related activity occur?
6. What are some aids measuring mounting behavior?
7. What are some aids measuring the secondary signs of heat?
8. In a milk progesterone kit, what does a reading of "high" indicate?
9. Compare and contrast OvSynch and HeatSynch.

ADDITIONAL RESOURCES

Articles

El-Zarkouny, S. Z., J. A. Cartmill, B. A. Hensley, and J. S. Stevenson. "Pregnancy in Dairy Cows after Synchronized Ovulation Regimens with or Without Presynchronization and Progesterone." *Journal of Dairy Science* 87 (April 2004): 1024–1037.

Lopez-Gatius, F., K. Murugavel, P. Santolaria, M. Lopez-Bejar, and J. L. Yaniz. "Pregnancy Rate after Timed Artificial Insemination in Early Post-Partum Dairy Cows After Ovsynch or Specific Synchronization Protocols." *Journal of Veterinary Medicine and Physiological Pathology of Clinical Medicine* 51 (February 2004): 33–38.

Navanukraw, C., D. A. Redmer, L. P. Reynolds, J. D. Kirsch, A. T. Grazul-Bilska, and P. M. Fricke. "A Modified Presynchronization Protocol Improves Fertility to Timed Artificial Insemination in Lactating Dairy Cows." *Journal of Dairy Science* 87 (May 2004): 1551–1557.

Internet

Estrus (Heat) Detection Guidelines: http://ianrpubs.unl.edu/Dairy/g952.htm.

Heat Detection Strategies for Dairy Cattle: http://pubs.caes.uga.edu/caespubs/pubcd/B1212.htm.

Publications and Articles on Dairy Reproduction: http://www.ansc.purdue.edu/dairy/repro/repropub.htm.

Protocols for Synchronizing Estrus in Yearling Heifers: http://www.uky.edu/Agriculture/AnimalSciences/extension/pubpdfs/asc164.pdf.

27

Assisted Reproductive Technologies

OBJECTIVES

- To describe the purpose and procedure of artificial insemination (AI).
- To describe embryo transfer and cloning technologies.

Natural service, the actual mating of a male and a female, is no longer the most common approach toward establishing a pregnancy in a dairy cow. Most cows in the United States are never exposed to a bull but instead become pregnant using one of several different breeding options. The term *assisted reproductive technologies* covers many areas, including artificial insemination, multiple ovulation embryo transfer (MOET) programs, in vitro fertilization, and cloning. There are many other potential ways to assist reproduction, but these are the most representative and will be the focus of the following discussion.

ARTIFICIAL INSEMINATION

Artificial insemination is the deposition of spermatozoa in the female genitalia by artificial rather than by natural means. Legend has it that this method had its origin in 1322, at which time an Arab chieftain used artificial methods to impregnate a prized mare with semen stealthily collected by night from a beautiful stallion belonging to an enemy tribe. However, the first scientific research relative to the artificial insemination of domestic animals was conducted with dogs by the Italian physiologist, Lazarro Spallanzani, in 1780.

In 1937, the North Central School of Agriculture and Experiment Station, Grand Rapids, Minnesota, conducted the first large-scale demonstration of artificial insemination of dairy cattle in the United States. Subsequent growth was rapid. To dairy producers, AI offered a way to enhance the genetics of the herd, control infectious diseases, and eliminate dangerous bulls on the farm.

In 1949, British scientists reported that the addition of glycerine to semen diluters permitted them to freeze certain semen at temperatures much below zero (they used dry ice to freeze at a temperature of $-79°C$ or $-110°F$) and still retain a high degree of fertility following thawing. Today, semen is frozen in liquid nitrogen at $-320°F$. Liquid nitrogen is the fourth coldest known substance. It allows uniform temperatures to be maintained for long periods of time and is more convenient for shipping and storing frozen semen. Frozen bull semen has been stored for more than thirty years and still produced viable pregnancies. There is a small reduction in the fertility of semen stored for long periods, but it is common to find semen stored one year or longer in routine use. Semen frozen with liquid nitrogen can be shipped to all parts of the world.

In the United States today, artificial insemination is more extensively practiced with dairy cattle than with any other class of farm animals. More than 65 percent of the nation's dairy cattle are artificially inseminated (Figure 27.1). Return over investment for

Figure 27.1 A cow is inseminated using frozen semen. *(Courtesy of Mark Kirkpatrick)*

the dairy producer averages about 50 percent on typical semen sales, and it may be as high as 100 percent for herds that are just starting AI. Genetic progress using AI sires has been phenomenal in recent years. The AI bulls are now advancing cows' genetics at the rate of nearly 250 pounds per year. Some of the other advantages of artificial insemination are listed here:

1. **It increases the use of outstanding sires.** Through artificial insemination, all producers have equal access to the best sires in the world.
2. **It alleviates the danger and expense of keeping a bull.** Significant hazard and expense are involved in keeping a bull.
3. **It decreases breeding costs.** Artificial insemination is usually less expensive than the ownership of a worthwhile bull together with the accompanying housing, feed, and labor costs.
4. **It helps to control diseases.** Artificial insemination is a valuable tool in preventing and controlling the spread of certain types of diseases. Artificial insemination organizations that are members of The National Association of Animal Breeders, Inc. are expected to follow the rigid Sire Health Code approved by The American Veterinary Medical Association and adopted by The National Association of Animal Breeders.
5. **It makes it feasible to prove more sires.** Because of the small size of the herds in which they are used, many bulls used in natural service are never proven. Through artificial insemination, it is possible to determine the genetic worth of a bull at an earlier age and with more certainty than in natural service. The best of the bulls proved at an early age are put into heavy use and have a longer period of usefulness than is possible under natural breeding methods.
6. **It increases pride of ownership.** The ownership of progeny of outstanding sires inevitably makes for pride of ownership.
7. **It alleviates distance and time as limiting factors.** The male and the female may be separated by thousands of miles, and with frozen semen, years may pass between the time of collection of the semen and insemination of the female.
8. **It increases profits.** The daughters of outstanding sires are usually higher producers and thus are more profitable. AI provides a means of using such sires more widely.

Like many other techniques, artificial insemination has limitations. A full understanding of such limitations, however, merely accentuates and extends its usefulness. Some of the limitations of artificial insemination follow:

1. **It requires training.** Artificial insemination requires some skills and training and much experience. While the cow breeding process is not complicated, a small percentage of people who attempt to learn it never succeed.
2. **It may accentuate the damage of a poor sire.** When a sire carries hidden genetic defects, his damage is potentially accentuated because of the increased number of progeny possible. For this reason, unproven bulls are seldom used extensively. Fortunately, suitable standards for evaluating sires have evolved through progeny testing. About 60 percent of dairy sires are proven, and these sires account for about 80 percent of all breedings.
3. **It may increase the spread of disease.** The careful use of artificial insemination decreases the risk of disease. To date, no outbreaks of disease traceable to the use of artificial insemination have been reported in the United States. However, potentially lethal genetic defects can be introduced into large numbers of females if the defects are difficult to identify.

Insemination Procedures

Frozen semen can be stored indefinitely if it is maintained constantly at a very low temperature. On many farms, however, semen quality is compromised severely prior to inseminating a cow by poor semen-handling practices. The critical temperature for sperm damage is $-112°F$; semen exposed to warmer temperatures (even for short periods) and then returned to the tank may be damaged. The extent of the damage depends on the length of exposure. Repeated exposures create further damage; it is critical to understand that sperm damage is additive in nature. Ice crystals in the sperm cells enlarge, damaging delicate cell membranes and decreasing the number of sperm surviving the freeze-thaw process.

The liquid nitrogen tank can maintain the tank temperature at $-320°F$ as long as at least 2 inches of liquid nitrogen is present, but producers should always maintain at least 4 inches of nitrogen for best results. To increase holding time, the tanks should be stored in a cool spot out of direct sunlight and in a clean, dry, well-ventilated area. Drafts from milk coolers, furnaces, and outside air should be avoided to prevent excessive nitrogen evaporation. Adequate ventilation in the tank storage area prevents the potential for suffocation: nitrogen can accumulate in the air in a closed area due to nitrogen evaporation.

The tank should be checked twice weekly with a measuring stick. The stick is inserted to touch the bottom of the tank for five to ten seconds and then removed, and the frost line can be read. To minimize nitrogen losses, the lid must be secure, and no frost should be apparent on the neck or vacuum fitting, which indicates loss of vacuum insulation. Under this condition, the nitrogen evaporates rapidly. Once the frost thaws to produce sweat, it is unlikely that any semen is still viable.

Within the tank, semen is stored in plastic straws. One-half-cubic-centimeter straws are stored, with five straws per goblet and two goblets per cane. Many canes can be stored per canister, depending on the capacity desired. Straws of semen should always be kept at least 4 inches down in the tank, unless they are going to be removed and used. The temperature in the neck of the tank increases dramatically at higher points than 4 inches. A semen inventory that indicates location of semen in the tank and number of units remaining is essential for minimizing semen exposure.

Distance from Top Temperature

Distance from Top	Temperature
0 inches	36° to 54°F
1 inch	5° to −8°F
2 inches	−40° to −51°F
3 inches	−103° to −116°F
4 inches	−148° to −184°F
5 inches	−220° to −256°F
6 inches	−292° to −313°F

Cows should be restrained and ready to breed prior to thawing semen (Figure 27.2). As previously described, the cane of straws should be kept as low as possible in the tank while an individual straw is removed with 12-inch tweezers (Figure 27.3). The straw is immediately immersed in a 90°F to 95°F insulated water bath for 40 seconds. The thermometer accuracy should be checked every month. After a straw is thawed, it must be dried thoroughly with a paper towel and protected from rapid cooling. Thawed straws need to be used within ten minutes; after this time, semen quality decreases and pregnancy rates drop (Figure 27.4).

Insemination guns should be warmed prior to loading a straw; a cold gun can also damage sperm cells (Figure 27.5). Inseminating guns consist of three portions: the barrel, plunger, and O-ring. French guns eliminate the O-ring by having sheaths that twist onto the barrel of the inseminating gun. Many sheaths also have a guide that minimizes semen leakage by sealing the straw into the sheath.

Cows are artificially inseminated by passing the inseminating gun past the vagina and into the cervix, stopping beyond the location where the cervix ends and the uterus begins (Figures 27.6 through 27.9). At

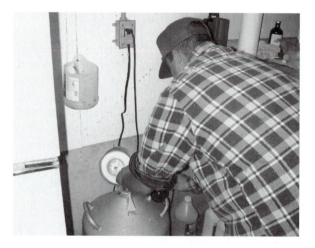

Figure 27.3   It is crucial to keep the canister below the frost line when removing straws of semen. (*Courtesy of Mark Kirkpatrick*)

Figure 27.2   This mobile inseminating vehicle allows semen to be selected and thawed as close as possible to the cow to be inseminated. (*Courtesy of Mark Kirkpatrick*)

Figure 27.4   Multiple insemination rods can be loaded at the same time as long as cows are inseminated within a ten-minute window. (*Courtesy of Mark Kirkpatrick*)

Figure 27.5 The straw of semen is carefully loaded into the insemination rod and then the sheath is placed over the rod to hold the straw in place. *(Courtesy of Mark Kirkpatrick)*

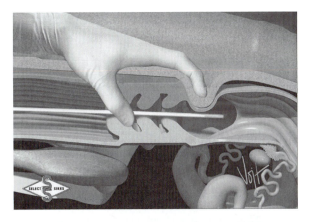

Figure 27.8 After passing through the cervix, the inseminating rod can be easily palpated through the uterine walls. *(Courtesy of Select Sires)*

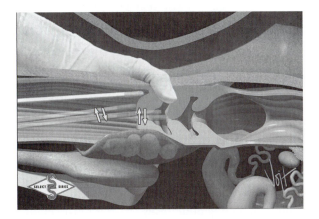

Figure 27.6 When the inseminating rod reaches the fornix, the rod is tipped down as the cervix is tipped up until the rod enters the cervical opening. *(Courtesy of Select Sires)*

Figure 27.9 A trained inseminator breeds a cow at a palpation rail. *(Courtesy of Mark Kirkpatrick)*

this point, the semen should be deposited slowly (Figures 27.10 and 27.11). Care should be taken not to retract the gun as the semen is deposited; otherwise, most of the semen is deposited in the cervix and vagina (Figure 27.12).

The success of artificial insemination is directly proportional to the cleanliness of the equipment, operators, and animals. Spermatozoa are highly sensitive to and quickly killed by dirt, water, urine, excess heat, cold shock, ultraviolet light, and by just about any other substance foreign to their natural environment.

MULTIPLE OVULATION EMBRYO TRANSFER (MOET)

Embryo transfer can hardly be considered a new technology. The first reported work was done in 1943 in Wisconsin, and in 1951, the first calf was born by surgically transferring an embryo from a donor cow to a recipient. In 1964, the first calf was born using nonsurgical techniques. In 1971, Cornell University transferred an embryo from a five-month-old heifer.

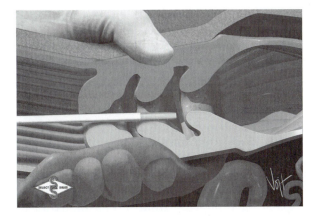

Figure 27.7 The inseminator should manipulate the cervix and gently pull it back over the inseminating rod to avoid damaging the cervical tissue. *(Courtesy of Select Sires)*

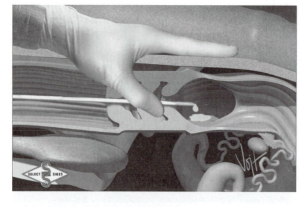

Figure 27.10 After proper positioning of the rod is determined, the semen is slowly deposited by pushing the plunger into the rod. *(Courtesy of Select Sires)*

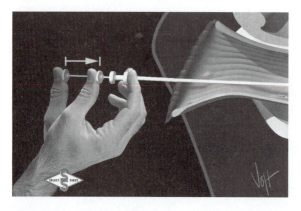

Figure 27.11 Semen should be deposited in the uterus by slowly depressing the plunger without retracting the inseminating rod. *(Courtesy of Select Sires)*

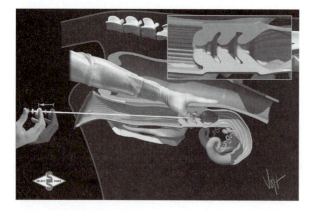

Figure 27.12 Retracting the insemination rod during the actual insemination process results in undesirable cervical and vaginal deposition of semen. *(Courtesy of Select Sires)*

Embryo transfer is a process whereby multiple embryos are transferred from a valuable donor cow to less valuable recipients. Full-term pregnancies result in full siblings (brothers and sisters) with the genetic traits of the donor cow and the bull to which she was bred. Recipients have no genetic influence on the calves they carry; they serve merely as incubators.

Figure 27.13 Offspring derived from embryo transfer and the donor cow from which they originated. *(Courtesy of Iowa State University)*

With this technology, producers can flush postpartum cows prior to recommended breeding dates. The normal waiting period can be exploited without delaying the breeding program. Producers can also flush older cows still exhibiting estrus and ovulating normally but incapable of carrying a pregnancy. The offspring produced must be valuable enough to warrant the procedure. Genetically infertile animals can be flushed, although the success rate is considerably lower and propagates the same problems in the offspring.

The implementation of a multiple ovulation embryo transfer (MOET) program can substantially enhance genetic progress in a herd (Figure 27.13). As with other technologies, the question is: Can the benefits outweigh the costs of this technology? Current estimates suggest that costs of $750 to $1,000 per procedure can be expected if multiple donor cows are flushed at the same time and over twice that amount if only a single donor is flushed. For this procedure to be profitable, donor cows must be truly genetically elite and not just the best cow in that particular herd. The use of MOET programs to improve the genetic status of a herd is typically not profitable unless some of the offspring can be marketed at a price that offsets the costs.

Superovulation consists of injecting the female with hormones [primarily follicle-stimulating hormone (FSH)] that cause the larger follicles (each of which contain one oocyte) to mature, rupture, and release the egg. The bull is capable of producing from several thousand to millions of sperm daily, whereas the cow normally produces one ovum (occasionally two ova) every seventeen to twenty-one days. It is possible, through the administration of hormones, to ob-

TABLE 27.1   Schedule for Superovulation of Donors and Synchronization of Recipients

Day	Donor	Recipient
1	25 mg $PGF_{2\alpha}$	
14	25 mg $PGF_{2\alpha}$	
17	Estrus	25 mg $PGF_{2\alpha}$
28	5 mg FSH—AM	
	5 mg FSH—PM	
29	4 mg FSH—AM	
	4 mg FSH—PM	
30	4 mg FSH—AM	
	4 mg FSH—PM	25 mg $PGF_{2\alpha}$
31	3 mg FSH + 25 mg $PGF_{2\alpha}$—AM	
	3 mg FSH + 15 mg $PGF_{2\alpha}$—PM	
32	2 mg FSH—AM	
	2 mg FSH—PM	
33	Estrus—breed—AM	Estrus
	Breed again—PM	Palpate
39	Flush	ET

tain several ova (five to fifty) from a cow at one estrous cycle. It is also feasible to obtain a large number of eggs from very young heifer calves by injection of hormones.

Eggs shed from the ovaries are stored in large follicles. The basic principle of superovulation is to stimulate extensive follicular development through the use of a hormone preparation, given intramuscularly or subcutaneously, with follicle-stimulating hormone (FSH) activity. Typically, six to twelve days after confirmed estrus, FSH is injected twice daily for four to five days, with prostaglandins provided on the next to last day of FSH injections. The total dose of FSH should be 33 to 37 mg given in decreasing doses to try to avoid any decrease in estrous intensity. Although FSH-treated animals and nontreated animals both have similar intensities and durations of estrus when treated in this manner, FSH-treated animals have a shorter and more variable interval from prostaglandin treatment to estrus (thirty-eight ± twelve hours for superovulated cows versus fifty-six ± three hours for nontreated animals).

Cows with multiple follicles release more progesterone after ovulation, which results in faster ova migration. Because ovulation occurs over a period of time, not all the eggs are fertilized unless the donor is inseminated repeatedly. Cows are usually inseminated twelve and twenty-four hours after the initiation of standing heat, and many procedures use four inseminations spread over two days. Embryos will be in the uterus four to five days later but are typically flushed on day 6 or 7 (morula stage). A yield of five to twelve good fertilized eggs per donor cow may be expected. Typically, the more eggs that

are flushed, the lower the percentage viability. (However, the total number of viable eggs increases with an increased total in flush.) The total number of eggs recovered can range from zero to over fifty, but the average is six to eight. With only 75 percent of transferred embryos resulting in a pregnancy, a total of four pregnancies per flush is considered successful.

The most common methods for recovering and transferring embryos are nonsurgical. During the collection process, the embryos are literally flushed from the donor's uterus using a three-way Foley catheter and less than a pint of flushing medium. The actual transfer of the embryo into the recipient is accomplished using techniques similar to artificial insemination (Figure 27.14). The embryo is drawn into a conventional semen straw, along with a small amount of flushing medium. The insemination gun is advanced fully into the uterine horn adjacent to the ovary containing a corpus luteum (Figures 27.15 through 27.17).

Synchrony between the donor cow and the recipients is critical for success. As little as a twelve-hour difference between the time of donor and recipient ovulations can reduce the possibility of a successful pregnancy becoming established. The lining of the uterus secretes many substances for the nourishment of the embryo that continually change during this critical period of development.

Embryos that are not immediately transferred can be frozen for later transfer. Typically, ethylene glycol or glycerol is used as a cryoprotectant. The embryos can then be thawed and transferred when a suitable recipient is available. Pregnancy rates are not as good following freezing and thawing as with fresh embryos.

GENETIC ENGINEERING

Building on the present knowledge and understanding of genes and the ability to manipulate DNA, scientists are now making ideas such as genetic engineering and cloning into realities. In 1977, scientists produced genetically altered bacteria capable of producing insulin, a valuable hormone previously extracted at slaughter from pigs, sheep, and cattle. This genetic wizardry has been extended to transplantation of genes into bacteria. The genes are responsible for many critical products in addition to insulin, among them, endorphin, somatotropin, interferon, and vaccines.

Genetic manipulations with the potential to create new forms of life make biologists custodians of great power. Despite contoversy, molecular biologists continue recombinant DNA studies, and breakthroughs continue to occur regularly.

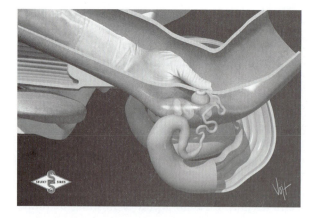

Figure 27.14 In the recipient cow, the ovary that has ovulated is detected by the presence of a corpus luteum. *(Courtesy of Select Sires)*

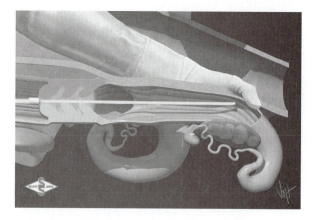

Figure 27.16 The goal is to deposit the embryo partway up the uterine horn on the same side where ovulation occurred (ipsilateral horn). *(Courtesy of Select Sires)*

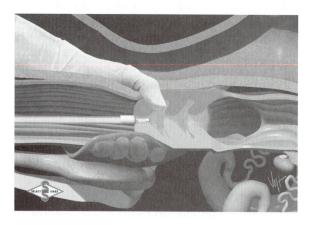

Figure 27.15 The cervix of the recipient cow has a cervical plug and is not dilated, making manipulation of the transfer rod difficult. *(Courtesy of Select Sires)*

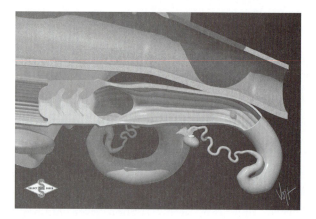

Figure 27.17 If properly transferred, the embryo is in the best position for survival and implantation. *(Courtesy of Select Sires)*

IN VITRO MATURATION (IVM) AND IN VITRO FERTILIZATION (IVF)

The ability to mature and/or fertilize ova in a culture dish (in vitro) allows even more human control over the reproductive processes of the dairy cow. Infertile or prepubertal animals can reproduce using such procedures. Immature oocytes are aspirated from the follicles of superovulated animals via guidance by transvaginal ultrasound imaging. These oocytes are cultured for about twenty-four hours; at maturity, they are fertilized by the addition of capacitated sperm to the culture media. These fertilized oocytes are then cultured further until they reach early blastocyst stage. At this stage, they can either be transferred to recipient animals or frozen for later transfer.

Animals derived from IVM and IVF procedures have a higher incidence of abnormal development patterns than do AI-derived fetuses. Gestation lengths are more variable, birthweights are more variable, and the incidence of both dystocia and stillbirth are increased.

PATENTED ANIMALS

In April 1987, the US Patent and Trademark Office (PTO) ruled that patents could be issued on genetically engineered animals. Thus, livestock producers must pay fees to those who patent genetically altered animals. The PTO also ruled that livestock producers must pay royalties to the patent holder on each generation of patented animals for the life of the patent, which may be as long as seventeen years. Stud fees and one-time payments for animals have always been part of livestock farming, but prior to this ruling, producers have possessed the rights to and complete control over subsequent generations without any additional payments.

Figure 27.18 Annie, born March 3, 2000, is a clone of a purebred Jersey calf. *(Courtesy of USDA-ARS)*

CLONING

Cloning of an animal is the production of an exact genetic copy (Figure 27.18). In a technical sense, identical twins are clones because they are derived from a single cell, a result of the embryo splitting early in development to yield two genetically identical copies. With few exceptions, all cells in the body of an animal contain the same genetic information. This information is contained in DNA located in the nucleus of the cell. Thus, within an animal, the DNA sequence in the nucleus of a liver cell is identical to that in a skin cell. These cells differ in appearance and function because they make use of different parts of the genetic information, not because the total amount of information differs. All of these cells have the genetic information that was present in the one-cell embryo that developed in the animal. Therefore, if the nucleus of any of these cells is used to replace the genetic information in any one-cell embryo, an exact genetic copy of the animal whose cells donated the nucleus would develop. With such an approach, thousands of cloned copies could be made.

Cloned calves are currently being produced routinely; however, the profitability of such production is limited, and the animals must have a function beyond simply producing milk. Through cloning, it is theoretically possible for all dairy animals in an entire herd to look alike, be genetically alike, have the same nutritive requirements, and produce the same quantity of milk of the same composition. In reality, environmental factors impinge on genetic potential, and often this reduction of genetic potential is permanent. Even altered embryonic environment (in vitro) induces long-term impairments of genetic potential. Most cloned calves are currently delivered by Cesarian section because these impairments are so severe that some calves cannot survive the birth process. Until the mechanisms of these cloning-induced impairments are understood, cloning will remain a technology of the future rather than of the present for most producers.

SUMMARY

- AI offers a way to enhance the genetics of the herd, control infectious diseases, and eliminate dangerous bulls on the farm.
- Limitations of AI are many: it requires training, it may accentuate the damage of a poor sire, and it may increase the spread of disease.
- Superovulation consists of injecting the female with hormones that cause the larger follicles to mature, rupture, and release the egg.

QUESTIONS

1. What does the term *assisted reproductive technologies* include?
2. Define *artificial insemination.*
3. What are some advantages of AI?
4. Thawed straws used in AI must be used within what amount of time?
5. Define *embryo transfer.*
6. How many pregnancies per flush are considered successful?
7. How are most cloned calves currently delivered?

ADDITIONAL RESOURCES

Articles

Fricke, P. M. "Bulls Are No Bargain." *Hoard's Dairyman* 142 (December 1997): 841.

Heersche, G., and R. L. Nebel. "Measuring Efficiency and Accuracy of Detecting Estrus." *Journal of Dairy Science* 77 (September 1994): 2754–2761.

Larson, L. L., and P. J. H. Ball. "Regulation of Estrous Cycles in Dairy Cattle: A Review." *Theriogenology* 38 (January 1992): 255.

Nebel, R. L., S. M. Jobst, M. B. G. Dransfield, S. M. Pansolfi, and T. L. Bailey. "Use of a Radio Frequency Data Communication System, HeatWatch™, to Describe Behavioral Estrus in Dairy Cattle." *Journal of Dairy Science* 80 (Supplement 1, 1997): 179.

Pursley, J. R., M. R. Kosorok, and M. C. Wiltbank. "Reproductive Management of Lactating Dairy Cows Using Synchronization of Ovulation." *Journal of Dairy Science* 80 (February 1997): 301–306.

Schmitt, E. J-P., T. Diaz, M. Drost, and W. W. Thatcher. "Use of a Gonadotropin-Releasing Hormone Agonist or Human Chorionic Gonadotropin for Timed Insemination in Cattle." *Journal of Animal Science* 74 (May 1996): 1084–1091.

Walker, W. L., R. L. Nebel, and M. L. McGilliard. "Characterization of Estrus Activity as Monitored by an Electronic Pressure Sensing System for the Detection of Estrus." *Journal of Dairy Science* 78 (Supplement 1, 1995): 468.

Xu, Z. Z., and L. J. Burton. "Reproductive Performance of Dairy Heifers and Estrus Synchronization and Fixed-Time Artificial Insemination." *Journal of Dairy Science* 82 (May 1999): 910–917.

Internet

Artificial Insemination in Dairy Cattle: http://astronaut.agoff.umn.edu/ansc3023/ai/.

Artificial Insemination in Dairy Cattle: http://edis.ifas.ufl.edu/DS089.

Benefits of Artificial Insemination and Estrus-Synchronization of Dairy Heifers: http://www.uky.edu/Ag/AnimalSciences/dairy/dairyrepro/rep011.pdf.

Embryo Transfer: http://www.wvu.edu/~exten/infores/pubs/livepoul/dirm26.pdf.

Factors Influencing Pregnancy Rate of Lactating Dairy Cows: http://www.wisc.edu/dysci/uwex/rep_phys/pubs/pregrate.pdf.

Reproductive Management of Dairy Heifers: http://www.wisc.edu/dysci/uwex/rep_phys/pubs/heifers502.pdf.

Timing of AI Relative to Estrus and Ovulation: http://www.wisc.edu/dysci/uwex/rep_phys/pubs/aitiming.pdf.

28

Pregnancy and Parturition

OBJECTIVES

- To list key concepts in pregnancy, diagnosis, and causes of early embryo losses.
- To identify the stages of labor.

Pregnancy is initiated at conception; the resulting embryo survives and thrives on uterine secretions until implantation around day 32 of gestation. Cattle develop a cotyledonary placenta, which means that the placenta attaches to specialized areas (caruncles) in the uterus. The dam nourishes her fetus through these specialized attachments until the fetus signals its readiness for birth.

Pregnancy loss can occur at any point in this process; however, the risk decreases as pregnancy progresses. Although near full-term abortions or stillbirths are most obvious to the producer and most costly on an individual cow basis, early embryo losses are by far the most costly form of pregnancy loss to the industry because of their frequency.

EARLY EMBRYO LOSSES

Early embryo losses are a significant problem in dairy cattle. It is often difficult to distinguish early embryo loss from failure of fertilization. Fertilization failure rates are approximately 10 to 15 percent in all cattle, while early embryo loss rates are 25 to 40 percent. In heifers, this rate is presumed to be much lower, but no accurate estimates are available. Most losses occur between days 8 and 19 after breeding, with subsequent losses being much lower. If embryo loss occurs early enough, the inter-estrus interval will be normal (eighteen to twenty-four days). Later losses appear as longer or missed cycles.

As with conception failure, many cases of early embryo loss are associated with inadequacies in nutritional management. Although clear-cut cause-and-effect relationships are difficult to es-

tablish, beta-carotene, selenium, phosphorus, and copper deficiencies have all been implicated as contributing factors. It is not only critical that the ration be properly balanced during this period, but once again poor transition cow feeding strategies can manifest themselves as early embryo losses.

Cytogenetic abnormalities also contribute to these losses. A critical requirement for normal embryo development is normal complement of properly expressed chromosomes. It has been estimated that 8 percent of twelve- to sixteen-day-old embryos may have abnormal karyotypes, and these chromosomal aberrations may account for one-third of lost conceptions in early pregnancy. These cytogenic abnormalities can be induced if sperm and egg are not united within an appropriate time frame; proper timing of insemination is critical to reduce losses from this source.

Immunological factors are also associated with early embryo loss. Following conception, the uterus is in contact with both sperm and embryonic antigens, and if immunosuppressive mechanisms are not functioning properly, the resulting antibodies formed in response to these antigens may reduce fertility. The trophoblast cells form the boundary zone around an embryo and are resistant to both cellular and humoral affectors of transplantation immunity. This is the reason for the survival of embryos and the maintenance of ectopic pregnancies.

In the uterus are critical time and steroid requirements for proper protein synthesis. These are extremely stringent and must be satisfied for normal development to proceed. The critical importance of donor synchronization for the success of embryo transfer is an example of the importance of uterine secretions for embryo survival. Uterine secretions have been shown to differ between normal cows and repeat breeders, with decreased protein concentrations and altered levels of phosphorus, zinc, and calcium in repeat breeders.

FETAL LOSSES

The term *fetal death* encompasses all the pregnancy losses between day 43 and term delivery. Early fetal deaths may often go undetected; the fetal tissues undergo autolysis and are expelled along with the fetal fluids. In some cases, the fetus is retained and mummified, the corpus luteum is retained, and the pregnancy continues. Often, the mummification is not discovered until the due date passes and the cow fails to show any indications of impending calving. The mummified mass is easily identified by palpation because the uterus remains tightly contracted around it; treatment with prostaglandins causes expulsion within two to three days.

Abortions are classified as calves that are expelled prior to 271 days of gestation. They may be born dead or alive; however, live calves survive less than twenty-four hours. Abortion rates that exceed 3 percent indicate a herd problem.

PREGNANCY DIAGNOSIS

Veterinary teaching hospitals and large veterinary practices often have sophisticated equipment available for the diagnosis of pregnancy in the cow. Most have access to ultrasound equipment that allows them to visualize the pregnancy and make a diagnosis earlier and more accurately than is possible by rectal palpation. Despite these advances, most pregnancies are still diagnosed by rectal palpation (Figure 28.1) because it is less expensive and is remarkably accurate when performed by an experienced palpator (Figure 28.2). Even when an early diagnosis of pregnancy is made using ultrasound imaging (Figure 28.3 and 28.4), hand palpation at a later date is still recommended because of the high risk of pregnancy loss between thirty and fifty days. The high cost of maintaining open cows makes this management practice critical for the progressive dairy farmer.

The best method to diagnose pregnancy by hand depends on the stage of pregnancy. The dates vary slightly between heifers and cows; heifers are easier to diagnose at an earlier date than are cows. From conception to thirty or thirty-five days, checking for the presence of a corpus luteum is the only option. From thirty to sixty days, the embryonic vesicle is large enough to palpate (the fetus is ⅜ to ⅞ of an inch and about the size of a lima bean). At sixty days, the pregnant horn is 4 to 7 inches long and bulging; it somewhat resembles a fat banana. At this stage, it is still possible to slip membranes as well. The fetus is the size of a mouse. At ninety days, the pregnant horn is 10 to 13 inches long. Small cotyledons are present, and the fetus can sometimes still be palpated (the fetal

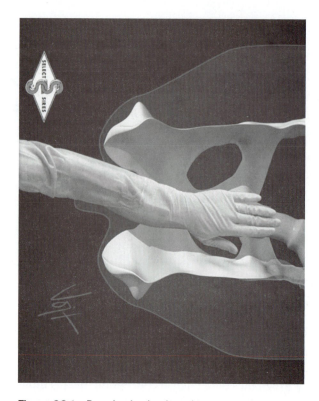

Figure 28.1 Rectal palpation is an important management tool for artificial insemination, pregnancy diagnosis, and embryo transfer. *(Courtesy of Select Sires)*

Figure 28.2 Experienced palpators can confirm the pregnancy status of cows in less than a minute. *(Courtesy of Mark Kirkpatrick)*

head is the size of a Ping-Pong ball). The uterus is the shape of an oversize boxing glove, with the nonpregnant horn representing the thumb of the glove. At 120 days, the uterus extends over the brim of the pelvis. Small cotyledons may be palpable, and the pregnant horn is 13 to 18 inches long (Figure 28.5).

Figure 28.3   An ultrasound monitor on the right indicates that this cow is forty-five days pregnant. *(Courtesy of USDA-ARS)*

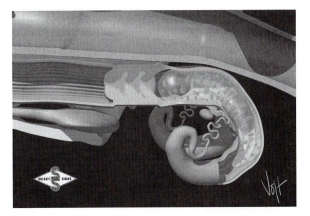

Figure 28.5   The developing fetus is surrounded by fluid-filled fetal membranes containing cotyledons that attach to the maternal uterine endometrium. *(Courtesy of Select Sires)*

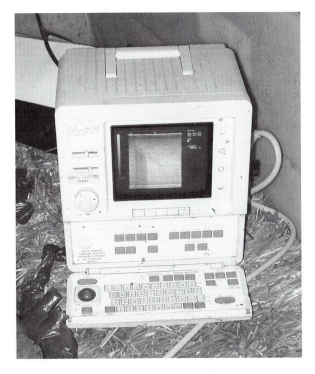

Figure 28.4   On-farm ultrasound is used for early pregnancy diagnosis and determination of the fetus's sex. *(Courtesy of Mark Kirkpatrick)*

Stage of Pregnancy	Length of Horn
45 days	4–6 cm
60 days	6–9 cm
90 days	12–16 cm
120 days	16–20 cm

At 150 days, the horn is 25 to 35 inches long, heavy pulsations are notable in the uterine artery, and distinct cotyledons are present. The fetus is the size of a rat terrier dog. By six to seven months, the uterus drops to the floor of the abdomen, and the middle uterine artery is pulsating heavily; the fetus is about 2 feet long (Figure 28.6). By eight to nine months,

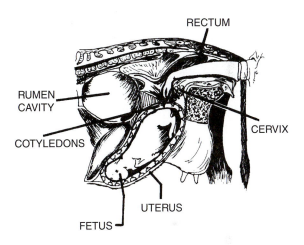

Figure 28.6   Rectal determination of pregnancy becomes more challenging in midgestation because of the location of the fetus. *(Courtesy of Iowa State University)*

cotyledons are very distinct, and you can often feel the head or legs of the fetus, which is close to 3 feet in length. There is more fluid than body mass in the uterus, especially during the early stages of pregnancy. In latter stages, much of the fluid is absorbed.

The most common technique employed for determining pregnancy status is palpating the uterine horn on the side of the corpus luteum. If the cow is pregnant, the embryonic vesicle may be felt sliding past the fingers and the chorioallantoic membranes can be slipped between the fingers near the greater curvature (ovarian end) of the pregnant uterine horn. The presence of intrauterine fluid is also indicative of either pregnancy or infection. An experienced technician can make a positive diagnosis as early as thirty-five days following breeding. These methods of pregnancy diagnosis are popular because they allow early diagnosis, they require no special equipment, and there is little risk of pregnancy loss if care is taken.

Other tests for pregnancy in cows are listed here:

1. Abdominal ballottement may be used from the fifth to the seventh month of pregnancy. This method consists of feeling the fetus by the following technique: place the hand or fist against the abdomen in the lower right flank region; execute a short, vigorous, inward-upward thrust in this region; and retain the hand in place. The hard fetus may be felt. Because of the amniotic fluid, this technique makes the fetus recede, but it will fall back in place almost immediately.

2. The fetal heartbeat can sometimes be detected after the sixth month of pregnancy, but this method is not as certain in the cow as in other classes of farm animals. Use of a stethoscope is preferred, although good results are sometimes secured by merely placing the ear against the right lower abdominal region and listening. The fetal heartbeat can be distinguished from that of the mother because of its greater frequency and lighter and higher pitch.

3. Fetal movements can sometimes be observed through the abdominal wall during the latter half of pregnancy. This method of detecting pregnancy requires much patience. The observer must simply wait until voluntary movement of the fetus on the right side of the cow is observed.

4. The presence of cotyledons (placentomes) is a definite indication of pregnancy. They are actually formed at thirty-five days but can't be easily palpated until the seventy-fifth day. They are about the size of a bean at seventy-five days and are more distinct by ninety days. By 120 days, cotyledons are about ½ inch in diameter, and after 150 days, they are about ¾ inch in diameter.

5. Experienced palpators can also feel pulsations in the middle uterine artery as pregnancy proceeds. It is located just forward of the cervix, and there is a stronger pulsation in the pregnant horn in the early stages. This pulsation can sometimes be detected as early as ninety days, but usually not until 120 days. The artery reaches up to a ½ inch in diameter and gives a buzzing sensation in later pregnancy due to the extreme velocity of blood flow through the vessel.

FETAL DEVELOPMENT

The developing embryo is considered a fetus when all its organ systems are present, at about forty-five days into pregnancy. Throughout pregnancy, the fetus is completely surrounded and protected by two membranes: a fluid-filled amnion and then the

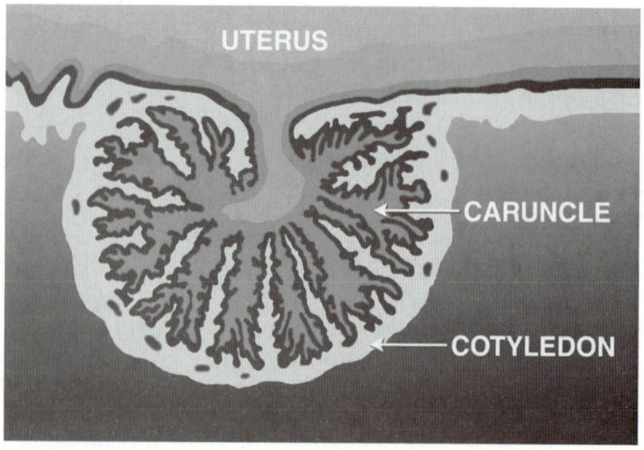

Figure 28.7 Placental cotyledonary tissue wraps around uterine caruncular tissue to create sites of nutrient and gas exchange between the fetus and dam. *(Courtesy of Select Sires)*

chorioallantois in a double-bag arrangement. Although there is less than a pint of fluid surrounding the fetus at forty-five days of pregnancy, there is over 3 gallons by the time the calf is born. About 2.5 gallons is allantoic fluid (the water bag), while the remaining half-gallon is amniotic fluid directly surrounding the calf. The allantoic fluid in the outer water bag is very thin and watery and mostly composed of the developing calf's urine and other waste products accumulated throughout pregnancy. The amniotic fluid directly around the calf is much thicker and more viscous (like honey) and acts as a lubricant during the delivery process to reduce friction between the calf and the birth canal.

The fetus develops a placenta to obtain nutrients and oxygen from the dam and remove carbon dioxide and some other waste products. The placenta is actually a fetal organ and does not exchange any blood directly with the cow. Cattle develop a cotyledonary placenta, a term derived from the specialized areas for nutrient exchange, called cotyledons, that connect the fetal placenta to the dam's specialized uterine areas for attachment, called caruncles (Figure 28.7). Typically, between eighty and 120 cotyledons develop in the placenta of the calf. The umbilical cord is the lifeline between the placental cotyledons and the fetus. A tremendous volume of blood normally passes through the umbilical cord for normal fetal development to continue. At birth, the umbilical cord is stretched and eventually ruptures, separating the calf from its placenta. The placenta is then no longer useful to the calf or the cow and is expelled from the cow's uterus shortly after the calf is delivered.

The age of the fetus can be estimated by its size, referred to as the crown-rump length. This estimate varies by breed and can also be greatly affected in the case of twins, but it is commonly used to estimate due dates for cows where the breeding date is

TABLE 28.1 **Average Weights and Lengths**
of Developing Fetal Calves

Stage of Gestation	Fetal Weight	Crown–Rump Length
30	0.01–0.02 oz.	0.3–0.4 in.
40	0.03–0.05 oz.	0.7–1 in.
50	0.1–0.2 oz.	1.4–2.2 in.
60	0.3–1 oz.	2.4–3.2 in.
70	0.9–3.5 oz.	2.8–4 in.
80	4.2–7 oz.	3.2–5.2 in.
90	7–14 oz.	5.2–6.8 in.
120	2–4.5 lb.	8.8–12.8 in.
150	6–9 lb.	12–18 in.
180	11–22 lb.	16–24 in.
210	18–40 lb.	22–30 in.
240	33–55 lb.	24–34 in.
270	44–110 lb.	28–40 in.

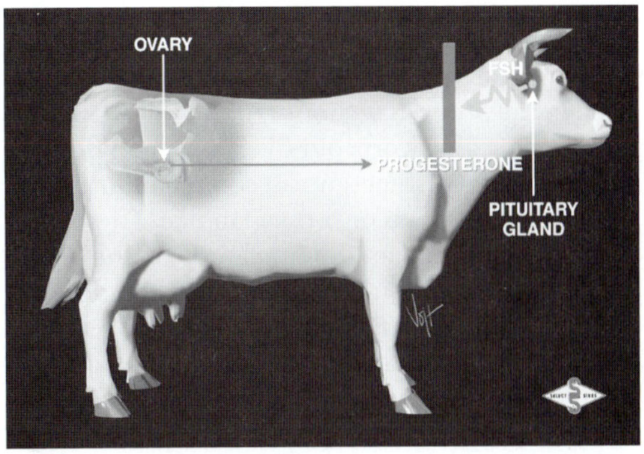

Figure 28.8 Progesterone produced by both the corpus luteum and the placenta maintains pregnancy in part by blocking the release of follicle-stimulating hormone (FSH) from the pituitary gland. *(Courtesy of Select Sires)*

unknown. Experienced palpators can estimate the length of the developing fetus and make fairly accurate predictions of the age of the fetus. Table 28.1 presents ranges of weights and lengths; the lower value represents a small Jersey calf, while the upper value would represent a large Holstein calf.

The metric version of the formula for estimating age from crown-rump length is:

$$\text{Fetal age} = 2.5 \times (\text{crown-rump length in cm}) + 21)$$

The US version is:

$$\sqrt{2 \times \text{crown–rump length in inches}}$$

The most rapid period of growth occurs at around day 230 of gestation, with the calf growing over half a pound per day.

The dominant hormone of pregnancy is progesterone (Figure 28.8). It is produced by luteal tissue from the ovary of the dam and by placental tissue from the fetus. Progesterone serves several important functions throughout the gestation period. It exerts a negative feedback control on the pituitary gland, suppressing the normal estrous cycle of the cow. Progesterone also stimulates the production of uterine milk, nourishing the embryo until the placenta is functional, and maintains the uterus in a quiet, relaxed state until near the time of parturition.

PREPARING FOR THE CALVING PROCESS

The timing of birth is ultimately decided by the fetus. As the fetal calf reaches the final stages of maturation prior to birth (about three weeks before birth), its adrenal gland produces increasing amounts of cortisol. This is a crucial turning point in the reproductive cycle of the dairy cow. The first secretions of cortisol by the calf trigger all the changes required for a normal delivery and a normal lactation. Cortisol causes the placenta to start producing estrogen from progesterone, which decreases circulating progesterone and increases circulating estrogen in the dam. The changing concentrations of progesterone and estrogen stimulate the mammary system to begin regenerating alveoli, the secretory cells of the udder. These hormonal changes also initiate colostrogenesis, or the production of antibody-rich colostrum by the mammary gland.

The dam's uterus is also sensitive to these hormonal changes and begins preparation for the upcoming birth process. This is all carefully synchronized with the final development of the calf to allow the delivery of a mature, healthy calf. Cortisol stimulates the final crucial maturation processes of the gut and the lungs of the calf so that these organs are fully functional at birth and can take over the functions that, until birth, have been the function of the calf's placenta. The placenta itself undergoes developmental changes that allow it to be expelled rapidly after birth.

The final surge of fetal cortisol the last few days before calving triggers a cascade of hormonal changes in the fetal placenta and maternal uterus that initiates the parturition process. Decreases in progesterone increase uterine muscular contractions in preparation for the work of labor, while a rapid increase in estrogen the last day before calving helps prepare the uterus to respond to the onslaught of hormones presented during the delivery process. Relaxin concentrations increase, causing relaxation of the pelvic ligaments as well as the ligaments around the pins (sacrosciatic and sacroiliac ligaments).

INDUCED CALVING

Inducing calving short-circuits the processes outlined in the previous section. Typically, the cow is administered a glucocorticoid (like dexamethasone or betamethasone), sometimes in combination with estrogenic compounds or prostaglandins. These compounds initiate a rapid stimulation of the dam's preparation for calving and stimulate the development of the calf's lungs and gut, much like fetal cortisol in a normal delivery. However, problems occur when calving is induced too early (especially before 260 days of gestation). In these cases, the calf is not fully mature; not only is survival at risk, but if the calf survives, it may also have compromised performance throughout its lifetime. Colostrum does not have a chance to be produced, further challenging the survival of the calf. In addition, the full regeneration of secretory tissue in the mammary gland may not be achieved, and milk production can be compromised for the next full lactation. The placenta is often retained if induction is initiated too early.

The incorporation of relaxin and a progesterone antagonist (RU486) into the induction protocol regimen alleviates some of these problems and creates a more nearly normal hormonal pattern for delivery. However, these compounds are not routinely used in most induction protocols. The natural variability in gestation length is the major obstacle in artificially timing the calving process. It is difficult to establish a time for induction that is safe 100 percent of the time yet does not have a high percentage of calves born naturally prior to that date. Cows must be treated as individuals, and the timing for induction should be based on the individual cow's progress toward readiness for calving.

DETERMINING THE TIME OF BIRTH

The best method for determining the time of calving for an individual cow is to observe the physical and behavioral changes that typically are apparent prior to calving. The position of the calf changes as it prepares for birth; this changes the shape of the cow when observed from behind. For most of the dry period, the cow appears to be barrel-shaped. During the final week of gestation, the calf shifts position and the cow appears pear-shaped. On the day of calving, the calf seems to disappear within the cow as it moves up into the birth canal. This is often the best close-up predictor of calving.

The vulva also starts to swell prior to calving. There are distinct individual and breed differences in the timing of this swelling; Jerseys often start to swell several weeks prior to calving, while other breeds may have little noticeable swelling until the

Figure 28.9 Mucus consistency and quantity both change dramatically shortly before calving. *(Courtesy of Howard Tyler)*

TABLE 28.2   Breed Effects on Gestation Length

Breed	Average Length of Gestation
Holstein	279 days
Jersey	279 days
Brown Swiss	291 days
Ayrshire	279 days
Guernsey	284 days

day of calving. The viscosity of the mucus produced in the reproductive tract also changes in response to hormonal changes. Normal discharge is scanty, opaque, and viscous; as parturition draws near, it becomes clear and free-flowing (Figure 28.9). Body temperature decreases as well, although this is not typically useful as an indicator of impending calving. The udder begins to fill several days to several weeks prior to parturition. It may become highly distended, especially in heifers. In addition, just prior to calving, the teats often become full and distended and may begin leaking colostrum. The action of relaxin is noted by a loosening of the sacrosciatic and sacroiliac ligaments at the tailhead. They start to relax the final week prepartum and become completely slack (more than 1 inch of movement at the back end) during the last twenty-four hours prior to the onset of parturition. Another physical reflection of this change is an increased tail flexibility as the ligaments between the vertebrae loosen. Ultimately, the length of gestation is controlled by a combination of genetic and environmental factors and varies greatly by breed and parity. See Table 28.2.

THE BIRTH PROCESS

The process of calving is arbitrarily divided into three stages, each of which will be discussed in this section.

Stage 1 Labor

The first stage of labor is the period of cervical dilation. Uterine contractions increase in frequency as the quiescent influence of progesterone is reduced. The fetus is moved by these contractions toward the birth canal. Pressure from the chorioallantoic sac that surrounds the fetus induces cervical relaxation. By the end of stage 1 labor, the cervix is dilated to approximately 3 to 6 inches in diameter. Behavioral changes in the cow are subtle and often go unnoticed. Increased restlessness and circling behaviors, copious mucus discharge from the reproductive tract, increased defecation and urination, high respiration rates, and relaxation of the pelvic ligaments are all associated with stage 1 labor (Figure 28.10). Stage 1 labor lasts for two to sixteen hours and varies with both breed and parity.

Stage 2 Labor

The second stage of labor is initiated as the fetus enters the cervix. Uterine contractions continue to increase in frequency and develop a rhythm that was previously lacking. Labor pains during this period occur as clusters of five to eight contractions. Initially, each individual contraction lasts one to two seconds, clusters of contractions last approximately one minute and are separated by two- to three-minute rest periods. Stretching of the birth canal induces release of oxytocin; concentrations increase thirty- to fiftyfold as the calf passes through the cervix. Abdominal contractions occur in combination with these clusters of uterine contractions to propel the fetus through the birth canal. The primary factor determining duration of parturition is the rate of cervical dilation. Pressure from the fetal head and shoulders passing through the cervix stimulates the final phase of cervical dilation (Figure 28.11). The fluid-filled amnion and chorioallantois protects the fetus from the forces exerted by the incompletely dilated cervix. The completely dilated cervix is totally confluent (no area of constriction) from the vagina to the uterus. Premature rupture of the chorioallantois (water bag) results in a temporary cessation of abdominal contractions because the effective space occupied by the fetus is suddenly reduced. The feet of the fetus typically appear about thirty minutes (seven to eight clusters of contractions) after the fetus enters the cervix (Figure 28.12), although this

Figure 28.10   The first stage of labor is characterized by circling behavior, a raised tail, and increased defecation and urination. *(Courtesy of Howard Tyler)*

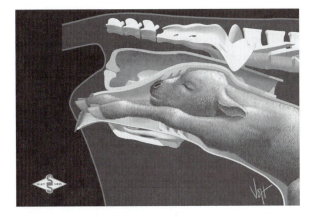

Figure 28.11   The normal birth presentation for a calf. *(Courtesy of Select Sires)*

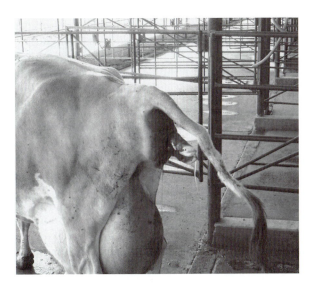

Figure 28.12   A Jersey cow in early stage 2 labor. *(Courtesy of Howard Tyler)*

phase is extended in calvings requiring assistance. Progress is slower after this point because expulsion of the fetal head and shoulders require further increases in cervical diameter. Approximately five to forty-five minutes after the appearance of the feet (depending on the breed and the size of the fetus), the intensity and frequency of contractions increase again (two- to three-second contractions, with shorter rest periods between clusters). The fetus is expelled within fifteen to thirty minutes (figure 28.13). Breeds with a higher incidence of dystocia require a longer period of time from the appearance of the chorioallantoic sac to the appearance of the feet, the appearance of the feet to the delivery of the head, and from the delivery of the head to the delivery of the calf. Stage 2 labor lasts fifteen minutes to three hours depending on breed and parity.

Stage 3 Labor

The third stage of labor is defined as the period leading to the expulsion of fetal membranes, which usually takes four to six hours. Fetal membranes that are not expelled within twelve hours are arbitrarily defined as retained placentae. Cows delivering male calves retain the placenta for longer periods of time. Dystocia increases the incidence of

Figure 28.13   The goal of a successful reproduction program is the delivery of a healthy calf. *(Courtesy of Mark Kirkpatrick)*

retained placenta by two- to threefold. Although the physiological mechanism leading to this increase is unknown, similar prepartum hormonal defects have been observed in cows retaining fetal membranes as in dystocic cows.

SUMMARY

- Many cases of early embryo loss are associated inadequacies in nutritional management.
- Diagnosing a pregnancy in a cow can be performed with ultrasound equipment or by rectal palpation, abdominal ballottement, fetal heartbeat after the sixth month, fetal movements, the presence of cotyledons, and pulsations in the middle uterine artery.
- Nutritional inadequacies, cytogenic abnormalities, and immunological factors are associated with early embryo loss.

QUESTIONS

1. How long is the normal gestation period of a Brown Swiss cow?
2. At what day does the embryo implant in the uterus?
3. What is the function of the placenta?
4. If a Jersey cow was bred on March 7, when is her due date?
5. What are the three stages of parturition?

ADDITIONAL RESOURCES

Articles

Anderson, R. "Problems Associated with Twin Pregnancies in Dairy Cows." *Veterinary Recommendations* 155 (July 2004): 127–128.

Santos, J. E., W. W. Thatcher, R. C. Chebel, R. L. Cerri, and K. N. Galvao. "The Effect of Embryonic Death Rates in Cattle on the Efficacy of Estrus Synchronization Programs." *Animal Reproductive Science* 82–83 (July 2004): 513–535.

Starbuck, M. J., R. A. Dailey, and E. K. Inskeep. "Factors Affecting Retention of Early Pregnancy in Dairy Cattle." *Animal Reproductive Science* 84 (August 2004): 27–39.

Internet

Getting Problem Cows Pregnant: http://muextension.missouri.edu/explore/agguides/dairy/g03030.htm.

Pregnancy and Calving: http://babcock.cals.wisc.edu/downloads/de/10.en.pdf.

The Physiology of Gestation and Parturition: http://www.wvu.edu/~exten/infores/pubs/livepoul/dirm3.pdf.

6

Concepts in Lactation Physiology

29

Anatomy and Physiology of the Mammary Gland

OBJECTIVES

- To outline the physiology of milk production and the method of milk harvesting.
- To describe the anatomy and physiology of the mammary gland.

Zoologically, cattle belong to the class *Mammalia*, which includes warm-blooded, hairy animals that give birth to living young and suckle them for a variable period on secretions, called milk, from the mammary glands. The number of mammary glands and their position on the body is peculiar to each species. For example, the cow has four glands (quarters), each with a passageway (teat) to the outside, whereas the sow generally has ten or more mammary glands. The ability of dairy cattle to produce large amounts of milk is the principal reason they are accorded a prominent place in American agriculture. It is important, therefore, that the physiology of milk production and the methods of milk harvesting be fully understood.

EXTERNAL STRUCTURE OF THE UDDER

The mammary glands, or udder, in the cow consist of four separate glands, called quarters. The teat hangs down from each quarter. It is hollow and more or less closed at the top and at the bottom. The bottom of the teat is closed by a circular (sphincter) muscle known as the teat canal, or streak canal. About 40 percent of cattle have extra nonfunctional teats called supernumerary teats (Figure 29.1). These teats are normally cut off when the calf is very young.

Teats vary in shape from cylindrical to conical. Rear teats are usually shorter than the fore teats. Each teat in the cow has one streak canal that allows the milk to exit the gland. In some mammals, such as the mare, each teat contains more than one streak

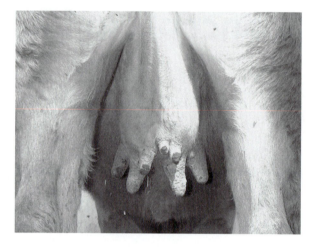

Figure 29.1 Some calves are born with several extra teats, which should be removed shortly after birth. *(Courtesy of Mark Kirkpatrick)*

canal. Because the primary function of dairy cows is the production of milk, the size and shape of the teats are quite important. They should be moderate in size and located on the udder in a way that facilitates the use of milking machines. The sphincter in each teat should be tight enough to prevent milk leakage without slowing the flow and causing difficult milking.

The udder of the cow is located in the inguinal region and is covered with fine hair, although teats are generally devoid of hair. The right and left halves of the udder are divided by a longitudinal groove called the intermammary groove. Occasionally, a cow may have a distinct groove that separates the front and hindquarters, and this characteristic is generally considered undesirable. The udder of a cow may weigh anywhere from 7 to 165 pounds. In many cases, it must also be able to support in excess of 80 pounds of milk. The rear quarters secrete about 60 percent of the milk. The udder continues to grow in size until the cow is about six years of age.

SUSPENSION OF THE UDDER

A well-attached udder fits snugly against the abdominal wall in front and on the sides and extends high between the thighs in the rear. Breaking away of the udder from the body occurs when the supporting ligaments weaken or stretch. The primary supporting structures of the udder are the skin, the median suspensory ligament, and the lateral suspensory ligament (Figure 29.2). The skin plays only a minor role in the support and stabilization of the udder. It is connected to loose connective alveolar tissue, cordlike tissue that keeps the surface of the forequarters close to the abdominal wall.

The median suspensory ligament is yellowish elastic tissue that separates the right and left halves of the udder. It connects the udder to the abdominal wall by way of a series of lamellae (plates) that are attached to the medial surface of the two halves of the udder. Because this tissue is elastic, it responds to the weight of milk in the udder. After milk ejection, the tissue tightens to provide more support to the udder.

In contrast to the median suspensory ligament, the lateral suspensory ligament is rather inflexible and consists of white fibrous tissue. These ligaments surround the outer wall of the udder and are attached to the prepubic and subpubic tendons, which are in turn attached to the ischium and the pubic bone. The intermammary groove is formed where the lateral suspensory ligament and the median suspensory ligament juncture (Figure 29.3).

INTERNAL ANATOMY OF THE UDDER

The streak canal functions to keep milk in the udder and bacteria out of the udder; the effectiveness of this barrier is based on the constrictive ability of the streak canal. Within the teat is a duct, with a capacity of 30 to 45 milliliters, called the teat cistern (Figure 29.4). It is separated from the streak canal by several folds of tissue, generally four to eight in number, that radiate in several directions, called Fürstenberg's rosettes. They serve as an additional means of preventing milk leakage.

The teat cistern is separated from the gland cistern by the cricoid fold. The gland cistern is capable of holding up to 400 milliliters of milk and acts as a collecting area for the mammary ducts. Branching from the gland cistern is an extensive, highly branched system of mammary ducts that may number anywhere from eight to fifty. Alveoli, the functional producers of milk, discharge their secretions into these ducts (Figure 29.5).

The dorsal part of the udder contains a network of connective tissue in which little milk is produced. It is rather hard and meaty.

Alveoli and Associated Ductwork

The basic milk-producing unit of the udder is a very small, bulb-shaped structure with a hollow center called the alveolus. It is estimated that each cubic inch of udder tissue contains 1 million alveoli; hence, the udder contains billions of alveoli. When

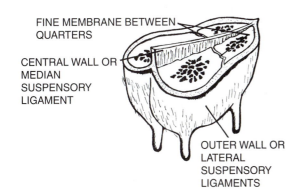

FINE MEMBRANE BETWEEN QUARTERS

CENTRAL WALL OR MEDIAN SUSPENSORY LIGAMENT

OUTER WALL OR LATERAL SUSPENSORY LIGAMENTS

Figure 29.2 The ligaments supporting the udder are shown in cross section. *(Courtesy of Iowa State University)*

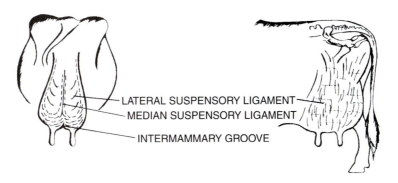

LATERAL SUSPENSORY LIGAMENT
MEDIAN SUSPENSORY LIGAMENT
INTERMAMMARY GROOVE

Figure 29.3 The ligaments that permit udder suspension. *(Courtesy of Iowa State University)*

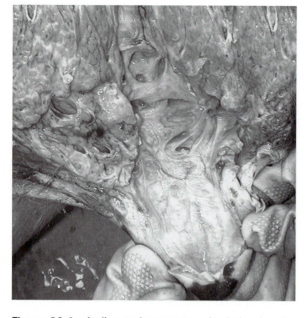

Figure 29.4 A dissected mammary gland showing the gland cistern, teat cistern, and streak canal. *(Courtesy of Leo Timms)*

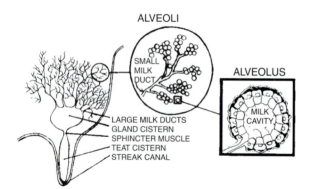

Figure 29.5 The mammary gland is comprised of a system of ducts connecting the milk-producing alveoli to the teats. *(Courtesy of Iowa State University)*

an alveolus is filled with milk, it is about 0.2 millimeters in diameter.

Alveoli are lined with a single layer of epithelial cells that are responsible for secreting milk. Their functions are threefold: remove nutrients from the blood, transform these nutrients into milk, and discharge milk into the center of the alveolus, or lumen. Each alveolus is surrounded by a network of capillaries, from which nutrients are extracted, as well as by a specialized type of muscle cell, called the myoepithelial cell, which is sensitive to the effects of oxytocin. When oxytocin is secreted into the blood, it stimulates contraction of these muscle cells, initiating milk ejection.

Groups of alveoli empty into a duct forming a functional unit called a lobule. Several lobules empty into another duct system forming a larger unit called a lobe. The ducts of lobes empty into what is referred to as a galactophore, which in turn empties into the gland cistern.

The ducts of the udder provide a storage area for milk and a means of transporting it to the outside. No milk secretion, per se, occurs within the ducts. The cells lining the cisterns and duct systems consist of two layers of epithelium. Myoepithelial cells are arranged in a longitudinal pattern that allows the ducts to shorten and increase the diameter to facilitate the flow of milk.

INNERVATION OF THE UDDER

Milk secretion is regulated primarily through hormonal mechanisms. However, milk letdown is initiated largely through neural mechanisms. The nervous system can be divided into two anatomical systems: the somatic nervous system and the autonomic nervous system. The somatic nervous system enables the body to adapt to stimuli from the external environment. Various stimuli, such as touch, are perceived by specialized receptors within this system, and the body responds accordingly. The autonomic system involves the maintenance of homeostasis, the internal environment of the body. The autonomic system can be further divided into the sympathetic autonomic nervous system and the parasympathetic autonomic nervous system. The sympathetic system is generally associated with the traditional fight or flight response, and the parasympathetic system is usually associated with routine integration of normal activity.

The udder has a network of afferent (sensory) and efferent (motor) nerves. Receptors in the udder are sensitive to touch, temperature, and pain. During preparation for milking, washing and cleaning the udder stimulates these receptors, and the process of milk ejection is initiated. Motor nerves transmit impulses from the brain and regulate blood flow and smooth muscle activity around the ducts and in the teat sphincter.

When the cow is startled or subjected to pain, adrenaline is released from the adrenal glands, and the sympathetic nervous system is stimulated. Blood vessels constrict so that blood can be shunted to other parts of the body (e.g., skeletal muscle), and the amount of oxytocin that can potentially be delivered to the gland is reduced proportionally. The final result is a diminished constriction of the myoepithelial cells surrounding the alveoli. Thus, milk ejection is slowed and production depressed.

There is no indication of parasympathetic innervation of the udder.

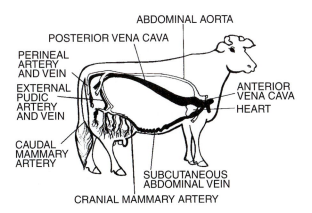

ABDOMINAL AORTA

POSTERIOR VENA CAVA

PERINEAL ARTERY AND VEIN

EXTERNAL PUDIC ARTERY AND VEIN

ANTERIOR VENA CAVA

HEART

CAUDAL MAMMARY ARTERY

SUBCUTANEOUS ABDOMINAL VEIN

CRANIAL MAMMARY ARTERY

Figure 29.6 Blood flow to and from the mammary gland determines the milk-producing capability of the cow. *(Courtesy of Iowa State University)*

CIRCULATION IN THE UDDER

Milk production places extremely high demands on the circulatory system. To produce 1 gallon of milk, more than 500 gallons of blood must pass through the udder. In the case of a low-producing cow, this ratio may increase to 1,000 to 1. The blood plasma of a lactating cow constitutes about 4.9 percent of the entire body weight, compared to 3.8 percent of the body weight for a nonlactating cow. Thus, it is imperative that the circulatory system within the udder be both extensive and efficient.

Blood enters the udder through a pair of arteries called the external pudic arteries (Figure 29.6). These arteries enter the udder dorsal to the rear teats and branch to form the cranial (forward) and caudal (rear) mammary arteries. The cranial and caudal arteries branch into increasingly smaller vessels, eventually forming a capillary system that surrounds each alveolus.

Blood is then picked up in the venules and anastomoses to form veins. These veins form a circle at the base of the udder. From this venous network, blood can travel through one of two routes. The first route is via the external pudic veins, which travel parallel to the external pudic arteries and empty into the vena cava, where the blood then travels to the heart. The second route is through the subcutaneous abdominal veins, commonly called the milk veins. Blood in these veins travels anteriorly along the abdomen and eventually empties into the anterior vena cava. The perineal veins travel through the udder carrying blood from the vulva and the anal region. However, blood from the udder does not enter this system.

LYMPHATIC SYSTEM

Together with the circulatory system, the lymphatic system helps to regulate the proper fluid balance within the udder and to combat infection. Unlike the circulatory system, however, lymph (a colorless fluid drained from tissue) only travels away from the udder. This unidirectional movement is facilitated by the following mechanisms:

1. Blood capillary pressure.
2. Contraction of muscles surrounding the lymph vessels.
3. Valves that prevent the backflow of lymph.
4. The mechanical action of breathing.

Lymph travels from the udder to the thoracic duct and finally empties into the blood system in the anterior vena cava. Flow rates of lymph depend on the physiological status of the cow. A nonlactating cow has a flow rate of from 15 to 250 milliliters per hour, while a lactating cow may have over a tenfold increase in flow rate.

UDDER DEVELOPMENT

The development of the udder starts early in the growth of the fetus and continues to change under hormonal influences through prepuberty, puberty, pregnancy, and lactation. The mammary glands develop from ectodermal tissue. At about thirty-five days of gestational age, the fetus develops two parallel ridges on each side of its ventral midline. These ridges develop into a series of nodules that in turn sink into the dermis of the fetus at about two months to form the mammary buds, four of which are formed in the cow. A cordlike sprout, called the primary mammary cord, forms at the apex of each bud and becomes canalized at about 100 days. The lumen (chamber) formed by this canalization eventually develops into the gland cistern. The teat cistern develops at the end of the cord. Secondary sprouts branch off from the primary sprout and eventually divide into tertiary sprouts to form what is later to become the duct system.

At birth, the duct system is restricted to a relatively small area surrounding the gland cistern. The gland and teat cisterns are rather well developed, increasing only in size, not in morphology, as postnatal growth occurs. During the period from birth to puberty, the amount of connective tissue and fat increases in the udder. A limited amount of glandular development also occurs, but the mammary glands, for the most part, do not radically change in size or cellular function until puberty, pregnancy, and lactation.

From birth to puberty, thyroxine, growth hormone, and the corticosteroid hormones are involved in the general growth and development of the animal, including the mammary gland. Anything that alters the concentrations of these hormones alters the development of the gland; for example, heifers that are overconditioned during this period have impaired secretion of growth hormone and mammary development is stunted. This impairment is not repaired after puberty, and milk production during the first lactation is decreased.

Extensive changes begin to take place in the mammary glands at the onset of puberty. Follicle-stimulating hormone (FSH) is secreted into the blood by the anterior pituitary and stimulates the growth and development of a follicle on the ovary. The newly functional ovary initiates estrogen secretion. Coinciding with the maturation of the follicle, luteinizing hormone (LH) is released by the anterior pituitary, thus triggering ovulation. After ovulation, a corpus luteum forms and subsequently produces another female sex hormone, progesterone. Estrogen is a potent stimulator of the development of the mammary duct system. Thus, with each recurring estrous cycle, more estrogen is released, and the duct system becomes more extensive.

Although the duct system of the mammary glands develops prior to pregnancy, the growth and development of alveoli is minimal. During the first three to four months of pregnancy, there is an extensive proliferation of the duct system. Throughout the remainder of the pregnancy, the lobule-alveolar system undergoes rapid development under the influence of progesterone. Progesterone and estrogen are secreted at this stage of development by two sources: the ovary and the placenta. Prolactin, growth hormone, thyroid hormones, insulin, gluco-corticoids, and placental lactogens are also critically important, along with estrogen and progesterone, in the development of the mammary system. Alveolar development starts around the gland cistern and branches out to the peripheral areas of the gland. By the fifth month of pregnancy, the lobular system is well formed. By the seventh month of gestation, the gland is fully capable of producing milk. If abortion occurs after this point, there is sufficient secretory tissue present to support a full lactation, although the quantity of milk produced is significantly reduced.

PARTURITION AND LACTATION

Immediately prior to the birth of the calf, colostrum synthesis is initiated, and the udder begins to enlarge greatly. At this time, prolactin increases greatly in the pituitary, and there is increased activity of the adrenal cortex. After lactation is initiated, prolactin and somatotropin drive the lactation process.

The functional development of the mammary glands is complete at the onset of the first lactation. The glands continue to grow, however, until the cow is about six years old. Hence, production of milk during each lactation should steadily increase to a peak at six years of age.

Between sixty and thirty days before calving, cows are dried off to allow them to recuperate from the heavy demands of lactation and to help prepare for the next lactation. During this period, alveolar epithelial cells regenerate and renew.

SUMMARY

- The bottom of the teat is closed by a circular muscle (sphincter) known as the teat canal, or streak canal.
- The right and left halves of the udder are divided by a longitudinal groove called the intermammary groove.
- The median suspensory ligament, a yellowish, elastic tissue; the lateral suspensory ligament, an inflexible, white fibrous tissue; and the skin are the primary supporting structures of the udder.
- The basic milk-producing unit of the udder is a very small, bulb-shaped structure with a hollow center called the alveolus.
- Groups of alveoli empty into a duct forming a lobule; lobules empty into another duct system forming a lobe; the ducts of lobes empty into a galactophore, which finally empties into the gland cistern.
- There is no indication of parasympathetic innervation of the udder.
- To produce 1 gallon of milk, more than 500 gallons of blood must pass through the udder.
- The lymphatic system helps to regulate the proper fluid balance within the udder and to combat infection.
- Edema is characterized by excessive accumulation of fluid in the intercellular spaces of the udder and forward of it.

QUESTIONS

1. How does the number of mammary glands in a cow differ from a sow?
2. How does the number of streak canals per teat in a cow differ from a mare?
3. Where is the intramammary groove formed?
4. Which cells line the alveoli? What are their functions?
5. What type of specialized muscle cell surrounds the alveolus?
6. What does oxytocin do when secreted into the blood?
7. What are the two divisions of the autonomic nervous system?
8. How is the process of milk ejection initiated?
9. What are the two venous routes blood can travel as it leaves the mammary gland?
10. What facilitates the unidirectional movement of lymph?
11. From what tissue do the mammary glands develop?

ADDITIONAL RESOURCES

Books

Turner, C. W. *The Mammary Gland. I: The Anatomy of the Udder of Cattle and Domestic Animals.* Columbia, MO: Lucas Brothers Publishers, 1952.

Vorherr, H. *The Breast: Morphology, Physiology, and Lactation.* New York: Academic Press, 1974.

Articles

Ballou, L. U., J. L. Bleck, G. T. Bleck, and R. D. Bremel. "The Effects of Daily Oxytocin Injections Before and After Milking on Milk Production, Milk Plasmin, and Milk Composition." *Journal of Dairy Science 76* (June 1993): 1544–1549.

Bar-Peled, U., E. Maltz, I. Bruckental, Y. Folman, Y. Kali, H. Gacitua, A. R. Lehrer, C. H. Knight, B. Robinzon, H. Voet, and H. Tagar. "Relationship Between Frequent Milking or Suckling in Early Lactation and Milk Production of High Producing Dairy Cows." *Journal of Dairy Science 78* (December 1995): 2726–2736.

Bruckmaier, R. M., and J. W. Blum. "Simultaneous Recording of Oxytocin Release, Milk Ejection and Milk Flow During Milking of Dairy Cows with and Without Prestimulation." *Journal of Dairy Research 63* (1996): 201–208.

Pamphlet

Turner, C. W. *The Comparative Anatomy of the Mammary Glands.* Columbia, MO: University Cooperative Store, 1939.

Internet

Lactation: http://nongae.gsnu.ac.kr/~cspark/teaching/chap10.html.

Mammary Gland Anatomy of Cattle: http://classes.aces.uiuc.edu/AnSci308/anatomycattle.html.

Oxytocin Release and Milk Removal in Ruminants: http://jds.fass.org/cgi/reprint/81/4/939.pdf.

The Purpose of the Milking Routine and Comparative Physiology of Milk Removal: http://www.uwex.edu/uwmril/pdf/MilkMachine/Liners/04_Costa_PurposeOfMilkingRoutine.pdf.

30

Lactogenesis

OBJECTIVES

- To outline the process of lactogenesis and protein synthesis.
- To describe the cytology of secretory cells and the chemical composition of milk.
- To understand the physiological factors affecting the amount and composition of milk.

The efficiency of milk secretion becomes apparent when one realizes a cow that produces 29,000 pounds of milk during one year manufactures 1,046 pounds of milk fat, 1,348 pounds of milk sugar, 954 pounds of milk protein, and 218 pounds of minerals and vitamins or a total of over 3,566 pounds of total solids. This is equivalent to the carcass weight produced by five steers in eighteen months' time. The cow is still alive and can repeat the productivity again and again, whereas the five steers must be slaughtered or spent.

The alveolus is, in effect, a milk factory. It has the ability to take nutrients from the blood and transform them into one of nature's most perfect foods. *Galactopoiesis* is the term used to describe the biosynthesis of milk. For the first one to two hours after milking, there is little intramammary pressure, and the rate of milk secretion is slowed. Once the gland has had a chance to recuperate from the last milking, however, galactopoiesis gets into full gear.

CYTOLOGY (CELL STRUCTURE)

To understand fully the process of milk secretion, one must become familiar with the anatomy of the secretory cells. As illustrated in Figure 30.1, the main parts of the cell are the nucleus, endoplasmic reticulum, Golgi apparatus, mitochondria, and lysosomes.

The nucleus is the informational center for the cell. It contains all the genetic messages required

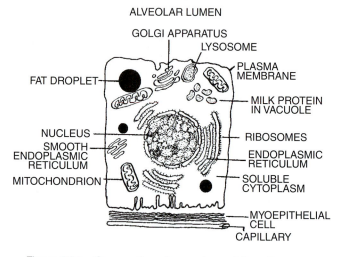

Figure 30.1 Cross section of a secretory cell from the mammary gland. *(Courtesy of Iowa State University)*

for milk synthesis and cellular metabolism. Two membranes surround the nucleus. The outer membrane is continuous with the endoplasmic reticulum and the Golgi apparatus. The inner membrane contains pores that allow material to pass from the nucleus to the cytoplasm. Several nucleoli may be present within the nucleus, and each nucleoli carries genetic information.

The cytoplasm is the area of the cell outside the nucleus and is contained by the outer cell membrane. The cytoplasm contains numerous nutrients and structures that carry out the messages sent by the nucleus and maintain energy production needed to drive the cell.

The mitochondria are often referred to as the powerhouses of the cell and, rightfully so, because most of the energy-liberating reactions occur in these organelles. As energy demands increase, the number of mitochondria likewise increases. Thus, mammary cells in nonlactating cows contain far fewer mitochondria than do cells from lactating cows.

Lysosomes produce enzymes, called lysozmes, which break large molecules into smaller ones. When these enzymes leak out into the cell, the cell is destroyed and digested; hence, a membrane normally surrounds the lysosomes to prevent leakage. When a cell becomes old and nonfunctional, the lysosomes provide one means of destroying the cell so another functional cell can take its place. Involution of the mammary gland during the drying off period and in some cases of mastitis has been linked to lysosomal activity.

In the lactating cow, the endoplasmic reticulum is well developed and abundant. These organelles are responsible for the synthesis of milk protein. Attached to the outer surface of the membrane of the endoplasmic reticulum are several dense particles called ribosomes, structures in which most protein synthesis takes place. The endoplasmic reticulum becomes smooth because it lacks ribosomes, as it leads to the Golgi apparatus. These smooth channels act merely as a means for transferring material throughout the cytoplasm of the cell. The Golgi apparatus in the secretory cells of lactating cows is an enlarged structure. Milk proteins are collected here for subsequent transport into the lumen of the alveolus.

CHEMICAL COMPOSITION OF MILK

Milk varies in composition by species as well as by breed. The composition of milk is also greatly affected by both physiological and environmental factors. Table 30.1 provides the typical range of concentrations of the primary components of milk from dairy cattle.

The first secretion of the mammary gland following parturition is known as colostrum. Colostrum is designed to give the newborn calf a good start in life. It is higher than milk in dry matter, protein, vitamins, and minerals. It also contains antibodies that give newborn animals protection against certain diseases. Following parturition, the synthesis of true milk is initiated (Figure 30.2), although the composition of milk harvested from the gland is in transition for a week or more.

Carbohydrates in Milk

Lactose, or milk sugar, is the most abundant form of carbohydrate in milk. Lactose is a disaccharide, a compound sugar formed from one molecule of glucose and one molecule of galactose. Glucose is found throughout the body and is one of the chief sources of energy. Most of the galactose in the body is produced in the milk, but some galactose is found in galactolipids, cerebrosides, and galactoproteins.

Lactose is synthesized solely from glucose. The secretory cell absorbs glucose and converts part of the supply to galactose, thereby providing the necessary units for lactose production. In the ruminant, most carbohydrates in the feed are broken down into the volatile fatty acids, acetic acid, propionic acid, and butyric acid. Acetic acid is used by the mammary gland primarily for the synthesis of milk fat. Propionic acid is converted to glucose that is subsequently used for the production of lactose. Butyric acid is equally divided among lactose, fat, and casein production. Thus, a ration producing high quantities of propionic acid should increase the production of lactose.

The synthesis of lactose is facilitated by an enzyme called lactose synthetase. It contains two subunits that, when complexed, cause glucose and galactose to combine. The first subunit is called galactosyltransferase and is found in the Golgi bodies of the epithelium. The second subunit, lactalbumin, is produced by the ribosomes along the endoplasmic reticulum and moves to the Golgi apparatus to complex with galactosyltransferase.

The amount of lactose in milk is constant, ranging from 4.5 to 5 percent of the milk, depending on the specific breed of cattle. Lactose in milk is largely responsible for the osmotic pressure exerted by milk. When the cell produces a large amount of lactose, fluids are drawn out into the milk to maintain a constant osmotic pressure. Thus, the rate of lactose synthesis drives milk volume to a great extent.

TABLE 30.1   Gross Composition of Milk

Constituent	Average Content (%)	Normal Variation
Water	87.2	82.4–90.7
Fat	3.7	2.5–6.0
Nonfat Solids	9.1	6.8–11.6
Protein	3.5	2.7–4.8
Casein	2.8	2.3–4.0
Lactalbumins Lactoglobulins	0.7	0.4–0.8
Lactose	4.9	3.5–6.0
Minerals	0.7	0.6–0.8
Total Solids	12.8	9.3–17.6

Source: Modified from Nutritional Quality and Safety of Milk and Milk Products, p. 1,
http://www.afns.ualberta.ca/drtc/dp472-4.htm.

Lipids in Milk

Only about 25 percent of the fatty acids found in milk fat are from dietary fat. The remaining portion is synthesized through numerous metabolic schemes.

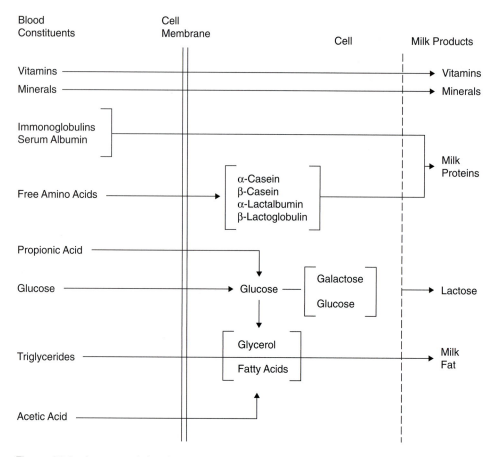

Figure 30.2   Lactogenesis involves the conversion of blood metabolites into milk components by secretory cells in the alveoli. *(Courtesy of Iowa State University)*

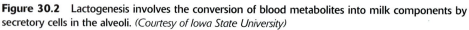

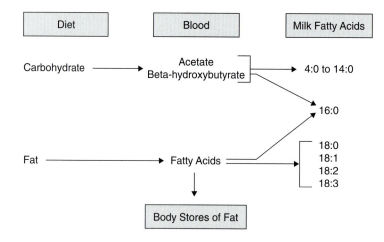

Figure 30.3   Biosynthesis of milk fatty acids.

Triglycerides compose almost all of the milk fat. The fatty acids in these triglycerides are, for the most part, short-chain fatty acids (C4 to C14). These short-chain fatty acids are generally odoriferous, thereby creating the strong smells characteristic of certain types of cheeses. A triglyceride is formed when three fatty acid molecules are combined with one molecule of glycerol.

Almost all of the short-chain fatty acids are synthesized within the cell (Figure 30.3). Two mechanisms have been implicated in fatty acid synthesis and elongation. The first mechanism involves acetic acid, one of the volatile fatty acids produced in the rumen. In this mechanism, successive two-carbon units are condensed until the proper length of the fatty acid has been attained. The second mecha-

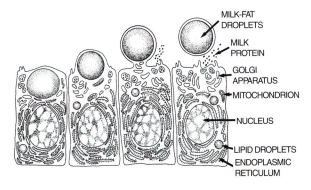

Figure 30.4 Milk-fat globules and other milk components are released from secretory cells by exocytosis. *(Courtesy of Iowa State University)*

nism involves the use of four-carbon molecules, beta-hydroxybutyric acid, as the building blocks for fatty acid synthesis. Long-chain fatty acids (C16 and C18) are absorbed from the blood and incorporated into milk triglycerides. Occasionally, some partial beta-oxidation of these long-chain fatty acids occurs, giving rise to C14 and C16 fatty acids.

The acetic acid pathway is used much more frequently than the beta-hydroxybutyric acid pathway. The prevalence of the acetic acid pathway helps explain why forages are so important in maintaining fat test. When forages are digested, acetic acid is synthesized in large amounts in the rumen. When high-energy concentrates are fed, propionic acid is synthesized in large amounts at the expense of acetic acid. Because increased propionic acid levels increase the production of lactose and therefore increase milk yield, acetic acid levels are too low to maintain high fat test.

Most of the glycerol used in the formation of triglycerides is synthesized in the cell from glucose. However, some glycerol is absorbed from the bloodstream as part of the triglycerides that are subsequently degraded in the cell. Because fat droplets formed in the secretory cells are too large to pass through pores in cell membranes, the cells have developed a specialized mechanism for secretion called emeiocytosis (Figure 30.4). In this mechanism, fat droplets migrate to the cell membrane leading to the lumen. The membrane then surrounds the droplet, and the outer portions of the membrane break down; the newly formed portion of the membrane that has surrounded the droplet prevents other cell contents from leaking out.

Proteins in Milk

The proteins that predominate in the milk protein fraction are alpha-casein, beta-casein, alpha-lactalbumin, and beta-lactoglobulin. These four proteins are found only in milk and constitute more than 90 percent of the milk protein. All of them are formed from free amino acid precursors in the blood. Casein is insoluble in water and forms micelles, small aggregates that are suspended in the milk. Kappa-casein prevents these micelles from forming curds. The primary function of casein is to provide a well-balanced protein for the calf. Alpha-lactalbumin and beta-lactoglobulin are soluble. Alpha-lactalbumin has a critical role in lactose production and has been discussed previously. Beta-lactoglobulin is the predominant protein in whey.

A second group of proteins constitutes the remaining fraction of milk protein. This group contains immune globulins, serum albumin, and gamma-casein. These proteins are not synthesized by mammary cells. Rather, they are absorbed from the blood intact and passed to the milk.

The synthesis of milk protein is not a random process whereby several amino acids are joined together; rather, it is a detailed, predetermined procedure. Within the cell, deoxyribonucleic acid (DNA) serves as the information center concerning the sequences of the various proteins to be synthesized in the cell. When DNA is decoded, amino acids are linked to form a specific protein that has its own particular physiological function.

For milk protein to be synthesized, all of its constituent amino acids must be available. If one amino acid is missing, the synthesis procedure is halted. When a particular amino acid is deficient, it is referred to as a limiting amino acid because it limits the synthesis of protein.

DNA is present in the nucleus and the mitochondria of the cell and acts as the genetic information source. It is composed of nucleotides containing the purines (adenine and guanine) and the pyrimidines (cytosine and thymine). Messages are relayed from DNA to the cytoplasm by ribonucleic acid (RNA), whereupon the sequences of amino acids in protein synthesis are dictated from the sequences of the various nucleotides transcribed from the DNA.

RNA acts as the messenger for DNA in the determination of the amino acid sequences of proteins. These nucleic acids consist of nucleotides containing the purines (adenine and guanine) and the pyrimidines (cytosine and uracil). There are three types of RNA, each one with a specialized function. Messenger RNA (mRNA) transcribes the sequences of DNA into a single strand of RNA. Transfer RNA (tRNA) consists of small molecules that act as carriers for the specific amino acids. The third type of RNA, ribosomal RNA (rRNA), is a major component of the ribosomes within the cell, but the functions of this type of RNA are not as clearly understood. Protein synthesis involves a series of reactions that are specific for each protein. An outline of this procedure follows:

1. Messenger RNA transcribes the sequence message from DNA to form a template for protein synthesis. The sequences of the nucleotides in the DNA are the keys to the sequence pattern forming this template. Triplets of the purine and pyrimidine bases in the DNA form codons that correspond to specific amino acids and are signals that control protein synthesis. For example, the codon of adenosine, guanine, and guanine signals the incorporation of arginine.

2. A specific transfer RNA combines with each of the respective amino acids to form an aminoacyl complex. There is at least one specific tRNA for each of the twenty amino acids.

3. The initiation of protein synthesis occurs when the ribosome (site of protein synthesis) recognizes a codon specific for initiation.

4. Once synthesis has been initiated, the protein is elongated through a series of successive additions of amino acids as determined by the mRNA template.

5. Eventually the procedure is terminated when a codon specific for terminating protein synthesis is reached. Then, the protein splits off the ribosome, and the procedure can be started again.

Vitamins and Minerals in Milk

Vitamins and minerals are passed from the blood through the epithelial cells and into the milk through filtration. The cell membrane of the epithelial cell acts as a barrier to regulate osmotic pressure within the milk and also as a carrier system for molecules that require active transport. Vitamins pass essentially unchanged from the blood into the milk. The concentration of the various vitamins, especially the fat-soluble vitamins, depends on their respective concentrations in the blood.

Many of the minerals pass from the blood into the cells and are complexed with other compounds. About 75 percent of the calcium in milk is complexed with phosphate, citrate, and caseinate. More than 50 percent of the phosphorus in milk is attached to caseinate. The primary minerals of milk are calcium, phosphorus, sodium, chlorine, and magnesium. In general, sodium and potassium concentrations are relatively constant because these minerals are involved in the regulation of osmotic pressure in the milk. When a cow has mastitis or is at the end of lactation, however, she may produce salty milk. This saltiness is caused by the decrease in concentration of lactose and potassium in the milk and the relatively high concentrations of sodium and chlorine.

FACTORS AFFECTING MILK YIELD AND MILK COMPOSITION

For most dairy producers, the milk check is the primary source of income. Most producers are paid essentially for milk components produced. Therefore, income depends on the volume of milk produced and the composition of that milk. Several physiological factors affect the amount and composition of milk; understanding these factors is crucial for maximizing the profitability of the herd. Factors that play a role in influencing milk composition include the following:

Breed and individual inheritance: Variation in the ability of cows to produce total milk, fat, and solids-not-fat is an inherited characteristic. There is both a breed difference and an individual difference. In general, total milk production decreases and butterfat content increases by breeds in the following order: Holstein, Brown Swiss, Ayrshire, Guernsey, and Jersey. Within the Holstein breed, a range in butterfat from 2.6 to 6 percent has been reported, within the Jersey breed, from 3.3 to 8.4 percent. Similar variation between breeds and individuals exists in total milk production.

Stage of lactation: The greatest variation in the composition of milk takes place immediately following parturition, within the first five days after freshening. Colostrum, synthesized in the udder prior to the time of calving, is far different than milk. It contains more globulins, vitamins A and D, iron, calcium, magnesium, chlorine, and phosphorus than does milk, but it contains less lactose and potassium than milk. Total milk production generally increases for the first month following calving, then decreases gradually thereafter. Conversely, the fat test is usually higher toward the end of the lactation period than soon after calving.

Persistency: This refers to the level at which milk production is maintained as lactation progresses. Generally speaking, following the peak lactation period, which is about a month after freshening, the total milk production each month is approximately 90 to 95 percent of that of the previous month.

Estrus; pregnancy: Milk and fat production may fluctuate, usually downward, on the day of or the day following a heat period. Pregnancy seems to have little effect on milk composition. Beginning about the fifth month of pregnancy, however, total production of gestating cows declines more rapidly than that of nonpregnant cows. It has been estimated that the energy requirement of the late-gestation fetus is equivalent to about 400 to 600 pounds of milk.

First- and last-drawn milk: The percentage of fat in last-drawn milk is higher than that in first-drawn milk.

Age: The age of a cow has a definite effect on production. Most cows reach maturity and maximum milk production at about six years of age, following which there is a decline in production. Records indicate that cows produce approximately 25 percent more milk at maturity than they do as two-year-olds. Also, after passing their prime, at six years of age, fat production gradually decreases with advancing age. Age adjustment factors have been developed to standardize 305-day lactation records to a mature equivalent basis and to minimize environmental variation due to the month of the year in which the record began. These factors remove, with considerable accuracy, the environmental effects from age and month of calving in individual breeds and regions.

Size: Within a breed, large cows usually produce more milk than do small cows. For each 100-pound increase in body weight, however, production typically increases only 70 percent of the proportional increase in body size.

Frequency of milking: Increasing the frequency of milking does result in more total milk produced. Cows milked three times a day consistently produce more milk than those milked twice a day, and cows milked four times a day produce more milk than those milked three times daily. Also, cows milked more frequently are more persistent in their production throughout the lactation; milk production declines less rapidly as lactation progresses. Of course, a decision about whether it pays to milk more than twice daily depends on whether the additional milk more than covers the added labor and other costs.

Irregular milking intervals: Unequal intervals between milkings affect both the quantity and composition of milk; more milk of slightly lower fat content is obtained following the longer intervals.

Environmental temperature; season: The fat percentage of milk varies with the season. It is higher in the fall and winter and lower in the spring and summer. It may vary up and down seasonally by an average of 0.3 to 0.5 percent. Solids-not-fat also show seasonable variation, with the low point in the spring and summer. Severe weather conditions usually decrease the amount of total milk produced and may influence the fat test either up or down. Temperatures above 85°F greatly affect cows, and the situation is accentuated when high temperatures are accompanied by high humidity. Cows calving in the fall months consistently produce more than those calving at other times of the year. Cows calving in the spring produce the least. This difference may be as much as 10 to 15 percent.

SUMMARY

- The nucleus is the informational center for the cell.
- The cytoplasm contains numerous nutrients and structures that carry messages and maintain energy production.
- Ribosomes are structures where most protein synthesis takes place.
- Colostrum is higher in dry matter, antibodies, protein, vitamins, and minerals than milk.
- Lactose is a disaccharide formed from one molecule of glucose and one molecule of galactose.
- In the ruminant, carbohydrates are broken down into the volatile fatty acids and acetic, propionic, and butyric acids.
- Almost all of the milk fat is composed of triglycerides.
- The primary minerals of milk are calcium, phosphorus, sodium, chloride, and magnesium.
- Increasing the frequency of milking results in more total milk produced.
- Disease affects milk secretion, in both total production and composition.

QUESTIONS

1. Define *galactopoiesis.*
2. Which structure is referred to as the powerhouse of the cell?
3. Which structure is responsible for the synthesis of milk protein?
4. What is the first secretion of the mammary gland following parturition?

5. What is the most abundant carbohydrate in milk?
6. What are the functions of acetic, propionic, and butyric acids in the mammary gland?
7. Which enzyme facilitates lactose synthesis?
8. What are the four proteins that predominate in the milk protein fraction? How are they formed?
9. What factors influence milk composition?
10. When is the butterfat test usually the highest?
11. At what age do cows reach maturity and maximum milk production?
12. What environmental factors affect the amount and composition of milk?

ADDITIONAL RESOURCES

Articles

Chakriyarat, S., H. H. Head, W. W. Thatcher, F. C. Neal, and C. J. Wilcox. "Induction of Lactation: Lactational, Physiological, and Hormonal Responses in the Bovine." *Journal of Dairy Science* 61 (December 1978): 1715.

Friggens, N. C., K. L. Ingvartsen, and G. C. Emmans. "Prediction of Body Lipid Change in Pregnancy and Lactation." *Journal of Dairy Science* 87 (April 2004): 988–1000.

Harness, J. R., R. R. Anderson, L. J. Thompson, D. M. Early, and A. K. Younis. "Induction of Lactation by Two Techniques: Success Rate, Milk Composition, Estrogen and Progesterone in Serum and Milk, and Ovarian Effects." *Journal of Dairy Science* 61 (December 1978): 1725.

Kensinger, R. S., D. E. Bauman, and R. J. Collier. "Season and Treatment Effects on Serum Prolactin and Milk Yield During Induced Lactation." *Journal of Dairy Science* 62 (December 1979): 1880.

Walsh, M. K., J. A. Lucey, S. Govindasamy-Lucey, M. M. Pace, and M. D. Bishop. "Coping Capacity of Dairy Cows During the Change from Conventional to Automatic Milking." *Journal of Animal Science* 82 (February 2004): 563–570.

Internet

Lactogenesis: http://animsci.agrenv.mcgill.ca/courses/460/topics/5/text.pdf.

Lactogenesis – Initiation of Lactation: http://classes.aces.uiuc.edu/AnSci308/lactogenesis.html.

Managing Milk Composition: http://www.cahe.nmsu.edu/pubs/_d/d-105.html.

Managing Milk Composition: Normal Sources of Variation: http://osuextra.okstate.edu/pdfs/F-4016web.pdf.

Milk Composition: http://classes.aces.uiuc.edu/AnSci308/milkcomp.html.

Modeling Extended Lactation Curve for Dairy Cattle: A Biological Basis for the Multiphasic Approach: http://jds.fass.org/cgi/content/full/86/3/988.

Nutrition Changes Milk Composition: http://www.ext.vt.edu/pubs/dairy/404-232/404-232.html.

31

Milk Letdown Reflex

OBJECTIVES

- To outline the process of milk letdown.
- To describe the role of oxytocin in milk letdown and milk ejection.

To remove milk from the udder, the streak canal must be opened. During machine milking, opening of this canal is accomplished by negative pressure (Figure 31.1). During hand milking, positive pressure is applied to the teat cistern, forcing milk through the streak canal. During suckling, both positive and negative pressures are applied to open the streak canal.

The milk ejection reflex must be activated to allow removal of the majority of the milk stored in the lumen of the alveoli and smaller ducts. Approximately 80 percent of the milk present in the udder at any point is stored in the collecting ducts,

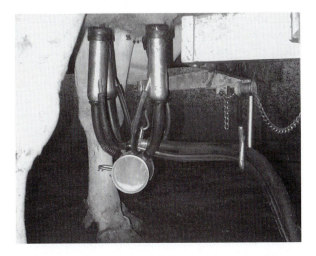

Figure 31.1 Proper techniques for stimulating milk letdown result in copious quantities of milk flowing into the milking unit. *(Courtesy of Mark Kirkpatrick)*

milk sinuses, and cisterns. In the absence of milk ejection, only the small portion of milk stored in the teat and gland cisterns and large lobular ducts can be removed.

Innervation of the mammary gland teats is abundant and pressure-sensitive receptors have been identified in the dermis of the teats. Mechanical stimulation of the teats activates these receptors, which lead to increased impulse transmission in the afferent nerves. For maximal release, a full ten seconds of teat-end stimulation is required. The afferent pathway is part of the spinothalamic system, which is usually activated by abrupt stimuli. Therefore, milk ejection occurs from stimuli that are of high intensity and relatively long duration (i.e., from a machine milking or suckling). On the other hand, there is little question that cues arising from the external environment or cerebral cortex may cause oxytocin release. For example, milk often leaks from the teats of animals waiting to be milked, and conditioned release of oxytocin has been reported. Direct stimulation of higher brain centers releases oxytocin and provides further evidence of involvement of the forebrain in milk ejection. A conditioned response of oxytocin to visual and auditory stimuli associated with milking occurred in over one-third of cows tested.

Following synthesis in cells' paraventricular or supraoptic nuclei, oxytocin and neurophysin I form neurosecretory (storage) granules. These granules are transported down the nerve axons and stored in herring bodies (swollen areas) along the neurons. In addition to storage in granules, there may be a readily releasable pool of oxytocin within the posterior pituitary.

To achieve maximal removal of milk with minimal teat-end damage from overmilking, it is critical to attach the milking machine to teats within forty-five to ninety seconds of teat stimulation (Figure 31.2). Following this period, the efficacy of oxytocin is reduced. Oxytocin has a half-life of only 1.5 to 2 minutes in

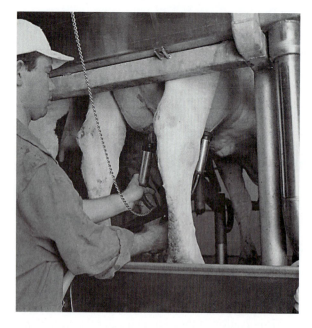

Figure 31.2 For maximal milk harvest, it is crucial that the milker be attached to coincide with milk letdown. *(Courtesy of Westfalia)*

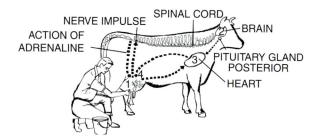

Figure 31.3 Stress-stimulated adrenaline release effectively blocks milk letdown. *(Courtesy of Iowa State University)*

blood. Major sites for the clearance of oxytocin are the kidneys and liver. In addition, an enzyme (oxytocinase) present in the blood of pregnant animals slowly inactivates oxytocin. Oxytocin measured directly in blood increases rapidly following udder stimulation and drops to very low concentrations within five minutes after the start of milking. Oxytocin concentrations in blood closely parallel intramammary pressure. Hence, once letdown has occurred, it is important that the milk be removed within approximately five to six minutes to obtain the greatest amount. A second stimulation cannot be obtained soon after the first.

Myoepithelial cells surrounding the alveoli and the smaller ducts of the mammary gland are not innervated. Oxytocin causes myoepithelial fibers to contract and also causes the dilation of smaller ducts. Oxytocin binds to specific receptors that increase in density throughout pregnancy and achieve greatest density during lactation. Receptors are not present on the smooth muscle of the ducts. Although oxytocin is the primary effector of myoepithelial cell contraction, the myoepithelial cells also contract in response to direct mechanical stimulation. Thus, massaging the udder may enhance expression of milk from the alveoli.

Suckling and milking are the most potent natural stimuli causing milk ejection. Suckling is the most potent stimulator. Washing the udder along with presentation of the calf to the mother is second, washing the udder alone is third, and presentation of the calf alone is fourth. Other stimuli inducing milk ejection include visual and auditory cues associated with the milking procedure. Milk ejection may also be induced by stimulation of the genital tract during artificial insemination, natural mating, or parturition.

Inhibition of the milk ejection reflex may be caused by various stressful stimuli. Stress stimulates release of epinephrine and norepinephrine, in turn causing vasoconstriction. This action minimizes the quantity of oxytocin reaching the myoepithelial cells by redirecting blood flow (Figure 31.3). In addition, epinephrine inhibits oxytocin activity directly at the level of the myoepithelial cell. Administration of exogenous oxytocin does not override this peripheral inhibition of the milk ejection reflex. The most common cause of inhibition of this reflex is associated with emotional disturbances of the animal. Under these conditions, release of oxytocin from the central nervous system is impaired. Emotional stress increases the release of norepinephrine, which blocks the firing rate of neurons in the paraventricular nuclei. Exogenous oxytocin elicits milk ejection in animals subjected to central inhibition stimuli. Shortly after parturition, heifers frequently exhibit central inhibition of milk ejection, and injection of exogenous oxytocin overrides the blockade. Injection of oxytocin is warranted because failure to remove milk, especially in early lactation, increases the risk of mastitis and results in a reduction in milk yield that persists for the duration of lactation.

Approximately 15 percent of the total amount of milk in the udder at the start of milking is not removed when milking is completed. This milk is called complementary or residual milk and may be harvested if oxytocin is administered and the cow is remilked. Twice daily injections of oxytocin (10 international units) immediately after a normal milking, followed by remilking, results in substantially greater total yields of milk. Over time, however, the percentage of milk collected at normal milking decreases, whereas the proportion of complementary milk increases. Furthermore, repeated injections of oxytocin interfere with the secretory activity of mammary epithelial cells and inhibit the normal milk-ejection reflex. Lactating heifers have less complementary milk than do older cows. Complementary milk decreases in proportion to milk yield as lactation progresses.

SUMMARY

- Machine milking is accomplished by negative pressure; hand milking creates a positive pressure.
- To achieve the maximum removal of milk, it is critical to attach the milking machine to teats within forty-five to ninety seconds of teat stimulation.
- Oxytocin concentrations in blood closely parallel intramammary pressure.
- Suckling, milking, washing the udder, presenting the calf, visual and auditory cues, and stimulation of the genital tract during artificial insemination, natural mating, or parturition can cause milk ejection.

QUESTIONS

1. What is the half-life of oxytocin in blood?
2. What is the primary effector of epithelial cell contraction?
3. What is the result of failure to remove milk?
4. Where is oxytocin stored and released?
5. Once letdown has occurred, the cow should be milked within how many minutes to obtain the greatest amount of milk?
6. What hormone is responsible for interfering with the action of oxytocin in a frightened cow?

ADDITIONAL RESOURCES

Articles

Gorewit, R. C., and K. B. Gassman. "Effects of Duration of Udder Stimulation and Milking Dynamics and Oxytocin Release." *Journal of Dairy Science* 68 (July 1985): 1813–1818.

Mayer, H., D. Schams, H. Worstorff, and A. Prokopp. "Secretion of Oxytocin and Milk Removal as Affected by Milking Cows with and Without Stimulation." *Journal of Endocrinology* 103 (December 1984): 355.

Rasmussen, M. D., and E. S. Frimer. "The Advantage in Milking Cows with a Standard Milking Routine." *Journal of Dairy Science* 73 (December 1990): 3472.

Internet

Milk Ejection: http://animsci.agrenv.mcgill.ca/courses/450/extra/450_milk.pdf.

Milk Ejection: http://classes.aces.uiuc.edu/AnSci308/milkejection.html.

Effect of Cow Prep on Milk Flow, Quality, and Parlor Throughput: http://www.ansci.umn.edu/dairy/dairyupdates/du119.htm.

Oxytocin: http://www.vivo.colostate.edu/hbooks/pathphys/endocrine/hypopit/oxytocin.html.

32

Milking Procedures and Processes

OBJECTIVES

- To identify milking equipment and functions of milking systems.
- To outline milking procedures and processes.

Milking is the act of removing milk from the udder. It is essentially the final harvest of the milk crop that determines whether or not a cow or a herd is profitable. The procedures and practices associated with milking the cow are critically important for harvesting the full complement of milk that the cow has produced and stored. Historically, it has been routinely carried out through three methods: suckling by the calf, hand milking, or machine milking (Figure 32.1).

SUCKLING

The fastest means of removing milk from the udder is through the use of the calf. The calf grasps the teat with its tongue and presses it against the soft palate on the roof of the mouth. Milk ejection is accomplished by the creation of a negative pressure in the mouth through widening of the jaws and the retrac-

tion of the tongue. This action causes the streak canal to open, releasing milk from the udder. When the calf swallows, a positive pressure is created, which acts as a resting and a massage phase for the teat.

HAND MILKING

Hand milking is still widely used in lesser developed countries where labor is cheaper than automation and in modern operations to milk out quarters that are infected or have been injured. In hand milking, the teat is grasped between the thumb and forefinger. By applying pressure with the other fingers (in a massaging motion down the length of the teat), milk is then forced from the teat cistern through the streak canal. Through this method, more milk can be obtained than by the use of a milking machine (Figure 32.2).

MACHINE MILKING

The history of machine milking in the United States goes back over 135 years (Figure 32.3). The first vacuum-type milking machine was patented by L. O. Colvin in 1865. In 1884, J. P. Martin devised a milking machine that had individual teat cups, connecting tubes, and a vacuum pump. The pulsator was

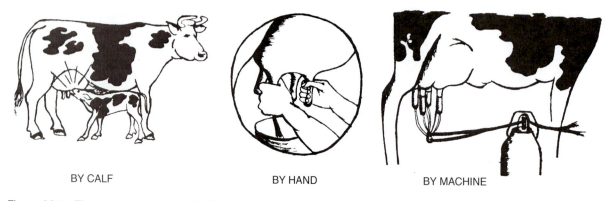

BY CALF   BY HAND   BY MACHINE

Figure 32.1 Three ways to harvest milk. *(Courtesy of Iowa State University)*

Figure 32.2 During hand milking, the milk is stripped from the teat cistern using positive pressure. *(Courtesy of Howard Tyler)*

Figure 32.4 Bucket system milking. *(Courtesy of Iowa State University)*

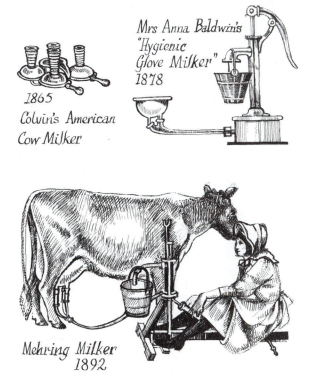

Figure 32.3 Milestones in the historical development of the modern milking machine. *(Courtesy of M. E. Ensminger)*

Figure 32.5 Higher vacuum levels are needed in systems that lift milk from the udder to an overhead milk line, as is the case in stanchion barn milking systems. *(Courtesy of Iowa State University)*

patented by Modestus Cushman in 1885. In 1903, an Australian, Alexander Gillies, developed the first prototype for what was eventually to develop into the modern milking machine. This machine has a vacuum source, a collection receptacle, pulsator, hoses, and individual teat cups and liners.

MILKING EQUIPMENT

Basically, there are two types of milking systems: the bucket system and the pipeline system. In the bucket system, the milk is received directly into a vacuumized portable bucket, which may be either of two types: a floor type (Figure 32.4) or a sus-

pended type. Conventional pipeline systems use a rigid heat-resistant glass or stainless steel sanitary pipe for carrying vacuum from the milk receiver to the individual milking units and for carrying the milk from the units to the receiver. Pipeline milkers may be used in either stanchion and tie-stall barns (Figure 32.5) or milking parlors (Figure 32.6). Regardless of make, the mechanical milking systems can be separated broadly into three major parts: vacuum supply, pulsation, and milking unit.

Vacuum Supply

Vacuum pumps are the engines that drive the milking system. They function to displace air, and the

Figure 32.6   A pipeline system in a milking parlor requires less vacuum with a low milk line. *(Courtesy of Westfalia)*

Figure 32.7   The milking claw is designed to harvest milk with the least amount of damage to teat-end tissues. *(Courtesy of Westfalia)*

controlled movement of air is crucial to the milking process. Two types of vacuum pumps are commonly used in milking operations: rotary and piston. Either type is readily adaptable to most milking operations, but it is generally recommended that the pump be of sufficient size to displace at least 25 percent more air than is required to operate the milking units and to lift and transport the milk to the cooling and storage area if a pipeline system is used. This additional displacement allows for decreased efficiency when the pump wears with age. The demands for a vacuum pump in a pipeline system are about 2.5 times as great as those for the bucket system.

Pulsation

Pulsators are installed in milking systems to direct the airflow leading to and from the milking unit. They cause the teat cup liners (inflations) to open and close, or pulsate, about once a second. Two types of pulsators are still being used: pneumatically controlled pulsators and electromagnetic pulsators. Pneumatic pulsators have a locking-screw valve that allows the milker to adjust the pulsation rate. Electromagnetic pulsators utilize a timing device to regulate pulsation rate and are commonly used in all newer milking systems. A master pulsation control box sends an electrical signal to an electromagnetically controlled plunger in each individual pulsator that controls air movement into the pulsation tubes leading to the milker.

  Older units often used a milking-time-to-collapse-time ratio of 50:50, where the milking phase and massage phase are each allotted about a half second. However, most new milking units are set to use a pulsation ratio of 60:40 or even 70:30 to

Figure 32.8   Teat-cup liners must be replaced at regular intervals to maintain optimal function of the milking system. *(Courtesy of Westfalia)*

allow for faster milk-flow rates. Pulsation ratios of 70:30 milk cows much faster than do lower ratios, but they are much more likely to damage teat ends.

Milking Unit

Each milking unit (Figure 32.7) has four individual teat cups attached by hoses to a unit pulsator. Each teat cup contains an inflation tube called a teat-cup liner (Figure 32.8) that is surrounded by a metal

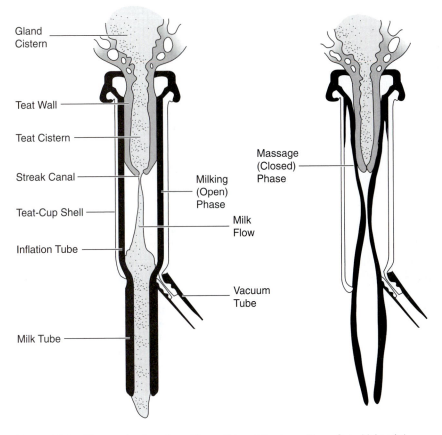

Figure 32.9 The mechanics of machine milking. *(Courtesy of Iowa State University)*

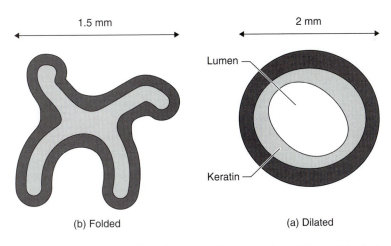

Figure 32.10 Proper milk letdown procedures stimulate dilation of the teat sphincter, permitting faster flow of milk from the teat end and more rapid milking times. *(Courtesy of Westfalia)*

outer shell generally made of stainless steel. Teat-cup liners come in various bore sizes, with the narrow bores generally causing less irritation to the teat than those with wide bores. The top of narrow bore liners tend to ride lower on the teat, avoiding contact with the sensitive, injury-prone area at the top of the teat. When teat liners lose their elasticity, teats become very susceptible to injury because the stiffer liner produces a sharp slapping action rather than a gentle massage on the teat.

Figure 32.9 illustrates the milking action within the teat cup. In the first phase of milking, called the expansion phase, pressure decreases in the space between the outer shell and the liner. This action dilates the teat canal (Figure 32.10) and promotes milk flow. The pressure inside the teat cistern is greater than the pressure outside the teat, and milk is forced through the streak canal (Figure 32.11). The vacuum level at the teat end is critically important; too little vacuum slows milking and reduces the quantity of

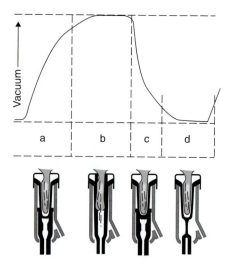

Figure 32.12 Vacuum diagrams display the inner workings of the teat-cup liner, allowing otherwise invisible problems to be detected: a) period of increasing vacuum, b) expansion phase, c) period of decreasing vacuum, and d) massage phase. During the expansion phase, milk flow rate is maximal, while milk flow rate becomes negligable during the massage phase. *(Courtesy of Westfalia)*

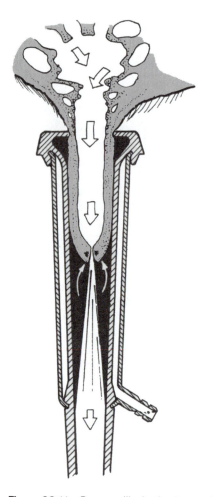

Figure 32.11 Proper milk ejection is a result of both milk letdown in the mammary gland and teat-end vacuum from the milking unit. *(Courtesy of Westfalia)*

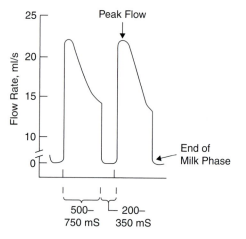

Figure 32.13 Changes in milk-flow rate during the milking cycle. *(Courtesy of Westfalia)*

milk harvested, while too much causes damage to the teat end and callus formation (Figure 32.12). The damage is caused by blood and lymph accumulation at the end of the teat, which results in swelling and congestion. Small amounts of swelling and congestion are inevitable side effects of the milking process, but either overmilking or excessively high vacuum levels result in teat-end damage that can increase in severity at each subsequent milking.

The massage phase, or nonmilking phase, is initiated when air is pumped into the space between the liner and the outer shell. The teat-cup liner then collapses around the teat. This massaging action promotes circulation in the teat and allows the teat a brief moment of rest. The teat sphincter is closed briefly, and blood and lymph is allowed to leave the teat. If the massage phase is too short and the expansion phase too long, circulation is impaired, and the teat again can become injured due to congestion. Ideally, milking machines operate at one massage cycle per second, or sixty cycles per minute. Some systems can be set to operate at dif-

ferent pulsation rates, anywhere from forty-five to sixty-five cycles per minute.

Milk-flow rate (Figure 32.13) is controlled by several factors:

1. **Pressure differential around the streak canal.** This pressure differential is determined by the pressure exerted by the milk in the udder and also by the pressure exerted by the vacuum pump and pulsator.
2. **Size and tautness of the streak canal.** More milk per unit time flows through a larger opening. Several factors affect the tautness of the streak canal, including intensity of stimulation for milk ejection, age of the cow, and effects of teat injury.

Figure 32.14 Whether using paper or cloth towels, it is important to dry teats thoroughly, especially teat ends, prior to applying the milking machine. *(Courtesy of Howard Tyler)*

MILKING PROCEDURES

Milking is the most important single job to be done on the dairy farm. It is important that cows be milked at regular times, preferably by the same milker, and that each milking be a pleasant experience. Cows like to be milked, if it is done properly. The physiology of the discharge of milk is a delicate process, and it requires the close cooperation of the milker and the cow if it is to be successful. A milking program consists of the following coordinated steps:

1. **Preparing the equipment:** Prior to milking, the equipment to be used in the milking process should be assembled and sanitized. Also, it should be checked and adjusted if necessary.

2. **Preparing the cow:** Under natural conditions, the cow is primed or stimulated by the suckling of the calf. This process can be simulated by cleaning the cow's teats with warm water (120° to 130°F) or a predip, then massaging and drying them with a paper towel (Figure 32.14). Following this process, remove two or three streams of milk from each quarter into a strip cup (Figure 32.15) (never strip milk onto the floor) and examine for visible evidence of mastitis. This process also washes out any debris adhering to the end of the teat and enhances milk letdown. It takes a full ten seconds of direct teat stimulation to provide the stimulus for full milk letdown. Inadequate stimulation (less than ten seconds) results in harvesting less than the full complement of milk (Figure 32.16). About forty-five seconds after the priming stimulus, the udder becomes full and firm (especially in early lactation), and milk occasionally leaks from the teats. This is evidence that the cow has let down her milk and is ready for the next step.

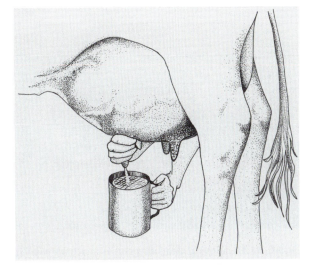

Figure 32.15 Strip cups can be used to check for the presence of clots in milk strippings. *(Courtesy of Iowa State University)*

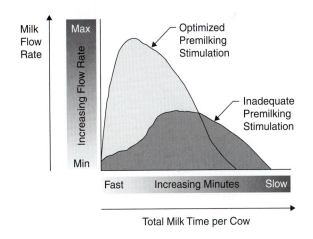

Figure 32.16 Optimizing time spent on premilking teat-end stimulation increases milk-flow rate and decreases milking time per cow. *(Courtesy of Westfalia)*

3. **Attaching the teat cup and beginning:** About forty-five seconds and not more than one minute after washing the udder, the teat cups should be attached (Figure 32.17), and milking should begin (Figure 32.18). Most cows milk out in three to six minutes, depending on the amount of milk and the characteristics of the cow.

4. **Removing the teat cup:** Both incomplete milking and overmilking should be avoided. The greatest cause of machine injury is leaving the teat cups on too long. Incomplete milking usually results because one or more quarters are more difficult to milk than the others. As soon as the udder is empty, and before the teat cups crawl up, they should be removed properly and gently (Figure 32.19). Automatic

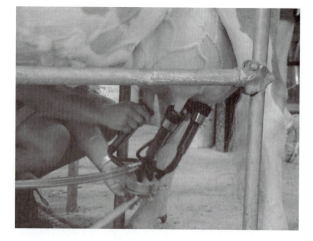

Figure 32.17 Kinking inflations prior to attaching milking units minimizes air leaks and vacuum fluctuations at the teat end. *(Courtesy of Howard Tyler)*

Figure 32.20 Automatic takeoff controllers allow cows to be milked consistently each milking. *(Courtesy of Howard Tyler)*

Figure 32.18 Following proper milk letdown, 60 percent of milk should be removed within the first two minutes after attaching the milking unit. *(Courtesy of Howard Tyler)*

Figure 32.21 George Kinner (left) and Stuart Greene watch as the automatic takeoff units prepare to remove the milking unit on this cow. *(Courtesy of USDA-ARS)*

Figure 32.19 As milk-flow rate decreases, liners tend to crawl up teats. Units should be removed prior to this time. *(Courtesy of Iowa State University)*

takeoffs (Figure 32.20) control the timing of removal by electronically monitoring the rate of milk flow and automatically shutting off vacuum to the milking unit when milk flow drops below a preset flow rate (Figure 32.21). Current recommendations suggest removing units when flow rates are less than 2 pounds per minute (Figure 32.22). The unit is then retracted away from the cow after vacuum levels at the teat end are allowed to diminish, thus minimizing stress on the teat-end tissue. The actual milking time per cow ranges from three to six minutes, with an average time of four minutes for cows in midlactation. But additional time must be allowed for letdown, adjustments, and interval between cows or groups of cows.

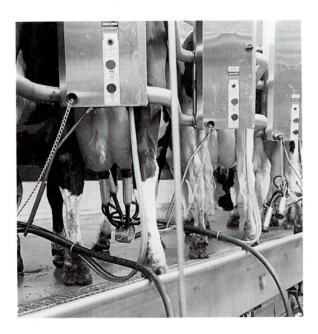

Figure 32.22   These chain-driven automatic detachers are triggered when milk flow drops below a set flow rate for a given period of time. *(Courtesy of Westfalia)*

Figure 32.24   Milking units are returned to the milk room for cleaning between milkings on this dairy facility. *(Courtesy of Mark Kirkpatrick)*

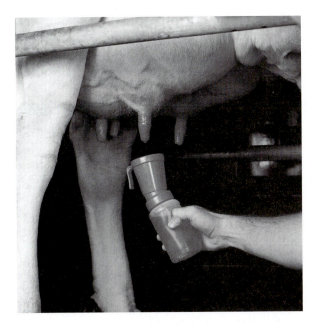

Figure 32.23   Teat dipping after milking is important to protect the mammary gland from pathogen entry until the teat sphincter closes. *(Courtesy of Westfalia)*

Figure 32.25   Clean-in-place systems allow milking units to be cleaned and sanitized in the parlor between milkings. *(Courtesy of Westfalia)*

Following milking, dip the teats in a fresh disinfectant solution (100 ppm iodophor, chlorine or other sanitizing agent) (Figure 32.23). This will remove the milk from the ends of the teats and prevent the invasion of bacteria into the udder. Also, it will avoid attracting flies.

5. **Cleaning equipment:** After milking the last cow, all milking equipment should be thoroughly cleaned and sanitized (Figures 32.24 and 32.25).

SUMMARY

- Milking is routinely carried out through three methods: suckling by the calf, hand milking, or machine milking.
- Milk is received directly into a nearby vacuumized portable bucket, which may be either of two types: floor type or suspended type.
- Conventional pipelines create vacuum from the milk receiver to the individual milking units and carry the milk from the units to the receiver.
- Pipeline milkers may be used in stanchion barns or milking parlors.
- Mechanical milking systems can be separated into three major parts: vacuum supply, pulsation, and milking unit.
- There are two phases of milking: the expansion phase, which dilates the teat canal and promotes milk flow and the massage phase, which promotes circulation and allows the teat a brief moment of rest.
- Incomplete milking and overmilking should be avoided.

QUESTIONS

1. What is the fastest means of removing milk from the udder?
2. What are the two types of milking equipment?
3. What are two types of vacuum pumps commonly used in milking operations?
4. What happens when the massage phase is too short and the expansion phase is too long?
5. What three factors control milk-flow rate?
6. What steps make up a milking program?
7. Why dip the teats with a fresh disinfectant solution after milking is completed?

ADDITIONAL RESOURCES

Articles

de Passille, A. M. "Sucking Motivation and Related Problems in Calves." *Applied Animal Behavioral Science* 73 (May 2001): 175–187.

Hogeveen, H., and W. Ouweltjes. "Sensors and Management Support in High-Technology Milking" (review). *Journal of Animal Science* 81 (Supplement 3, January 2003): 1–10.

Internet

Milking Procedures: http://extension.usu.edu/files/agpubs/milktech.htm.

Proper Milking Procedures: http://edis.ifas.ufl.edu/DS129.

Recommended Milking Procedures: http://nyschap.vet.cornell.edu/module/mastitis/section1/Recommended%20Milking%20Procedure%20Fact%20Sheet.pdf.

33

Milk Quality

OBJECTIVES

- To understand the important factors in milk quality.
- To distinguish the factors affecting milk flavor.
- To learn about regulatory programs, including federal, state, and local programs.

Consumers and health departments all have a vested interest in the quality of milk. Milk quality ultimately determines consumer satisfaction, and a bad experience with poor quality milk can result in the loss of a customer for life. Quality milk can be produced only when the producer pays special attention to several factors:

1. **Health of the herd:** The herd should be free from diseases that might be spread to people through the milk. Bacteria in milk coming from cows must be eliminated. Mastitis is the most important herd health problem.
2. **Clean animals:** Clean flanks and udders help prevent dirt from getting into the milk (Figure 33.1). Clean floors, plentiful bedding, and a well-drained yard make for clean cows.
3. **Clean equipment:** All milking equipment should be kept as clean and free from bacteria as possible. Bacteria grow in cracks and rough spots on equipment if it is not washed properly.
4. **Cool and store milk properly:** Proper cooling and storage of milk on the dairy farm require facilities that cool the milk promptly from the in-the-cow temperature of about 101°F down to 40°F and then hold it at that temperature until it is collected (Figures 33.2 through 33.4). Bacteria reproduce (divide) once every thirty minutes in an environment kept between 70° and 90°F; thus, in twelve hours, one bacterium can

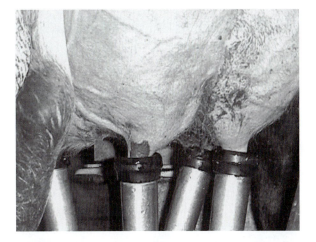

Figure 33.1 Milk-quality issues often trace back to dirty cows. *(Courtesy of Monsanto)*

Figure 33.2 Plate coolers decrease the milk temperature prior to the time that milk enters the bulk tank. *(Courtesy of Howard Tyler)*

Figure 33.3 Bulk milk storage tanks cool milk rapidly and efficiently to maintain milk quality. *(Courtesy of Westfalia)*

Figure 33.4 On this dairy, milk is pumped from the cows into storage silos and then pipelined directly to a cheese-manufacturing plant. *(Courtesy of Howard Tyler)*

reproduce 16 million. Cooling controls this growth.

5. **Keep the barn and milk house clean:** The milking barn should be clean and have a concrete floor. Barn odors may be eliminated by effectively ventilating the building. A milk room is important to the convenience of the operator and an aid to the production of high-quality milk.

6. **Control flies:** Fly-control measures are important to dairy producers. Flies add to the bacterial count of milk; cases are on record of flies carrying as many as 1.25 million bacteria. They can carry typhoid, dysentery, and other contagious diseases. Breeding places for flies, such as manure piles and mud holes, should be eliminated.

MILK FLAVOR

Most consumers base the quality of any product on its flavor, and milk is no exception. They want milk that tastes good. Off-flavors result in unhappy consumers and lost markets. The off-flavors found most often in milk and their causes and prevention are listed here:

1. **Feed and weed flavors:** Off-flavors from rancid feed, wild onions, and other weeds can carry into milk.

2. **Oxidized flavors:** These flavors are sometimes described as cardboard flavors. Some causes of oxidized flavors are metallic contamination from copper and iron, which may be alleviated by using stainless steel; exposure to sunlight or just daylight; foaming; and drylot feeding. Feeding vitamin E to the milking herd reduces or eliminates oxidized flavors.

3. **Rancid flavor:** This flavor is caused by a breakdown of the butterfat, which releases strong-flavored fatty acids. This action is caused by the enzyme lipase, which is present in all milk. The primary causes of rancid milk are excessive agitation of milk, due to high lifts and sharp turns in pipeline milking, and slow cooling with foaming (Figure 33.5).

4. **Barny:** This flavor is caused by dirty stables, poor ventilation, unclean milking, and unclean cows, all of which can be alleviated.

5. **Salty:** This flavor, which masks the slightly sweet flavor of milk, is caused by mastitis. Milk from cows that have mastitis should not be marketed.

6. **Malty:** Malty flavor is primarily due to high bacteria count. The remedy is to keep bacteria out of milk as much as possible and to prevent growth of bacteria in milk. Rapid cooling practically eliminates malty flavor.

7. **High-acid, sour milk:** This flavor is due to very high bacterial count. In these days of mechanical refrigeration, there is no excuse for sour milk. Simply cool milk as rapidly as possible to 40°F.

8. **Unnatural or foreign:** This category refers to flavors that come from medicinal agents and disinfectants. The control of such off-flavors consists in handling medicines and disinfectants so that the flavor or odor from them does not get into the milk and in using chemical sanitizers only in the concentrations indicated by the directions. Do not market milk from drug-treated cows for the entire withdrawal period prescribed on the drug label or by the veterinarian.

Figure 33.5 Milk foaming in the receiver jar leads to decreases in milk quality. *(Courtesy of Mark Kirkpatrick)*

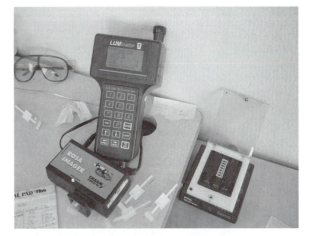

Figure 33.6 On-farm antibiotic residue tests help prevent antibiotic contamination of the milk supply. *(Courtesy of Mark Kirkpatrick)*

For good-tasting milk, the producer should keep it clean, keep it cold, use good-quality feed, and not ship milk from problem cows.

REGULATORY PROGRAMS

Because of the essential nature of milk, plus the fact that it is easily contaminated and a favorable medium for bacterial growth, it is inevitable that numerous regulatory programs have evolved. These regulatory programs include federal, state, and local programs.

Producers are issued permits allowing them to ship Grade A milk. The permit is revoked if either the bacteria count or somatic cell count of raw milk exceeds current standards.

In addition to cleanliness and freedom from mastitis, temperature is important in processing quality milk. Most bacteria cannot reproduce effectively below 40°F; dairy farmers should cool milk below 40°F as quickly as possible. By law, all fluid milk sold for human consumption must be pasteurized. At dairy processing plants, milk is pasteurized to kill disease-causing organisms. It may be pasteurized at either 145°F for thirty minutes or 161°F for fifteen seconds.

In addition, antibiotic residues are a major concern to producers and processors. Processors must dump any milk found to have antibiotic residues from tankers or storage silos, and the financial consequences to producers can be severe. Therefore, producers should be especially careful in following withholding periods for antibiotics administered to lactating cattle. On-farm tests are available to verify that milk is antibiotic-free prior to leaving the farm (Figure 33.6).

The Food and Drug Administration (FDA) is charged with inspecting dairy products and processing plants for contamination and adulteration. Presently, many cooperatives and some milk dealers pay a premium to dairy farmers for producing high-quality milk. Various premiums and penalty bases are in use to reward good production practices that lead to higher-quality product.

SUMMARY

- Mastitis is the most important herd health problem affecting milk quality and milk flavor.
- The milking barn should be clean and have a concrete floor.
- Clean facilities, clean cows, and rapid cooling are all critical elements of milk quality.
- Feeding vitamin E to the milking herd may help reduce or eliminate oxidized flavors.
- Malty flavor is primarily due to high bacteria count; high-acid, sour milk is due to very high bacterial count.
- By law, fluid milk sold for human consumption must be pasteurized at either 145°F for thirty minutes or 161°F for fifteen seconds.

QUESTIONS

1. What factors influence how quality milk is produced?
2. How can the bacterial count in milk be kept down?
3. What are the flavors often found in milk?
4. Which enzyme causes a breakdown of the butterfat? Which releases strong-flavored fatty acids?

ADDITIONAL RESOURCES

Articles

Boor, K. J. "Fluid Dairy Product Quality and Safety: Looking to the Future." *Journal of Dairy Science* 84 (January 2001): 1–11.

Jones, G. M. "Preventing Drug Residues in Milk and Cull Dairy Cows." Virginia Tech, Virginia Cooperative Extension, Publication Number 404–403.

Lau, K. Y., D. M. Barbano, and R. R. Rasmussen. "Influence of Pasteurization of Milk on Protein Breakdown in Cheddar Cheese During Aging." *Journal of Dairy Science* 74 (March 1991): 727–740.

Ma, Y., C. Ryan, D. M. Barbano, D. M. Galton, M. A. Rudan, and K. J. Boor. "Effects of Somatic Cell Count on Quality and Shelf-Life of Pasteurized Fluid Milk." *Journal of Dairy Science* 83 (February 2000): 264–274.

Internet

Bacteria in Milk Sources and Control: http://ianrpubs.unl.edu/dairy/g1170.htm.

Milk Quality Resources: http://www.uwex.edu/milkquality/.

Producing Milk with a Low Bacteria Count: http://ianrpubs.unl.edu/dairy/g678.htm.

7

Concepts in Dairy Biosecurity

34

Dairy Farm Biosecurity

OBJECTIVES

- To define the importance, purpose, and goals of biosecurity programs.
- To describe diagnostic approaches and identify diagnostic tests available.

Biosecurity has become a high-profile issue among dairy producers in recent years. Biosecurity is not simply a matter of reducing the risk of introducing new pathogens to a dairy herd. It also includes management practices that reduce the spread of pathogens within a herd. The primary return on investment in a good biosecurity program comes in the form of reduced involuntary culling rates and decreased veterinary treatment costs. Persistent subclinical infections by many pathogens reduce milk yield and increase reproductive problems. These two parameters account for most of the involuntary culls on dairy operations. Reducing the number of cows exiting herds for these reasons can substantially increase profit potential for dairy producers utilizing biosecurity measures as a management tool to increase herd health.

When pathogens overwhelm the ability of the animal's immune system to respond effectively, infectious disease results (Figure 34.1). It follows that the only methods of reducing the level of infectious disease within a herd are to reduce pathogen load or to increase the ability of the immune system to respond to the pathogen (Figure 34.2). Most health programs are based on the latter, focusing specifically on vaccination programs. Vaccination programs are essential components of herd health programs; however, vaccines are costly and do not confer absolute immunity. Effective plans for reducing pathogen load and exposure to pathogens can dramatically affect the outcome of a herd health program. Biosecurity plans are based on pathogen eradication and control.

Figure 34.1 If the pathogen load is great enough to overwhelm the immune system, disease results. *(Courtesy of Dana Boeck)*

It is estimated conservatively that annual U.S. losses from diseases, parasites, and pests of livestock and poultry are equivalent to about 15 percent of the cash receipts from marketing of livestock and livestock products. Dairy producers may suffer even greater economic losses from decreased growth of young stock, lowered milk production, and increased production costs. Considerable cost is also involved in disease prevention.

A biosecurity program should include plans for reducing the risk of new pathogen introduction to a herd as well as a plan for controlling or eliminating pathogens already on the farm. Very few producers can maintain complete isolation from all sources of outside contamination. Contamination can potentially occur even in closed herds from commonly overlooked sources; a shared fenceline with another cattle herd or a borrowed cattle trailer can become a source of infection for exposed cattle. Visitors can carry potentially devastating pathogens on their boots or clothing, especially if they have visited other farms. All visitors should wear clean boots (or plastic boot covers) and pass through a decontaminating footbath prior to gaining access to cattle.

Purchased animals are common sources of exposure to new pathogens. Nearly half of all producers purchase animals for their operation each year. The risk of contamination from this source may be reduced by requiring a vaccination history prior to purchase and avoiding animals that have been mixed with other cattle, such as at a cattle sale, prior to purchase (Figure 34.3). Cattle should be quaran-

Figure 34.2 In healthy cows, the response of the immune system is greater than the pathogen load to which they are exposed. *(Courtesy of Dana Boeck)*

tined for thirty days following arrival; this requires isolation from shared feeders, waterers, or other equipment. Purchasing heifers allows easier quarantine after purchase. If milking cows are purchased, milk these animals last to avoid cross contamination with other cows via the milking parlor. Vaccination of all purchased animals prior to introduction to the herd is also recommended.

People and equipment are the most common carriers of pathogens between production units on a farm. Calves are the most susceptible animals to pathogens; a primary goal should be to reduce their exposure to all farm pathogens. Calf feeders should complete calf chores prior to working with other cattle on a farm to reduce transfer of pathogens from cows to calves. Feeding adequate high-quality colostrum is critical for optimizing immunity for the first few weeks after birth. However, both colostrum and milk from infected cattle are also potential vectors for infection, especially for bovine lymphoma virus and Johne's disease. Producers should discard these potential sources of contamination from positively identified adult carriers. Persistently infected animals should be sent to slaughter.

As always, sanitation is an important key to protecting animals at any age (Figures 34.4 and 34.5). The mere presence of pathogens is not necessarily a herd health threat; if pathogen load can be maintained at a low enough level, most animals can effectively avoid infectious disease. In addition to reducing pathogen load, good sanitation measures have the added benefit of keeping animals clean and dry. Clean and dry animals are less likely to become infected by pathogens, regardless of the level of exposure.

The relationship of cattle diseases and parasites to other classes of animals and to humans is also of

Figure 34.3 Cattle trucks that haul cows from multiple facilities are a high risk for transferring pathogens from facility to facility. *(Courtesy of Monsanto)*

Figure 34.4 No vaccination program can overcome exposure to overwhelming levels of pathogens. *(Courtesy of Monsanto)*

Figure 34.5 Pathogens that spread via fecal-oral pathways typically flourish in wet, muddy conditions. *(Courtesy of Mark Kirkpatrick)*

critical importance because many are transmissible between species. For example, over 200 different types of infectious and parasitic diseases, including brucellosis (undulant fever), tuberculosis, cryptosporidia, and salmonella, can be transmitted from animals to people. Thus, stringent on-farm sanitation procedures as well as rigid milk and meat inspection is necessary for the protection of human health. This is an added expense that the producer, processor, and consumer all share.

The success of a dairy health program is the responsibility of all members of the dairy team, and it depends on the genetics of the animal, its nutritional status, the environment in which it is maintained, and the preventive medicine program. The herd manager and the veterinarian typically play the most active roles in the development and implementation of a herd biosecurity program; however, the nutrition and sire selection programs are both important to the ultimate success of the plan.

Dairy producers should inform their veterinarians of all treatments that were administered since the veterinarian's last visit, as well as any changes in feeding and management practices. Each cow should be individually identified and have an individual health record indicating all vaccinations, tests, past diseases, and treatments. This information assists the veterinarian in establishing a more accurate diagnosis when illness is encountered.

Veterinarians are responsible for providing an effective disease prevention program that meets the needs and requirements of each particular farm. Visits should be scheduled routinely to prevent problems rather than simply to diagnose diseases or confirm pregnancies. Although the exact herd health program varies according to the specific conditions existing on each individual dairy farm, the basic principles for maintaining biosecurity remain the same.

BIOSECURITY GOALS

Producers must address four primary goals in any biosecurity program:

1. Minimize the use of treatments, especially antibiotics.
2. Minimize pathogen load to the animal.
3. Maximize the immune function of the animal.
4. Minimize the potential for disease transmission.

Minimize Antibiotic Use

It is important that producers use antibiotics judiciously. The appropriate use of antibiotics provides both humane treatment for the infected animal and increased profit for the producer. Inappropriate use of antibiotics has been directly implicated in the development of antibiotic-resistant strains of bacteria.

Microbes naturally vary in their susceptibility to an antibiotic; some microbes (the most adaptable) adapt to a period of antibiotic exposure and develop resistance, especially if the full course of antibiotics is not administered. These resistant microbes then replicate to form a new population after antibiotic use has been discontinued. The best ways to reduce the formation of these resistant populations are to use antibiotics only when warranted by appropriate diagnostic techniques and to administer the recommended amounts for the entire recommended period of administration.

Minimize Pathogen Load to the Animal

Sanitation may be the most important aspect of maintaining animal health. Animal areas should be designed to be as well-drained and as dry as practical to prevent breeding places for pathogens and parasites (Figures 34.6 and 34.7). Fence cattle out of pasture mud holes and ponds for the same reason (Figure 34.8).

The high concentration of dairy animals and the continuous use of buildings often result in pathogen buildup. As pathogens accumulate in the environment, the risk of disease problems becomes more severe. These diseases can also be transmitted to each succeeding group of animals raised on the same premises. Cleaning and disinfection are critically important in breaking the life cycle of many pathogens. Effective disinfection depends on five factors:

1. Thorough cleaning before application.
2. The phenol coefficient of the disinfectant, which indicates the killing strength of a disinfectant compared to phenol (carbolic acid). It is determined by a standard laboratory test in which the typhoid fever germ is often used as the test organism.
3. The dilution at which the disinfectant is used.
4. The temperature—most disinfectants are much more effective if applied hot.
5. Thoroughness of application and time of exposure.

Disinfection should always be preceded by a very thorough cleaning. The presence of organic matter directly interferes with the activity of the disinfecting agent.

Sunlight possesses disinfecting properties, but it is variable and superficial in action. Heat and chemical disinfectants are more effective. The application of heat by steam, hot water, burning, or boiling is an effective method of disinfection. In many cases, however, it may not be practical to use heat.

Maximize Immune Function of the Animal

Effective vaccines exist for the control of many common infectious diseases of dairy cattle. Producers and their veterinarians should develop a vaccination program that is specific for the management skills and pathogens associated with their specific herd, and they should follow it routinely. Programs vary from herd to herd, depending on:

1. Risk of the disease.
2. Costs associated with a disease outbreak.
3. Cost of the vaccine.
4. Efficacy of the vaccine.

The loss of just one animal or the decreased productivity that usually accompanies a disease outbreak often exceeds the cost of vaccinating the entire herd. Some vaccinations can also be used to control certain diseases once they have been accurately diagnosed.

Figure 34.6   Freestall alleys should slope to the center of the alley to avoid accumulation of contaminated water at the base of stalls. *(Courtesy of Monsanto)*

Figure 34.7   Taildocking is often a response to poor sanitation and/or design flaws in freestall alleys. *(Courtesy of Monsanto)*

Figure 34.8   Stagnant water is a potential biosecurity risk for dairy farms. *(Courtesy of Iowa State University)*

Ideally, vaccinations stimulate the immune system to produce large quantities of antibodies against a specific pathogen. In addition, some vaccines activate cellular mechanisms of immunity, such as B- and T-lymphocytes. The resultant increased immunity raises the threshold that pathogens must overcome to cause clinical disease.

There are two basic types of vaccines: modified live and killed virus products. Modified live products are based on an attenuated live virus. Attenuated viruses can replicate normally but cannot cause disease. Most important, they can stimulate the immune system in the same way as a normal virus. The replication process stimulates activation of the cellular immune system (including killer T-cells) in addition to the humoral (antibody) system.

Viruses in killed products do not replicate and therefore do not stimulate cellular immunity. They do stimulate a humoral response; however, the initial antibody response is not very strong. A second dose, administered three weeks after the initial vaccination, stimulates an anemnestic response (greatly heightened antibody production response) and provides a longer-lasting immunity.

If modified live products provide superior immune protection, why would producers ever use a killed product? Some modified live virus vaccines increase the risk of abortion due to the replication of the attenuated virus. Therefore, killed vaccines are typically recommended for pregnant cattle. Again, without a second dose at the proper time (three weeks later), the killed product provides little protection. In addition, some marketed vaccines are a combination of modified live and killed viruses that package a compromise between safety and efficacy.

Minimize Potential for Disease Transmission

The risk of disease transmissions on dairy farms is affected by a combination of facility design, people-and-cow flow patterns, and animal management practices (Figures 34.9 through 34.11). Pets and pests are a common source of pathogen movement around and between farms (Figures 34.12 and 34.13). Dairy producers must also consider outside sources of infection, including sales, shows, visitors, potential runoff from neighboring land, streams, and a myriad of other sources. For producers with registered stock, the economic benefits of transporting the most valuable animals on a farm to various county and state fairs must be balanced with the risk associated with increased exposure to potentially devastating pathogens.

When disease problems arise, affected animals should be isolated as much as possible to minimize the potential for disease transmission.

Figure 34.9 Automated alley scrapers help keep this freestall barn as clean and sanitary as possible. *(Courtesy of Mark Kirkpatrick)*

Figure 34.10 By reducing the risk of disease, provision of ample amounts of clean, dry bedding often reduces the need for other expenses, such as antibiotics. *(Courtesy of John Smith)*

Figure 34.11 Maintaining separate buckets for feeding and for scraping lots reduces the spread of pathogens on farms. *(Courtesy of Iowa State University)*

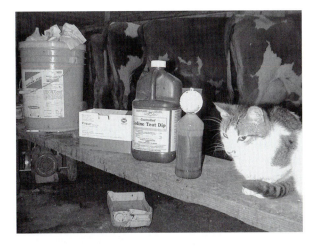

Figure 34.12 Pets and pests, such as cats, mice, and rats, are common sources of potential pathogens on many dairy farms. *(Courtesy of Mark Kirkpatrick)*

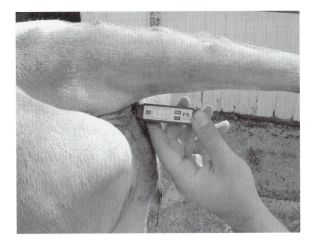

Figure 34.14 An important component of any physical examination is the determination of rectal temperature. *(Courtesy of Mark Kirkpatrick)*

Figure 34.13 Birds that feed in cattle feed bunks also spread pathogens that can cause disease. *(Courtesy of Monsanto)*

DIAGNOSING THE SICK ANIMAL

The key to successful treatment is early diagnosis. Recognizing the signs of disease is critical not only for initiating an appropriate treatment regimen and insuring a speedy recovery, but also for minimizing any long-term losses. Dairy cattle are milked two or three times daily, and milk production is a good barometer of the health of the cow. This daily observation and inspection often enables the dairy producer to detect illness in the early stage in these animals. Other signs of illness include loss of appetite (although this can be difficult to assess in group feeding systems), lack of rumination (resting cows should be ruminating), lameness, fever (Figure 34.14), fecal consistency, appearance of urine, appearance of milk, or a change in temperament or habits.

A thermometer and stethoscope are useful and inexpensive diagnostic tools to have at hand. The temperature is measured by inserting the mercury thermometer full length in the rectum, where it should be left a minimum of three minutes. Prior to inserting the thermometer, a long string should be tied to the end. Many infectious diseases are accompanied by a rise in body temperature, but body temperature is also affected by outside temperature, exercise, excitement, age, feed, and so forth. It is lower in cold weather in older animals and at night.

Veterinary assistance should be obtained immediately to identify the causative agent definitively. Early diagnosis is essential to initiate prompt treatment and rapid return to normal function. All animals entering the herd from shows or sales (or simply returning from such events) should be isolated for three weeks. Many highly infectious diseases are prevented from spreading by quarantine. If possible, divert drainage from adjacent farms and avoid common fencelines that permit direct contact with the neighbor's cattle. Pathogens may be transmitted from farm to farm on shoes, clothing, or vehicles. Travel patterns for feed delivery, milk pickup, and all other routine visitors should be planned carefully to avoid the transmission of pathogens from these sources to animals housed at the facility. Feed samples should be obtained and frozen to facilitate testing if a disease outbreak occurs.

The pulse rate indicates the heart rate. The pulse is taken either on the outside of the jaw just above its lower border, on the soft area immediately above the inner dewclaw, or just above the hock joint. The younger, smaller, and more excitable the animal, the higher the pulse rate. Also, the pulse rate increases with exercise, excitement, and high outside temperature.

The breathing rate can be determined by placing a hand on the flank, by observing the rise and fall of the flanks, or in the winter by watching the breath condense as it exits the nostrils. Rapid breathing due to recent exercise, excitement, hot weather, or poorly ventilated buildings should not be confused with disease. Respiration is accelerated with pain and in febrile conditions.

DIAGNOSTIC TESTING

Diagnostic testing is a necessary expense to monitor the health status of the herd. It is far easier to treat or cull a few infected animals than it is to manage a clinical outbreak after the pathogen has had the opportunity to infect a high percentage of the herd. The most common means of testing is to look at serum titer levels against specific pathogens (Figure 34.15). For the most part, however, serum titer levels tell you very little about the herd health status because titer levels are confounded by vaccination status and previous exposure. Serum samples can be frozen and banked, however, and if an infectious outbreak is suspected, they can be tested. The resulting values are used as a baseline to compare with current titers. This process of serum banking is inexpensive and provides insurance against future problems (Figure 34.16).

In addition to testing for serum antibody titers, producers have other options for detecting the presence of a pathogen. Virus or bacterial isolation procedures look for actual growth of the organism,

and they are quite specific. Success depends on being able to obtain a sample that contains adequate concentrations of the pathogen and on the ability of the organism to grow in culture. Strict sanitary conditions must be observed when obtaining the sample and when developing the culture to eliminate introduction of the pathogen from another source, which would provide a false positive result. Similarly, false negative results can occur if the sample is handled improperly. Enzyme-linked immunosorbent assays (ELISAs) use antibodies to determine the presence of the organism or the presence of specific antibodies. The advantage of this procedure over serum titers is the confirmation of the actual presence of the pathogen or an increased sensitivity to determining the presence of antibody. There are limitations on how these tests can be performed, and most can be run only on fluid samples. Tissue immunohistochemistry tests again look for the actual presence of the pathogen in a tissue sample. These tests are typically very specific, but they are also more expensive than other options.

Regardless of the method of testing, pathogens must be rapidly and accurately identified so appropriate action can be taken. These practices are extremely important in herds undergoing expansion or in new herds. Culling rates in expansion herds can exceed 50 percent in some cases, and the accuracy of testing procedures is a critical component to the ultimate profitability of the herd.

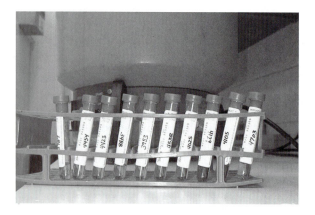

Figure 34.15 Antibody titres provide a snapshot in time that describes exposure to various pathogens. *(Courtesy of Monsanto)*

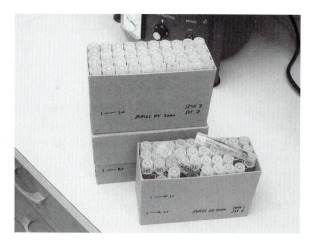

Figure 34.16 Serum banking is a useful method for maintaining the health history of a herd. *(Courtesy of Mark Kirkpatrick)*

SUMMARY

- Biosecurity involves reducing the risk of introducing new pathogens, reducing the spread of pathogens through management practices, reducing involuntary culling rates, and decreasing veterinary treatment costs.
- The only ways to reduce the level of infectious diseases are to reduce pathogen load or to increase the efficacy of the immune system.
- Contamination can occur from a shared fenceline, borrowed trailer, visitors, shoes, clothing, vehicles, runoff from adjoining land streams, cow flow patterns, facility design, sales, shows, and management practices.
- Sanitation, such as keeping clean and dry animals, is an important key to protecting animals and may be the most important aspect of maintaining animal health.
- Cleaning and disinfection are critically important in breaking the life cycle of many pathogens.
- Each cow should be individually identified and have an individual health record indicating all vaccinations, tests, past diseases, and treatments.
- The key to successful treatment is early diagnosis.

QUESTIONS

1. What can visitors do to reduce the spread of pathogens?
2. How long should cattle be quarantined following arrival?
3. What are the most common carriers of pathogens between production units on a farm?
4. How many different types of infectious and parasitic diseases can be transmitted from animals to people?
5. What are four primary goals in any biosecurity program?
6. Disinfection depends on what five factors to be effective?
7. What are the two types of vaccines?
8. If modified live products provide superior immunity protection, why would a producer use a killed virus product?
9. What are some signs of illness?
10. What are some common means of diagnostic testing to monitor herd health?

ADDITIONAL RESOURCES

Articles

Dinsmore, R. P. "Biosecurity for Mammary Diseases in Dairy Cattle" (review). *Veterinary Clinicians of North America Food Animal Practices* 18 (March 2002): 115–131.

Rauff, Y., D. A. Moore, and W. M. Sischo. "Evaluation of the Results of a Survey of Dairy Producers on Dairy Herd Biosecurity and Vaccination Against Bovine Viral Diarrhea." *Journal of the American Veterinary Medical Association* 209 (November 1996): 1618–1622.

Internet

Biosecurity Basics for Cattle Operations and Good Management Practices (GMP) for Controlling Infectious Diseases: http://ianrpubs.unl.edu/animaldisease/g1411.htm.

Biosecurity Guide for Livestock Farm Visits: http://cvm.msu.edu/extension/Biosecurity/Livestock_farm2.pdf.

Biosecurity on Dairies: http://cvm.msu.edu/extension/Biosecurity/BAHMDairy.pdf.

Biosecurity: Protecting Your Health and the Health of Your Animals: http://ianrpubs.unl.edu/animaldisease/nf484.htm.

35

Infectious Diseases

OBJECTIVE

• To identify some specific infectious diseases in cattle and their causes, symptoms, and treatment.

Losses due to infectious disease are the primary source of preventable economic loss in the dairy industry. Innumerable pathogens, infectious diseases, and parasites may affect dairy cattle; however, this section will focus on those with the greatest current economic impact on the dairy industry in the United States. Infections of the udder and hoof will be covered in Topics 37 and 38.

JOHNE'S DISEASE (PARATUBERCULOSIS)

The causative agent involved in Johne's disease is *Mycobacterium paratuberculosis.* Johne's disease is an extremely slow onset, chronic, progressive, incurable, and fatal disease. Estimates for a 100-cow herd with a 20 percent infection rate suggest an annual loss of nearly $25,000 due to reduced milk yield, reduced cull value, increased culling, and wasted feedstuffs. It is seen chiefly in cattle, but it is also found in sheep and goats and more rarely in swine and horses. It resembles tuberculosis in many respects. The disease is widespread, having been observed in practically every country where cattle are raised on a large scale. The prevalence in the United States is estimated to be approximately 20 percent of all herds. The clinical incidence in infected herds ranges from 2 percent to 25 percent in severely infected herds, although this is only a small percentage of the total infected animals. It is one of the most difficult diseases to eradicate from a herd. Transmission is through ingestion of contaminated feces from an infected animal; the organism re-

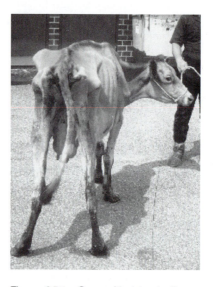

Figure 35.1 Cows with Johne's disease typically lose tremendous amounts of body weight, although feed consumption may be normal. *(Courtesy of Mark Kirkpatrick)*

mains viable in feces, soil, or water for six to eleven months.

Johne's disease involves calfhood exposure with no apparent evidence of infection for six months to several years. The rate of progression of infection to clinical disease depends on age, genetic background, nutritional status, management, and stresses due to lactation and parturition, as well as level of exposure to the organism. At the onset of clinical symptoms, the animal loses condition (Figure 35.1) and displays intermittent diarrhea and constipation, the former becoming more prevalent. Affected animals often retain a good appetite and normal temperature. The feces are watery but contain no blood and have a normal odor. The disease can be fatal; however, the animal often continues living from a month to two years and may go into remission. Infected animals also have increased susceptibility to other diseases.

Figure 35.2 Enzyme-linked immunosorbent assays (ELISAs) are useful diagnostic tools for estimating the prevalence of Johne's disease in infected herds. *(Courtesy of Leo Timms)*

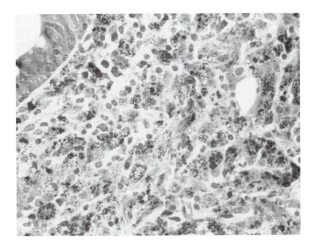

Figure 35.3 To confirm infection with Johne's disease, tissues can be stained for the immunohistological detection of *Mycobacterium paratuberculosis*. *(Courtesy of Mark Kirkpatrick)*

The organism primarily affects the gastrointestinal tract and accumulates in macrophages in the intestine, lymphoid tissues, and other organs. At necropsy, the thickening of the infected part of the intestines, covered with a slimy discharge, is all that is evident. This thickening prevents the proper digestion and absorption of food and explains the emaciation. The disease is caused by the ingestion of the bacterium, *Mycobacterium paratuberculosis*. Apparently healthy animals can spread the disease. Organisms have been isolated from fetuses and uteri of infected animals and are also present in milk and semen, although their importance as disease vectors is unknown.

Testing at regular intervals of three to six months, removing reactors, disinfecting housing, and raising young stock away from mature animals should be practiced in infected herds. Diagnostic tests are notoriously inaccurate; some affected animals fail to react to the more rapid tests (Figure 35.2). Young animals rarely mount an adequate immune response to allow accurate diagnostic testing; therefore, testing is typically limited to springing heifers and cows. Direct visualization of the organism is possible using tissues obtained from necropsy (Figure 35.3). Fecal culture is the most accurate method in live animals; however, false negatives are still a problem due to low sensitivity. The four-month culture period limits the usefulness of this technique.

No satisfactory treatment for Johne's disease has yet been found. Effective prevention is accomplished by keeping the herd away from infected animals. If it is necessary to introduce new animals into a herd, they should be purchased from reputable breeders, and the owner should be questioned regarding the testing history of the herd.

Control programs are based on testing, culling, management practices to limit exposure, and vaccination. The emphasis of these control programs is on eliminating exposure of susceptible animals and reducing environmental contamination. Control measures for infected herds include the following:

1. Calve cows in clean, uncontaminated pasture or in a maternity pen.
2. Clean the perineal region of the cow prior to calving to avoid fecal transmission during the birth process.
3. Remove the calf from the dam prior to nursing, wash the dam's udder thoroughly before milking out colostrum for feeding, and preferentially use colostrum from test-negative cows.
4. Raise young livestock in an uncontaminated environment, separate from mature animals.
5. Test mature animals every six months by fecal culture; remove infected animals promptly.
6. Cull any animal showing clinical symptoms of Johne's disease.
7. Reduce environmental contamination by good sanitation; avoid fecal contamination of feed and water, and fence off or drain stagnant water.
8. Do not spread manure on pasture land.
9. Purchase any replacements from herds free of Johne's disease.
10. Vaccinate cows. The effectiveness of vaccination is controversial; while it reduces the incidence of clinical disease, it does not confer absolute immunity. Vaccines are administered during the first month of life, and the effectiveness decreases over time. Revaccination, however, is associated with decreased immunity.

BOVINE VIRUS DIARRHEA

Bovine virus diarrhea (BVD) is not new, having first been described in 1946. The disease is widespread in the United States. The greatest dairy losses are in feed and milk production. Mortality is low and rarely exceeds 5 percent.

The incubation period is seven to nine days following exposure to the virus. The disease is characterized by high temperature (104°F to 107°F) for two to five days, nasal discharge, rapid breathing, depression, and loss of appetite. Some animals make a prompt recovery. In other cases, signs persist, including nasal discharge and diarrhea. Sometimes blood flecks occur in the feces. Coughing, eye lesions, and lameness may affect 10 percent of the herd. In pregnant cows, abortions generally appear three to six weeks after infection; in lactating cows, a marked loss in milk production occurs.

As indicated by its name, this disease is caused by a virus. The most effective preventive measures consist in avoiding contact with affected animals and in keeping away from contaminated feed and water. Also, all incoming animals should be isolated for at least thirty days. Once the disease makes its appearance, sick animals should be isolated, and rigid sanitary measures should be initiated.

When virus diarrhea is a constant problem, cows should be vaccinated. Immunity against BVD can be achieved by the intramuscular administration of modified live or inactivated vaccines. One vaccination should last a lifetime. Two precautions are important: do not use the vaccine on pregnant cows because of possible abortions and birth defects and do not vaccinate calves under six months of age because it may be ineffective due to the temporary immunity from colostrum of immune dams. Replacement dairy heifers should be vaccinated at nine to twelve months of age. Antibiotics or sulfonamides effectively combat the secondary bacterial invaders that accompany the disease. Administration of balanced electrolytes and fluid is indicated to rehydrate animals with diarrhea.

A class of BVD-infected animals should be considered separately because diagnostic and management strategies are different. Persistently infected (PI-BVD) animals are of great concern in the dairy industry and can be a tremendous threat to an individual herd. These animals are infected *in utero* and do not recognize the virus as foreign; therefore, they do not mount an immune response. Subsequently, they shed the organism in great amounts into the environment and are a major source of infection for their herdmates. Because they do not recognize the virus as foreign, antibody-based tests are ineffective for identifying these animals; a direct test for the virus is necessary (Figures 35.4 and 35.5).

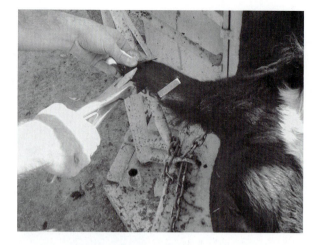

Figure 35.4 Obtaining an ear-notch tissue sample for immunohistochemical diagnosis of BVD infection. *(Courtesy of Leo Timms)*

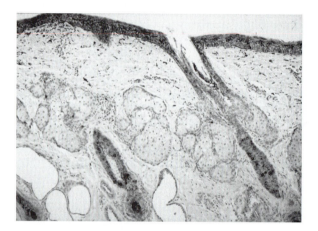

Figure 35.5 Immunohistochemical techniques help veterinarians visualize the BVD virus in ear tissue. *(Courtesy of Mark Kirkpatrick)*

SALMONELLA

Salmonella is a pathogen associated with stress and immunocompromised animals. As such, it is a major risk in expansion units, new facilities, and calf-raising facilities. Salmonella infections are characterized by a rapid onset, severe diarrhea. The affected animal rapidly becomes weak and dehydrated. Death losses can be extremely high.

This pathogen is transmitted in feces; thus, maintaining high sanitation standards on a daily basis is critically important. Particular care should be taken to avoid infecting the most immunocompromised animals on the farm: periparturient cows and newborn calves. The maternity barn should be carefully maintained and be separate from any sick cows.

Serological tests provide information on exposure to five different serogroups of salmonella and are useful for identifying chronically infected cattle.

Cattle infected with salmonella are responsive to antibiotics; animals often require supplemental fluids to prevent dehydration.

BOVINE RESPIRATORY DISEASE COMPLEX

Bovine respiratory disease complex (BRDC) is an acute respiratory disease of cattle. The disease is most common in calves and following shipment. It occurs widely throughout the world, especially among thin and poorly nourished young animals that are subjected to shipment by truck or rail during periods of inclement weather, though it may occur in animals in good condition.

The first sign of the disease (which may appear within two to twenty one days after moving cattle) is a tired appearance and reduced appetite. The affected animal may show signs of depression, watery to slimelike nasal discharge, increased body temperature (rising to 105°F to 107°F), occasional soft or hacking cough, rapid breathing, and loss of appetite followed by loss of body weight and drop in milk production. In very acute forms, animals may die showing no symptoms. Death losses may be high in untreated cases. Calves are more susceptible than older animals, but cattle of all ages are affected.

Bovine respiratory disease complex is caused from multiple infection due to the interaction of viruses and bacteria and is accentuated by environmental conditions creating physical tension or stress. Change in weather and feed, overcrowding, lack of rest, and improper shelter (Figure 35.6) all help usher in the disease. The three viruses that cause most bovine respiratory infections are infectious bovine rhinotracheitis (IBR), bovine virus diarrhea (BVD), and parainfluenza 3 (PI3). Other viruses that may cause respiratory problems include adenovirus, syncytial virus (Figure 35.7), rhinovirus, and rotavirus. Two or more of these organisms may infect a herd at the same time. But viruses are not the only agents of respiratory infection; bacteria can also cause problems, especially in cattle already weakened by infections. For example, infection by *Pasteurella multocide* and *Pasteurella haemolytica* is thought to be a major cause of shipping fever (hemorrhagic septicemia). Other pathogens that may infect weakened cattle include *Haemophilus somnus* and species of salmonella, pseudomonas, mycoplasma, and leptospira. As a preventive measure, newly purchased animals should be isolated for two to three weeks before being placed in the herd.

Immunity against IBR, BVD, and PI3 can be achieved by administration of modified live or inactivated vaccines, in single or combination forms. The routes of administration of these vaccines are intramuscular (IBR, BVD, and PI3) or intranasal

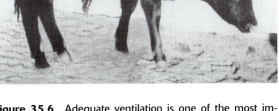

Figure 35.6 Adequate ventilation is one of the most important considerations for the prevention of bovine respiratory disease complex in dairy cattle. *(Courtesy of USDA)*

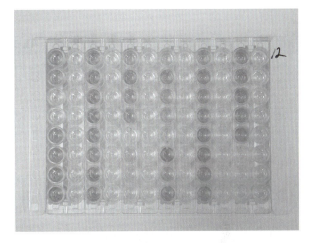

Figure 35.7 For this ELISA test for bovine respiratory syncytial virus (BRSV), the intensity of the blue color is proportionate to the titer of specific antibody in the sample. *(Courtesy of Leo Timms)*

(IBR and PI3 only). Both intramuscular and intranasal vaccines provide adequate immunity. Antibiotics and sulfa drugs are effective treatments if they are given early in the course of the disease. Treatment after BRDC develops is often ineffective.

PNEUMONIA

Pneumonia is an inflammation of the lungs in which the alveoli (air sacs) fill up with an inflammatory exudate or discharge. The disease is often secondary to many other conditions. It is difficult to describe and classify because the lung is subject to more forms of inflammation than any other organ in the body. It affects all animals. In cattle, it is seen most commonly as calf pneumonia, and it frequently accompanies shipping fever. If it is left untreated, 50 to 75 percent of affected animals die.

Figure 35.8 Fever, dullness, lack of appetite, coughing, and nasal discharge are the most common symptoms of pneumonia in calves. *(Courtesy of University of Illinois)*

The disease is characterized by elevated temperature. There is quick, shallow respiration, with discharge from the nostrils and perhaps from the eyes. A cough may be present. The animal appears distressed, stands with legs wide apart (Figure 35.8), drops in milk production, shows no appetite, and is constipated. There may be crackling noises with breathing, and gasping for breath may be noted. If the disease terminates favorably, the cough loosens, and the appetite picks up.

The causes of pneumonia are numerous. Many microorganisms found in other acute and chronic diseases, such as mastitis and metritis, have been incriminated, and pneumonia can be caused by several different viruses. Also, it is generally recognized that changeable weather during the spring and fall and poorly ventilated, damp barns are conducive to pneumonia.

Prevention includes providing good hygienic surroundings and practicing good, sound husbandry. Sick animals should be segregated and placed in quiet, clean quarters away from drafts. Calves can be treated with a broad-spectrum antibiotic for four to five days. Secondary bacterial pneumonia may also be treated with sulfonamides or an antibiotic.

PINKEYE (KERATITIS)

Pinkeye is the common name for an infectious disease that affects the eyes of cattle. It may be caused by several different infectious agents. Of the two most common forms of the disease, one is caused by a bacteria and the other by a virus. It attacks animals of any age, but it is more common in young animals. It seems to become more virulent in certain years and in certain communities. The disease is widespread throughout the United States.

The first symptoms one may notice in bacterial pinkeye are the liberal flow of tears and the tendency to keep the eyes closed. Redness and swelling of the lining membrane of the eyelids and sometimes of the visible part of the eye also appear. There may also be a discharge of pus. Ulcers may form on the cornea. If unchecked, they may cause blindness and even loss of the eye. The attack may also be marked by slight fever, reduction in milk flow, and slight digestive upset.

In viral pinkeye, the eyeball itself is only slightly affected. Infectious bovine rhinotracheitis (IBR), a virus infection of the eyes of cattle, mainly affects the eyelids and the tissues surrounding the eyes. It causes a severe swelling of the lining of the lids.

Bacterial Pinkeye

The most prevalent bacterial form of the disease is caused by *Moraxella bovis*. This organism produces a toxin that irritates and erodes the covering of the eye. Bacterial pinkeye occurs mainly during warm weather. Bright sunlight, wind, and dust may contribute to the cause of the disease. Cattle with white faces or lack of pigment around the eyes are rather susceptible. Transmission is mainly by flies and other insects that feed on eye discharges of infected animals and then carry the infection to susceptible animals. The disease is also spread by direct contact, from animal to animal. Prevention of bacterial pinkeye consists of the following: controlling face flies and other insects that feed around the eyes; good nutrition, including adequate vitamin A; and isolation of affected animals.

The most common treatment for bacterial pinkeye is the application of antibiotics or sulfa drugs to the affected eye as ointments, powders, or sprays, preferably, with treatment made twice daily. Foreign protein therapy, which is the subcutaneous or intramuscular injection of, for example, sterile milk, has been used to treat pinkeye for years, with some success. Cortisone is sometimes combined with antibiotics and injected under the covering (at the outer edge) of the eyeball. The cortisone aids in reducing inflammation and pain and lessens the tears. Recovery is speeded up by keeping the infected animals in a dark barn. A commercially produced protective eye patch is also available. It completely covers the infected eye, holding the medication in place; protects the eye from insects and bright sunlight; and reduces the work and expense of handling and isolation. Held in place by a special adhesive, the eye patch drops off and decomposes after about seven to ten days.

Viral Pinkeye

The most common virus causing pinkeye is the infectious bovine rhinotracheitis (IBR) virus. It is much less common than bacterial pinkeye. When this organism infects the eyes of cattle, there may or may not be other signs of disease, such as respiratory infection, vaginitis, or abortion commonly associated with IBR. It occurs most frequently in the winter, but it may be seen at other times of the year. The disease is highly contagious by direct and indirect contact of infected animals with susceptible animals.

Prevention can be accomplished by proper vaccination of animals prior to onset of the disease. The herd should not be vaccinated once the disease appears, nor should pregnant cows be vaccinated. Affected animals should be isolated. Treatment of IBR conjunctivitis is seldom of value, although antibiotics sometimes help reduce the incidence of secondary bacterial infection.

RINGWORM

Ringworm, or barn itch, is a contagious disease of the outer layers of skin caused by certain microscopic molds or fungi (trichophyton, achorion, or microsporon). All animals are susceptible. Ringworm is widespread throughout the United States. Though it may appear among animals on pasture, it is far more prevalent as a barn disease. Affected animals may experience considerable discomfort, but the actual economic losses attributed to the disease are not too great.

The period of incubation for this disease is about one week. The fungi form seeds or spores that may live eighteen months or longer in barns or elsewhere. Ringworm is usually a winter disease, with recovery the following summer when the animals are on pasture. Round, scaly areas almost devoid of hair appear mainly in the vicinity of the eyes, ears, side of the neck, or the root of the tail. Crusts may form, and the skin may have a gray, powdery, asbestos-like appearance. The infected patches, if not checked, gradually increase in size. Mild itching usually accompanies the disease.

The organisms are spread from animal to animal or through the medium of contaminated fence posts, curry combs, and brushes. Thus, prevention and control consists of disinfecting everything that has been in contact with infected animals. The infected animals should also be isolated. Strict sanitation is essential in the control of ringworm.

COCCIDIOSIS

Coccidiosis, a parasitic disease affecting cattle, sheep, goats, swine, pet stock, and poultry, is caused by microscopic protozoan organisms known as coccidia, which live in the cells of the intestinal lining. Each class of domestic livestock harbors its own species of coccidia; hence, there is no cross infection between animals.

Cattle are affected by twenty-one species of coccidia, but only *Eimeria bovis* and *Eimeria zuerni* are important in the United States, with the latter tending to cause the most serious infection. The distribution of the disease is worldwide. Except in very severe infections or where a secondary bacterial invasion develops, infested animals usually recover. The chief economic loss is in lowered growth of young stock. It is very severe in young dairy calves.

Infected animals may eliminate in their droppings thousands of coccidia organisms (in the resistant oocyst stage) daily. Under favorable conditions of temperature and moisture, coccidia sporulate to maturity in three to five days, and each oocyst contains eight infective sporozoites. The oocyst then gains entrance into an animal by being swallowed with contaminated feed or water. In the host's intestine, the outer membrane of the oocyst, acted on by the digestive juices, ruptures and liberates the eight sporozoites within. Each sporozoite then attacks and penetrates an epithelial cell, ultimately destroying it. While destroying the cell, the parasite undergoes sexual multiplication and fertilization, with the formation of new oocysts. The parasite (oocyst) is then expelled with the feces and is again in a position to reinfect a new host. The coccidia parasite abounds in wet, filthy surroundings; resists freezing and ordinary disinfectants; and can be carried long distances in streams.

A severe infection with coccidia produces diarrhea, and the feces may be bloody. The bloody discharge is due to the destruction of the epithelial cells lining the intestines. Ensuing exposure and rupture of the blood vessels then produces hemorrhage into the intestinal lumen. In addition to bloody diarrhea, affected animals usually show pronounced unthriftiness and weakness. Coccidiosis can be prevented by strict sanitation practices. Prompt segregation of affected animals is important and should be done if practical. Manure and contaminated bedding should be removed daily. Low, wet areas should be drained. If possible, segregation and isolation of animals by age should be used in controlling the disease. All precautions should be undertaken to keep droppings from contaminating the feed. The oocysts resist freezing and certain disinfectants and may remain viable outside the body for one or two years, but they are readily destroyed by direct sunlight and complete drying.

FOOT AND MOUTH DISEASE (FMD)

This highly contagious disease of cloven-footed animals (mainly sheep, cattle, and swine) is characterized by the appearance of watery blisters in the mouth (and in the snout in the case of hogs), on the skin between and around the claws of the hoof, and on the teats and udder. Humans are mildly suscepti- ble but are very rarely infected, while the horse is im- mune. One attack does not induce permanent immunity, but the disease has a tendency to recur perhaps because there are several strains of the causative virus. The disease is not present in the United States, but there were at least nine outbreaks (some authorities claim ten) in this country between 1870 and 1929, each of which was stamped out by the prompt slaughter of every affected and exposed ani- mal. No U.S. outbreak has occurred since 1929, but the disease is greatly feared. Drastic measures are ex- ercised in preventing the introduction of the disease into the United States or in eradicating it in the case of actual outbreak. Foot and mouth disease is con- stantly present in Europe, Asia, Japan, the Philip- pines, Africa, and South America. It has not been reported in New Zealand or Australia.

The disease is characterized by the formation of blisters (vesicles) and a moderate fever three to six days following exposure. These blisters are found on the mucous membranes of the tongue, lips, palate, and cheeks; on the skin around the claws of the feet; and on the teats and udder (Figure 35.9). Presence of these vesicles, especially in the mouth of cattle, stimulates a profuse flow of saliva that hangs from the lips in strings (Figure 35.10). Complicating or sec- ondary factors are infected feet, caked udder, abor- tion, and great loss of weight. The mortality of adult animals is not ordinarily high, but the usefulness and productivity of affected animals is likely to be greatly damaged, thus causing great economic loss.

The infective agent of this disease is one of the smallest of the filterable viruses. In fact, it now ap- pears that there are at least six strains of the virus. In- fection with one strain does not protect against the other strains. The virus is present in the fluid and cov- erings of the blisters and in the blood, meat, milk, saliva, urine, and other secretions of the infected an- imal. The virus may be excreted in the urine for over 200 days following experimental inoculation. The virus can also be spread through infected biological products and by the cattle fever tick.

Neither live cloven-hoofed animals nor their fresh, frozen, or chilled meats can be imported from any country in which it has been determined that foot and mouth disease exists (meat imports from these countries must be canned or fully cured). In the United States, two methods have been applied in control: the slaughter method and the quaran- tine procedure. If the existence of the disease is con- firmed by diagnosis, the area is placed immediately under strict quarantine; infected and exposed ani- mals are slaughtered and buried, with owners being paid indemnities based on their appraised value. Everything is cleaned and thoroughly disinfected.

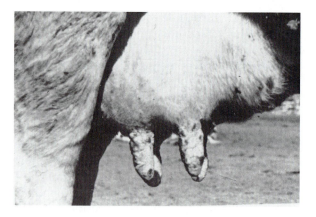

Figure 35.9   Foot and mouth disease (FMD) is characterized by blisterlike vesicles on the tongue, lips, mouth, and teats and between the hooves. *(Courtesy of USDA)*

Figure 35.10   Blisters in the mouth and on the tongue caused by foot and mouth disease result in excessive slob- bering. *(Courtesy of USDA)*

ANAPLASMOSIS

Anaplasmosis is an infectious disease whose etiology and symptoms are similar to cattle tick fever, except that more carriers are involved. It is caused by a minute parasite, *Anaplasma marginale,* that invades the red blood cells. The parasite is transmitted from infected to healthy animals by ticks, horseflies, sta- ble flies, mosquitoes, deer flies, and probably by other biting insects. Lack of proper sanitation and management practices during dehorning or vacci- nation procedures can also be a significant vector.

The disease is widely distributed in warm cli- mates throughout the world. In the United States, it has been prevalent throughout the southern states, but it is slowly spreading to the northern states. The mortality rate may vary from 2.5 percent to as high as 50 to 60 percent. The most severe losses occur in older animals and during hot weather.

In infected animals, the causative parasite, *Anaplasma marginale,* lives in the red blood cells.

The parasite and consequently the disease may be transmitted from animal to animal by means of biting insects and by such mechanical agencies as needles, dehorning instruments, and so forth. Any animal that has once contracted the disease retains the parasite in the blood permanently, though no signs of ill health may be evident. Such animals are carriers and are potential sources of danger to others. In addition to carrier animals, another reservoir of anaplasmosis infection in the western range states is the wood tick. This insect is a biological vector because the disease overwinters in its body.

Cattle are susceptible at all ages, although clinical symptoms generally do not appear until eighteen months of age. The symptoms may be those of a mild, acute, or chronic condition. Calves usually have the mild type of infection, simply becoming dumpy for a few days and then apparently recovering, though their blood remains the permanent abode of the parasite. The more characteristic symptoms in mature animals include anemia, with pale, nonpigmented skin areas, and jaundice, causing yellow appearance of nonpigmented skin areas. Animals also show rapid, pounding heart action; labored and difficult breathing; rise in temperature (up to 107°F); dry muzzle; marked depression; muscle tremors; loss of appetite; and a great reduction in the milk flow. Animals usually show yellowing of the eye and other mucous membranes and of the skin, as in jaundice. Depraved appetite, evidenced by the eating of bones or dirt, is not uncommon. Sick animals may also show brain symptoms and an inclination to fight. Unlike cattle tick fever, bloody urine is not common in anaplasmosis. In severe acute cases, death may follow in one to a few days. Recovery is usually very slow, and although no clinical symptoms remain, such animals continue as permanent carriers of the parasite. Once an animal is infected with anaplasmosis, it becomes a permanent carrier, harboring the disease agent in its bloodstream for life. This animal then becomes a continuous source of infection in the herd.

SUMMARY

- Infectious diseases affecting dairy cattle cause enormous economic losses.
- The causative agent in Johne's disease is *Mycobacterium paratuberculosis.*
- The most effective preventive measures for BVD consist of avoiding contact with affected animals and keeping away from contaminated feed and water.
- Persistently infected animals are infected in utero and thus do not recognize the BVD virus as foreign. A direct test for the virus is necessary for diagnosis.
- Salmonella is associated with stress and immunocompromised animals.
- Bovine respiratory disease complex is an acute respiratory disease of cattle ushered in by a change in weather and feed, overcrowding, lack of rest, and improper shelter.
- Pneumonia is an inflammation of the lungs in which the alveoli (air sacs) fill up with inflammatory exudates or discharge.
- Pinkeye affects the eyes and is caused by either a bacteria, most commonly *Moraxella bovis,* or a virus.
- Ringworm, or barn itch, is a contagious disease of the outer layers of skin caused by molds or fungi, usually occurring in the winter.
- Coccidiosis is a parasitic disease caused by protozoan organisms that live in the cells of the intestinal lining.

QUESTIONS

1. What is the primary source of preventable economic loss in the dairy industry?
2. How long does the organism in Johne's disease remain viable in feces, soil, or water?
3. What are some clinical symptoms of Johne's disease?
4. How often should a herd be tested for Johne's disease? What is the most accurate method?

5. What are some control measures for infected herds?
6. How is BVD characterized?
7. How can immunity against BVD be achieved? What are two precautions?
8. How is salmonella characterized?
9. How is salmonella transmitted, and what is the treatment?
10. What are the signs of a cow with BRDC?
11. What is the cause of BRDC?
12. What are the three most common viruses and two bacteria causing BRDC?
13. How is pneumonia characterized?
14. How is pinkeye characterized?
15. When does bacterial pinkeye mainly occur? What can contribute to the cause? How is it transmitted? What is the treatment?
16. How is metritis characterized?
17. How is ringworm characterized?
18. How is coccidiosis characterized?

ADDITIONAL RESOURCES

Articles

Ames, T. R. "Dairy Calf Pneumonia: The Disease and Its Impact" (review). *Veterinary Clinicians of North America Food Animal Practices* 13 (November 1997): 379–391.

Bitsch, V., and L. Ronsholt. "Control of Bovine Viral Diarrhea Virus Infection Without Vaccines" (review). *Veterinary Clinicians of North America Food Animal Practices* 11 (November 1995): 627–640.

Biuk-Rudan, N., S. Cvetnic, J. Madic, and D. Rudan. "Prevalence of Antibodies to IBR and BVD Viruses in Dairy Cows with Reproductive Disorders." *Theriogenology* 51 (April 1999): 875–881.

Breeze, R. "Respiratory Disease in Adult Cattle" (review). *Veterinary Clinicians of North America Food Animal Practices* 1 (July 1985): 311–346.

Das, P. "Infectious Diseases Surveillance Update." *Lancet: Infectious Diseases* 4 (March 2004): 134.

Fleischer, P., M. Metzner, M. Beyerbach, M. Hoedemaker, and W. Klee. "The Relationship Between Milk Yield and the Incidence of Some Diseases in Dairy Cows." *Journal of Dairy Science* 84 (September 2001): 2025–2035.

Groenendaal, H., and D. T. Galligan. "Economic Consequences of Control Programs for Paratuberculosis in Midsize Dairy Farms in the United States." *Journal of the American Veterinarian Medical Association* 223 (December 2003): 1757–1763.

Internet

Johne's Disease: http://www.extension.iastate.edu/Publications/PM1548.pdf.

Keeping Johne's in Check: http://www.traill.uiuc.edu//dairynet/paperDisplay.cfm?ContentID=200.

Bovine Viral Diarrhea Virus: http://cvm.msu.edu/extension/docs/bvdv.htm.

Bovine Virus Diarrhea: http://www.oznet.ksu.edu/library/lvstk2/mf2435.pdf.

Salmonella in Michigan: http://www.msue.msu.edu/vanburen/salmone.htm.

Infectious Bovine Rhinotracheitis IBR (Red Nose): http://edis.ifas.ufl.edu/VM051.

Infectious Bovine Rhinotracheitis (IBR/Red Nose): http://us.merial.com/veterinary_professionals/veterinarians/dairy/disease_pdf/IBR_RN.pdf.

36

Reproductive Diseases and Disorders

OBJECTIVE

- To understand the symptoms and signs, causes, prevention, and treatment of the following reproductive diseases: brucellosis, leptospirosis, vibriosis, bovine trichomoniasis, metritis, bovine protozoal abortion, and foothill abortion.

BRUCELLOSIS (BANG'S DISEASE)

Brucellosis, which occurs throughout the world, is an insidious (hidden) disease in which the lesions frequently are not evident. Brucellosis derives its name from a British Army surgeon, Sir David Bruce, who discovered the bacteria, later named *Brucella melitensis,* in 1887. In cattle, it is called Bang's disease after Professor Bang, noted Danish research worker, who first discovered the organism, *Brucella abortus,* responsible for bovine brucellosis, or contagious abortion, in cattle in 1896. In swine, it is Traum's disease, or infectious abortion, caused by *Brucella suis.* In goats, it is Malta fever, or abortion, caused by *Brucella melitensis.* In humans, it is Mediterranean fever, or undulant fever.

Although brucellosis remains a threat to the U.S. cattle industry, much progress has been made toward its eradication. The first national testing program, which was initiated in connection with the cattle reduction program necessitated by the drought of 1934, revealed a cattle infection level of 11.5 percent. Today, the infection level nationally is less than one-half of 1 percent. The average infection rate within infected herds is about 20 percent. Control and eradication of the disease are important to lessen economic loss. We also need to alleviate the danger of human infection because brucellosis (undulant fever) is still an occupational risk for certain people.

The symptoms of brucellosis are often rather indefinite. While abortion is the most readily observed symptom in cows, not all animals that abort are affected with brucellosis and not all animals affected with brucellosis necessarily abort. On the other hand, every case of abortion should be regarded with suspicion until proved noninfectious. The infected animal may give birth prematurely to a dead fetus, usually during the last third of pregnancy. On the other hand, the birth may be entirely normal. But the calf may be weak, or there may be retention of the afterbirth, inflammation of the uterus, and/or difficulty in future conception. Milk production is usually reduced. There may be abscess formation in the testicles of the male and swelling of the joints (arthritis). The observed symptoms in humans include weakness, joint pains, undulating (varying) fever, and occasionally orchitis (inflammation of the testes).

The disease is caused by a bacteria called *Brucella abortus* in cattle, *Brucella suis* in swine, and *Brucella melitensis* in goats. The *suis* and *melitensis* types are seen in cattle, but the incidence is rare. Swine are infected with both the *suis* and *melitensis* types and horses may become infected with all three types. Humans are susceptible to all three species of brucellosis. In most areas, the vast majority of undulant fever cases in humans are due to *Brucella suis.* The swine organism causes a more severe disease in humans than does the cattle organism, although not so severe as that induced by the goat type (*Brucella melitensis*). Far fewer people are exposed to the latter simply because of the limited number of goats and the rarity of the disease in goats in the United States. Dairy producers are aware of the possibility that humans may contact undulant fever from handling affected animals, especially at the time of parturition; from slaughtering operations or handling raw meats from affected animals; from consuming raw milk or other raw by-products from cows or goats; and from eating uncooked meats infected with brucellosis organisms. The simple precautions of pasteurizing milk and cooking meat, however, make these foods safe for human consumption.

The *Brucella* organism is quite resistant to drying but is killed by common disinfectants and by pasteurization. It is found in immense numbers in the various tissues of the aborted young and in the discharges and membranes from the aborted animals. It is harbored indefinitely in the udder and may also be found in the sex glands, spleen, liver, kidneys, bloodstream, joints, and lymph nodes.

Brucellosis appears to be commonly acquired through the mouth in feed and water contaminated with the bacteria or by licking infected animals, contaminated feeders, or other objects to which the bacteria may adhere. Venereal transmission by infected bulls to susceptible cows through natural service may occur, but it is rare.

Freedom from disease should be the goal of all control programs. Testing; the removal of infected animals; strict sanitation; proper and liberal use of disinfectants; isolation at the time of parturition; and the control of animals, feed, and water brought into the premises are the keys to the successful control or eradication of brucellosis. The nationwide cooperative federal-state brucellosis eradication program, which was initiated in 1934, has been effective in reducing the incidence of bovine brucellosis in the United States. Four principles are involved: (1) finding infected animals and eliminating them from the herd; (2) vaccinating where there is a disease problem; (3) certifying brucellosis-free herds and areas, with the certification progressing from an individual herd, out to an area or county, finally to a state; and (4) providing indemnity to farmers whose animals are condemned under the program.

LEPTOSPIROSIS

Leptospirosis was first reported in cattle in the United States in 1944, although it had been found in dogs in the United States since 1939. Bovine leptospirosis has been reported in Europe and Australia, in addition to the United States. Human infections may be contracted through skin abrasions when handling or slaughtering infected animals, by swimming in contaminated water, through consuming raw beef or other uncooked foods that are contaminated, or through drinking unpasteurized milk.

In most herds, leptospirosis is a mild disease. However, the symptoms may vary from herd to herd or even within a herd. In general, the symptoms noted in cattle include high fever, poor appetite, abortion at any time (Figure 36.1), bloody urine, anemia, and ropy milk. All ages of cattle and both sexes (including steers) are affected. The disease is caused

Figure 36.1 Leptospirosis can cause abortions during the last half of gestation. *(Courtesy of USDA)*

by several species of corkscrew-shaped organisms of the spirochete group, primarily *Leptospira pomona* in cattle, although *Leptospira hardjo* and *Leptospira grippotyphosa* are becoming more common.

The following preventive measures are recommended:

1. Test animals prior to purchase, isolate them for thirty days, and then retest prior to adding them to the herd.
2. Keep premises clean, and avoid used of stagnant water.
3. Control rodents and other vectors such as canines.
4. Vaccinate susceptible animals annually if the disease is present in the area.

Carrier animals, that is, animals that have had leptospirosis and survived, may spread the infection by shedding the organism in the urine. The infected urine may then either be inhaled as a mist in cow barns, or it may contaminate feed and/or water and thus spread the infection. Recovered animals may remain carriers for two to three months or longer after getting over the clinical symptoms. The organisms seldom survive for more than thirty days outside the animal; however, stagnant water and mild temperatures favor their survival.

VIBRIOSIS

Vibriosis is an infectious venereal disease of cattle that causes infertility and abortion. Infected herds are characterized by abortions in the middle third of pregnancy, increased services per conception, and irregular heat periods.

The disease is caused by the microorganism *Campylobacter fetus*, which is transmitted at the time of natural service. Prevention consists in avoiding contact with diseased animals and contaminated feed, water, and materials. Vaccination, repeated annually, is also effective in preventing the disease. Artificial insemination is a rapid and practical method of stopping infection from cow to cow. Aborting cows should be isolated, and aborted fetuses and membranes should be burned or buried. Contaminated housing should be cleaned and disinfected thoroughly.

BOVINE TRICHOMONIASIS

Trichomoniasis is a protozoan venereal disease of cattle characterized by early abortions (usually between the second and fourth months of pregnancy) and temporary sterility. The protozoa that cause the disease, known as *Trichomonias foetus*, are one-celled, microscopic in size, and capable of movement. They are found in aborted fetuses, fetal membranes and fluids, vaginal secretions of infected animals, and the sheaths of infected bulls. Diagnosis can be confirmed microscopically. The infected bull is the source of the infection. On the other hand, the disease appears to be self-limiting in the cow. This disease is found throughout the United States and is a serious problem in infected herds. The economic loss is primarily due to the low percentage calf crops in infected herds.

The protozoa that cause the disease are one-celled microscopic organisms with three thread-like whips (flagella) at the front and one at the rear. Evidence indicates that the disease is spread from the infected to the clean cow by an infected bull at the time of service and that other types of contact infection do not occur. Following one or perhaps two abortions, cows appear to be immune to reinfection.

Infected bulls appear normal (Figure 36.2). There may be some mucous discharge from the sheath, and the sheath may be slightly inflamed. The only clinical evidence of infection is the transmission of the disease to the females serviced. Infected cows frequently show a whitish vaginal discharge, abortion in the first third of pregnancy, uterine infections, irregular heat periods, and increased numbers of services per conception. Early abortions or erratic heat periods in individuals or herds that are known to be free of Bang's disease should lead one to suspect the presence of the trichomonad infection. Definite diagnosis of infection in the bull is made by means of microscopic examination of smears taken from the prepuce of the bull or the vagina of the cow.

Prevention lies in the use of clean bulls or artificial insemination. Slaughter, rather than treat-

Figure 36.2 This bull, infected with trichomoniasis, appears normal and will breed normally, but he can infect an entire herd through natural service. *(Courtesy of University of Illinois)*

ment of bulls, is recommended. If natural service is essential, only young bulls should be used because they lack susceptibility to trichomonad infection. All bulls over four years of age should be eliminated. In cows, the disease appears to be self-limiting; that is, cows appear to acquire an immunity after about three months' sexual rest. Infected cows should be tested for a minimum of three months and then bred by artificial insemination to a bull known to be clean.

METRITIS

Metritis is an inflammation of the uterus, usually caused by various bacteria, that affects cattle, horses, sheep, and swine. Metritis usually develops soon after the animal has given birth. It is characterized by a foul-smelling discharge from the vulva that becomes thick and yellow or white and finally brownish or blood-stained. Chilling, high temperature, rapid breathing, marked thirst, loss of appetite, and lowered milk production also occur. Pressure on the right flank may produce pain. The animal may lie down and refuse to get up. Affected animals may die in one to two days, or the acute infection may develop into a chronic form, producing sterility.

Metritis can be caused by several different bacteria with *Escherichia coli* the most common. Laceration of the uterus at the time of calving and/or retention of the afterbirth are the principal predisposing causes. However, almost all cows harbor microorganisms in their uterus after calving. The most common of these, in addition to coliforms, are streptococci and staphylococci species and *Actinomyces pyogenes*.

Preventive measures consist in alleviating as many of the predisposing factors as possible, including bruises and tears while giving birth, exposure to wet and cold, and the introduction of disease-causing bacteria during delivery or the manual removal of the afterbirth. Clean, well-bedded maternity stalls should be provided. If assistance at calving time becomes necessary, producers should first disinfect their hands and arms as well as the animal's external genitals.

BOVINE PROTOZOAL ABORTION

Bovine protozoal abortion is a relatively recently discovered disease (first described in 1985) but was likely an undiscovered problem for many years prior to being discovered. Not all infected cows abort, and calves born from infected cows often experience nervous system disease. The causative agent is *Neospora*, and it appears to be transmitted through congenital infection as well as through fecal-oral transmission.

FOOTHILL ABORTION (EPIZOOTIC BOVINE ABORTION)

Foothill abortion is an infectious disease of cattle manifested primarily by abortion. It is epizootic in California, where it is known as foothill abortion because of its high prevalence in cows pastured in foothill terrain. The disease has been reported in the western United States and in Europe. Cows may abort when three to six months pregnant, with the abortion rate frequently reaching 65 percent. Some calves are stillborn, while others are weak at birth.

The disease is caused by a virus (psittacoid virus). The soft-bodied pajaroello tick, *Ornithodoros coriaceus,* is the vector. Prevention consists in moving cattle out of tick-infested areas (dryland brush areas) during the three- to six-month gestation period. Animals that have aborted usually are immune and can be retained safely in the herd.

SUMMARY

- Brucellosis, which occurs throughout the world, is an insidious disease in which the lesions frequently are not evident. It causes contagious abortions.
- Testing; the removal of infected animals; strict sanitation; proper and liberal use of disinfectants; isolation at the time of parturition; and the control of animals, feed, and water brought into the premises are the keys to the successful control or eradication of brucellosis.
- Vibriosis is an infectious venereal disease of cattle that causes infertility and abortion.
- Bovine trichomoniasis is a protozoan venereal disease of cattle characterized by early abortions (usually between the second and fourth months of pregnancy) and temporary sterility.
- Metritis is an inflammation of the uterus, usually caused by various bacteria, that affects cattle, horses, sheep, and swine.

QUESTIONS

1. Humans are susceptible to which three strains of brucellosis?
2. What are the symptoms of leptospirosis in cattle?
3. What are three measures for preventing leptospirosis in cattle?
4. Name the signs and symptoms of vibriosis in cattle.
5. What are the symptoms of a cow infected with trichomoniasis?
6. Characterize a cow infected with metritis.

ADDITIONAL RESOURCES

Articles

Bennett, R. M. "Decision Support Models of Leptospirosis in Dairy Herds." *Veterinary Record* 132 (January 1993): 59–61.

Bennett, R. M., K. Christiansen, and R. S. Clifton-Hadley. "Estimating the Costs Associated with Endemic Diseases of Dairy Cattle." *Journal of Dairy Research* 66 (August 1999): 455–459.

Culter, S., and A. Whatmore. "Progress in Understanding Brucellosis." *Veterinary Record* 153 (November 2003): 641–642.

Patterson, R. M., J. F. Hill, M. J. Shiel, and J. D. Humphrey. "Isolation of Haemophilus Somnus from Vaginitis and Cervicitis in Dairy Cattle." *Australian Veterinary Journal* 61 (September 1984): 301–302.

Internet

Brucellosis: http://www.aphis.usda.gov/vs/nahps/brucellosis/.

Infectious Abortions in Dairy Cows: http://www.vetmed.ucdavis.edu/vetext/INF-DA/Abortion.html.

Leptospirosis: http://osuextra.com/pdfs/F-9130web.pdf.

Leptospirosis in Cattle: http://www.aces.edu/pubs/docs/A/ANR-0858/ANR-0858.pdf.

Metritis and Endometritis: http://www.wvu.edu/~exten/infores/pubs/livepoul/dirm22.pdf.

37

Infections of the Mammary Gland

OBJECTIVES

- To identify the signs, symptoms, causes, and treatment of mastitis.
- To outline mastitis prevention and testing programs.

Mastitis is an infectious inflammation or irritation in the udder that interferes with the normal flow of milk and/or its quality. It takes a heavier toll on the dairy industry than any other single disease, costing the industry more than $2 billion per year. Losses are primarily from decreased milk production and decreased milk quality (Table 37.1). Mastitis causes a yearly loss of $225 per afflicted cow, according to the National Mastitis Council. Studies at Michigan State University have shown an overall average mastitis infection rate of 35 percent in dairies, with some individual operations running as high as 75 percent. Approximately 10 percent of all cows in the United States produce abnormal milk at any given time, and 40 percent of these cows are infected with pathogenic bacteria in two or more quarters. On average, every cow has two new infections per lactation.

SYMPTOMS AND SIGNS

In acute mastitis, the udder is hot, very hard, and tender. The animal has an increase in temperature, refuses to eat, and has dull eyes and a rough coat. The inflammatory response results in an increase in blood proteins and leukocytes in mammary tissue and milk; the purpose of the response is to destroy or neutralize the irritant, repair tissue damage, and return the udder to normal function.

Inflammation is the normal process of leukocytes (white blood cells) multiplying and converging at the site of an infection to combat disease and infection. Leukocytes in milk can phagocytose the invading organism and probably account for a large portion of the mastitis cases that undergo spontaneous recovery. Within the udder, neutrophils and macrophages phagocytize a wide variety of particles including microorganisms, milk-fat globules, and casein. Phagocytosis is a process of recognition, ingestion, and digestion of foreign particles by neutrophils and macrophages. In the lactating udder, neutrophils predominate, while the dry udder has more macrophages. In cattle, billions of neutrophils are mobilized to fight infection. Over 50 million neutrophils per milliliter may be found in infected quarters. Despite this capacity to mobilize tremendous numbers of neutrophils, the mammary gland remains highly susceptible to mastitis. Ingestion of milk-fat globules alters the surface structure of neutrophils and reduces phagocytic ability. Casein phagocytosis is also inhibitory to the neutrophil's ability to destroy microorganisms.

Slight increases in somatic cell scores indicate abnormal secretion but not necessarily a case of mastitis. However, a high somatic cell count in milk indicates an udder infection and probable mastitis. Flakes in milk are congealed neutrophils, secretory cells, and protein. Cowside screening tests are milk-quality indicators and are based on the congealing. The somatic cell content of milk is composed of both neutrophils from the blood and epithelial secretory cells. Neutrophils are present as a response to infection or injury, while epithelial cells are present as a result of infection or injury.

The quantity of milk produced is often dramatically reduced, and the milk that is produced may be lumpy or watery. Mastitic milk has poor flavor; is typically lower in fat, lactose, protein, minerals, and vitamins; and is higher in water and chlorides. The extent of the changes in composition varies with the severity and extent of infection. Severely infected milk may have a total solids content of less than one-third that of normal milk. Abscesses may appear on

TABLE 37.1   National Trends for Herd Test Days for Milk Yield and Somatic Cell Count (SCC) of Milk by Year

Year	Herd test days (no.)	Cows+ per herd (no.)	Average daily milk yield (lb)	Average SCC (cells/ml, 1000s)	Herd Test Days‡ with SCC Greater Than			
					750,000 Cells/ml (%)	600,000 Cells/ml (%)	500,000 Cells/ml (%)	400,000 Cells/ml (%)
1995	265,844	50.0	65.3	304	4.1	9.3	16.0	27.2
1996	255,039	55.5	64.7	308	4.1	9.2	16.1	27.8
1997	287,789	57.4	66.4	314	4.2	9.5	16.6	28.8
1998	283,695	60.8	66.8	318	4.5	10.1	17.8	30.3
1999	273,364	67.0	68.2	311	4.3	9.7	17.1	29.8
2000	260,139	73.3	69.1	316	4.1	9.4	16.8	29.5
2001	244,940	79.1	69.0	322	4.9	10.6	18.2	31.1
2002	267,809	77.5	69.5	320	5.6	11.0	18.1	30.0
2003	251,182	80.5	69.6	319	5.6	11.2	18.4	30.4
Annual trend	−2,160	4.1	0.7	1.8	0.2	0.2	0.3	0.4

Source: Modified from USDA-AIPL: http://www.aipl.arsusda.gov/publish/dhi/current/.

the udder. Death often occurs in untreated, acute mastitis. In chronic mastitis, the only symptom that may be noted is that the milk is thick or lumpy.

CAUSES OF MASTITIS

Mastitis may be either infectious or noninfectious. Infectious mastitis, resulting from the invasion of bacteria in the gland (Figure 37.1), may be from several different types of bacteria. Noninfectious mastitis is the result of injury, chilling, bruising, or rough or improper milking. Infectious mastitis may be present either as a subclinical (hidden) infection or as an overt clinical infection. In subclinical cases, bacteria are present in the gland, but both the udder and milk seem normal. However, certain changes in the milk, such as an increase in chloride content, can be detected.

There are between fifteen and forty cases of subclinical mastitis for every case of clinical mastitis occurring in the herd. The losses from subclinical mastitis account for two-thirds of the total economic losses from mastitis. In mild clinical or chronic cases of mastitis, a few flakes or clots are apparent in the foremilk, but no swelling or fever appear in the udder. These cases usually involve a long-standing subclinical infection with periodic flare-ups into the clinical form. In severe clinical or acute cases, milk is abnormal, and quarters are hot, swollen, and painful. These infections occur suddenly and may become systemic, with signs of fever, rapid pulse, depression, weakness, and loss of appetite.

At least twenty different microorganisms can cause mastitis, but most of these cases are due to a few major pathogens.

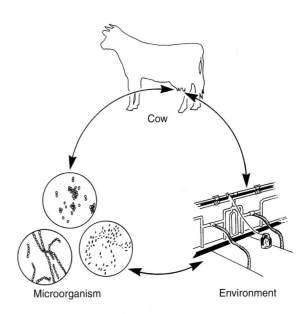

Figure 37.1 The risk of contracting mastitis between milkings depends on the level of pathogen exposure and several environmental factors. *(Courtesy of Iowa State University)*

Contagious Mastitis

Pathogens that are spread from cow to cow primarily during milking are commonly referred to as contagious pathogens. Typically, they affect a high percentage of the herd when they are present. The primary contagious pathogens are listed here:

Staphylococcus aureus: The principal reservoirs of infection are udder and teat skin and infected milk. These infections can be spread during milking by machines or personnel. The organism penetrates deeply into glandular epithelium and is therefore difficult to eradicate with conventional antibiotic treatments during lactation. The immune response of the cow is not typically successful in eliminating established infections with this pathogen.

Streptococcus agalactiae: These pathogens are readily eradicated with antibiotics. They multiply in milk and on mammary epithelial surfaces, although they will not survive in the environment. This is not the case with other streptococcal organisms.

Mycoplasma: Mycoplasma may infect many tissues or organs, including the mammary gland. This type of infection is spread easily from animal to animal and is difficult to eradicate. Cows infected with mycoplasma typically shed in high enough amounts that bulk tank detection is an option. This allows producers to test for the presence of this pathogen in their herd without resorting to testing individual animals. The natural reservoir for *Mycoplasma bovis* is the upper respiratory tract of cattle. Mycoplasma mastitis is characterized by a severe, purulent mastitis affecting multiple quarters. There are usually no systemic manifestations of the disease. There is also no effective treatment, and clinically infected cows should be culled.

Environmental Mastitis

Pathogens that primarily infect cows between milkings are referred to as environmental pathogens. When present in a herd, they typically affect a much smaller percentage of cows than is the case with outbreaks of contagious organisms. The primary environmental pathogens are the following:

Streptococcus dysgalactiae and *Streptococcus uberis:* These organisms are commonly called environmental strep organisms. Symptoms of infection with these organisms are typically confined to the mammary gland.

Coliforms, for example, *Klebsiella* **or** *Escherischia:* These organisms are ubiquitous in the environment. Coliform infections are relatively rare probably because the organisms are susceptible to humoral and cellular factors in milk. They frequently appear when no other pathogens are present (in other words, they are opportunistic). Trends toward larger herds and confinement housing favor a higher incidence of coliforms. No practice can insure freedom from coliforms, but sanitation throughout the operation is the key.

Pathways for Infection

As stated previously, *Streptococcus agalatiae* and *Staphylococcus aureas* are thought to be transmitted into the udder via the streak canal primarily during

Figure 37.2 Backflush systems reduce the spread of contagious pathogens, such as *Staphylococcus aureus,* by sanitizing milkers between groups of cows. (*Courtesy of John Smith*)

Figure 37.3 Uncomfortable freestalls can increase the incidence of mastitis by increasing exposure to environmental pathogens. (*Courtesy of Monsanto*)

milking (Figure 37.2), while coliforms are believed to be transmitted between milkings (Figure 37.3). All these pathogens can spread from cow to cow, although proper sanitation reduces the problem. These organisms require damage or irritation to udder tissue to gain entry to the gland. Cows with teat erosion are three times more susceptible to new infections, and mastitis occurs in 75 percent of quarters with damaged teats.

The establishment of disease depends on the ability of the bacteria to multiply in the udder. Some infections disappear spontaneously, while some require antibiotic treatment for elimination. Some can never be eliminated.

The most common route of entry for pathogens is the streak canal at the end of the teat. A sebum-like mass of cells and lipid (waxy) substances (keratin) line the streak canal, which provides a seal in

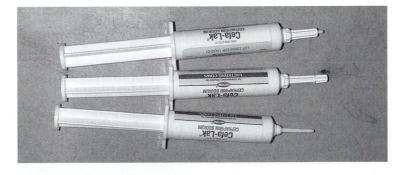

Figure 37.4 Cannulas on commercial mastitis treatments should be inserted only one-quarter inch into the teat end to minimize keratin removal. *(Courtesy of Leo Timms)*

the teat and also exhibits bactericidal properties. The bactericidal properties are associated with long-chain fatty acids. Furstenberg's rosette produces the keratin in the streak canal, which is the primary barrier to infection. Cows with small-diameter streak canals are more resistant to mastitis. Conversely, cows that tend to leak milk have an increased susceptibility to mastitis. Insertion of teat dilators or antibiotic tubes may remove keratin from the streak canal (Figure 37.4), leaving cows more prone to mastitis. After damage to these cells occurs, it takes about one month for full regeneration of the keratin plug in the streak canal to occur.

Once bacterial infection is established, bacteria multiply at a rapid rate. As they multiply, they compete with milk-secreting cells for nourishment and destroy them. They also produce toxins that destroy more milk-secreting cells. In addition, they may mechanically block normal circulation and change the filtering ability of membranes. In chronic infections, bacteria multiply and cause clotting of milk; continued infection causes inflammation of the ducts, often blocking them off completely. If treatment is not begun promptly or is unsuccessful, scar tissue forms, and the ducts become permanently blocked. Loss of function in these glands may be permanent (Figure 37.5). In acute mastitis, blood vessels become greatly dilated, causing stagnation. Milk ducts become compressed so that little or no milk is formed, and the milk that is formed cannot be withdrawn. Because of these changes, medication cannot be effectively introduced into the gland.

Each case of mastitis results in the loss of more milk-secreting cells and the blockage of more milk ducts and further decreases in production potential. Therefore, even though the treatment of a clinical case of mastitis may eliminate the pathogens involved, economic losses from those infections continue until the following lactation.

Many factors increase the susceptibility of cows to infection by pathogenic organisms. The inci-

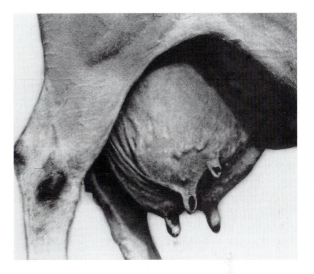

Figure 37.5 Severe mastitis can result in the loss of function of individual quarters. *(Courtesy of USDA)*

dence of mastitis increases with increasing age of the cow. Nonetheless, it is possible for the udders of first-calf heifers to be infected at parturition. The level of milk production is probably not directly related to incidence of mastitis. However, other factors that affect milk yield, such as pendulous udders and milking rate, may be related to incidence of mastitis.

Although bacteria can infect an udder at any time, the majority of new infections become established during the first three weeks of the dry period and during the first month following parturition. The length of time that milk is retained within the mammary gland is critically important in allowing pathogenic organisms to multiply and ultimately overwhelm the immune system. Numbers increase exponentially over time in a high-quality growth medium such as milk; the increase in susceptibility in the first few weeks of the dry period depends on the number of infectious organisms present in the mammary system at dry-off and the quantity of milk present for growth of these organisms. During the period immediately following parturition, the immune system of the cow is compromised, so even a low pathogen load can overwhelm host defense and result in a clinical case of mastitis. During both these periods, sanitation during the milking procedures and cleanliness of the cow's environment is of utmost importance in reducing the incidence of clinical infections.

Frequency of milking also affects mastitis rate. Pathogen load in the gland can be reduced simply by evacuating the milk more frequently, thus reducing the time available for pathogen growth before they are removed from the gland. Thus, increasing milking frequency from twice daily to three times a day not only increases milk production but also simultaneously reduces the risk of clinical mammary gland infections.

The position of the gland makes it extraordinarily accessible to pathogens in the environment. The more exposed the teats become to the environment, the greater the risk. Chilling of the udder on cold ground in the fall or spring also increases the incidence of clinical mastitis. Adequate bedding is recommended to protect the udder from extreme variations in temperature. Housing, as it relates to the degree of udder and teat injuries and to improper ventilation and dampness, influences the incidence of mastitis. Fewer teats are stepped on and injured in loose housing or freestall systems than in stanchion barns.

Susceptibility of a cow to mastitis is inherited, but heritability is low. Heritability estimates for susceptibility to mastitis may be related more to conformation of the cow than to genetics of the immune response. The length of the leg in proportion to udder size and relative strength of udder attachments are examples of conformational factors affecting pathogen load at the teat end. Furthermore, milking rate, which depends in part on dilatability of the teat orifice and teat length, is an inherited characteristic that is related to susceptibility to infection. Selection of bulls that produce daughters with increased resistance to mastitis may become more widespread in the future.

PREVENTION

From a herd health standpoint, a milking program should embrace the following:

1. **Udder and teat protection:** Most mastitis is predisposed by injuries to the udder and teats (Figure 37.6). The best injury preventive measure consists in using liberal amounts of clean, dry bedding.

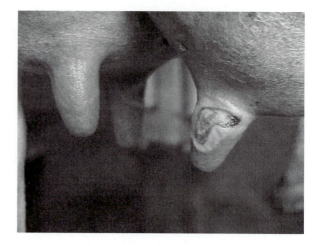

Figure 37.6 This erosive teat condition is caused by a bacteria called *Dermatophilus congolensis,* and it occurs in wet conditions that soften teat skin. *(Courtesy of Mark Kirkpatrick)*

2. **Sanitation:** Clean udders and teats are essential to the production of clean milk of high quality. This requires adequate space and a clean barn; clean, dry corrals (free from low, wet, muddy, or swampy areas); and clean, sterile milking equipment.

3. **Proper milking procedure:** First, the cow must be stimulated to release her milk (letdown). This can be accomplished by keeping the cow content in a quiet, peaceful atmosphere; keeping regular milking hours; and thoroughly washing and drying the teats, which both stimulates and cleanses. After stimulation, rapid and complete milking must follow immediately. Within about forty-five seconds following the stimulation, pressure builds up in the cow's udder due to the action of the hormone oxytocin, which is released by the pituitary gland. This hormone action lasts for about five to six minutes. It causes the muscles to contract around the milk cells. Within one minute following stimulation, attach the teat cups and begin milking. The milking machine should be removed promptly when milk flow ceases because overmilking may contribute to increased infection rates, clinical mastitis, and elevated somatic cell counts. Incomplete milking increases the incidence of mastitis, but machine stripping also actually increases susceptibility to mastitis. There appears to be an increased risk to infection near the end of a milking. Retrograde flow of milk from teat cistern to gland cistern may occur at the time the teat-cup liner collapses. This may provide the means for transfer of pathogens from the teat lining to the gland cistern. Vented inflations and low

Figure 37.7 Teat dips are effective only if adequate coverage of all teats is accomplished. *(Courtesy of Mark Kirkpatrick)*

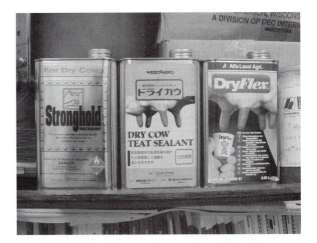

Figure 37.8 Barrier teat dips are useful for preventing mastitis in dry and transition cows. *(Courtesy of Mark Kirkpatrick)*

milk lines reduce retrograde impact frequency. If vacuum is not broken prior to removing the claw of the milking machine, air may propel bacteria back against the teat orifice at the speed of approximately 50 feet/second. By the same token, hand milking the first few squirts flushes the streak canal without allowing retrograde flow back into the gland cistern. This technique may help reduce the number of new infections.

4. **Postmilking teat dip:** Use a postmilking teat dip to lessen the incidence of new infections during lactation (Figure 37.7). Dip the ends of the teats with an approved and effective teat dip after milking. Dipping is the most effective preventive measure against *Streptococcus agalactae* and also helps reduce *Staphylococcus aureas* infections. Iodine, chlorine-based, and chlorhexidine dips are not particularly effective against coliform infections. Teat dips lose most of their germicidal properties within fifteen minutes after application; therefore, they are least effective against environmental pathogens and most effective against those infections occurring at the end of milking. Infections at this time are usually the result of milking very wet cows, having faulty or worn liners, or vacuum fluctuations. Chapping or drying of teats can result from dipping. Allow cows to drip dry prior to leaving the parlor during extremely cold weather. At least two-thirds of the teat should be immersed, and the dip should be used up within five or six cows (organic matter can decrease effectiveness). Teat dips containing more than 10 percent lanolin or glycerine as emollients should never be used. Barrier

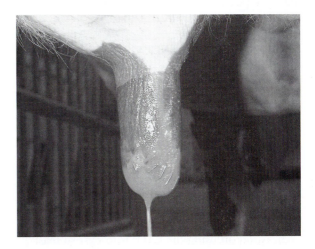

Figure 37.9 Barrier dips simply block bacterial access to the animal and can also be used to protect cut areas where the skin barrier has been compromised. *(Courtesy of Leo Timms)*

dips form a physical barrier over the teat end (Figure 37.8). They are not used routinely because of cost and the difficulty in removing the barrier from teats prior to milking. They are used most often at drying off to protect from environmental pathogens, and they should remain on teats for one to three days under normal conditions (Figure 37.9).

5. **Milking machine repairs:** Keep milking machines, the most used machines on the dairy farm, in good operating condition at all times. Milking machines do not cause mastitis if the system is properly installed, regularly maintained, and periodically tested and deficiencies are corrected. Most machine-related issues are related to improper use of the equipment.

Improper use of the milking machine is related to tissue damage and therefore to the incidence of mastitis. Teat erosions, cyanosis, and edema of teat ends develop more rapidly when vacuum levels reach 18 to 20 mm Hg. Teat erosions provide a place for bacteria to multiply, thus increasing susceptibility to new infections. Irregular fluctuation of vacuum level is an important contributing factor to the susceptibility of cows to mastitis. However, pulsation rates or ratios are of little importance in the etiology of mastitis provided vacuum levels and teat-cup assemblies meet the manufacturer's recommendations. The use of narrow-bore inflations reduces the incidence of clinical mastitis probably because they reduce teat-end vacuum fluctuations.

TESTING

Although mastitis is often apparent, it may be a hidden disease. Therefore, several different tests have been developed for detecting the presence of the causative microorganisms in lactating cows; among them are several screening tests, or presumptive tests. These are performed either at the side of the cow or at the bulk tank. The California Mastitis Test (CMT) is the most widely used of these tests. It involves the mixing of one or two streams of milk with an equivalent amount of reagent in a four-compartment paddle. The greater the gel formation, the higher the number of somatic cells in the milk (Figures 37.10 through 37.14). A reasonable goal, based on using the CMT, is to have at least 75 percent of the milk samples score negative or trace. Less than 75 percent negative and trace readings in-

dicates a milking management problem. On an individual quarter-of-the-udder basis, 90 percent of the samples scoring negative or trace indicates a well-managed herd.

Other tests for abnormal milk or infection, usually conducted in a laboratory or dairy plant, include the Wisconsin Mastitis Test (WMT), catalase test, electronic counters (Foss-o-matic or Coulter counters), or specific laboratory tests designed to detect the causative organism. Except for the direct laboratory tests for determination of the organism, each of these techniques is subject to high levels of variation on any individual sample and should not be the sole basis for decision-making. The reliability

Figure 37.11   Step two of the CMT is to add reagent to each cup. *(Courtesy of Mark Kirkpatrick)*

Figure 37.10   The first step in performing the California Mastitis Test (CMT) is to strip a few squirts of milk into each cup of the paddle. *(Courtesy of Mark Kirkpatrick)*

Figure 37.12   The amount of reagent added to each cup should be approximately equal to the amount of milk present in each cup. *(Courtesy of Mark Kirkpatrick)*

of these tests depends on the sanitary manner in which milk samples are collected and handled after collection, as well as variability in operator proficiency and machine variability. In addition, somatic cells of individual cows can be measured monthly by the DHIA tester, which can be an additional tool for monitoring udder health. Its limitations need to be kept in mind, however, and decisions should never be based strictly on somatic cell counts.

When interpreting individual herd or cow SCCs, several concepts should be kept in mind. Uninfected cows at the beginning and end of lactation frequently secrete milk that contains more than 200,000 cells per milliliter. The foremilk and strippings of unin-

fected cows may also be especially high in cell numbers, and older cows have more cells in their milk than do younger cows. Because these are normal responses, they should not be considered to be inflammatory responses in the usual sense. Somatic cells may also increase in the summer, although whether this increase is due to an increase in infections or is merely a normal physiological response is unclear. In general, producers should strive for bulk tank SCCs of 200,000 per ml of milk. SCCs over 400,000 would definitely indicate room for improvement, while counts over 750,000 render milk unsalable.

TREATMENT

All of these tests are effective tools only when they are followed with action when a problem quarter is located. The veterinarian should be consulted about treatment. Success depends on treating promptly and appropriately for the specific pathogen that is present. For treatment purposes, the specific organisms should be identified by milk cultures. Samples need to be taken aseptically. Sensitivity tests can also be conducted to determine which antibiotics will kill the pathogens that are present most effectively. Withholding periods need to be strictly observed; they contribute to the economic loss incurred. Milk from all quarters must be withheld because infusion in one quarter produces detectable quantities of antibiotics in all other quarters.

Mastitis is usually treated by the intramammary injection of antibiotics. Acute cases should also be treated systemically. Infections that have become systemic may also be treated with intravenous solutions of 7 percent saline, which greatly stimulates water consumption and subsequent renal clearance of endotoxins. More frequent milking schedules for mastitic cows reduce pathogen loads while simultaneously reducing the amount of growth medium (milk) available for the remaining pathogens. The use of oxytocin can be helpful for fully clearing the gland of residual milk, especially in cases where painful infections may otherwise inhibit milk letdown. Dry cow treatments are long-lasting antibiotic preparations and the most effective treatments available for *Staphylococcus aureus* infections. These should not be used if the cow will not have a full dry period.

All intramammary infusions should be conducted aseptically. Teat ends should be cleaned using individual alcohol pads, and the infusion tube should be inserted only as far as necessary (partially through the streak canal; about 2 to 3 millimeters). This prevents excessive dilation of the teat canal and removal of keratin and assures drug contact with any bacteria colonizing the keratin. Disposable sterile syringes should always be used, and antibiotics should be stored as suggested on their labels.

Figure 37.13 After the reagent is added, the paddle should be gently swirled to mix the reagent thoroughly with the milk. *(Courtesy of Mark Kirkpatrick)*

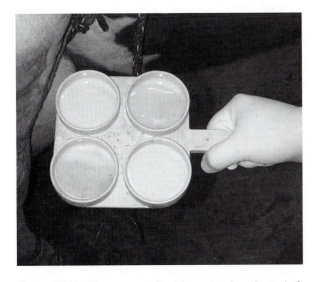

Figure 37.14 The amount of gel formation is estimated after about ten seconds of mixing. *(Courtesy of Leo Timms)*

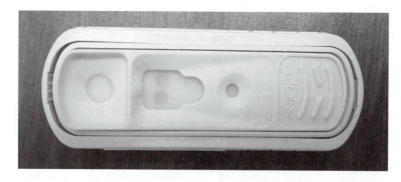

Figure 37.15   Simple on-farm tests have been developed to reduce the risk of antibiotic residues in milk. *(Courtesy of Mark Kirkpatrick)*

There are many reasons to scrupulously avoid antibiotic contamination of our milk supply (Figure 37.15). First, antibiotics in milk are illegal, and contamination results in stiff financial penalties. In addition, antibiotic-contaminated milk may result in development of resistance to or reactions to antibiotics in babies. Over 20 million people in the United States are hypersensitive to antibiotics. Antibiotics are not destroyed during milk processing,

and they reduce cheese yield. Milk from treated cows needs to be separated completely from that of normal cows, which means using completely separate equipment. Vacuum should be from the pulsation line, not the milk line. Installation of manual stopcocks may be required in some systems. Careful attention to these details and testing of shipped milk has led to milk being the safest animal-based food in our nation's food supply.

SUMMARY

- On average, every cow has two new infections per lactation.
- Leukocytes are present as a response to infection or injury, while epithelial cells are present as a result of infection or injury.
- Mastitis may be either infectious or noninfectious.
- Infectious mastitis may be present as a subclinical infection or as an overt clinical infection.
- *Streptococcus agalactiae* and *Staphylococcus aureus* are thought to be transmitted into the udder via the streak canal primarily during milking, while coliforms are believed to be transmitted between milkings.
- Increasing milking frequency from twice daily to three times a day not only increases milk production but also simultaneously reduces the risk of clinical mammary gland infections.
- Several tests, including screening tests and specific laboratory tests, have been developed for detecting the presence of microorganisms in the mammary glands of lactating cows.
- Producers should strive for a bulk tank SCC of 200,000 per ml or less.

QUESTIONS

1. Define *mastitis.*
2. What are the signs and symptoms of a cow with acute mastitis?
3. Compare and contrast infectious and noninfectious mastitis.
4. What microorganisms cause nearly all cases of mastitis?
5. What is the treatment for mammary glands infected with mycoplasma?
6. What is the most common route of entry for pathogens into the mammary glands?
7. When do the majority of new mastitis infections occur?

8. What prevention methods should a milking program embrace?
9. Under what conditions do milking machines not cause mastitis?
10. What are some recommended preventive measures for mastitis?
11. How is mastitis treated?

ADDITIONAL RESOURCES

Articles

Bayoumi, F. A., T. B. Farver, B. Bushnell, and M. Oliveria. "Enootic Mycoplasmal Mastitis in a Large Dairy During an Eight-Year-Period." *Journal of the American Veterinary Medical Association* 192 (April 1988): 905–909.

Brown, M. B., J. K. Shearer, and F. Elvinger. "Mycoplasmal Mastitis in a Dairy Herd." *Journal of the American Veterinary Medical Association* 196 (April 1990): 1097–1101.

Bushnell, R. B. "Mycoplasma Mastitis." *Veterinary Clinicians of North America Large Animal Practices* 6 (July 1984): 301–312.

Gonzales, R. N., and P. M. Sears. "Persistence of Mycoplasma Bovis in the Mammary Gland of Naturally Infected Dairy Cows." *Bovine Processes* 26 (February 1994): 184–186.

Gonzalez, R. N., P. M. Sears, R. A. Merrill, and G. L. Hayes. "Mastitis Due to Mycoplasma in the State of New York During the Period 1972–1990." *Cornell Veterinarian* 82 (January 1992): 29–40.

Internet

Dairy 10-Point Quality Control Program—Mastitis Treatment Records: http://ianrpubs.unl.edu/dairy/g1101.htm.

Mastitis Control Programs: http://www.ext.nodak.edu/extpubs/ansci/dairy/as1129w.htm.

Mycoplasma Mastitis: http://cvm.msu.edu/extension/docs/mycomast.htm.

Streptococcus agalactiae Fact Sheet: http://www.aabp.org/strep.pdf.

38

Health and Care of the Hoof

OBJECTIVES

- To understand the anatomical structure of the hoof.
- To be familiar with hoof infections and proper hoof care.

The best way to increase profitability for any dairy enterprise is to identify the sources of economic loss. Most dairy producers can readily identify two of the top three of these sources: mastitis and reproductive failure. But the disorder causing the third greatest economic loss to the dairy industry is lameness, which usually comes as a surprise. These losses are significant; research suggests that the average 1,000 cow herd will lose roughly $25,000 to $35,000 annually. This estimate includes treatment costs as well as costs associated with decreases in milk production, impaired reproductive performance, and increased risk of culling.

STRUCTURE OF THE HOOF

The hoof is the structural foundation of the cow. Consequently, any problem that affects the structural integrity of the cow puts the entire animal at risk. Hoof problems compromise the ability of the animal to move, making it more difficult for the animal to lay down, to stand up, and even to stand still comfortably. Every person that works with cows should have an understanding of the normal structure of the hoof so he or she can recognize hoof-related problems before they compromise the survivability of the cow.

The external anatomy of the hoof is composed of the coronary band that separates the leg and the hoof itself, the wall and the sole of the hoof, and the bulb (soft horn tissue that extends from the coronary band to the sole). The sole of the hoof is a concave structure that increases in thickness as it approaches the heel. The hoof itself has two digits, or toes.

Internally, the coronary cushion is an area composed of elastic tissue. The vasculature supplying blood to the hoof runs through this area. The laminae and papillae (or digital cushion) are arranged in folds that act as the shock absorbers of the hoof. In normal hooves, these tissues greatly reduce the force associated with the impact of the hoof as it contacts a walking surface. These tissues also produce the horny tissue of the hoof wall and sole. The digital cushion and the bulb together comprise the heel of the cow. The lamellae are the ridges on the lower portion of the inside wall of the hoof; strong fibers run between the lamellae and the coffin bone. In this manner, the hoof is attached directly to the skeletal structure of the cow.

INFECTIONS OF THE HOOF

The hoof is typically exposed to a higher concentration of pathogens on a daily basis than any other part of the cow, although it is surprisingly resistant to infections. Any condition that compromises the structural integrity of the hoof or surrounding tissues, however, dramatically increases the risk of infection. The two most common infectious diseases affecting the hoof area are foot rot and hairy heelwart. These diseases are potential hazards wherever cows are kept, but especially in wet, muddy areas.

Hairy Heelwart

Hairy heelwart is a biosecurity concern when bringing new stock into a facility. Hairy heelwart is a highly contagious infection characterized by a strawberry-colored growth between the digits of the hoof (Figures 38.1 and 38.2). Hair often grows from this growth, providing the basis for the name (Figure 38.3).

The cause of hairy heelwart is not yet definitively identified, but it is suspected to be a spirochete bacterium. It is a common problem in both new and expansion units. Producers bringing new animals into

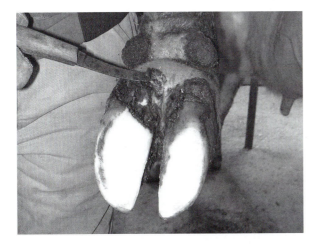

Figure 38.1   Hairy heelwart is a common problem in expansion herds. *(Courtesy of Leo Timms)*

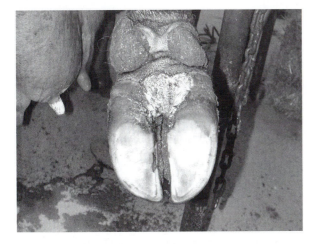

Figure 38.2   Recent surveys have suggested that over half of all lame cows are afflicted with hairy heelwart. *(Courtesy of Mark Kirkpatrick)*

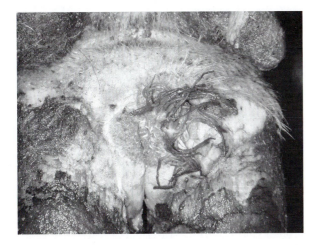

Figure 38.3   Hooves with hairy heelwart develop characteristic fingerlike projections. *(Courtesy of Mark Kirkpatrick)*

Figure 38.4   Topical hoof treatments can be applied conveniently with a sprayer. *(Courtesy of Leo Timms)*

Figure 38.5   Direct application of topical treatments is accomplished easily in the milking parlor. *(Courtesy of Leo Timms)*

a herd should design an incoming treatment program using parlor sprays (Figures 38.4 and 38.5) or footbaths for all cattle. Treatment programs require at least two weeks to ensure getting this condition down to a manageable level. The problem is rarely eradicated, but it can be controlled with vigilance and determination. Hairy heelwart is another reason for the isolation of all incoming stock prior to mixing new cows with the rest of the herd.

Foot Rot

Foot rot is a less obvious infection than hairy heelwart. A shrewd observer first notices a reddening and swelling of the skin just above the hoof, between the toes, or in the bulb of the heel. Lameness becomes apparent as the infection progresses. If not arrested, the infection invades the soft tissue and causes a discharge of pus from the infected breaks in the skin. At this stage, a characteristic foul odor is present. Later, the joint cavities may be involved, and the animal may show fever and depression characteristic of a general infection. Affected animals lose weight, and if lactating, produce less milk; they may die if left untreated.

Foot rot is a contagious, infectious disease caused by the organism *Bacterioides nodosus* in conjunction with *Fusobacterium necrophorum*. Possibly other organisms such as *Spirocheata penortha* and *Corynebacterium pyogenes* may be involved. Walking animals over contaminated areas where infected animals have been is the principal means of spreading the disease, although its spread is also influenced by weather conditions. New animals should be isolated on arrival, their hooves should be trimmed, and then they should be walked through a 3 percent formalin foot bath or a 5 percent copper sulfate foot bath. Oral iodides have been beneficial as preventatives in some cases. Control of foot rot in cattle is best achieved by eliminating muddy or abrasive walking areas, regular trimming of hooves, sanitation, isolating affected animals, and use of a suitable disinfectant.

Systemic and local treatment with antibiotics is recommended. Footbaths with drying agents are often used (Figures 38.6 through 38.8). Other procedures that may speed recovery are cleaning the foot, applying a protective dressing (Figure 38.9), wiring the claws together, and removing the necrotic in-

Figure 38.7   Various treatments can be applied topically to hooves via hoof baths. *(Courtesy of Mark Kirkpatrick)*

Figure 38.8   Copper sulfate is often used in footbaths to reduce the risk of hoof infections. *(Courtesy of John Smith)*

Figure 38.6   An empty foot bath showing the correct sizing and layout. *(Courtesy of Howard Tyler)*

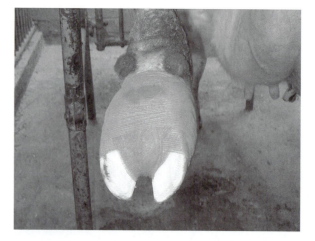

Figure 38.9   Treated hooves should be bandaged properly to protect them from excessive pathogen exposure. *(Courtesy of Leo Timms)*

terdigital mass. Zinc methionine has been recommended for both treatment and prevention. In advanced stages, best results are obtained by surgical amputation of the affected claw.

LAMINITIS

Laminitis is broadly defined as an inflammation of the laminae and the papillae in the hoof. Cases of laminitis can be classified as acute, subacute, or chronic. Experts suggest that 60 to 90 percent of cows are affected by laminitis, although only a small percentage of these may be lame at any time. There is a genetic component to laminitis; cows with more nearly correct foot angle are more structurally sound and have decreased hoof problems. The ideal hoof angle is about 45°. Hooves that are steeper or shallower reduce cow longevity. Hoof angle is only lowly heritable, however, and many management factors affect this trait. Many of these same management-related factors also contribute to the incidence of laminitis, including inappropriate walking surfaces, cows spending excessive time standing or walking, or acute physical trauma. In addition, fiber deficiencies in lactating cows' diets often result in acidosis, which can either cause an acute form of laminitis or exacerbate an existing subacute case of laminitis.

High-energy rations with inadequate amounts of forage are a common problem in herds with high-producing cows. In the rumen, these diets produce acid at a faster rate than they produce buffers, resulting in rumen acidosis. Many strains of rumen bacteria die under these conditions; endotoxins released by these dying microbes are absorbed and stimulate histamine release, which ultimately affects the microvasculature of the corium of the hoof. This leads to abnormal growth (overgrowth) of the hoof wall and sole and results in a softer horn reaching the surface of the hoof. In cows that have experienced multiple bouts of acidosis, this is seen as a series of white or yellowish parallel lines in the hoof. The timing of the acidosis can be estimated by the spacing of the lines; hoof growth occurs at the rate of about ⅕ in (5 mm) per month. There is also swelling of the laminae and papillae. Cows with acidosis-induced laminitis are often in severe pain, which falls under the classification of acute laminitis.

Factors that adversely affect cow comfort result in cows remaining on their feet for longer periods than is desirable. Breakdowns in the structural integrity of the hoof result from cows spending too much time walking on concrete that is wet or excessively rough. Wet conditions soften the horny tissue of the hoof wall and sole, allowing abnormal wear patterns and increasing the risk of injury or other damage. As the laminae and papillae of the hoof become softer, the pedal bone may begin to separate from the wall and sole.

HOOF CARE

Hoof trimming should be a routine practice on all dairy operations. Obviously, numerous management practices affect the extent of trimming required in an individual herd, but regular hoof care and professional trimming extends the productive life of cows in all management systems (Figure 38.10). A good hoof management program extends the useful life of the average cow by a full lactation.

The primary goal of the trimming procedure is to produce a balanced claw that allows equal distribution of weight between the digits (Figures 38.11 through 38.14). This is accomplished by careful removal of horny tissue from the sole of the hoof to reestablish a properly shaped hoof. The trimming process itself stimulates production of new, healthy horn tissue.

Figure 38.10 A hoof knife is a useful tool during hoof trimming. *(Courtesy of Leo Timms)*

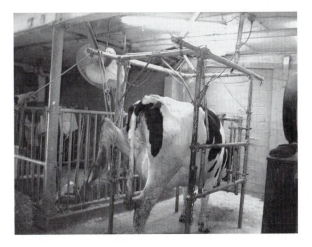

Figure 38.11 Hoof examination can be conducted in simple chutes by hoisting the rear leg after properly tying up the hock. *(Courtesy of Leo Timms)*

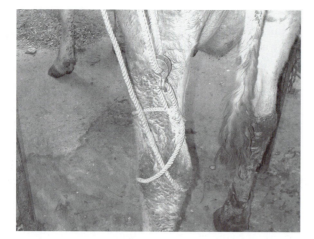

Figure 38.12 A close-up of the leg restraint method to permit hoof examination of standing cows. *(Courtesy of Leo Timms)*

Figure 38.14 The goal of hoof trimming is to reshape and rebalance the hoof. *(Courtesy of Mark Kirkpatrick)*

Figure 38.13 A tilt table allows access to all hooves simultaneously. *(Courtesy of Mark Kirkpatrick)*

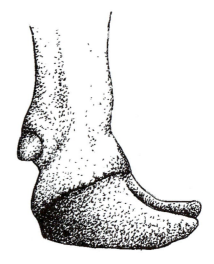

Figure 38.15 Hoof overgrowth is a common problem that can be corrected by hoof trimming. *(Courtesy of Nathan Klein)*

Cows in confinement tend to grow long toes (Figure 38.15) and build up excessive tissue on the soles of the feet. As a result, more weight is carried by the heels and the hocks, and the pasterns are subjected to extra stress. If these conditions are not corrected by proper hoof trimming, permanent damage may result in the form of crooked legs and weak pasterns, and the productive life of the cow will be shortened. If the hoofs are trimmed properly, the animal stands squarely and walks properly, with each leg directly under the weight it supports. The reason for and the technique of trimming hoofs are the same whether a cow is in a commercial production string or being fitted for show.

SUMMARY

- The hoof is the structural foundation of the cow.
- The hoof has two digits.
- Laminae and papillae are arranged in folds that act as the shock absorbers of the hoof.
- The hoof is exposed to a higher concentration of pathogens on a daily basis than is any other part of the cow.
- The ideal hoof angle is about 45°. Steeper or shallower hooves reduce cow longevity.
- Hoof trimming should be a routine practice on all dairy operations.

QUESTIONS

1. What are the top three economic losses for dairy producers?
2. What are the two most common infectious diseases affecting the hoof area?
3. Describe hairy heelwart, include the treatment time to get the condition down to a manageable level.
4. Define *foot rot* and its causes.
5. What is the best way to control foot rot?
6. Define *laminitis*.
7. What is the primary goal of hoof trimming?

ADDITIONAL RESOURCES

Articles

Clark, A. K. and A. H. Rakes. "Effect of Methionine Hydroxy Analog Supplementation of Dairy Cattle on Hoof Growth and Composition." *Journal of Dairy Science* 65(August 1982): 1493.

Greenugh, P. R., and J. J. Vermunt. "Evaluation of Subclinical Laminitis in a Dairy Herd and Observations on Associated Nutritional Management Factors." *Veterinary Record* 128(January 1991): 11.

Whitaker, D. A., J. M. Kelly, and B. E. Smith. "Incidence of Lameness in Dairy Cows." *Veterinary Record* 113(July 1983): 60.

Internet

Hoof Health and Dietary Interrelationships in Lactating Dairy Cows: http://www.ads.uga.edu/annrpt/1997/97_121.htm.

Hoof Health and Feeding Relationships: http://www.traill.uiuc.edu/dairynet/paperDisplay.cfm?Type=currentTopic&ContentID=6689.

Lameness and Hoop Health: http://www.moomilk.com/archive/a_health_35.htm.

Lameness in Dairy Cattle: Are Activity Levels, Hoof Lesions and Lameness Connected?: http://www.agsci.ubc.ca/animalwelfare/publications/documents/amanda_thesis.pdf.

Laminitis and Foot Health: Guidelines for Dairy Cows: http://www.ansc.purdue.edu/dairy/health/diseases/lam.htm.

39

External Parasites

OBJECTIVES

- To outline methods of identifying the propagation of parasites.
- To identify external parasites.

Dairy cattle are attacked by a wide variety of external parasites. The prevention and control of parasites is one of the quickest, cheapest, and most dependable methods of increasing milk production with no extra cows, no additional feed, and little additional labor.

TICKS

The lone star tick (*Amblyomma americanum*), the Gulf Coast tick (*Amblyomma maculatum*), the Rocky Mountain wood tick (*Dermacentor andersoni*), the Pacific Coast tick (*Dermacentor occidentalis*), and the American dog tick (*Dermacentor variabilis*) are three-host species that attack cattle during the summer months. The black-legged tick (*Ixodes scapularis*) is also a three-host tick that is common in later winter and early spring. The winter tick (*Dermacentor albipictus*) is a one-host species found on cattle and horses in fall and winter. In addition, larvae and nymphs of the so-called spinose ear tick (*Octobius megnini*), a one-host species, attach deep in the ears of cattle and feed there for several months.

Ticks are widely distributed, especially throughout the southern part of the United States. But they are usually seasonal in their activities. Ticks suck blood (Figure 39.1). They create economic losses by transmitting diseases; causing restlessness, anemia, and inefficient feed utilization; and necessitating expensive treatments. Among the diseases transmitted to or produced in cattle by ticks are Texas fever, anaplasmosis, Q fever, tick paralysis, and piroplasmosis.

Figure 39.1 After feeding on cattle, engorged female ticks drop to the ground, where they can lay up to 5,000 eggs. *(Courtesy of USDA)*

Generally, all species of ticks have similar stages of development. The females lay eggs that hatch into six-legged larvae (seed ticks). The larvae attach to a host, engorge on blood, and molt to eight-legged nymphs. The nymphs attach to a host, engorge on blood, and molt to eight-legged adults. Mating usually occurs on the host. The female then engorges fully, drops off the host, lays several thousand eggs, and dies.

Injury to cattle from tick parasitism varies directly with the number of parasites. Ticks feed exclusively on blood. Thus, when several hundred ticks feed, the host becomes anemic and unthrifty and loses weight. In addition, some female ticks generate a paralyzing toxin. The spinose ear tick, commonly called the ear tick, takes up residence along the inner surfaces of the ears and in external ear canals, where it is extremely annoying. Cattle heavily infected by spinose ear ticks droop their heads, rub and shake their ears, and turn their heads to one side.

FLIES

Several kinds of flies live around dairy cattle in and around buildings, in pastures, and on the animals (Figures 39.2 through 39.7). Whether feeding on the animals (Figures 39.8 and 39.9) or just annoying them, flies can reduce milk production and slow the growth of young stock (Figure 39.10). Flies are also unsanitary and may contaminate milk. Figure 39.11 describes a typical life cycle of flies that attack or annoy dairy cattle.

Blowfly

The flies of the blowfly group include several species that find their principal breeding ground in dead and putrefying flesh, although they sometimes infest wounds or unhealthy tissues of live animals and fresh or cooked meat (Figure 39.12). Black blowfly larvae frequently infest dehorning wounds during winter months and occasionally the navel of newborn animals. All the important species of

Figure 39.2 Horn flies are blood-feeders that feed up to thirty times per day. In extreme cases, several thousand horn flies may feed on a single animal. *(Courtesy of USDA)*

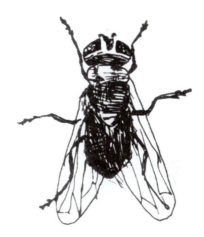

Figure 39.5 Face flies cause direct irritation to eye tissue and can transmit pinkeye. *(Courtesy of Iowa State University)*

Figure 39.3 The female screwworm fly lays her eggs on the edges of a wound. The larvae hatch within twelve to twenty hours and burrow deep into tissue to cause extensive damage. *(Courtesy of USDA)*

Figure 39.6 House flies are nonbiting flies that are common around barns and lots. *(Courtesy of Iowa State University)*

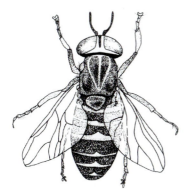

Figure 39.4 Horse flies and deer flies are biting flies that attack cattle. *(Courtesy of Iowa State University)*

Figure 39.7 Stable flies are blood-feeding flies that can be vectors for certain blood-borne diseases, such as anaplasmosis. *(Courtesy of Iowa State University)*

Figure 39.8 Bunching in response to feeding stable flies damages pastures, causes heat stress, and increases injuries, especially to calves. *(Courtesy of USDA-ARS)*

Figure 39.9 Cattle bunch together to attempt to avoid face flies that feed on secretions from the nose and eyes. *(Courtesy of Iowa State University)*

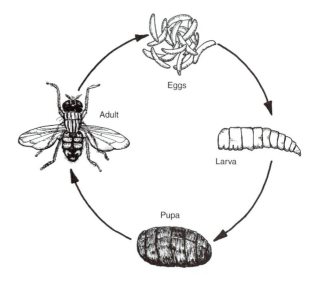

Figure 39.11 The typical life cycle of a fly. *(Courtesy of Iowa State University)*

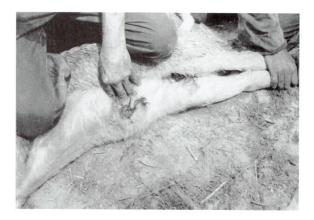

Figure 39.10 Screwworms are most dangerous when they infest the navel area in newborn calves. *(Courtesy of USDA)*

Figure 39.12 Black blowfly larvae most commonly infest dehorning wounds or umbilical stumps of newborn calves. *(Courtesy of USDA)*

blowflies except the flesh flies, which are grayish and have three dark stripes on their backs, have a more or less metallic luster. Although blowflies are widespread, they present the greatest problem in the Pacific Northwest and in the South and southwestern states. Death losses from blowflies are not excessive. But they cause much discomfort to affected animals and they lower production.

With the exception of the group known as gray flesh flies, which deposit tiny living maggots instead of eggs, the blowflies have a similar life cycle to the screwworm. The only difference is that the cycle is completed in about one-half the time. The blowfly causes its greatest damage by infesting wounds and the soiled hair of cattle. Most damage is inflicted by the black blowfly, and the damage is similar to that caused by screwworms. Infested animals rapidly become weak and fevered; although they recover, they may remain in an unthrifty condition for a long period.

Prevention of blowfly damage consists of eliminating the pest and decreasing the susceptibility of animals to infestation. Because blowflies breed principally in dead carcasses, the most effective control is effected by promptly destroying all dead animals by burning, deep burial, or sending to a rendering plant. The use of traps, poisoned baits, and electrified screens is also helpful in reducing trouble from blowflies.

Grubs (Heel Flies, Warbles)

Cattle grubs are the maggot stage of insects known as heel flies, warble flies, or gadflies. Two species of cattle grubs are present in the United States. The northern cattle grub (*Hypoderma bovis*) occurs mainly in

Figure 39.13 Adult female heel flies deposit eggs on the hair of cattle early in the spring. *(Courtesy of USDA)*

the north, though it is found as far south as southern California, northern Arizona, Oklahoma, Tennessee, South Carolina, and Hawaii. The common cattle grub (*Hypoderma lineatum*) occurs throughout the forty-eight contiguous states and in Hawaii and Alaska. The cattle grub or heel fly is probably the most destructive insect attacking dairy animals.

The species *Hypoderma lineatum* is widely distributed throughout the United States, whereas *Hypoderma bovis* is confined chiefly to the northern states (Figure 39.13). The damage inflicted by cattle grubs affects dairy producers, packers, tanners, and consumers. The fly does not bite or sting, but when it lays its eggs on an animal's lower legs, it usually terrifies the animal, causing it to run with tail hoisted, seeking relief (Figure 39.14). It may run through fences or over cliffs or become hopelessly bogged down in a mud hole. Milk production from dairy cows may be reduced from 10 to 25 percent during the period when heel flies are laying their eggs.

The grubs also create problems. Many cattle carcasses are damaged by grubs. The yellowish, watery patches caused by the migration of the larvae under the skin require the removal of two to three pounds of jellied beef from the loins and ribs of each grubby animal. As a result, the damaged cut of meat is devalued because of the ragged appearance. Approximately one-third of all cattle hides produced in the United States are damaged by grubs. This loss is caused by the migration of the grub through the back, which leaves a scar in the most valuable part of the hide (Figure 39.15). Commonly as many as 40 and occasionally 100 or more grub holes are found in a single hide. Hides of the latter quality are not considered worth tanning and are sold for by-products. In certain older animals that have been previously sensitized, the breaking of a grub under its skin may cause anaphylaxis (an allergic reaction). The area may be greatly swollen and form an abscess, and there may be such a general reaction that the animal may die from shock.

Figure 39.14 Heifers trying to escape from heel flies. *(Courtesy of Iowa State University)*

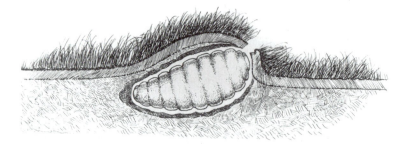

Figure 39.15   Cattle grubs cause hide damage by burrowing under the hide and by opening holes to obtain air. *(Courtesy of USDA)*

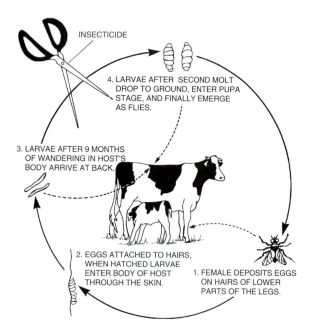

Figure 39.16   The life cycle of the cattle grub can be effectively disrupted by timely application of insecticides. *(Courtesy of Iowa State University)*

INSECTICIDE

4. LARVAE AFTER SECOND MOLT DROP TO GROUND, ENTER PUPA STAGE, AND FINALLY EMERGE AS FLIES.

3. LARVAE AFTER 9 MONTHS OF WANDERING IN HOST'S BODY ARRIVE AT BACK.

2. EGGS ATTACHED TO HAIRS, WHEN HATCHED LARVAE ENTER BODY OF HOST THROUGH THE SKIN.

1. FEMALE DEPOSITS EGGS ON HAIRS OF LOWER PARTS OF THE LEGS.

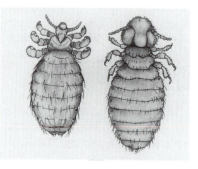

Figure 39.17   Cattle lice infest during periods of winter weather stress, and infestations (over ten lice per square inch) adversely affect performance. *(Courtesy of Iowa State University)*

Basically, the two species of cattle grubs have a similar life cycle (Figure 39.16). The female flies, called heel flies, attach their eggs to the hairs of the legs and bodies of cattle. The eggs hatch into larvae after about three days and enter the animals at the bases of the hairs. Once inside, the common cattle grub migrates from the point of entry to the gullet; the northern cattle grub migrates to the spinal column (both migrations take two to four months). After some additional months in the gullet or spinal column, grubs of both species migrate to the animals' backs, cut breathing holes in the hide, and remain there for about six weeks while they increase greatly in size. The resultant swellings are often called wolves or warbles. Fully grown grubs leave the hide through the breathing holes and drop to the ground where they pupate. Then, in a few weeks, they transform to nonfeeding adult flies that

emerge and mate. On bright sunny days, the female then seeks cattle for egg laying, which causes the cattle to roam. The entire life cycle takes about one year, and the same stages are usually found at about the same time each year in any given area.

LICE

The louse is a small, flattened, wingless insect parasite (Figure 39.17). Cattle are attacked by four species of bloodsucking lice, the short-nosed cattle louse (*Haematopinus eurysternus*), the cattle tail louse (*Haematopinus quadripertusus*), the long-nosed cattle louse (*Linognathus vituli*), and the little blue louse (*Solenopotes capillatus*), and by one species of biting louse (*Bovicola bovis*). Cattle lice do not remain on other farm animals, nor do lice from other animals infest cattle. Lice are always more abundant on weak, unthrifty animals and more troublesome during the winter months than during the rest of the year. The presence of lice on animals is almost universal, but the degree of infestation depends largely on the state of animal nutrition and the extent to which the owner tolerates parasites. The irritation caused by the presence of lice retards growth and milk production.

Figure 39.18   Blood-sucking lice seek sheltered parts of the body to feed, such as the neck area of this infested cow. *(Courtesy of University of Illinois)*

Figure 39.19   Tunneling by mange mites injures the skin and eventually forms characteristic scabs. *(Courtesy of Iowa State University)*

Lice spend their entire life cycle on the host's body. They attach their eggs or nits to the hair near the skin, where they hatch in about two weeks. Two weeks later, the young females begin laying eggs, and after reproduction they die on the host. Lice do not survive more than a week when separated from the host, but under favorable conditions eggs clinging to detached hairs may continue to hatch for two to three weeks.

Infestation shows up most commonly in winter in ill-nourished and neglected animals. There is intense irritation, restlessness, and loss of condition. Many lice are bloodsuckers, and they devitalize their host. There may be severe itching, and the animal may be seen scratching, rubbing, and gnawing at the skin. The hair may be rough and thin and lack luster and scabs may be evident. In cattle, favorite locations for lice are the root of the tail, on the inside of the thighs, over the fetlock region, and along the neck and shoulders (Figure 39.18). In some cases, the symptoms may resemble that of mange, and the two may occur simultaneously. With the coming of spring, when the hair sheds and the animals go to pasture, the problem of lice is greatly diminished.

MITES (MANGE, SCABIES)

Scabies in cattle, also known as scab, mange, or itch, is caused by mites living on or in the skin. Each species of domesticated animals has its own peculiar species of mange mites, and with the exception of the sarcoptic mites, the mites of one species of animals cannot live and propagate permanently on a different species. The sarcoptic mites are transmissible from one class of animals to another, and in the case of the sarcoptic mite of the cow and horse, from animals to humans. Three of the more impor-

tant species are the psoroptic, or common, scab mite, the sarcoptic scab mite, and the chorioptic, or symbiotic, scab mite. The sarcoptic form is most damaging because, in addition to their tunneling, the mites secrete an irritating poison that results in severe itching.

Injury from mites is caused by bloodsucking and the formation of scabs and other skin affections (Figure 39.19). In a severe attack, the skin may be much less valuable for leather. Growth is retarded, and the production of milk and meat is lowered. The mites that attack cattle breed exclusively on the bodies of their hosts and will live for only two or three weeks when removed. The female mite that produces sarcoptic mange, the most severe form of scabies, lays from ten to twenty-five eggs during the egg-laying period, which lasts about two weeks. At the end of another two weeks, the eggs hatch, and the mites have reached maturity. A new generation of mites may be produced every fifteen days. Mites are more prevalent during the winter months, when animals are confined and in close contact with each other.

When the mite pierces the skin to feed on cells and lymph, there is marked irritation, itching, and scratching. Exudate forms on the surface and then coagulates, crusting over the surface. The crusting is often accompanied or followed by the formation of a thick, tough, wrinkled skin. Frequently, there are secondary skin infections. The only certain method of diagnosis is to demonstrate the presence of mites.

MOSQUITOES

Mosquitoes, particularly species of the genera *Aedes, Psorophora,* and *Culex,* are a severe nuisance to dairy cattle in many areas. Mosquitoes are rather widely distributed, but they are most numerous in the southeastern United States, especially in swampy regions that have permanent pools of water or that are exposed to frequent flooding. Sometimes they kill cattle, although this is rare.

Almost all female mosquitoes must take a blood meal before they can lay eggs. The males do not suck blood but instead feed on nectar and other plant juices. Eggs are laid singly or in rafts on the surface of the water or on the ground in depressions that are flooded by tidal waters, seepage, overflow, or rainwater. The larvae and pupae are aquatic.

Mosquitoes may occur in such abundance that cattle refuse to graze. Instead, they bunch together or stand neck deep in water for protection from the attack. Mosquitoes annoy cattle day and night, and they can cause serious losses in milk production or even death in extreme cases. They may also be disease carriers.

Mosquitoes can be controlled by elimination of breeding places, by chemical destruction of larvae through treating the relatively restricted breeding areas with proper larvicides, and by chemical destruction of adults. Elimination of breeding sites is by far the most satisfactory and effective method of control. However, either this method or chemical destruction of larvae may not be economically practical if the breeding area is extensive.

SUMMARY

- Ticks cause economic losses by transmitting diseases; causing restlessness, anemia, and inefficient feed utilization; and necessitating expensive treatments.
- The louse is a small, flattened, wingless insect parasite of which there are several species.
- Lice spend their entire life cycle on the host's body.
- Prevention of mites consists of avoiding contact with mangy animals or infested premises.
- Mosquito larvae and pupae are aquatic.

QUESTIONS

1. What are the most effective methods for controlling ticks?
2. What are the four stages in the life cycle of a fly?
3. What types of damage are inflicted by cattle grubs?
4. How long does it take for a cattle grub to arrive at the host's back?
5. What causes scabies?
6. What are some ways mosquitoes can be controlled?

ADDITIONAL RESOURCES

Internet

External Parasites of Dairy Cattle: http://edis.ifas.ufl.edu/IG050.

Lice on Beef and Dairy Cattle: http://www.uky.edu/Agriculture/Entomology/ entfacts/pdfs/entfa512.pdf.

Pest Management Recommendations for Dairy Cattle: http://www.nysipm.cornell .edu/publications/dairycattlerec00.pdf.

Suggestions for Managing External Parasites of Texas Livestock and Poultry: http:// insects.tamu.edu/extension/bulletins/b-1306.html.

8

Concepts in Dairy Cow Comfort

40

Fundamentals of Dairy Cow Behavior

OBJECTIVES

- To define the types of learning processes and behavioral patterns of cows.
- To identify social relationships and communication between cattle.

A high-producing milk cow is a composite of unique behavioral traits blended with a favorable environment and skillful management. Knowledge of cattle behavior, or cow sense, is necessary to manage and handle dairy animals successfully. Bridging the gap between the principles and application of dairy cattle behavior allows maximal productivity of cows in the herd.

Animal behavior is the reaction of animals to certain stimuli or the manner in which they react to their environment. Through the years, dairy cattle behavior has received less attention than the quantity and quality of milk produced. But recently there is renewed interest in behavior, especially as a factor in obtaining maximum production and efficiency. With increasing confinement of cows in herds, many abnormal behaviors have emerged, including loss of appetite, pica, stereotyped movements, poor maternal behavior, excessive aggressiveness, and a host of other behavioral disorders. Animal behavior is the result of three forces: genetics, simple learning (training and experience), and complex learning (intelligence).

We need to breed dairy cattle adapted to artificial environments. Confinement has not only limited space, but it has also interfered with the habitat and social organization to which cattle became adapted and best suited through thousands of years of evolution. Producers need to concern themselves more with understanding the natural habitat of animals.

SIMPLE LEARNING (TRAINING AND EXPERIENCE)

In general, the behavior of animals depends on particular instinctive reaction patterns. These instincts and reflexes are unlearned forms of behavior. Cattle learn by experience. However, the training is only as effective as the inherited neural pathways will permit. Several types of simple learning processes are described here.

Habituation

Habituation is adapting to, or ignoring, certain stimuli.

Operant Conditioning

Conditioning is a learned response to a specific stimulus. For example, after hearing the rolling of a barn door, a cow may lick her tongue and moo, even though she can see no feed; after hearing the vacuum pump, she may let down her milk. Another example of conditioning is the use of an electric fence. When an electric fence is installed, the immediate instinct of cattle is to investigate, to touch it with their noses. After receiving a shock, they back off and leave it alone. Thereafter, the electricity can be shut off for a considerable period of time before some animal tests it again.

Operant conditioning, or operant learning, refers to the cow learning an act that has some consequence, that is, one that operates the environment, like pressing a bar that supplies feed or turns off a light. Broadly speaking, training is operant conditioning because it is an attempt to modify an animal's behavior. There are two types of training: reinforced training, usually with positive rewards, and forced training, in which the animal is compelled to complete certain tasks.

Insight Learning (Reasoning)

Insight learning is the sudden adaptive reorganization of experience or sudden production of a new adaptive response not arrived at by overt trial-and-error behavior. Of course, it is difficult to be certain in such cases that the animal did not previously have a similar type of problem. Even so, the immediate application of past experience to a new situation is a noteworthy capacity. The most important single factor to remember in training animals is that an animal's mind functions by intuition, not logic. It also has no conscious sense of right and wrong. Thus, it is one of the trainer's tasks to teach an animal the difference between right and wrong, between good and bad. Although the animal cannot utilize pure reason, it can remember, and it has the ability to use the memory of one situation as it applies to another.

Imprinting (Socialization)

This form of early social learning has been observed in some species. The pioneering work in this field used goslings and was performed by the Austrian zoologist Konrad Lorenz. He found that if baby goslings were exposed immediately after hatching to some moving object, especially if the object emitted sound, they would adopt that object as a parent or companion. Further studies revealed that goslings would adopt any other moving object in the same manner. Also, it was found that the same principle applies to mammals. Newborn calves can become imprinted to allow human interaction that normally would not be allowed. Apparently, genetics controls the time and the length of the critical period when an individual can be imprinted, the type of object to which it can be imprinted, the tendency to respond to the first object to which it is exposed, and the permanence of the attachment to the object following imprinting.

Memory

Memory is the ability to remember and the capacity to retain or recall that which is learned or experienced. The existence of dominance in cattle is evidence that cattle do remember (recognize) each other.

COMPLEX LEARNING (INTELLIGENCE)

Complex learning is the capacity to acquire and apply knowledge, the ability to learn from experience and to solve problems. It is the ability to solve complex problems by something more than simple trial and error, habit, or stimulus-response modifications. We recognize this capacity in humans as the ability to develop concepts, behave according to general principles, and put together elements from past experience into a new organization. Animals learn to do some things, whereas they inherit the ability to do others. The latter is often called instinct.

BEHAVIORAL PATTERNS

Animals behave differently, according to species. Also, some behavioral patterns are better developed in certain species than in others. Ingestive and sexual behavior patterns have been most extensively studied because of their importance commercially. Nevertheless, most cattle exhibit the following nine general behavioral systems:

1. Ingestive (eating and drinking)
2. Eliminative
3. Shelter-seeking
4. Investigative
5. Sexual
6. Gregarious
7. Agonistic (combat)
8. Allelomimetic (mimic)
9. Maternal-newborn (caregiving and care-seeking)

Ingestive Behavior

This type of behavior includes eating and drinking; therefore, it is characteristic of animals of all species and all ages. The first ingestive behavior trait, common to all young mammals, is suckling. For high production, animals must have aggressive eating habits and consume large quantities of feed (Figure 40.1).

Figure 40.1 Ingestive behaviors are dictated by the anatomy of the mouth and the digestive tract. *(Courtesy of Iowa State University)*

Each species has its own particular method of ingesting feed. The natural feeding (grazing) position of cattle is heads down. In this position, they produce more saliva, which aids digestion. When grazing, cattle wrap their tongues around grass and then jerk their heads forward to cut off the vegetation with the lower incisor teeth. There are no upper incisor teeth, only a thick, hard dental pad. When grazing, cattle also move their heads from side to side. This movement, aided by protuberant eyes and thin legs, gives them a continuous view of their entire surroundings, essential for wild cattle in an environment containing dangerous predators. It is important that artificial feeding devices and arrangements not depart too far from this natural pattern because cows have a built-in antipathy to being more or less blindfolded while eating.

Rumination is the act of chewing the cud, characteristic of herbivorous animals with split hoofs. It involves regurgitation of ingesta from the reticulorumen, swallowing of regurgitated liquids, remastication of the solids accompanied by reinsalivation, and reswallowing of the bolus. Rumination occupies about eight hours of the cow's time each day. In addition, grazing time may take another eight hours. This means that cows may ingest feed over a sixteen-hour day.

Eliminative Behavior

In recent years, elimination has become a most important phenomenon, and pollution has become a dirty word. Nevertheless, nature ordained that if animals eat that they must defecate. A full understanding of eliminative behavior makes for improved animal building design and help in handling manure. Cattle deposit their feces in a random fashion. Although cows can defecate while walking, with the result that feces become scattered, they generally deposit their so-called chips in neat piles.

Shelter-Seeking Behavior

All species of animals seek shelter: protection from the sun, wind, rain, snow, insects, and predators. Cattle are not as sensitive to extremes in temperature (heat and cold) as are swine. Nevertheless, they do seek shelter under severe weather conditions. Shelter may consist of hills, valleys, timber, and other natural windbreaks, or they may simply group closely together.

Cattle seem to be able to sense the coming of a storm, at which time they race about and act up. During a severe rain or snowstorm, they turn their rear ends to the storm and tend to drift away from the direction of the wind. By contrast, bison (buffalo) face a storm head on. During the hot summer

Figure 40.2 Jeff Dailey places a calf in a T-maze while Julie Morrow-Tesch collects data and records video. *(Courtesy of USDA-ARS)*

months, cattle seek either shade or a waterhole during the heat of the day. Then they graze in the cool of the evening or early morning. There are well-known breed differences in tolerance to heat, with Jerseys generally considered the most heat-tolerant dairy breed.

Investigative Behavior

All animals are curious and have a tendency to explore their environment. Investigation takes place through seeing, hearing, smelling, tasting, and touching. Whenever an animal is introduced into a new area, its first reaction is to explore it (Figure 40.2). Experienced producers recognize that it is important to allow animals time for investigation, either when they are placed in new housing or when new animals are introduced into the herd, before attempting to work them.

If they are not afraid, cattle investigate a strange object at close range. They proceed toward it with their ears pointed forward and their eyes focused directly on it. As they approach the object, they sniff and their nostrils quiver. When they reach the object, sniffing is replaced by licking; if the object is small and pliable, they may chew it or even swallow it.

Cattle exhibit investigative behavior when placed in a new pasture or in a new barn. As a result, if there is an open gate in a pasture or a hole in a fence, they usually find it and explore the new area. Calves are generally more curious than are older cattle.

Figure 40.3   Chin-resting behaviors are exhibited by this image of a bull courting a cow in estrus. *(Courtesy of Iowa State University)*

Sexual Behavior

Sexual behavior involves courtship and mating. The extent of this behavior is affected by various hormones, although males that are castrated after reaching sexual maturity usually retain considerable sex drive and exhibit sexual behavior. This observation suggests that learned behaviors, as well as hormonal factors, may be involved in sexual behavior. Males in most species of farm animals detect females in heat by sight or smell. Courtship is more intense on pasture than in confinement, and captivity has the effect of producing many distortions of sexual behavior compared to that observed in wild animals.

Experienced producers can usually detect in-heat cows through one or more of the following characteristic symptoms: restlessness, mounting other cows, standing to be mounted by another cow (standing heat appears to be the best single indicator of the proper time to breed), a noticeable swelling of the labia and lips of the vulva, frequent urination, switching and raising the tail, and a mucous discharge. Dry cows and heifers usually show a noticeable swelling or enlargement of the udder during estrus, whereas in lactating cows a rather sharp decrease in milk production is often encountered. When kept alone, some cows become restless, walk the fence, and bawl when they are in heat.

A bull can often detect a cow that is coming in heat twenty-four to forty-eight hours before she will mate. Courtship of the bull consists of following the cow; licking and smelling the external genitalia, with the head extended horizontally and the lip upcurled; and chin resting, with the chin and throat resting on the cow's rump (Figure 40.3).

Gregarious Behavior

Gregarious behavior refers to the flocking, or herding, instinct of certain species. It is closely related to allelomimetic behavior. If animals imitate each other,

Figure 40.4   Cattle perform agonistic behaviors such as butting heads and bulling to sort the dominance hierarchy. *(Courtesy of Iowa State University)*

they will stay together. If they stay together as a mobile group, they must use allelomimetic behavior to do so. All such behavior arises out of the process of social attachment. Cattle tend to roam as groups of various sizes when a large herd is placed on a pasture. However, there is usually considerable space between the members of the herd. On close observation, one can detect several small groups within a herd, each group ranging from three to five head.

Agonistic Behavior

This type of behavior includes fighting, flight (and flight zones), and other related reactions associated with conflict. Among all species of farm mammals, males are more likely to fight than are females (Figure 40.4). Nevertheless, females may exhibit fighting behavior under certain conditions. Castrated males are usually quite passive, which indicates that sex hormones (especially testosterone) are involved in this type of behavior. Farmers have for centuries used castration as a means of producing docile males, particularly cattle, swine, and horses.

Figure 40.5 Allelomimetic behaviors developed as an effective predator detection system. *(Courtesy of Iowa State University)*

In combat, bulls paw the ground and bellow, followed by head butting. Also, breeds of cattle differ in their agonistic behavior. High-yielding cows generally have excellent temperaments; well-managed herds have tame cows, with small flight zones.

Allelomimetic Behavior

Allelomimetic behavior is mutual mimicking behavior. Thus, when one member of a herd of cattle does something, another tends to do the same; because others are doing it, the original individual continues (Figure 40.5). In the wild state, this trait was advantageous in detecting and escaping predators. Under domestication, animals are usually protected from predators. Nevertheless, the allelomimetic behavior still has important consequences. Cows moving across a pasture toward a milking barn often display allelomimetic behavior. One cow starts toward the barn, and the others follow. Because the rest of the herd is following, the first cow proceeds on. Exploiting this behavioral response can be profitable. For example, there is usually higher per animal feed consumption among a group of calves than by one calf alone.

Maternal-Newborn (CareGiving and Care-Seeking) Behavior

Caregiving behavior is largely confined to females among domestic animals, where it is usually described as maternal; the care-seeking behavior is normal for young animals. This behavior begins shortly after birth and extends until the young are removed from their dam. Caregiving and care-seeking vary widely among different species of farm animals.

Cows seek isolation at calving time. Where possible, they will hide. Following birth, the caregiving behavior of the new mother becomes evident al-most immediately. She gets up and begins to dry her newborn calf by licking it. Simultaneously, some cows "talk" to their newborn. They may become quite concerned and nervous as their baby first attempts to stand, takes a few footsteps, and falters. Aided by its mother's licking and encouraged by her "talking," eventually the calf makes it to its unsteady feet and commences to search for a teat. A newborn calf cannot see too well, but it can smell, touch, and taste. If on pasture, the new mother usually hides her calf. During the first day or two, the calf sleeps a great deal, while the mother grazes nearby, taking great pains not to disclose the hiding place of her calf. At intervals, she returns to feed it.

Recognition between mother and a calf is by smell (olfactory), sight (visual), and sound (auditory). Cows usually sniff their calves after being away for a time, and the calf recognizes its mother's call. The attachment of the mother to her calf is very strong. Calves that are removed from their mothers during the critical period of an hour or so after birth, then resubmitted to their mothers, are frequently rejected. Dairy calves are normally removed from their mothers at birth or shortly after, with the result that the tie between mother and offspring is soon severed.

SOCIAL RELATIONSHIPS OF CATTLE

Social behavior may be defined as any behavior caused by or affecting another animal, usually of the same species but also of another species in some cases. Social organization may be defined as an aggregation of individuals into fairly well-integrated groups in which the unity is based on the interdependence of the separate animals and on their responses to one another.

The social structure and infrastructure in the dairy herd are of great practical importance. Older cows generally dominate the younger ones, and heavier animals (usually the older animals) tend to dominate the lighter. For this reason, two-year-old heifers should be segregated from older cows. Among cows of similar age and breed, however, the more aggressive ones are most dominant. Also, cows with more seniority in the group and cows with horns tend to be of higher social rank. Aggression in cows appears to be ritualized, with most encounters taking place in the following sequence: approach, threat, and physical contact (or fighting).

A relationship exists between the social status of cattle and spacing, or social distance. The higher the social rank, the more likely cows are to be found near other members of the herd. Also, dominant cows tend to allow close approach by other cows more often than do subordinates.

Figure 40.6 When moving from area to area, submissive cows consistently bring up the rear of the herd, dominant cows vary in position within the group, and the leader is at the front. *(Courtesy of USDA)*

When moving from the paddock to the milking parlor, dairy cows travel in a consistent order (Figure 40.6). Mid-dominant cows tend to be in front of the group. However, the same individuals are seldom consistent leaders; instead, a pool of animals tends to be in or near the lead. More consistency is found in the cows bringing up the rear of the moving herd. So subordination is a more distinctive feature than leadership. The animals at the rear are usually the younger, subordinate heifers.

In most species of farm animals, the alpha animal in the herd or flock is dominant over all other individuals, and the omega animal is subordinate to all. In between, some animals are subordinate in some relationships and dominant in others. Once these relationships are established, they seldom change. The social order is usually important only in females because mature bulls are seldom run together in groups.

When several cows are brought together to form a herd, there is a substantial period during which there is much butting and threat posturing to establish a dominance hierarchy. This is disruptive to a dairy herd and results in reduced production. Usually the older and larger cows come out at the top of the hierarchy. Once the social rank order is established, it results in a peaceful coexistence of the herd. After this point, when the dominant one merely threatens, the subordinate animal submits and avoids conflict. If strange animals are introduced into such a group, social disorganization results in the outbreak of new fighting as a new social rank order is established.

Social rank among dairy animals is of little consequence as long as they are on pasture and there is plenty of feed and water. But it becomes of very great importance when animals are placed in con-

finement. When cows are moved into limited quarters, social dominance decrees that replacement heifers be sorted out and fed separately and that old cows with poor teeth be fed separately; otherwise, these animals will not get enough feed. When self-feeders and central water tanks are used, care must be taken to provide both adequate space and proper placement; otherwise, submissive animals may find it difficult to get out of eye contact of dominant animals and eat and drink in peace. Of course, social rank becomes doubly important if limited feeding space is available. Under such circumstances, the dominant individuals crowd the subordinate ones away from the feed bunk, reducing feed intake, growth rates, and production.

Several factors influence social rank, including age (both young and geriatric animals rank toward the bottom), early experience (once a subordinate in a particular herd, usually always a subordinate), weight and size, and aggressiveness or timidity. In dairy confinement operations, social facilitation is of great practical importance. Dominant animals should be grouped together if possible. Of course, they will establish a new social order; in the meantime, both feed efficiency and milk production will suffer. But as a result of removing the dominant cows, the feed intake of the rest of the animals will be improved, followed by greater feed efficiency, production, and profit. Among the more settled animals, social facilitation will become more evident. After the dominant cows have been removed, the rest of the animals will settle down into a new hierarchy, but within the limits of their dominance. Their interaction or social facilitation will be far more likely to have a calming effect on this group.

Leader-follower relationships are also important in cattle. The young follow their mothers; as they age, they continue to follow their elders. The leader is the cow that is usually at the head of a moving column and often seems to initiate a new activity. It is important to distinguish leader-follower relationships from dominance; in the latter, the herd is driven, rather than led. After the dominant cows have been removed from the herd, the leader-follower phenomenon usually becomes more evident. The dominant animal is not necessarily the leader; in fact, it is very rarely the leader. When a string of cows moves from the pasture into the milking parlor, the dominant animals are generally in the middle of the procession, the leader is in front, and the subordinate ones bring up the rear.

Social relationships can also be transferred to human beings. As a result, a young calf associates everything good or bad with humans. This is the period in life during which calves are dehorned, castrated, branded, and vaccinated. To minimize the problem, calves should be worked as little as possible, with all such jobs done at one time. Good dairy caretakers usually form a care-dependency relationship with the animals under their care, with the result that the cows readily come to them.

COMMUNICATION BETWEEN CATTLE

Communication between cattle involves one individual giving some signal that influences the behavior of others.

Sound

Sound is an important means of communication among cattle. They use sounds in many ways, including a way to express hunger (bawling) by the young, distress calls like the bellowing of a bull (Figure 40.7), sexual behavior and related fighting, mother-young interrelations to establish contact and evoke care behavior, and maintenance of the group in its movements and assembly. Cattle have a very acute sense of hearing, perceiving higher and fainter noises than the human ear can.

Figure 40.7 Vocalizations, such as bellowing, are used to communicate anger, fear, or other emotions. *(Courtesy of Iowa State University)*

Smell

Cattle can smell at a greater distance than people can. On a day with a 5-mile wind and a humidity of 75 percent, a cow can smell up to 6 miles away; as wind and humidity increase, she can smell even further. In cattle, females in estrus secrete a substance (pheromones) that attracts males. Bulls locate cows that are in heat by the sense of smell.

Visual Displays

When several strange cows are brought together, there is threat posturing as well as butting to establish a dominance hierarchy. Also, bulls strike a hostile stance prior to fighting. Visual displays during courtship are less evident among cattle than other species, but they do occur to some extent.

Sight

The eyes of most animals are on the side of the head (the cat is an exception). This location gives them an orbital (or panoramic) view: to the front, to the side, and to the back almost at the same time. Also, this is a rounded, or globular, type of vision, which leads to a different interpretation than that of the binocular type of vision of humans.

The wideset eyes of cattle enable them to have a large panoramic field of vision, even to the extent of seeing everything around them with slight head movements. Only what is immediately behind their hindquarters is outside their field of view. A cow does not see in color; she sees in shades of grays and blacks. If a cow sees movement, her instinct is to escape; movements around cattle should thus be made very quietly and slowly.

Homing Behavior

Through sound, scent, or some sense we do not understand, cattle often find their way back home when they are moved to distant places.

SUMMARY

- A high-producing milk cow is a composite of unique behavioral traits blended with a favorable environment and skillful management.
- Operant conditioning, or operant learning, refers to animal operation of some aspect of the environment to obtain access to feed or other animals.
- An animal's mind functions by intuition, not logic, and it has no conscious sense of right and wrong.
- Cattle seem to be able to sense the coming of a storm, at which time they race about.
- Sexual behavior is largely controlled by hormones.
- Gregarious behavior refers to the flocking, or herding, instinct of certain species.
- Allelomimetic behavior is mutual mimicking behavior.
- Older and larger cows rank at the top of the hierarchy.
- Cows do not see in color.

QUESTIONS

1. What are five types of learning processes?
2. Define *habituation* and *conditioning*.
3. Give an example of operant conditioning.
4. Define *insight learning, memory,* and *complex learning*.
5. What are nine general behavioral systems?
6. What is the first ingestive behavior trait?
7. Through what characteristics can experienced producers detect in-heat cows?
8. When can a bull detect a cow coming in heat?
9. What type of behavior includes fighting, flight, and other related reactions associated with conflict?
10. Define *social behavior* and *social organization*.
11. What is the name of the animal that is dominant over all other individuals? What is the name of the animal that is subordinate to all?
12. How do cows communicate with each other?

ADDITIONAL RESOURCES:

Articles

Ceballos, A., D. Sanderson, J. Rushen, and D. M. Weary. "Improving Stall Design: Use of 3-D Kinematics to Measure Space Use by Cows When Lying Down." *Journal of Dairy Science* 87 (July 2004): 2042–2050.

DeVries, T. J., M. A. G. von Keyserlingk, and D. M. Weary. "Effect of Feeding Space on the Inter-Cow Distance, Aggression, and Feeding Behavior of Free-Stall Housed Holstein Dairy Cows." *Journal of Dairy Science* 87 (May 2004): 1432–1438.

Fregonesi, J. A., C. B. Tucker, D. M. Weary, F. C. Flower, and T. Vittie. "Effect of Rubber Flooring in Front of the Feed Bunk on the Behavior of Dairy Cattle." *Journal of Dairy Science* 87 (May 2004): 1203–1207.

Internet

Behavior: http://www.cowdoc.net/pages/manage_practices/behavior.html.

Understanding Animal Welfare: Science, Values and Emerging Standards: http://www.agsci.ubc.ca/animalwelfare/publications/documents/FAO_Seminar2004.pdf.

41

Animal Welfare and Animal Stress

OBJECTIVES

- To become familiar with the concerns of animal welfare and animal rights.
- To detail the fundamentals of animal stress.

In recent years, the behavior and environment of animals in confinement have come under the increased scrutiny of animal welfare and animal rights groups all over the world. For example, Sweden passed legislation in 1987 designed to phase out layer cages as soon as a viable alternative can be found; to discontinue the use of sow stalls and farrowing crates; to provide more space and straw bedding for slaughter hogs; and to forbid the use of genetic engineering, growth hormones, and other drugs for farm animals except for veterinary therapy. Violators can either be fined or imprisoned. Some animal welfare activists see many modern practices as unnatural and not conducive to the welfare of animals. In general, they view animal welfare as the well-being, health, and happiness of animals. They believe that certain intensive production systems are cruel and should be outlawed. The animal rights activists go further; they maintain that all animals should be accorded the same moral protection as is provided for humans. They contend that animals have essential physical and behavioral requirements that, if denied, lead to stress and suffering. They conclude that all animals have the right to live.

Animal welfare issues tend to increase in intensity with urbanization. Fewer and fewer urbanites have farm backgrounds. As a result, the animal welfare gap between town and country widens. Also, both the news media and legislators are increasingly from urban centers. It follows that the urban views that are propounded will have a greater impact in the years ahead.

WELFARE ISSUES IN THE DAIRY INDUSTRY

For the most part, the dairy industry has escaped serious scrutiny by animal rights activists. Historically, dairy farmers have recognized that compassionate care leads to greater profits. Because dairy cattle are raised to produce milk rather than for slaughter, the animal-human bond is more apparent than with some other species. Dairy producers know that the abuse of animals in intensive and/or confinement systems leads to lowered production and income, and they maintain welfare of the animal and profit on the same side of the ledger. They recognize that husbandry that reduces labor and housing costs often results in physical and social conditions that increase animal problems. Nevertheless, means of reducing behavioral and environmental stress are needed so that lower labor and housing costs are not offset by losses in productivity. Activists claim that the evaluation of animal welfare must be based on more than productivity; they believe that there should be behavioral, physiological, and environmental evidence of well-being, too. Most dairy producers recognize that they should provide as comfortable an environment as is feasible for their animals, for both humanitarian and economic reasons. They must pay attention to environmental factors that influence the behavioral welfare of their animals as well as their physical comfort, with emphasis on the two most important influences of all in animal behavior and environment: feed and confinement. Despite this understanding, several areas of dairy management have come under more intense scrutiny regarding their relationship to cattle welfare.

Veal Production

Easily the most scrutinized practice involving dairy animals is the raising of young calves for veal. It involves several emotional hot buttons: removal of young animals from their mothers, raising young animals in confinement, and eventually the slaughter of these young calves for their meat. The veal industry has become the most acceptable industry for other aspects of animal agriculture to target as inhumane. Neither the beef industry nor the dairy industry provides their full support for the veal industry, preferring to distance themselves politically from such an obvious target for animal welfare reforms.

Without doubt, the greatest issue regarding the production of veal is the degree of confinement in the housing system. Housing systems that prevent calves from assuming normal resting and sleep positions are targets for animal welfare organizations. However, systems that allow more freedom of movement are more difficult to design economically, which presents a challenge to the industry: to create housing systems that cater to the welfare of the calf and still allow for a profitable enterprise. Housing systems that allow social interaction, although desired by some welfare organizations, permit free pathogen transfer between animals that have an innately weak ability to respond immunologically to new pathogens. The result is more disease, higher mortality rates, and a less humane situation for the animal.

The second issue regarding veal production is the focus of the industry on producing white veal. To attain this goal, iron intake must be closely monitored. Although the levels of iron fed to veal calves are adequate for their needs, there is only a small margin for error. Complicating this matter is the fact that most calves, whether destined for veal production or for dairy production, are somewhat anemic at birth. This condition is simply a physiological result of poor iron transfer across the placenta. Because hemoglobin production takes a matter of months, it is not possible for veal producers (or dairy producers, for that matter) to ensure that none of their calves are anemic during the milk-feeding period. Although the issue of anemia has been a historical area of contention (and an area of abuse in some cases), current industry regulations attempt to ensure that all calves are provided adequate levels of iron for their needs.

Certainly, current veal management practices do not resemble past practices that aroused the wrath of the public. The perception persists, however, and the biggest challenge to the veal industry today is not improving management practices as much as it is improving its public image.

Heifer Calf Management

To outside observers, the practice of separating the calf and dam at birth seems an inhumane practice. Few institutions are more sacred than the bond between mother and offspring, and the idea of terminating that bond immediately after birth is not very palatable to many people outside the industry.

The act of separation is both a management issue and a welfare issue for producers. Cows that are separated from their calves at birth are more easily integrated into the milking herd, and they let down their milk more readily than those that are still associated with their calves. In addition, isolating calves immediately after birth limits their exposure to pathogens, resulting in healthier calves and fewer losses from death. As with most issues on the farm, a trade-off must be considered. Is it more harmful to the calf to lose the opportunity to experience the mother-calf bonding experience or to be exposed to a greater risk of morbidity and mortality?

Similar issues surround calf housing decisions. Most calves are raised in individual housing, limiting their opportunities for socialization with other calves, but also limiting their exposure to pathogens and their risk of illness or death.

Cow Housing

Issues regarding mature cow housing have been less scrutinized but in reality have probably been a greater problem. Confinement is a real issue, and cows spending extended time on concrete and its effects on hoof health is a growing issue of real concern to the industry. The sizing and design of stalls are still being debated and researched, and the development of a housing system that meets the behavioral needs of the cow and the economic needs of the producer is a concern for producers. Great progress has been made in increasing cow comfort, and this progress has manifested itself in greater profits for producers. More progress will no doubt be made in the near future, and the impact of these innovations on animal welfare will continue to grow.

Other issues regarding dairy cattle management include the handling of downer animals. More downer animals originate from the dairy industry than any other industry (with the possible exception of veal). Downer cows should be euthanized humanely on the farm rather than transporting them to a slaughter facility. The lost income is of little consequence when compared to the damage done to the public image of the dairy industry.

Tail docking is another management practice that has come under close scrutiny. The primary reason that producers amputate tails is to improve conditions for milkers by reducing the number of times they are hit by manure-laden cows' tails. The practice of removing tails, while not necessarily painful to the cow, does present issues with fly

Figure 41.1 Tail docking is a largely unnecessary procedure typically implemented to overcome poorly designed or maintained facilities. *(Courtesy of Monsanto)*

control for the cows (Figure 41.1). The real issue is sanitation and facility design, and in this case, the procedure is an unwarranted and unnecessary response to an environmental problem that should be solved by changes in management.

The producer's charge in all these situations is to make the right decision based on the welfare of the animal and the profitability of the operation and to communicate the reasons for those decisions to the general public. The role of producer as educator is not totally unfamiliar; dairy farms have historically been a favorite visiting place for educational field trips for school-age children. Producers should use these opportunities to educate the children, as well as their teachers, on the welfare issues associated with dairy farms and the considerations for animal welfare that are made on a day-to-day basis.

FUNDAMENTALS OF ANIMAL STRESS

Abnormal behavior of domestic animals is not fully understood. As with human behavior disorders, more work is needed. However, we have learned from studies of captured wild animals that when the amount and quality of the surroundings of an animal are reduced that abnormal behavior will likely develop.

Also, confinement of animals often leads to unfavorable changes in habitat and social interactions.

Abnormal behavior in animals develops with a combination of confinement, excess stimulation, and forced production and also a lack of opportunity to adapt to the situation. Homosexual behavior is common among all species where adult mammals of one sex are confined together. Pica (consumption of dirt, hair, bones, or feces) may develop perhaps because of boredom, nutritional inadequacies, or physiological stress. Milk cows may kick because they are in pain, they are frightened, or they have been mistreated.

Stress

Stress is defined as physical or psychological tension or strain. Among the external forces that may stress animals are abrupt ration changes, change of water, space, level of production, number of animals housed together, changing housing or herdmates irregular care, transporting, excitement, presence of strangers, fatigue, previous training, illness, management, weaning, temperature, and abrupt weather changes. Animals experience many periods of stress. For example, high-producing dairy cows are constantly under stress. As a result, they are usually very sensitive to any changes in their environment or their routine, even changes such as a change in milkers, an unequal interval between milkings, or overmilking.

The principal criteria used to evaluate stress in cattle are growth rate or production, efficiency of feed use, efficiency of reproduction, body temperature, pulse rate, breathing rate, mortality, and morbidity. Stress is unavoidable. Wild animals were often subjected to great stress; there were no caretakers to modify their weather; often their range was overgrazed; and sometimes malnutrition, predators, diseases, and parasites took a tremendous toll. Domestic animals are subjected to different stresses than their wild ancestors, were, especially stresses relating to more restricted areas and greater animal density. To be profitable, however, producers must ensure that stresses are minimal.

SUMMARY

- Dairy producers know that the abuse of animals in intensive and/or confinement systems leads to lowered production and income.
- Animal welfare issues tend to increase with urbanization.
- When the amount and quality, including variability, of the surroundings of an animal are reduced, abnormal behavior will likely develop.
- Stress is defined as physical or psychological tension or strain.
- High-producing dairy cows are usually very sensitive to any changes in their environment.

QUESTIONS

      1. When can abnormal behavior develop in animals?

      2. Define *pica*.

      3. List five external forces that may stress animals.

      4. What are the principal criteria used to evaluate stress in cattle?

ADDITIONAL RESOURCES

Books

Coats, C. D. *Old MacDonald's Factory Farm.* New York: The Continuum Publishing Co., 1991.

Fox, M. W. *Farm Animals.* Baltimore, MD: University Park Press, 1984.

Rollin, B. E. *Farm Animal Welfare.* Ames: Iowa State University Press, 1995.

Singer, P. *Animal Liberation,* 2nd ed. New York: Random House Inc., 1990.

Articles

Cohen, R. D. H., B. D. King, L. R. Thomas, and E. D. Jangen. "Efficacy and Stress of Chemical Versus Surgical Castration of Cattle." *Canadian Journal of Animal Science* 70 (1990): 1063–1072.

Dickrell, J. "They Don't Waver on Waffles." *Dairy Today* 7 (1991): 26–28.

Friend, T. H. "Teaching Animal Welfare in the Land Grant Universities." *Journal of Animal Science* 68 (October 1990): 3462–3467.

Lay, D. C., T. H. Friend, R. D. Randel, C. L. Bowers, K. K. Grissom, and O. C. Jenkins. "Behavioral and Physiological Effects of Freeze or Hot-Iron Branding on Crossbred Cattle." *Journal of Animal Science* 70 (February 1992): 330–336.

Rollin, B. E. "Animal Welfare, Animal Rights, and Agriculture." *Journal of Animal Science* 68 (October 1990): 3456–3461.

Trunkfield, H. R. and Broom, D. M. "The Welfare of Calves During Handling and Transport." *International Journal of the Study of Animal Problems* 28 (1990): 135–152.

Turner, F., and J. Strek. "Farm Animal Welfare: Some Economic Considerations." *International Journal of the Study of Animal Problems* 2 (1981): 15–18.

Internet

Animal Rights and Animal Welfare: http://www.sover.net/~lsudlow/ARvsAW.htm.

Animal Rights versus Animal Welfare: http://www.rce.rutgers.edu/pubs/pdfs/fs753.pdf.

Dehorning of Dairy Cattle: Necessity, Methods, and Implications: http://www.rce.rutgers.edu/pubs/pdfs/fs769.pdf.

Stress and Dairy Calves: http://www.vetmed.ucdavis.edu/vetext/INF-AN/INF-AN_StressDairyCalves.pdf.

42

Dairy Cattle Facilities and Housing

- To outline the purpose, design, and factors influencing dairy buildings and facilities.
- To list the various facilities for calves and replacement heifers.
- To compare and contrast stall barns and loose housing.

Modern dairy cattle buildings and equipment should be designed to facilitate the comfort, health, and productivity of cows; minimize maintenance needs and labor requirements; minimize bedding costs; and facilitate manure handling and disposal. Environmental concerns should be a primary consideration when designing a new facility or expanding an existing facility. The amount of land needed is dictated by the amount of manure produced; the plans for handling the manure; and the current regulations regarding manure storage, handling, and disposal. Because of the high costs associated with these processes, it is also critical to anticipate the potential for future changes in these regulations and plan for these changes in current construction plans.

In addition to addressing environmental concerns, facilities must be engineered to enhance the profitability of the dairy business, and minimizing labor costs and maximizing cow welfare and comfort are the primary considerations in that regard. The most critical considerations for optimizing cow performance are adequacy of ventilation, accessibility and quality of feed and water, lighting, and cow comfort (including both stall configuration and bedding quality). To minimize labor costs, the entire facility should be designed to enhance cow movement and minimize the time required for daily procedures such as feeding, bedding, and milking (Figure 42.1).

FACILITIES

The decisions regarding facility design are critical; cutting corners on facilities can easily result in substantial decreases in milk production. However, facilities providing optimum environments can only provide the means for animals to express their full genetic potential of production; they cannot compensate for poor management, health problems, or improperly balanced or delivered rations.

With the shift to confinement structures and high-density production operations, building design and environmental control are more critical. Animals are more efficient; they produce and perform better and require less feed if they are raised under ideal conditions of temperature, humidity, and ventilation. Controlled environment increases milk yield, fat percentage, fat yield, 4 percent fat-corrected milk, and conception rates. The primary reason for having livestock buildings, therefore, is to modify the environment. Properly designed barns and other shelters, shades, insulation, ventilation, and air conditioning can be used to approach the ideal environment. However, the cost per head is much higher for environmentally controlled facilities. Thus, the decision on whether or not confinement and environmental control can be justified should be determined by economics. Will the cows in environmentally controlled quarters produce sufficiently more milk on less feed to justify the added cost? Of course, manure disposal and pollution control should also be considered. Environmental control is rather common in poultry and swine housing; it is becoming more common for dairy facilities. In hot climates, the use of shades, fans, sprinklers, sprayers, foggers, mechanical ventilation systems, and windbreaks is increasing.

When planning an entirely new dairy farm, the choice of location for the buildings is the first consideration. When appraising the desirability of an

Figure 42.1 Facility design should consider animal movement and labor requirements. *(Courtesy of Iowa State University)*

existing facility, the same factors should be considered. These factors are listed here:

1. **Climate:** The comfort of cattle is maximized in a thermoneutral environment. Cattle are more suited to colder environments because of the heat generated by fermentation in the rumen. Cow comfort is generally maximized between 15° and 70°F with humidity levels between 50 and 75 percent. Environments that vary from these ideals for significant lengths of time require buildings or special equipment (misters, fans) that modify the environment accordingly. Thus, building costs vary in accordance with the severity of the climate.

2. **Critical mass of other dairy producers:** Unlike many other livestock operations, a dairy facility depends on easy access to a processor for survival and therefore depends on the stability and size of other nearby dairy farms to maintain a stable milk market.

3. **Labor supply:** Labor costs have tripled or quadrupled over the last twenty-five years, and the availability of a well-trained, appropriately educated supply of labor for the farm staff can determine the success or failure of a dairy operation.

4. **Access to feeds:** A high percentage of the cost of milk production is attributed to feed costs. The more low-cost feed options that are available in an area, the more attractive it becomes as a location for a dairy operation.

5. **Water supply:** Water must be available year-round, and it must be plentiful and inexpensive. The cost of water is a significant cost to a dairy producer in areas where water supplies are limited. In irrigated areas, it is also desirable to be near an irrigation turnout.

6. **Environmental concerns:** Land requirements are dictated by the amount of manure pro-

Figure 42.2 Farm layout should be influenced by the topography of the land, cow-flow patterns, labor efficiencies, and ventilation and odor considerations. *(Courtesy of Iowa State University)*

duced; storage and handling systems; and current federal, state, and local regulations.

7. **Topography:** The topography should be high and level with no abrupt slopes. A relatively level area requires less site preparation, thus lowering building costs (Figure 42.2).

8. **Drainage:** The soil should be porous, and the slope should be gentle. Dairy cattle health is much more easily maintained when the yards and buildings are well drained; manure disposal considerations are also easier.

9. **Access to utilities:** Ease of access and cost of access to utilities can be a significant barrier to creating a new facility or expanding an existing facility.

10. **Neighbors:** Facilities located near populated areas often face more scrutiny than those that are more isolated. Regardless of the population density, it is important to have good relationships with the people living near any animal facility.

ARRANGEMENT OF FARM BUILDINGS

Orientation: Although the farm plan will be developed in general to present the front to the road, most buildings can be turned, quarter-turned, or reversed, as may be necessary to take advantage of the prevailing winds, sunlight, view, and so forth. When possible, sheds in colder climates are faced to secure direct sunlight and yet to face away from the direction of the prevailing winds. In warmer climates, sheds are usually oriented for maximum shade and storm protection.

Efficiency: The buildings should be located to facilitate ease of animal movement from one area to another without compromising biosecurity. Facilitating ease of animal movement also minimizes labor associated with these tasks. Routine access by milk haulers, feed suppliers, and other visitors should be planned to minimize the potential introduction of new pathogens. The potential for future expansion should be planned into the spacing, orientation, and sizing of newly constructed buildings.

Appearance: Avoiding complaints from neighbors and nearby communities starts with careful planning during the building stage. In addition, careful attention to the dairy headquarters arrangement can add to the attractiveness of the entire unit (Figure 42.3). Manure storage areas and unsightly objects should not be visible from the main highway or house; appropriate landscaping can screen unsightly objects and diffuse or redirect objectionable odors (Figure 42.4); fences and buildings should be repaired and painted regularly; and yards, driveways, and corrals should be kept free from rubbish, scattered farm machinery, and so forth.

HOUSING OPTIONS FOR CALVES

The biggest challenge in providing a functional housing facility for calves is in balancing the needs of the calves with the comfort of the calf feeders. Calves have very strict requirements for ventilation and are extraordinarily susceptible to respiratory problems if ventilation is inadequate (Figure 42.5). They also need strict isolation to prevent pathogen transfer from sick calves to healthy calves (Figure 42.6). It is difficult to design an indoor facility that meets both needs, and outdoor facilities have the disadvantage of exposing feeders to severe weather situations. As a result, some producers prefer to house preweaned calves in indoor facilities (Figures 42.7 through 42.10), while others prefer cold housing (Figures 42.11 through 42.15). The preferred pen temperature is within the range of 20° to 75°F. However, if calves are allowed to become wet and/or muddy, the lower critical temperature increases dramatically to over 40°F. If a warm barn is desired, it should be extremely well ventilated; up to sixty air changes per hour are required in extremely hot conditions. In cold weather, air speed must be less than 50 feet per minute over the calves, but they still require a minimum of fifteen air changes per hour. Solid partitions between individual pens reduce drafts and chilling but impair ventilation. The fronts of pens should be wire or slatted. A minimum of 24 square feet of pen space should be provided for individual calves, and 30 square feet should be provided for calves in groups without outside runs.

Calves should be housed separately (in individual pens or huts) from birth until at least one week

Figure 42.3 Farm facilities can be attractive as well as functional. *(Courtesy of Iowa State University)*

Figure 42.4 Some simple landscaping additions add tremendous eye appeal to a dairy farm. *(Courtesy of Iowa State University)*

Figure 42.5   Indoor facilities pose a difficult choice between maintaining isolation and compromising ventilation. *(Courtesy of Iowa State University)*

Figure 42.8   Using wire panels between calves enhances ventilation but allows nose-to-nose contact. *(Courtesy of Mark Kirkpatrick)*

Figure 42.6   Open stalls for calves enhance ventilation but permit disease transmission via nose-to-nose contact. *(Courtesy of Iowa State University)*

Figure 42.9   An open-air calf facility in southern California. *(Courtesy of Iowa State University)*

Figure 42.7   An elevated floor in the calf pens enhances ease of cleaning for this calf facility. *(Courtesy of Iowa State University)*

Figure 42.10   Pens should be spaced 18 to 24 inches apart to minimize the transmission of airborn pathogens. *(Courtesy of South Dakota State University, Brookings)*

Figure 42.11 Plastic calf hutches are a popular choice for housing preweaned calves. *(Courtesy of South Dakota State University)*

Figure 42.13 Although less expensive than individual hutches, calf condos are more difficult to move and to sanitize between calves. *(Courtesy of Mark Kirkpatrick)*

Figure 42.12 Old tires can be used as weights to keep plastic hutches in place during periods of high winds. *(Courtesy of Mark Kirkpatrick)*

Figure 42.14 Dome-style plastic hutches are another popular choice for housing individual calves. *(Courtesy of Mark Kirkpatrick)*

Figure 42.15 Wood hutches are less expensive than plastic ones, but they are also heavier and more difficult to sanitize properly. *(Courtesy of Monsanto)*

Figure 42.16   Animal size and animal density are important factors when determining the ventilation requirements of a facility. *(Courtesy of Mark Kirkpatrick)*

Figure 42.18   The design of this barn, with its open back wall and sloped roof, allows for efficient use of natural ventilation. *(Courtesy of Mark Kirkpatrick)*

Figure 42.17   After weaning, calves should be housed in small groups. *(Courtesy of Mark Kirkpatrick)*

Figure 42.19   Retrofitted sheds are often difficult to ventilate properly. *(Courtesy of Mark Kirkpatrick)*

after weaning. After weaning, they may be raised in groups (Figure 42.16), with a maximum of ten head per group (preferably six to eight per group). Each group should feature a maximum age difference of two months between calves and a weight difference of 100 pounds between calves. Calf housing must be clean, dry, and well ventilated at all times.

HOUSING OPTIONS FOR REPLACEMENT HEIFERS

Heifers reared in an open shed require 20 to 30 square feet per head for small cattle and 30 to 40 square feet per head for large cattle (Figures 42.17 through 42.19). These facilities should be bedded to keep heifers dry at all times. Replacement heifers do not need a closed barn; the main requisite of their

housing is that they be protected from drafts, rain, snow, and winds. Artificial shade is required in hot climates if natural shade is not available (Figure 42.20). Adequacy is critical; shades should be sized at 20 to 30 square feet per animal, and they should be built 8 to 10 feet high.

Feed bunks should be 24 to 30 inches above the ground (to the top of the bunk, with the height determined by the size of cattle), 8 to 12 inches deep (12 inches deep for silage), and 18 to 24 inches wide when feeding from one side (36 inches wide when feeding from two sides). In addition, 1 linear foot of water tank space is necessary for each ten animals, or one automatic watering bowl is needed for each twenty-five animals. Water temperature may range from 35° to 80°F; warming to 50°F in the winter is desirable.

Figure 42.20 Holstein heifers in open pens. Barriers are used to provide shade and a windbreak. *(Courtesy of Iowa State University)*

Figure 42.21 Stanchions restrict normal cow behaviors and are rarely used in new facilities as a routine housing option. *(Courtesy of Iowa State University)*

HOUSING SYSTEMS FOR COWS

Two basic facility systems are used for cows: stall barns and loose housing. Similar milk production can be achieved with either system. The stall barn allows the cows to be displayed to greater advantage, which is particularly important in the purebred herd. Generally speaking, cows can be observed more frequently in stall barns than they can in loose housing systems. On the other hand, loose housing systems require less labor, typically save bedding, and usually cost less to construct.

Stall Barns

The stall barn consists of one or two rows of cows in stanchions (Figures 42.21 and 42.22) or in tie or comfort stalls. In tie stalls, each cow is individually tied with a strap or chain, which offers cows considerably more freedom than in stanchions but requires more labor. Comfort stalls are a special type of tie stall and are designed to give cows more freedom than is afforded by ordinary tie stalls. Each cow is secured with a strap or chain fastened to a curb at the front. Horizontal bars at the front force the cow to stand near the rear of the platform. The tie rail should be 48 inches above the mattress and 8 inches in front of the manger curb. Adjacent stalls are separated by stall dividers. Typically, concrete floors are used (Figures 42.23 and 42.24); sometimes they are covered with rubber mats or mattresses. The floor slopes into a gutter, which is usually 16 inches wide and 8 inches deep on the stall side and 6 inches deep on the alley side.

In stall barns, cows may be faced inward toward the center of the barn or faced outward toward the walls (Figure 42.25). A time and motion study revealed that, when caring for a dairy herd, 60 per-

Figure 42.22 Holstein cows in a stanchion barn at feeding time. *(Courtesy of Iowa State University)*

cent of the time is spent behind the cows, 15 percent in front of the cows, and 25 percent in other parts of the barn and in the milk house. Because four times as much time is spent behind the cows as in front of them, this finding should be a major design consideration.

Facing the cows out has the following advantages. It reduces labor, and the wide middle alley in the center of the barn facilitates milking and cleaning. This configuration also alleviates manure-splattered sidewalls. The advantages of facing the cows in are that it is easier to get them into their stalls, and feeding is facilitated because both rows of

Figure 42.23   Time spent laying down is dramatically decreased in this stanchion barn with concrete stall platforms, which reduces cow performance. *(Courtesy of Iowa State University)*

Figure 42.26   Open barns, such as this transition cow facility, are designed primarily to provide shade. *(Courtesy of John Smith)*

Figure 42.24   Older facilities were not designed for cow comfort but instead focused on worker comfort and cow cleanliness. *(Courtesy of Iowa State University)*

Figure 42.25   A stall barn with cows facing out. *(Courtesy of Iowa State University)*

cows can be fed without backtracking. Regardless of configuration, during the winter months, the temperature in stall barns should not be permitted to drop below 40°F.

The standard milk house for a stall-type barn is located on the side of the barn opposite the feed room and silos. It has provision for the cooling and storage of milk and for washing the utensils, and it has machinery for operating milkers and refrigeration units.

Loose Housing

In loose housing, the herd is handled on a group basis except at milking time. It may be open lot or enclosed.

There are two types of resting or loafing arrangements:

1. **Group housing:** In this system, the cows rest in a common area, with or without bedding (Figure 42.26), on a manure pack. About 60 to 70 square feet of bedded area should be provided per cow (Figure 42.27). The bedding material should be deep to begin with (Figure 42.28); each day 10 to 12 pounds of fresh bedding per cow should be added, although more may be required in wetter climates. The advantage of group housing over individual stall housing is that the initial cost per cow for construction is less.

2. **Freestall housing:** This is a modification of the group housing system. It consists of individual open stalls, which are bedded. For small breeds, use 3.5- by 7-foot stalls with neck rails

Figure 42.27 Cows housed on a bedded pack should have 60 to 70 square feet of bedded area per cow to avoid overcrowding. *(Courtesy of Iowa State University)*

Figure 42.29 Well-designed freestalls with sand bedding. *(Courtesy of Monsanto)*

Figure 42.28 This bedded-pack barn is used to house transition cows comfortably. *(Courtesy of Mark Kirkpatrick)*

Figure 42.30 This four-row freestall barn has curtained sidewalls and a large ridge opening to take advantage of natural ventilation. *(Courtesy of John Smith)*

that are 40 to 42 inches high and 60 inches from the rear of the stall. The brisket board should be 60 inches from the rear of the stall for small-breed cows. For large breeds, use 4- by 8-foot stalls (Figure 42.29). Longer stalls are required in side-lunging stalls (stalls where front partitions prevent lunging forward). Neck rails should be positioned between 48 and 50 inches high and 66 inches from the rear of the stall, and the brisket board should also be 68 to 70 inches from the rear of the stall. Freestall partition loops should be 37 to 39 inches high, and the top loop should be 48 to 50 inches high in side-lunging stalls. Open-front stalls allow a shorter curb to curb length (16 feet) because cows can utilize the stall in front of them for lunge space.

The advantages of freestall housing over group housing are that less space per cow is necessary, bedding costs may be reduced by as much as 75 percent, less labor is required to bed the cows, and the cows are cleaner, with the result that less cow-washing time is required.

Regardless of the type of housing, adequate ventilation is critical for cow productivity. Ventilation needs are typically determined by the amount of air changes needed for moisture removal. The removal of large quantities of water, especially in the winter when barns are closed, can be challenging. To facilitate ventilation, freestall barns should be built with an open ridge (a minimum of 2 inches per 10-foot barn width), high sidewalls (14 to 16 feet), and an adequate roof slope (a 4-inch drop every 12 inches or a 4 in 12 slope) (Figure 42.30).

Figure 42.31   These two-row barns are separated so that air movement is not restricted between barns and natural ventilation can be used within barns. *(Courtesy of Howard Tyler)*

Figure 42.32   Two-row barns in Kansas with curtained sidewalls on the upwind side and a feeding alley on the downwind side. *(Courtesy of Howard Tyler)*

Figure 42.33   Four-row freestall barns have an alley between the outer row of stalls and the outside wall on each side of the barn. *(Courtesy of Mark Kirkpatrick)*

Figure 42.34   Six-row freestall barns have three rows on each side of the feeding alley, with the outside row facing the outside walls of the barn. *(Courtesy of John Smith)*

Figure 42.35   Wide crossover with two water troughs. Water intake is higher from the trough placed outside the normal cow traffic pattern. *(Courtesy of John Smith)*

Sidewalls are typically open, with curtains to control airflow. Building freestall facilities on an east/west orientation reduces radiant energy load when compared to barns built with a north/south orientation. In addition, two- or four-row barns have improved ventilation and increased bunkspace per cow when compared with three- or six-row barns (Figures 42.31 through 42.34).

Crossovers are generally placed every twenty to twenty-five stalls to facilitate cow movement between resting areas and feeding areas (Figure 42.35). Crossovers should be at least 12 feet wide, and a water trough should be accessible in each crossover (Figure 42.36). Shallow (less than 10 inches deep) water troughs allow rapid filling and rapid draining for ease of cleaning. Tip waterers allow rapid draining by easy tipping of the entire water trough (Figure 42.37). Other strategies used on some facilities include using a large drain hole with a bowling ball for a drain plug (Figures 42.38 and 42.39). Whatever type is used, water troughs should be able to be cleaned rapidly and frequently (at the minimum once weekly) (Figure 42.40).

Figure 42.36   Water troughs should be placed outside cattle travel patterns in freestall crossovers. *(Courtesy of Mark Kirkpatrick)*

For larger herds, cows are grouped to permit faster movement of each group through the milking parlor. This reduces stress, increases feed intake, and improves profitability. Group sizes are determined by parlor size and parlor throughput so that each group is a multiple of the parlor size and each group can be milked in less than one hour. For example, for a double-ten parlor milking 160 cows per hour, group sizes of 120 cows per group allow three full parlor turns per group and a forty-five minute milking time per group.

Maternity Stalls

Box-type maternity and isolation stalls should be provided for close observation at this critical time (Figure 42.41). It is recommended that 200 square

Figure 42.37   Tip waterers are mounted on swivels for easy draining and cleaning. *(Courtesy of Monsanto)*

Figure 42.39   Bowling balls are used as drain plugs for this water trough. They are easy to grasp and remove, and they allow rapid draining. *(Courtesy of Monsanto)*

Figure 42.38   An old bowling ball is used to plug a large drain hole in this customized water trough. *(Courtesy of Monsanto)*

Figure 42.40   To minimize the need for cleaning, some producers stock their water troughs with catfish. *(Courtesy of Monsanto)*

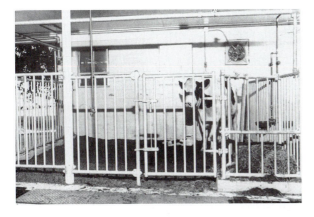

Figure 42.41 Maternity pens should provide at least 200 square feet of space. *(Ohio Agricultural Research and Development Center, Wooster)*

feet be allowed per stall, with one stall per twenty milk cows in the herd, unless seasonal calving or group synchronizations occur. The temperature of the maternity stall should be about the same as is maintained for the milking herd. Dairy cattle have a thermal neutral zone that ranges between 40° to 70°F, but the newborn calf will chill rapidly at temperatures under 55°F.

FEED AND WATER

Feed and water accessibility is critical for lactating cows to maximize feed intake and milk production of these animals. Most large dairies feed complete rations in fenceline or drive-through feeding systems (Figures 42.42 through 42.45), with the herd separated into production groups. Feed needs to be readily accessible to cows and protected from the weather (Figures 42.46 through 42.50). Mangers should provide 30 linear inches per head for roughage feeding for the larger breeds of dairy cows and 24 to 28 inches

for the smaller breeds. Bunks and roughage racks should be 30 inches wide when cattle are fed from one side, and 36 inches wide when they are fed from both sides. Feed bunks for total mixed rations require less space (20 linear inches per head) because of increased density of the feeds offered.

Adequate water trough space for lactating cows depends on the level of milk production desired (Figure 42.51); minimum requirements would be 1 linear foot of open tank per eight to ten head or one automatic bowl per fifteen head. Individual water cups in tie stalls or stanchions require a flow rate of at least 3 gallons per minute. Water temperature should be maintained within the range of 35° to 80°F and warmed to 50°F in the winter.

MILKING FACILITIES

The system of milking utilized on a particular operation is often dictated by the housing system present. Freestalls and loose housing arrangements lend themselves to parlor milking, while cows housed in stanchion or tie-stall barns may be milked at the stall (Figure 42.52).

Holding Area

This area is for the purpose of confining cows in preparation for milking (Figures 42.53 through 42.57). It should be paved, easy to clean daily, and funneled to the parlor. About 16 to 17 square feet per cow should be allowed for large-breed cattle and 14 to 15 square feet for smaller breeds. This is often the area where cows are exposed to the most heat stress; heat abatement in this area is crucial to maximizing cow productivity. A combination of natural ventilation and forced-air movement is typically required because of the high animal density.

Figure 42.42 Fenceline feeding saves labor costs, but leaving feeding areas exposed to the elements shortens the bunk life of feed. *(Courtesy of Iowa State University)*

Figure 42.43 Self-locking headcatches in this drive-through feeding alley make it easy to restrain these Brown Swiss cows for breeding purposes or routine veterinary checks. *(Courtesy of Mark Kirkpatrick)*

Figure 42.45 The self-locking headcatches in this facility are widely spaced to prevent overcrowding at the feed bunk and reductions in feed intake. *(Courtesy of Monsanto)*

Figure 42.44 The high rear barrier in this feed bunk helps minimize feed losses. *(Courtesy of Iowa State University)*

Figure 42.46 The shade over the feedbunk on this two-row barn in Kansas was installed to protect the feed from sun damage more than to protect the cows. *(Courtesy of Howard Tyler)*

Figure 42.47 The side view of this open facility for transition cows shows the protection provided for the feedbunk area as well. *(Courtesy of John Smith)*

Figure 42.48 Cows ingest feed for eight hours daily; therefore, feedbunk design is a crucial aspect of profitability for the facility. *(Courtesy of Iowa State University)*

Figure 42.50 Tile-lined bunks increase ease of cleaning and reduce the risk of feed contamination through spoilage. *(Courtesy of Iowa State University)*

Figure 42.49 This transition cow facility has plenty of bunk-space in the feeding area to facilitate the feeding of bulky forages. *(Courtesy of Monsanto)*

Figure 42.51 Water should be available for cows exiting the parlor. *(Courtesy of John Smith)*

Milking Parlors

Freestall systems lend themselves particularly well to the use of a milking parlor. The milking parlor improves labor efficiency, working conditions, and sanitation surrounding the milking operation. Although there is a wide range of choices in milking parlor designs, the most common parlor configurations currently in use include the following:

Tandem-type (side-opening): Cows stand single file in line and broadside to the operator's pit (Figure 42.58).

Herringbone: Cows stand at an angle to the operator's pit, and the angle may vary greatly from parlor to parlor. This parlor design, which was developed in New Zealand, takes its name from the arrangement of the cows in the parlor (Figures 42.59 through 42.62).

Trigon or polygon: These are three- and four-sided herringbone parlors. They were initially developed to maximize the labor-saving capabilities of equipment used in mechanized parlors. They are best adapted for use in herds of more than 400 cows (Figures 42.63 and 42.64).

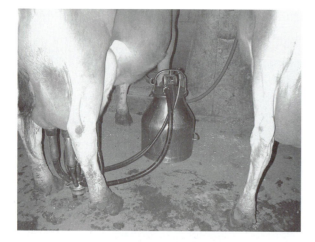

Figure 42.52   Cows in stall barns are typically milked at the stall. *(Courtesy of Mark Kirkpatrick)*

Figure 42.55   View of a holding pen from the back of a milking parlor as cows wait to be milked. *(Courtesy of John Smith)*

Figure 42.53   A cow's view from the holding pen into a parallel parlor. *(Courtesy of Monsanto)*

Figure 42.56   The crowd gate (in a raised position) drops down behind cows after they enter the holding pen and pushes cows toward the parlor. *(Courtesy of Westfalia)*

Figure 42.54   A crowd gate moves cows toward the parlor in this large, well-ventilated holding pen in South Dakota. *(Courtesy of Mark Kirkpatrick)*

Figure 42.57   Crowd gates should never be used to push cows through holding pens. They should be used only to shorten the length of the pen gradually. *(Courtesy of Westfalia)*

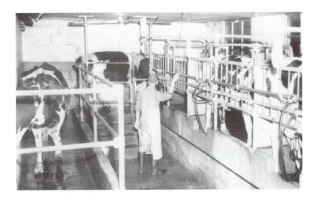

Figure 42.58   An example of a side-opening milking parlor. *(Courtesy of Iowa State University)*

Figure 42.59   Jet washers spray down the parlor floor between groups of cows. *(Courtesy of Howard Tyler)*

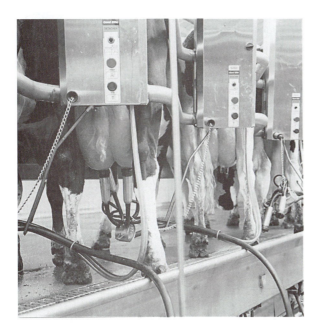

Figure 42.60   These chain-driven automatic detachers used in this herringbone parlor are triggered when milk flow drops below a set flow rate for a given period of time. *(Courtesy of Westfalia)*

Figure 42.61   Rapid exit parlors require a larger building space to accommodate the exit area adjacent to the milking area. *(Courtesy of Westfalia)*

Figure 42.62   A group of cows completes milking in this herringbone milking parlor. *(Courtesy of Westfalia)*

Figure 42.63   Polygon parlors have decreased in popularity because of the development of more labor-efficient options, such as parallel parlors. *(Courtesy of Iowa State University)*

Figure 42.64 A view from above a polygon parlor. *(Courtesy of Iowa State University)*

Figure 42.66 Some rapid-exit parlors allow the operator to release a single animal that requires individual attention. *(Courtesy of Westfalia)*

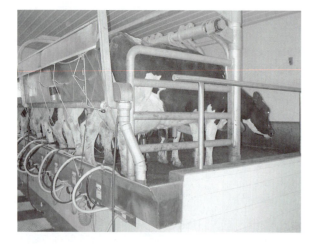

Figure 42.65 In parallel parlors, cows are milked from between their rear legs. *(Courtesy of Monsanto)*

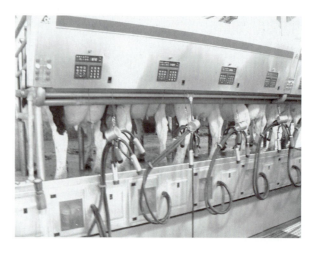

Figure 42.67 Cows waiting to be milked in a parallel parlor in Kansas. *(Courtesy of Howard Tyler)*

Parallel: Cows stand perpendicular to the operator pit (Figures 42.65 through 42.69). This arrangement reduces the building space required, relative to that required by a herringbone configuration, and the time required for milking is also reduced.

Rotary (or carousel): The parlor, which most often incorporates either a herringbone (Figure 42.70) or parallel (Figure 42.71) configuration, rotates around while the operators stay in place (Figures 42.72 through 42.76).

Other designs, such as the inexpensive step-up parlors (Figure 42.77), are also in operation, but all parlor designs are similar because they bring cows to milkers rather than having milkers move to cows. Each of these designs differs in terms of initial cash investment, maintenance costs, labor requirements, and parlor efficiency. Parlor efficiency is typically described as turns per hour. The term *turns*

per hour is defined as the number of times that both sides of the parlor are milked in an average hour of milking time. Obviously, rotary parlors must be evaluated differently, and the steady-state throughput per hour or average number of cows milked per hour is a good measure of parlor efficiency for all parlors.

Parlor efficiency can be maximized by altering the design of the parlor or by changing the groupings of the cows. For example, cows that either move slowly or milk slowly impair parlor efficiency

Figure 42.68 Cows wait for the rear gate to lift in a rapid-exit parallel milking parlor. *(Courtesy of John Smith)*

Figure 42.71 A milker attaches a milking unit to a cow in a parallel-style carousel parlor. *(Courtesy of Westfalia)*

Figure 42.69 Milkers work together in this double-thirty-six parallel parlor to keep parlor efficiency high. *(Courtesy of Mark Kirkpatrick)*

Figure 42.72 A cow enters a stall on a large carousel parlor in Australia. *(Courtesy of John Smith)*

Figure 42.70 Cows are accessed from the side for milking in a herringbone carousel parlor. *(Courtesy of Westfalia)*

Figure 42.73 A large carousel parlor in Australia stands empty between milkings. *(Courtesy of John Smith)*

Figure 42.74 A twenty-four stall parallel carousel parlor. *(Courtesy of Westfalia)*

Figure 42.76 Cows are rotated on an elevated platform that allows workers to stand comfortably while milking. *(Courtesy of John Smith)*

Figure 42.75 The drive system turns the carousel parlor at a specific speed that varies with the size of the carousel. *(Courtesy of Westfalia)*

Figure 42.77 For some producers, a step-up parlor is an inexpensive alternative to a traditional milking parlor. *(Courtesy of Mark Kirkpatrick)*

dramatically. Even if they slow milking by only two minutes per turn, milking lame cows or slow milkers as a separate group can often save from one to three hours daily in labor costs.

Parlor size should be calculated to maximize parlor usage in situations where milking is accomplished through hired labor. Parlors are a major investment on most dairies, and this practice allows maximum return on the investment. In these situations, the parlor should be in use twenty to twenty-four hours per day. Farms with limited labor availability or very expensive labor, such as those relying strictly on family labor, typically use a different strategy, designing parlors to minimize time spent milking to allow increased labor availability for other purposes.

When calculating the parlor size needed for a given facility, the number of cows to be milked is obviously a major consideration. The number of cows and the number of daily milkings determine the parlor size, assuming the goal is to milk twenty hours per day. Thus, our assumptions are that each milking will be ten hours if we are milking twice a day and about 6.5 hours if we milk three times daily. This leaves time for special-needs cows and for parlor cleaning. We can conservatively estimate achieving three to four turns per hour when milking twice daily and four to five turns per hour when milking three times daily. Turns per hour is the most variable number and is based on premilking hygiene, adequacy of stimulation of milk letdown, and speed of cow movement in and out of the parlor stalls.

Build the parlor with the future in mind. Allow room for expansion or for the ability to retrofit a design change. For example, building a herringbone parlor with a wider than normal exit lane allows easy conversion to a parallel parlor if so desired.

Return lanes are designed to allow cows exiting the parlor to reenter their group housing areas. They should be at least 16 feet wide, and they are often designed even wider (up to 24 feet wide) for very large parlors with forty or more cows exiting at a time. These extra wide lanes reduce stress, reduce crowding, and facilitate cow flow.

SUMMARY

- Cattle buildings and equipment are designed to facilitate the comfort, health, and productivity of cows; minimize maintenance needs and labor requirements; minimize bedding costs; and facilitate manure handling and disposal.
- To optimize cow performance, ventilation, accessibility and quality of feed and water, lighting, and cow comfort should be considered.
- The following factors should be considered when planning a new dairy farm or appraising an existing facility: location, climate, other dairy producers, labor supply, access to feeds, water supply, environmental concerns, topography, drainage, access to utilities, and neighbors.
- Calves can be reared indoors or in outdoor wooden pens, plastic hutches, polydomes, or calf condos.
- The main requisite of replacement heifer housing is protection from drafts, rain, snow, and wind.
- Feedbunk dimensions and water tank space depend on the size and number of animals.
- There are two basic facility systems: stall barns and loose housing.
- A stall barn consists of tie or comfort-type stalls.
- Loose housing, either an open lot or enclosed system such as freestalls, is a system in which the herd is handled on a group basis.
- Adequate ventilation is critical for cow productivity.
- The most common milking parlors are tandem-type, herringbone, trigon or polygon, parallel, and rotary.
- Parlor designs differ in terms of initial cash investment, maintenance costs, labor requirements, and parlor efficiency.

QUESTIONS

1. What is the primary function of dairy buildings?
2. What is the preferred number of calves in a group?
3. What are the advantages of freestall housing over group housing?
4. Define the design of each of the five common milking parlors.
5. Define *turns per hour* or parlor efficiency.

ADDITIONAL RESOURCES

Articles

Longenbach, J. I., A. J. Heinrichs, and R. E. Graves. "Feed Bunk Length Requirements for Holstein Dairy Heifers." *Journal of Dairy Science* 82 (January 1999): 99–109.

Rotz, C. A., C. U. Coiner, and K. J. Soder. "Automatic Milking Systems, Farm Size, and Milk Production." *Journal of Dairy Science* 86 (December 2003): 4167–4177.

Schnier, C., S. Hielm, and H. S. Saloniemi. "Comparison of Milk Production of Dairy Cows Kept in Cold and Warm Loose-Housing Systems." *Preventive Veterinary Medicine* 61 (December 2003): 295–307.

Sisto, A. M., and T. H. Friend. "The Effect of Confinement on Motivation to Exercise in Young Dairy Calves." *Applied Animal Behavioral Science* 73 (July 2001): 83–91.

Thomas, C. V., M. A. DeLorenzo, and D. R. Bray. "Factors Affecting the Performance of Simulated Large Herringbone and Parallel Milking Parlors." *Journal of Dairy Science* 79 (November 1996): 1972–1980.

Internet

Animal Housing Systems: http://server.age.psu.edu/extension/Factsheets/g/G76.pdf.

Benefits of Calf Hutches for Housing Young Dairy Calves: http://www.calfnotes.com/pdffiles/CN056.pdf.

Building and Managing Super Calf Hutches: http://www.extension.umn.edu/distribution/livestocksystems/DI0416.html.

Designing Efficient Animal Handling Facilities: http://www.wisc.edu/dysci/uwex/mgmt/pubs/DesigningFacilities.html.

Flat Barn Milking Systems: http://cecommerce.uwex.edu/pdfs/A3567.PDF.

Milking Parlor Cleaning: http://edis.ifas.ufl.edu/DS151.

Natural or Tunnett Ventilation for Freestall Structures: http://www.ansci.cornell.edu/tmplobs/doc225.pdf.

Newborn Housing for Dairy Calves: http://www.ansci.cornell.edu/tmplobs/doc213.pdf.

Planning a Milking Center: http://www.oznet.ksu.edu/library/LVSTK2/MF2165.pdf.

43

Assessing Cow Comfort

OBJECTIVES

- To define proper cow rest and sleep behaviors.
- To describe appropriate stall designs and bedding types.

Cow comfort is a topic that integrates facility design with animal behavior and animal welfare. When troubleshooting a dairy operation, comfort should be the first on-farm management area to assess because it affects almost every other aspect of cow and herd performance. Lack of comfort affects feed intake, growth rates, milk production, herd health, reproductive efficiency, and almost every other area of production. Several factors are critical to maintaining cow comfort, including adequacy of stalls, choice of bedding, feed and water accessibility, quality of walking surfaces, adequacy of ventilation, heat abatement, adequacy of lighting, and control of stray voltage.

One of the most important aspects of herd management is the ability to assess cow comfort accurately. Assessments of cow comfort are often made indirectly, through measurements of air changes, stall dimensions, bedding quality, or other facility issues that affect the cow. Even though these measurements are valuable, our observations should not be limited to measurements.

The most accurate clues to cow comfort still come from the cows themselves. Contented, comfortable cows are eating, drinking, milking, or waiting to be milked; moving to a specific destination; exhibiting estrus; or lying down and ruminating. Herds with a high percentage of cows that are simply standing around are herds with a cow-comfort issue. In many cases, we have become so accustomed to abnormal behavior patterns that they appear normal. Therefore, it is important for producers to understand which behaviors are truly normal and which are indicative of a comfort issue.

REST AND SLEEP

Normal behavior in sleep should be recognized especially because it differs widely between species. Cattle always lie into a slope and lay uphill. Cattle typically lie on the stomach (sternal recumbancy), or they may tilt to one side, with the forelimbs folded under the body and one hind limb extends forward, while the other protrudes toward the outside. Although cattle rest in this manner, they do not sleep in the sense that the term usually suggests. While lying down, they do shut their eyes for short periods of time. Calves commonly spend up to a half-hour at a time with their heads turned back in the flank position.

Cattle assume four common resting positions. The wide position, although not assumed for long periods of time, is still important in the rest sequence. In this position, cows are in partial lateral recumbency (on their sides) with their head and one or both front legs stretched out (Figures 43.1 and 43.2). The narrow position has the cow in sternal recumbency (on her chest), her head held high

Figure 43.1 Cows assuming the wide resting position extend both rear legs and forelegs. *(Courtesy of Howard Tyler)*

Figure 43.2 A variation on the wide resting position—the dead cow position. *(Courtesy of Howard Tyler)*

Figure 43.3 Cows spend more time in the narrow resting position than any other resting position. *(Courtesy of Howard Tyler)*

and one or both front legs tucked under the chest (Figure 43.3). In the short resting position, the cow is in sternal recumbancy, both front legs are tucked under, and her head is laid back against her flank (Figures 43.4 and 43.5). Cows in the long position extend both front legs while in sternal recumbency (Figure 43.6). This stretches the neck forward more than it is stretched in other positions. Cows also assume combinations of the above positions for short periods of time (Figure 43.7).

When rising, the cow must lunge forward, using the knees as a pivot point. This motion transfers much of her weight off the rear legs. It allows the animal to raise her hindquarters first, then rise to a full standing position.

Stall Design

The ideal stall design integrates an understanding of cow behavior, cost considerations, air movement and the ventilation needs of the cow, and labor requirements for maintenance. In ideal conditions,

cows lie down for over fourteen hours daily, and stall comfort is tremendously important for optimal cow performance. Reductions in lying time by as little as two hours significantly affect cow performance and profitability. Critical considerations in stall design include the provision of a spacious yet defined resting space, adequate lunge and bob room so the cow can easily rise (Figure 43.8), and an adequate amount of bedding for cushion and friction reduction. Cows must also be able to assume all four normal rest positions.

The choice of bedding is a critical one both for the producer and for the cow. While sand is often considered the gold standard among beddings types (Figure 43.9), in many situations, it is either not an option or not the best option. Other bedding choices include straw (Figure 43.10), shavings, rice hulls, shredded newspaper, and many others. The factors that make a bedding work include provision of adequate cushion, reduction in friction, resistance to microbial growth, and removal of moisture from the cow.

Regardless of the type of bedding, provision of adequate cushion requires a bedding depth of at least 6 inches. Bedding should be maintained level with the outside curb height (Figures 43.11 through 43.14). Medial hock lesions are a frequent occurrence in facilities with inadequate bedding (Figure 43.15). It is also important to remember that cows avoid lying downhill, and stalls that slope from back to front, whether through inappropriate maintenance or through digging out, will not be used (Figure 43.16). Cows dig out the bedding in uncomfortable stalls more than they do in comfortable stalls (Figure 43.17).

Friction reduction is important to reduce skin abrasions as cows shift positions or rise from a resting position. This can be accomplished with minimal amounts of bedding and is typically a problem only in facilities that use mattresses. Mattresses are often highly abrasive without the provision of bedding, an important point that is often overlooked in facilities so equipped. Mattresses are often a popular alternative to other beddings (Figures 43.18 through 43.20). Stalls with mattresses still require some bedding to reduce friction and minimize leg abrasions. In addition, if gaps are present between stalls, they alter lunging behavior.

Minimization of microbial growth is much easier when inorganic beddings, such as sand, are used. However, organic beddings that are well maintained and frequently changed easily outperform inorganic beddings that are contaminated with manure. Moisture removal can be accomplished either through highly absorbent beddings or nonabsorbent beddings that drain exceptionally well.

Figure 43.4 A cow in the short resting position tucks its head back against its flank. *(Courtesy of Howard Tyler)*

Figure 43.5 Cows restrained by rope or chain ties must have an adequate tie length for assuming the short resting position. *(Courtesy of Howard Tyler)*

Figure 43.6 In the long resting position, the forelegs are extended in front of the cow, as shown in this crossbred cow. *(Courtesy of Howard Tyler)*

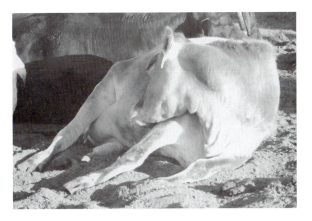

Figure 43.7 Cows also assume unique combinations of resting positions, such as the wide short position. *(Courtesy of Howard Tyler)*

Figure 43.8 The challenge for these cows was not entering a stall but will be apparent when they attempt to leave. *(Courtesy of Iowa State University)*

Figure 43.9 Sand is an excellent inorganic bedding option. It provides excellent cushioning characteristics without promoting bacterial growth. *(Courtesy of Monsanto)*

Figure 43.10   Straw is an excellent bedding material if it is kept clean and deep. *(Courtesy of Mark Kirkpatrick)*

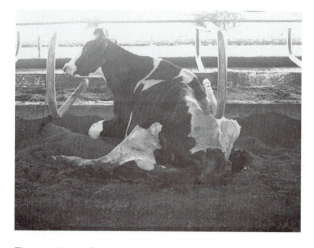

Figure 43.13   Cows can get trapped under stall partitions if bedding levels get too low. *(Courtesy of Monsanto)*

Figure 43.11   In most facilities, fresh bedding should be added to stalls on a daily basis. *(Courtesy of Monsanto)*

Figure 43.14   This stall, although well-groomed, is not bedded adequately. The low level of bedding creates lunging barriers for cows and restricts stall use. *(Courtesy of Monsanto)*

Figure 43.12   Inadequate bedding exposes the curb and increases the risk of serious leg abrasions. *(Courtesy of Monsanto)*

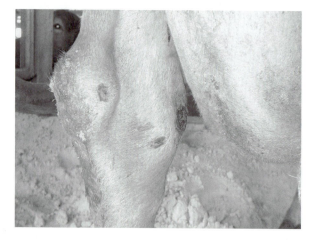

Figure 43.15   Medial hock lesions typically indicate issues with stall comfort. *(Courtesy of Mark Kirkpatrick)*

Figure 43.16 Cows will lay uphill, even if it requires backing into stalls. *(Courtesy of Mark Kirkpatrick)*

Figure 43.19 Presewn tubes in mattresses prevent shifting of materials inside the mattress and uncomfortable stall surfaces. *(Courtesy of Mark Kirkpatrick)*

Figure 43.17 Digging behaviors are responses to uncomfortable stalls. *(Courtesy of Mark Kirkpatrick)*

Figure 43.20 Mattresses are typically installed directly on a concrete platform, making it extremely challenging to change to a deep-bedded option. *(Courtesy of Monsanto)*

Figure 43.18 Well-designed stall mattresses provide adequate cushion, but they still require a minimal amount of bedding for friction reduction as well as keeping cows dry. *(Courtesy of Monsanto)*

In freestall facilities that use sand, daily usage rates per stall can reach 70 pounds. Bedding savers, such as sandtraps, may reduce this usage by 75 percent; however, in poorly maintained facilities, they are much more difficult to clean (Figures 43.21 and 43.22). The type of sand used in stalls is also an important consideration; fine sand is preferable to coarse sand.

In facilities with well-designed and well-maintained stalls, 90 percent of cows in a group lie down within two to three hours after milking (Figures 43.23 through 43.25). At any time, more than 80 percent of cows that are not eating or drinking should be lying down. Hock lesions and joint swelling thus are minimal, cows are cleaner, the incidence of lameness and mastitis is lower, detection of estrus is easier, and milk production is higher.

Figure 43.21   Sand savers are simply a series of linked, stiff rubber belts or strips from used tires that restrict sand movement out of the stall. *(Courtesy of Monsanto)*

Figure 43.22   Sand savers can greatly reduce sand use in stalls without reducing comfort if they are installed and maintained properly. *(Courtesy of Monsanto)*

Figure 43.23   Cows in a comfortable environment lie down within a few hours of milking. *(Courtesy of Mark Kirkpatrick)*

Figure 43.24   It is easy to recognize facilities with well-designed, comfortable stalls. *(Courtesy of Monsanto)*

Figure 43.25   The cows themselves provide the best indication of the adequacy of these freestalls. *(Courtesy of Howard Tyler)*

The major reasons that stalls are not used properly include a lack of adequate lunge space (Figures 43.26 and 43.27), improper neck rail placement (Figure 43.28), a lack of cushion, and a lack of ventilation at the stall itself (Figure 43.29). Signs of uncomfortable stalls include refusal to use the stalls, a high percentage of cows standing (often with their front feet in the stall and their back feet in the alley) (Figures 43.30 through 43.32), or observations of abnormal behaviors as cows enter or leave stalls. These behaviors include difficulty or an inability to assume normal positions of rest (Figure 43.33), difficulties in standing from a lying position, backing into stalls, lying on the stall curb (Figure 43.34), or apprehensive actions prior to entering the stall. These intention behaviors or hesitations prior to entering an uncomfortable situation can last for several minutes. Cows attempting to stand after lying in stalls with inadequate lunge room (Figures 43.35 through 43.37) will

Figure 43.26 A poorly designed outside row in a freestall barn that does not provide an opportunity for either forward or side lunging. *(Courtesy of Monsanto)*

Figure 43.27 Forward lunge space is limited, so partitions between stalls in the outside rows in six-row freestall barns should permit side lunging. *(Courtesy of Mark Kirkpatrick)*

Figure 43.28 A front view clearly demonstrates how low neck rails can impede access to stalls. *(Courtesy of Monsanto)*

Figure 43.29 The concrete barrier between these stalls impairs stall usage and also impairs airflow to the stalls. *(Courtesy of Mark Kirkpatrick)*

Figure 43.30 Cows standing half in and half out of stalls clearly indicate a defect in stall design. *(Courtesy of Monsanto)*

Figure 43.31 These cows are prevented from fully accessing the stall because of the neck rail, which is too low to permit easy access. *(Courtesy of Monsanto)*

Figure 43.32 This retrofit for neck rails is easily adjustable and will not harm cows if they are positioned too low. *(Courtesy of Monsanto)*

Figure 43.35 The cow has no place to lunge when rising in this poorly designed and bedded stall. *(Courtesy of Monsanto)*

Figure 43.33 Stanchions restrict cow movement because the cows cannot assume the short resting position. *(Courtesy of Iowa State University)*

Figure 43.36 The pain that a cow encounters when lunging to rise eventually discourages stall use in this facility. *(Courtesy of Monsanto)*

Figure 43.34 This cow is resting half in the stall to avoid lying downhill in this poorly maintained stall. *(Courtesy of Monsanto)*

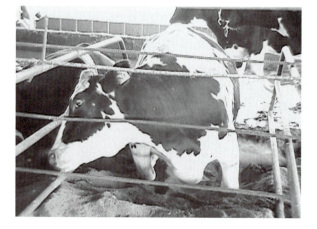

Figure 43.37 Barriers that restrict lunge space result in painful experiences associated with stall use, and eventually stall use by cows declines. *(Courtesy of Monsanto)*

often end up dog-sitting after attempting to rise front-end first.

Cows with inadequate lunge space often lie at an angle in the stall (Figure 43.38), creating space issues for her neighbors and increasing the likelihood that she will defecate and urinate in the corners of the stall rather than in the gutter. Narrowing the stall, while solving the defecation issue, forces the cow to lie further back in the stall so that she can eventually rise (Figure 43.39). Cows are often forced to dangle one rear leg (along with their tail) off the back of the stall (Figure 43.40). The movement of the rear leg on and off the stall platform can cause medial hock lesions (Figure 43.41) and also creates cleanliness issues (Figure 43.42). Opening the front of the stall (Figures 43.43 and 43.44) allows the cow to lie further forward and solves the cow-comfort issue.

Figure 43.40   Cows rest with one leg dangling off the back of the stall platform if it is designed too short or too narrow. *(Courtesy of Mark Kirkpatrick)*

Figure 43.38   This short stall with a high brisket board has forced this cow to lie at an angle in the stall. *(Courtesy of Monsanto)*

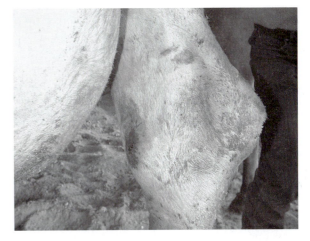

Figure 43.41   Cows that drag one leg on and off the stall platform in uncomfortable stalls often develop medial hock lesions. *(Courtesy of Mark Kirkpatrick)*

Figure 43.39   These freestalls are too narrow, which forces cows backward and results in cows hanging into the alley. *(Courtesy of Mark Kirkpatrick)*

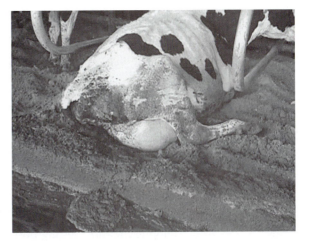

Figure 43.42   Cows in uncomfortable stalls drag manure from the alley into the stall bed as they drag their rear legs on and off the stall platform. *(Courtesy of Monsanto)*

Figure 43.43 Removing barriers between stalls ensures adequate space for lunging and improves stall usage. *(Courtesy of Mark Kirkpatrick)*

Figure 43.45 Brisket boards are used to position the cow in the stall. *(Courtesy of Monsanto)*

Figure 43.44 Freestalls should be designed to minimize barriers between stalls, which minimizes impediments to lunging. *(Courtesy of Monsanto)*

Figure 43.46 Brisket boards can be retrofitted easily into most facilities with a little ingenuity. *(Courtesy of Monsanto)*

Brisket boards (Figures 43.45 and 43.46), while effectively placing the cow in a stall, can restrict her from assuming the long resting position, especially if they are installed too high or bedding is inadequate (Figures 43.47 and 43.48). Inadequate bedding exposes more brisket board and creates a barrier. In this situation, cows often extend the front leg laterally rather than forward, leading to discomfort and increased restlessness (especially hind leg movements). The increased restlessness leads to a greatly increased incidence of hock lesions.

Tie stalls present an additional complication. The chain length in these stalls is often too short and prevents cows from assuming the short resting position (Figure 43.49). Cows in these stalls shift position more frequently to attempt to find a more comfortable option, but there is no option that allows them

Figure 43.47 Brisket boards in poorly bedded stalls provide a barrier to the long resting position. *(Courtesy of Monsanto)*

Figure 43.48   In a properly bedded stall, the brisket board should be barely visible. *(Courtesy of Monsanto)*

Figure 43.50   Crosscut groove patterns enhance drainage and prevent slipping. *(Courtesy of Monsanto)*

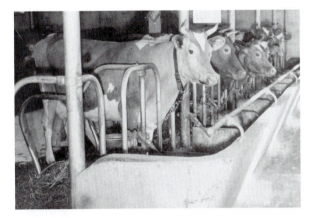

Figure 43.49   Extending the chain length in older tie stalls allows cows to assume the short resting position. *(Courtesy of Iowa State University)*

Figure 43.51   Grooves that are cut at about 3-inch intervals optimize the footing surface. *(Courtesy of Monsanto)*

complete comfort. Short chains in tie stalls are also associated with increased incidence of silent heats and increased difficulties in estrus detection.

Walking Surfaces

Ease of cow mobility is important for cow comfort, hoof health, and ultimately for cow profitability. Cows that are not confident in their footing (poorly grooved or slippery floors) or that experience pain when walking (poorly maintained, rough floors) have decreased feed intake, have decreased milk production, show minimal signs of estrus, and are at increased risk for developing hoof and joint problems.

Methods to improve footing include grooving concrete surfaces (Figures 43.50 and 43.51) or adhering rubber belting to such surfaces. Rubber belts (Figures 43.52 through 43.54) reduce the time spent on concrete, improve foot health, and improve cow longevity. However, they are expensive and must be securely fastened in some fashion (Figure 43.55). Slotted floors with gaps that are too wide not only reduce cow movement and estrus behavior, but can also be a safety hazard.

Feed and Water

Cow comfort is an important consideration for maximizing feed intake. Facility designs that enhance free and easy access to feed and water improve cow productivity. Many facility designs that are based on labor concerns inadvertently affect cow comfort, thereby decreasing feed and/or water intake. For every aspect of feedbunk design, the impact on feed accessibility and feed intake should be considered.

Figure 43.52 The rubber belting installed in this cow alley provides greater cushioning for the hoof, thus potentially decreasing the incidence of lameness. *(Courtesy of Monsanto)*

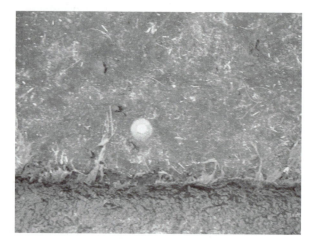

Figure 43.55 Rubber mats must be fastened securely to a concrete surface. *(Courtesy of Howard Tyler)*

Figure 43.53 Rubber belting provides a comfortable surface for walking. *(Courtesy of Monsanto)*

Figure 43.56 Self-locking headcatches need to be spaced properly and installed at the proper height to avoid affecting feed intake. *(Courtesy of Iowa State University)*

Figure 43.54 Cows show a marked preference for walking on belted areas rather than on concrete. *(Courtesy of Monsanto)*

Fenceline feeders are often designed with self-locking headlocks (Figure 43.56). The use of such headlocks greatly enhances the accessibility of cows for routine procedures such as vaccinations and breeding, but there have been concerns that they may also inhibit the dry-matter intake of lactating cows. The truth is that it depends on the design and installation of the headlocks. The feed barrier created by the base of the headlock should be no more than 18 inches high. Properly designed and maintained headlocks do not adversely affect either feed intake or milk production. However, headlocks that force overcrowding or otherwise decrease cow comfort during eating can be expected to decrease the cow's willingness to access them for the amount of time required for maximum dry-matter intake. For larger breeds, four headlocks per 10 feet should be

Figure 43.57   The cables in front of this feedbunk limit cow access to feed and ultimately limit milk production. *(Courtesy of Iowa State University)*

Figure 43.59   The closely spaced vertical bars at this feedbunk provide a barrier to maximal feed intake. *(Courtesy of Iowa State University)*

Figure 43.58   Properly placed cables at feedbunks do not hinder feed intake. *(Courtesy of Iowa State University)*

Figure 43.60   The widely spaced and slanted bars on this elevated feedbunk are not a barrier to feed intake and also reduce feed waste by cows. *(Courtesy of USDA-ARS)*

the maximum designed density, although five in 10 feet is acceptable for smaller breeds, such as Jerseys.

Any barriers that affect the cow's ability to reach feed should be removed and redesigned (Figures 43.57 through 43.60). To increase feed accessibility and minimize the potential for neck abrasions, the curb should be slanted down to the feed (18 inches on the cow side decreasing to 14 inches on the feed side). Sloping the barrier increases the distance a cow can access feed by 5.5 inches. Cows develop avoidance behaviors when feedbunk barriers are poorly designed (especially those that are designed too high) or poorly maintained; these behaviors directly affect feed intake and milk production. Cows prefer eating at ground level to eating from elevated bunks (Figure 43.61). Cows eating with their heads down produce more saliva, which increases

Figure 43.61   Cows eating from elevated feedbunks waste more feed than those eating at ground level. *(Courtesy of Iowa State University)*

Figure 43.62 Bunk liners increase dry-matter intake, leading to increases in milk yield. *(Courtesy of Mark Kirkpatrick)*

Figure 43.64 Water troughs at crossovers in a freestall barn. *(Courtesy of John Smith)*

Figure 43.63 The cow side of this feedbunk allows space for cows to eat without disruption by cows exiting and entering freestalls. *(Courtesy of John Smith)*

their ability to buffer the rumen from excess acidity. Feed tossing is also more common in feedbunks that are too high. In addition, rough bunk surfaces can reduce intake; providing smooth plastic liners for bunks improves the bunklife of the feedstuff and is less irritating to the cow's tongue during feed ingestion (Figure 43.62). The space provided on the cow side of the feedbunk is also critical. Cows forced to eat while in the traffic pattern of other cows exhibit decreased intake when compared to cows given ample room to minimize contact with cows behind them (Figure 43.63).

Water is not only an essential nutrient for the dairy cow, but it may also be the most essential nutrient for the lactating cow. Lack of appropriate accessibility to water can severely limit milk production. For stanchions or tie-stall barns, cows can be pro-

vided water through individual water bowls or through a water trough running the length of the stalls. In situations where a water bowl is shared, intake by submissive cows is reduced. In freestall barns, water troughs should be placed in alley crossovers (every twenty to twenty-five stalls). There should be a minimum of three inches of trough space per cow. Crossovers should be at least 12 feet wide to allow cows to feel safe while standing and drinking at a water trough (Figure 43.64). Narrower crossovers (many older barns have 8-foot crossovers) lead to avoidance behaviors by cows; they often drink only from the ends of the water troughs to avoid conflict with other cows moving through the crossover area. Placing the water trough on the side of the crossover with the least traffic flow also increases accessibility, water consumption, and cow productivity.

Holding pens, because of their high animal density and high propensity for heat buildup, should be built so cows have free access to water. After milking, cows immediately head for water. Exit alley water troughs should be large enough to allow all cows exiting the parlor to have simultaneous access (two feet per cow). Therefore, a double-ten parlor would require a 20-foot-long water trough in the exit alley. Water intake of lactating cows from exit lane water troughs exceeds 3 gallons per day in well-designed facilities.

Cows indicate problems with water sources by reducing intake (impossible to measure in most situations) and by altering drinking behavior. Cows typically drink fifteen times per day and about 1 to 1.5 gallons per drinking bout. If water quality is poor or water troughs are poorly maintained, cows lap at water rather than drink and often splash more than cows at clean, fresh water sources.

Figure 43.65 The spacing between these two-row barns in Kansas permits adequate airflow for maintaining proper building ventilation. *(Courtesy of Howard Tyler)*

Ventilation

Facility designs should always maximize ventilation. Providing 100 feet of space between buildings optimizes airflow but may impair efficiency of animal movement (Figure 43.65). High sidewalls, curtains, and ridge openings all facilitate natural ventilation (Figure 43.66). It is important to assess whether the air changes are adequate at the places where you would like cows to spend time. Stalls and feeding areas especially need adequate air movement to ensure full utilization by cows. Barriers between stalls or in front of stalls can create poorly ventilated areas in an otherwise well-ventilated barn.

Poorly ventilated barns are characterized by cows that congregate near doorways or near other areas with improved air movement (especially near the downwind side of a barn). Cows may use stalls in poorly ventilated barns, but they don't stay in these stalls for as long as they would if ventilation were better. Tunnel ventilation is an excellent method for improving ventilation in tie-stall or stanchion barns or in penning areas. It also serves to cool cows in heat-stress situations by moving air directly across the confined cows. The system should be designed to meet a minimum desired air velocity of 220 feet per minute. The requirements, therefore, depend on the cross-sectional area of the barn to be ventilated.

Heat Abatement

Although cold weather is rarely a major cow comfort issue, hot weather quite often impinges on cow comfort, reducing feed intake, impairing reproductive function, decreasing milk production, and adversely affecting profitability. Heat stress becomes apparent in dairy cattle when the total heat load exceeds the cow's capacity to lose heat. The

Figure 43.66 Airflow coming in through the high sidewalls and the thermal buoyancy from heat generated by cows move air up and out the open ridge vent. *(Courtesy of Monsanto)*

heat load includes the heat generated by the cow (including the heat generated by rumen fermentation) and the heat imposed on the cow from her environment. The environmental load includes any number of factors; however, the temperature-humidity index (THI) incorporates several of them to provide an estimate of environmental heat stress. For dairy cattle, the formula for calculating THI is:

$$THI = T_{db} - (0.55 - 0.55 \times RH/100) \times (T_{db} - 58)$$

T_{db} is ambient air temperature.
RH is relative humidity (%).
When RH is 1,005, then THI = T_{db}.

As a general rule, when THI exceeds 70, mild heat stress is present. Heat stress is considered severe when THI exceeds 90. The response of the cow to severe environmental heat stress depends on the heat abatement measures in place on the facility. The response of the cow in stress situations is dictated by the intensity and duration of the stressor. In a heat-stress response, cows can exchange heat with the environment in several ways, including evaporation, convection, conduction, and radiation.

Evaporative heat losses are accomplished through panting and sweating. In the cow, the primary route of evaporative heat loss is through panting; the rapid, shallow breathing maximizes airflow across the moist mucous membranes in the nasal passages. Evaporation of the moisture in the membranes serves to cool the inspired air. The heat loss in this case is accomplished across the pulmonary tissue. Sweating is the other route for evaporative heat loss. Although cows have numerous sweat glands, the glands have a low rate of secretion, limiting the effectiveness of this method for heat abatement.

Convective heat exchange involves the exchange of heat with the air. This loss, although not the primary route of heat loss for dairy cattle, is improved dramatically in moving air. The movement of heat from the cow to a physical surface is termed conductive heat loss. This is not a major route of heat loss for dairy cattle unless they have a wet area to lie in or they can partially immerse themselves in water.

Radiant heat exchange depends on the radiant energy from the sun. Although it is not a mechanism for heat loss, it is the most important means of increasing heat stress in cattle. Reducing the exposure to radiant energy, therefore, is an important mechanism for reducing this heat load.

The physiological response of a dairy cow to heat stress includes an increase in peripheral blood flow (which increases rectal temperature over 102.5°F), increased respiration rate (over eighty breaths per minute) (Figure 43.67), and sweating. Water intake may double in extreme heat. These responses impair performance by decreasing feed intake, milk production, immune resistance to pathogens, and reproductive performance. Water loss in mature animals can double in extremely hot situations. This loss is much greater in very young animals that have a high surface area to body mass ratio. The potential for dehydration is great unless great amounts of water are consumed.

Although the physiological mechanisms whereby heat stress affects cow performance is similar in most situations, the strategies for heat abatement can vary dramatically depending on regional factors. The strategies available for reducing heat stress in dairy cattle include cooling lanes or ponds, sprinklers or misters, the provision of extra shaded areas, and fans and other evaporative cooling systems.

Figure 43.67   Panting by heat-stressed cows cools inspired air and removes heat across the lung tissue by evaporative cooling. *(Courtesy of Iowa State University)*

Figure 43.68   Properly installed fans help cool cows via convective heat exchange. *(Courtesy of John Smith)*

Fans move air across cattle, increasing convective heat exchange. Therefore, fans are effective only if the ambient air temperature is less than the body temperature of the cow or if the cow is wet. Fans should provide an air velocity of 400 to 600 feet per minute. Overhead fans that are 3 feet in diameter can be placed 30 feet apart, while 4-foot fans should be spaced 40 feet apart. Fans should be tilted downward at a 30° angle to optimize air/cow heat exchange potential (Figure 43.68).

To optimize air movement across the cow, fans should be placed over feedbunks, stall areas (Figure 43.69), exit alleys, and holding pens. Holding pens require an increased fan density because cow density is increased (Figure 43.70). Air should be directed away from the parlor, into the faces of the cows. Fan maintenance is crucial to

Figure 43.69   Although sidewalls are not required in hot climates, fans must be used to move air because natural ventilation is ineffective in hot weather. *(Courtesy of Monsanto)*

Figure 43.71   Poorly maintained fans lose much of their ability to move air effectively. *(Courtesy of Monsanto)*

Figure 43.70   Fan density is increased proportionally with animal density, as in this holding pen. *(Courtesy of John Smith)*

Figure 43.72   Sprinklers at feeding areas are effective for both cooling cows and increasing feed intake. *(Courtesy of Iowa State University)*

maintain proper air movement (Figure 43.71). In dry climates, foggers can be added to the fan to cool the air and distribute water over the cows, thus increasing evaporative cooling. In very humid conditions, this method is ineffective for improving cooling and can actually be detrimental to cow health.

Foggers and sprinklers are both used in cooling systems on dairy farms. Sprinklers pump water through low-pressure pumps, while foggers or misters use high-pressure pumps to greatly reduce water use. They are typically set to start functioning at temperatures over 78°F. They then operate for one minute out of every fifteen-minute time period while the fans run continuously. Sprinklers are relatively inexpensive and are easy to install, operate, and maintain. Sprinklers cool the cow directly (evaporative heat loss) rather than cooling the air (Figures 43.72 and 43.73). They also use a lot of water, which can be a disadvan-

Figure 43.73   This cow soaker is designed to wet cows in a short period and then allow a period of intense evaporative heat loss. *(Courtesy of Howard Tyler)*

tage if water availability is limited or water is expensive. Ideally, sprinklers are positioned to wet the cow directly and use low-pressure nozzles (10 psi) that deliver large droplets of water to penetrate the haircoat. A fine mist does not penetrate the haircoat properly; the nozzles should deliver large droplets at the rate of about 1 gallon per minute. If sprinklers are positioned too high above the cow, they also wet the area rather than the animal directly, and they can wet down stalls, feedbunks, and flooring. This increases pathogen growth in stalls and increases the risk of foot problems. Foggers, on the other hand, do not wet the cow or the surrounding area, and therefore, the risk of sprinkler-associated problems is greatly reduced (Figures 43.74 through 43.76). Foggers act to cool the air through evaporative cooling, and then the cooler air acts to cool the cow. They use a similar amount of water as do sprinklers. They are more expensive ini-

tially, however, and require a much greater level of maintenance due to the filtration system.

Another important consideration in using wet systems for cooling is preventing water from building up on the cow. Water that does not evaporate rapidly from the cow actually increases the heat load of the animal by creating an insulating effect. Water droplets remaining in the hair coat block convective heat loss.

Shade is used to reduce radiant heat load reaching the cow. Natural shade, in the form of mature trees, is occasionally available in pastures and on smaller operations but must be protected from overuse; excessive defecation and urination near the roots can compromise the survival of even well-established trees (Figure 43.77). Mechanical shade

Figure 43.74   Foggers operate by applying water to the air, where it vaporizes and absorbs the heat and cools the air. The air is then blown across the cow. *(Courtesy of Monsanto)*

Figure 43.76   Misting systems cool air without wetting cows or their surroundings if the misting systems are designed and maintained properly. This feature allows their use where sprinklers are problematic. *(Courtesy of Monsanto)*

Figure 43.75   Misters are designed to cool air rather than cool cows. *(Courtesy of Iowa State University)*

Figure 43.77   These large trees provide a ready source of natural shade, but the high animal density around their base can damage or kill the trees if access is not controlled. *(Courtesy of Monsanto)*

should be considered for areas where cows are otherwise routinely unprotected from radiant heat for extended periods (Figures 43.78 and 43.79). Shade cloth is inexpensive, and mobile shade units can be created to increase flexibility of use (Figure 43.80).

Lighting

In addition to the physiological effects of adequate lighting, there are also beneficial behavioral effects. The visual acuity of the cow can be greatly affected by the intensity of lighting in cattle facilities. Unfortunately, cows cannot see well through areas of extreme shadow and light. They display this characteristic behaviorally with an unusual walking gait through such areas or reluctance to enter these

areas. For similar reasons, deep gutters (such as cattle crossings) provide a formidable barrier to cow movement. Thus, areas that are either too bright (shafts of bright sunshine in otherwise dark barns) or too dark can induce avoidance behaviors and affect cow movement or stall use.

Stray Voltage

Stray electrical voltage has been suggested as a cause of serious problems on many dairy farms, affecting animal behavior and lowering milk production. It may affect other animal species also. Contrary to popular belief, stray voltage is not new; it is as old as electricity itself. However, it has become a problem on many farms recently because

Figure 43.78 Providing shade is a remarkably simple, cheap, and effective method of heat abatement. *(Courtesy of Monsanto)*

Figure 43.79 Thorough planning is critical to success in any endeavor. This permanent shade structure is shading the alley rather than the stalls for most daylight hours. *(Courtesy of John Smith)*

Figure 43.80 The provision of temporary shade over feed areas encourages feed intake and reduces heat stress, both of which enhance animal performance. *(Courtesy of Monsanto)*

there is more electrical load on today's farms. In the last thirty years, we have used more equipment grounding for safety purposes.

Stray voltage is excessive voltage between two animal contact points. The conditions that cause stray voltage are electrically quite simple; if sufficient voltage is present, it may force a current through any available conductor, including a cow's body. Cows are good conductors because of their body design (the length from mouth to front and rear legs). Cows bridge the gaps between electrically grounded objects and true earth. People seldom feel the current for several reasons. Usually, caretakers wear rubber-soled shoes when in the barn. Also, humans have only two legs instead of four like the cow, and a human's legs touch the floor near the same vicinity.

Any electrical condition that creates large enough voltage between two animal contact points may create a stray voltage problem. The source of stray voltage may be either on-farm or off-farm. On-farm voltage problems stem from defective equipment, faulty wiring, bad connections, or having several 120-V motors on the same line. On-farm stray voltage can be minimized by maintaining good electrical wiring systems that meet the requirements of the National Electric Code. Also, properly balanced 120-V circuits and conversion of larger 120-V meters to 240 V reduces the effect of secondary neutral voltage drops at the farm service entrance. Off-farm voltage comes onto the farm through the electrical supplier's lines. Voltage varies with the load and the natural grounding ability of the area. As usage increases, so may stray voltage. Heavier loads occur at milking time and in the fall when grain dryers may be running on many farms.

The following signs may indicate that stray voltage exists in a dairy:

1. Cows are reluctant to enter the parlor (avoidance behaviors). When cows are subjected to stray voltage in the parlor stalls, they soon become reluctant to enter the parlor. Cows show apprehension prior to entering the parlor by increased defecation and urination in the holding area.

2. Cows are nervous in the parlor. Cows that are exposed to stray voltage often dance or step around almost constantly while in the milking parlor.

3. Uneven milk letdown and milkout occur. When milk letdown and milkout are uneven, longer milking times become apparent.

4. The incidence of mastitis increases. When milkout is incomplete, more mastitis is likely to occur. All that is required is the presence of an infectious bacteria. In turn, this results in increased somatic count.

5. Reduced feed intake occurs in the parlor. If cows encounter stray voltage while eating from the grain feeders, a reluctance to eat and reduced feed intake usually follow.

6. Cows are reluctant to drink water. If stray voltage reaches the cows in stall barns through the water supply or metal drinking cups, the animals soon become reluctant to drink. Lapping at water is a common observation in these situations.

7. Lowered milk production occurs. Each of the symptoms listed here is associated with stress and reduced feed intake, followed by a drop in daily milk production.

Other factors, such as mistreatment of animals, milking machine problems, disease, sanitation, and nutritional disorders can create problems that manifest themselves in the seven symptoms mentioned previously. Cows that routinely receive treatments or injections during the milking process develop avoidance behaviors.

The only sure method of determining if significant stray voltage is present is to have a qualified person perform a stray voltage survey, using approved equipment and monitoring the voltage through one and preferably two milkings. Point-to-point measurements between cow and contact points determine if the voltage is actually getting to the cow. Generally, stray voltage is not constant throughout the day, and readings should be taken over a long period.

SUMMARY

- Ideal stall designs integrate an understanding of cow behavior, cost considerations, air movement, and labor requirements for maintenance.
- Comfortable cows lie down for over fourteen hours daily.
- In well-designed and well-maintained facilities, hock lesions and joint swelling are minimal, cows are cleaner, the incidence of mastitis is lower, the detection of estrus is easier, and milk production is higher.
- Cows prefer walking on rubber belted surfaces and eating at ground level.
- Tunnel ventilation is an excellent method for improving ventilation in tie-stall or stanchion barns or in penning areas.
- Stray voltage is excessive voltage between two animal contact points.

QUESTIONS

1. What are the effects of a lack of cow comfort?
2. What are some factors critical to the maintenance of cow comfort?
3. What are the four common resting positions?
4. Which bedding is considered the gold standard among bedding types?
5. What are some signs of an uncomfortable stall?
6. Tie stalls introduce what complication to cow comfort?
7. How many inches of water trough space should be present per cow?
8. How do cows characterize poorly ventilated barns?
9. At what THI is heat stress considered severe?
10. List ways that cows exchange heat with the environment.
11. What is the primary route of evaporative heat loss in the cow?
12. Compare convective heat exchange with radiant heat exchange.
13. What are some signs that stray voltage may exist in a dairy?
14. What is the only sure method for determining whether stray voltage is present?

ADDITIONAL RESOURCES

Book

Rodenburg, J., H. K. House, and N. G. Anderson. "Free Stall Base and Bedding Materials: Effect on Cow Comfort." In *Dairy Systems for the 21st Century,* edited by R. Bucklin, 159–164. St. Joseph, MI: Society of Agricultural Engineers, 1994.

Articles

Albright, J. L., D. K. Stouffer, and B. L. Coe. "Cow Comfort and Preference in Free Stalls with Reference to Flooring and Bedding." *Journal of Dairy Science* 72 (January 1989): 67.

Grant, R. J., and J. F. Keown. "Managing Dairy Cattle for Cow Comfort and Maximum Intake." NebGuide G95-1256-A (1995). Lincoln: University of Nebraska, Cooperative Extension Service.

Natzke, R. P., D. R. Bray, and R. W. Everett. "Cow Preference for Free Stall Surface Materials." *Journal of Dairy Science* 65 (January 1982): 146.

Internet

Cozying up to Cow Comfort: www.wisc.edu/dysci/uwex/brochures/madison10.pdf.

44

Manure Management Practices

OBJECTIVES

- To outline issues associated with manure handling.
- To describe methods of manure handling.

The term *manure* refers to a mixture of animal excrements, consisting of undigested feeds plus certain body wastes and bedding. Planned manure management is an important part of a dairy production system. The collection, transport, storage, and use of manure must meet sanitary and pollution-control regulations. Nitrogen and phosphorus are the nutrients of primary concern for meeting land application regulations.

Most state regulations are directed toward maintaining the quality of surface water and groundwater. Nitrogen can have adverse effects on both ground and surface water sources, but phosphorus is primarily a surface water contaminant. Direct discharge of untreated wastewater, effluents, or runoff into surface water can have dramatic consequences on water quality. The phosphorus load from these sources stimulates excessive growth of algae and eutrophication of surface waters. Precipitation that falls on or flows across manure-covered areas or manure stacks can cause severe pollution in streams, lakes, or ponds. This runoff must be kept from reaching usable private or public waters. State and local regulations usually govern runoff control systems.

In addition to water and soil quality concerns, many states also address air quality considerations. Several other compounds besides ammonia and methane are volatilized from both fresh and stored manure, including volatile fatty acids, phenols, and sulfides. Odor issues are more difficult to define and are related to the nuisance value of the odor. In other words, unlike phosphorus in groundwater, odors are issues only if they affect the quality of life of the people exposed to them. Some states have legislated the amount of odor that can reasonably be expected to be a public nuisance (measured as odor intensity at the property line). Proper facility design, especially in regard to the location of manure-handling and storage facilities, can minimize problems associated with nuisance odors. For example, the direction of prevailing winds should be considered when selecting a site for manure storage, and windbreaks can be used to diffuse and deflect the airborne emissions away from populated areas. Mechanical aeration in lagoon systems can reduce some odors. Incorporating wastes directly into soils rather than surface application also helps minimize odor emissions. In addition to controlling odors, fly populations can be dramatically affected by the manure-handling and storage practices on a dairy operation.

AMOUNT, COMPOSITION, AND VALUE OF MANURE

The quantity, composition, and value of excrement produced varies according to the species, weight, kind, and amount of feed and kind and amount of bedding. The manure recovered and available to spread where desired is considerably decreased when animals are kept on pasture part of the time. Losses often run as high as 60 percent when manure is exposed to the weather for a considerable time. In addition, almost one-fourth of the total nitrogen of cow manure may be lost in twelve hours of drying at high temperature. Nitrogen losses can be minimized by the addition of nitrification inhibitors. About 75 percent of the nitrogen, 80 percent of the phosphorus, and 85 percent of the potassium contained in animal feeds are returned as manure. In addition, about 40 percent of the organic matter in feeds is excreted as manure. It is commonly estimated that 80 percent of the total nutrients consumed in feeds are excreted as manure.

Naturally, manure from well-fed animals is higher in nutrients and worth more than that from poorly fed ones. The urine, or liquid manure, contains nearly 50 percent of the nitrogen, 6 percent of the phosphorus, and 60 percent of the potassium of average manure—roughly one-half of the total plant food of manure. Also, the nutrients in liquid manure are more readily available to plants than the nutrients in the solid excrement, which explains why it is important to conserve urine.

Determination of the actual monetary value of manure can and should be based on equivalent cost of a commercial fertilizer. Of course, the cost of application of manure as compared to the cost of application of fertilizer must also be considered in these calculations.

The value of manure cannot be measured strictly in terms of increased crop yields and equivalent cost of a like amount of commercial fertilizer. It has additional value because of the organic matter that it contains, which almost all soils need. Also, due to the slower availability of its nitrogen and to its contribution to the soil humus, manure contributes lasting benefits, which may continue for many years. Approximately one-half of the plant nutrients in manure are available to the crops in the immediate cycle of the rotation to which the application is made. Of the unused remainder, about one-half is taken up by the crop in the second cycle of the rotation, one-half of the remainder in the third cycle, and so on. Likewise, the continuous use of manure through several rounds of a rotation builds a backlog that brings additional benefits and a measurable climb in yield levels.

BEDDING CATTLE

Bedding is used primarily for the purposes of keeping animals clean, dry, and comfortable. But bedding also soaks up the urine that contains about one-half the total plant food of manure and makes manure easier to handle. In addition, it absorbs plant nutrients, fixing both ammonia and potash in relatively insoluble forms that protects them against losses by leaching. This characteristic of bedding is especially important in peat moss but of little significance with sawdust and shavings.

Other facts of importance relative to certain bedding materials and bedding uses follow:

1. **Wood products (sawdust, shavings, tree bark, chips, etc.):** Shavings and sawdust decompose slowly, but this process can be expedited by the addition of nitrogen fertilizers. Also, when plowed under, they increase soil acidity, but the change is both small and temporary. Softwood (on a weight basis) is about twice as absorptive as hardwood, and green wood has only 50 percent the absorptive capacity of dried wood.

2. **Cut straw:** Cut straw absorbs more liquid than long straw, but chopped straws may be dusty. From the standpoint of the value of plant food nutrients per ton of air-dried material, peat moss is the most valuable bedding, and wood products are the least valuable.

3. **Sand:** Sand is a very popular inorganic bedding source that can create severe problems in some manure management systems. It contributes little value to the use of manure as a fertilizer, feed source, or source of biogas. It is highly abrasive to mechanical systems, however, and tends to settle and clog pumps and drains in gravity-flow systems.

The minimum desirable amount of bedding to use is the amount necessary to absorb completely the liquids in manure. The minimum daily bedding requirement of a dairy cow is about 10 pounds per twenty-four-hour period of confinement, based on uncut wheat or oat straw. Under average conditions, about 500 pounds of bedding are used for each ton of excrement.

SYSTEMS FOR MANAGING WASTE

Planned manure management is an important part of a dairy production program. Dairy buildings and equipment should be designed to handle the manure produced efficiently, with a minimum of labor and pollution; retrieve the maximum value of the manure; and maximize animal sanitation and comfort. Any proposed waste management system should be approved by the appropriate regulatory officials prior to construction.

Manure may be stacked, stored in a separate tank, stored in a nearby cement-lined lagoon or pit, composted, or left to accumulate in a pit under slotted floors. Water is often added to a pit to replace evaporation losses from wastes. Thus, if the manure is to be pumped, 20 to 40 percent of the storage volume may be needed for the extra water. For irrigation, there should be at least 95 percent water and less than 5 percent manure, and often the target is less than 3 percent total solids. By contrast, water should be kept to a minimum if the manure is field spread with a tank wagon. If available land is inadequate for the amount of manure produced, then export of manure is required and a dry scraping system is necessary to keep water content low.

Manure is typically handled in one of three ways:

1. As a dry product, which actually runs 20 to 30 percent solids and which refers to feces plus bedding or feces after liquid separation.
2. As a slurry, which may be up to 15 percent solids and which refers to feces, urine, and some dilution water.
3. As a liquid, with less than 5 percent solids, including feces, urine, and large amounts of dilution water.

Dry Systems

Manure can be handled as a solid if it is mixed with bedding or if the liquids are allowed to evaporate or drain away. Manure from a stall barn is usually loaded directly into a spreader with a barn cleaner and spread on the land daily (Figure 44.1). However, provisions should be made for manure storage capacity for a period of 180 days or more. To calculate the size of storage areas, multiply the number of 1,000-pound cow units by 2.5 cubic feet per day (this figure includes bedding) and then multiply by the number of days of storage desired. Adjustments for annual rainfall and potential runoff into storage areas should also be included in the final size of the storage facility.

Scrape, Stack, and Spread

Storage by stacking works best with manure containing bedding. It is well suited for use with stall barns and up to eighty cows in the herd. The investment in facilities is usually lower than with liquid or slurry storage systems. Moving stacked manure to the land requires a manure loader (Figure 44.2), spreader, and tractor.

Composting

Composting is of interest to many producers as a means of reducing odor, facilitating export from the farm, and potentially earning income from manure. It provides a means of stabilizing raw manure aerobically, rapidly converting the biodegradable products in manure to stable end products. Composting systems often require the addition of other materials to remove water and provide additional fermentable carbohydrate sources for optimal aerobic fermentation. Many different materials can fill this need, including shavings, hay, grain hulls, and yard waste. The most common means of aerating the compost is the windrow method, where the manure is stacked into windrows and turned at defined intervals (Figures 44.3 and 44.4). Compost is often bagged and sold through retail outlets, which requires a stable market to justify the significant additional expenses involved in this management system.

Figure 44.2 Manure solids loaded onto a transport truck. *(Courtesy of Mark Kirkpatrick)*

Figure 44.1 A manure spreader disperses a high solids content product on a field. *(Courtesy of Iowa State University)*

Figure 44.3 A mechanical compost turner is used to rotate compost piles. *(Courtesy of Mark Kirkpatrick)*

Figure 44.4 This equipment turns the compost pile, which facilitates the composting process. *(Courtesy of Mark Kirkpatrick)*

Figure 44.5 An earthen storage basin is agitated prior to being emptied. *(Courtesy of John Smith)*

Slurry Systems

Yearlong storage of manure is a practical goal for dairy producers. It permits incorporating manure into the soil at the best time to preserve fertilizer value. Milk house and parlor wastes can also be stored in slurry storage systems. In comparison with storing solid manure, storage of slurry is generally more costly. Odors can be a problem, especially when agitating and spreading (Figures 44.5 and 44.6), and labor requirements may interfere with field work.

Several types of storage are used in slurry systems: storage tank under the barn; outside, belowground storage tank; earthen storage basin; and aboveground storage (Figure 44.7). Manure is moved to storage by dropping it through slots into a storage area below, moving it with automatic floor scrapers (Figure 44.8), tractor scrapers (Figure 44.9), and barn cleaners (Figures 44.10 and 44.11), or by pumping it from a hopper or tank into the storage structure. Generally, twelve months' storage capacity is required. When determining storage requirements, the amount of manure produced can be estimated, but the final storage requirement is significantly affected by the climate of the region. While evaporative losses in arid regions reduce needs, rainwater accumulation in wetter areas of the country can dramatically increase storage needs.

Figure 44.6 Storage basins are agitated prior to being pumped out so that the solids settled at the bottom are forced back into suspension. *(Courtesy of John Smith)*

Slotted Alleys

The scraping of alleys takes time and effort, and it can easily be neglected during busy seasons, resulting in undesirable conditions. Slotted alleys eliminate the labor and cost of scraping and the cost of scraping equipment because wastes pass directly through the slots into the storage area below. Manure does not build up on the floor. As a result, cows' hooves remain comparatively clean, and less

Figure 44.7 An aboveground manure slurry storage unit. *(Courtesy of University of Minnesota)*

Figure 44.8   A cable pulls this mechanical scraper down a freestall alley at set times. *(Courtesy of Mark Kirkpatrick)*

Figure 44.11   A vacuum truck literally sucks manure into a storage and transport vehicle. *(Courtesy of Mark Kirkpatrick)*

Figure 44.9   Manual removal of manure from freestall alleys using a skid steer loader is labor intensive, but a manual system requires a lower initial investment than automatic scrapers or flush systems. *(Courtesy of John Smith)*

Figure 44.12   Flush tanks are sized by the area of the alley to be flushed. *(Courtesy of Howard Tyler)*

Figure 44.10   A manure transport truck that vacuums manure directly from freestall alleys. *(Courtesy of Mark Kirkpatrick)*

manure is tracked into the milking parlor from the slotted holding area. The disadvantage of this system is that it is significantly more expensive to construct, and failure to remove manure from beneath the facility, even for short periods of time, can significantly threaten cow health if manure gases are allowed to accumulate.

Liquid Systems

Most flushing systems contribute too much water to permit the economical use of the same haul and spread systems used in either solid or slurry systems. Flush systems wash the manure out of the barn (Figures 44.12 and 44.13). The amount of water contributed by the flushing system depends on the type of system in use and the frequency of flushing. Some

Figure 44.13 Water valves are opened to release flush water down this freestall alley. *(Courtesy of Mark Kirkpatrick)*

Figure 44.15 Flush water enters a drainage ditch on the way to a settling basin. *(Courtesy of John Smith)*

Figure 44.14 Flush water drops through grates into a belowground gravity-flow system. It reemerges in drainage ditches. *(Courtesy of John Smith)*

Figure 44.16 A gravity system directs manure-laden flush water into a settling basin, where the solids can be separated from the liquid. *(Courtesy of John Smith)*

dairies use a combination of scraping followed by flushing for cleaning to minimize the solids in the flush water. The amount of total water utilized in these systems can be reduced significantly without compromising facility cleanliness standards by collecting and recycling flush water. After the manure is flushed from the facility, it may be transported either by gravity (Figures 44.14 through 44.16) or by pumps (Figure 44.17) into settling (sedimentation) basins (Figures 44.18 through 44.20). The solids are eventually removed from the settling basin (Figure 44.21) and recycled as bedding or feed sources, spread onto cropland, or hauled away.

Water can also be removed via solids separation (Figures 44.22 and 44.23) and pumped into a lagoon (Figure 44.24). Lagoons can be either aerobic (Figure 44.25) or anaerobic (Figure 44.26). From the lagoon, the liquid is then dispersed onto cropland.

Figure 44.17 Pumps that move low solids flush from the facility to a lagoon must be maintained properly for the system to function optimally. *(Courtesy of John Smith)*

Figure 44.18 The sidewalls of settling basins are interlaced with a series of screens to allow for water removal while retaining solids. (*Courtesy of John Smith*)

The best method of applying liquid manure to the land is through an irrigation system. Most irrigation systems can handle fluid waste with up to 4 percent solids, which are typical of lot runoff and effluent from a lagoon or parlor. The solids separation system concentrates the solids, especially if they are to be recycled for bedding or feed. These systems do not remove sand as effectively as a settling basin, and for facilities using sand bedding, the combination of a settling sand trap (Figure 44.27) with solids separation is most effective. Solids separators remove 10 to 30 percent of the nitrogen and minerals, along with about 20 to 30 percent of the organic matter.

There are other alternatives to these systems. Any system that separates the solids from the liquid fraction enhances ease of handling relative to handling the original slurry (Figures 44.28 and 44.29).

Figure 44.19 Water is removed from manure in a settling basin, while the solids are retained in the basin by screens. (*Courtesy of John Smith*)

Figure 44.20 Once the water has been removed from the basin, the solids are removed manually, and the basin is reloaded with wet manure. (*Courtesy of John Smith*)

Figure 44.21 Dried manure is removed from a settling basin. (*Courtesy of Mark Kirkpatrick*)

Figure 44.22 A mechanical manure separator. *(Courtesy of Mark Kirkpatrick)*

Figure 44.24 The low-solids liquid from a mechanical solids separator is pumped into a storage lagoon. *(Courtesy of Howard Tyler)*

Figure 44.23 A mechanical solids separator forms a pile of solids; the liquid portion is pumped directly into a storage lagoon. *(Courtesy of Mark Kirkpatrick)*

Figure 44.25 An aerobic lagoon uses a series of agitators to supply oxygen to the system, which increases costs but reduces odors. *(Courtesy of Mark Kirkpatrick)*

Figure 44.26 Manure can be stored in a lagoon until the weather is suitable for pumping and removal to fields. *(Courtesy of Mark Kirkpatrick)*

Figure 44.27 A simple sand-trap recovery system at the end of a dairy barn removes a high percentage of sand from the slurry by sedimentation. *(Courtesy of Howard Tyler)*

Figure 44.28 Shallow settling basins in California rely on solar energy to dry manure. *(Courtesy of Howard Tyler)*

Figure 44.29 An extruder processes manure using a combination of pressure and heat. *(Courtesy of Mark Kirkpatrick)*

MANURE GASES

When stored inside a building, gases from liquid wastes create a potential health hazard for people and animals as well as undesirable odors. Most (95 percent or more) of the gases produced by manure decomposition are methane, ammonia, hydrogen sulfide, and carbon dioxide. Several have undesirable odors or possible animal toxicity, and some promote corrosion of equipment.

Animals and people can be killed (asphyxiated) because methane and carbon dioxide displace oxygen. Methane is odorless, and a lack of odor is not an indication of safety. Most gas problems occur when manure is agitated or when ventilation fans fail. No one should enter a storage tank unless the space over the wastes is first ventilated with a fan and another person is standing by to give assistance if needed. Workers should wear self-contained breathing equipment, the kind used for firefighting or scuba diving. It is important that maximum building ventilation be provided when agitating or pumping wastes from a pit. Also, an alarm system to warn of power failures in tightly enclosed buildings is important because there can be a rapid buildup of gases when forced ventilation ceases.

Methane emissions also pose a potential threat to the environment through contribution to global warming. The gases that surround the earth allow shortwave radiation to enter the atmosphere and warm the earth. Some of this energy is radiated back to the atmosphere as longwave radiation. Some of these gases, including methane, create a greenhouse effect that absorbs longwave radiation rather than allowing it to be radiated away from the earth, which creates a condition of global warming. The effect of gases from manure and from cows on global warming is a highly debated subject. Although carbon dioxide is the most abundant greenhouse gas and is expected to contribute to about half of future global warming, methane traps one-quarter more radiant heat than carbon dioxide on a molecule-by-molecule basis. Atmospheric methane is increasing about 1 percent per year, and animal-related methane contributes about 3 percent to the greenhouse gas total.

MANURE AS A NONFEED ENERGY SOURCE

Manure can also serve as a source of nonfeed energy, a concept that, of course, is not new. The pioneers burned dried bison dung, which they called buffalo chips, to heat their sod shanties. In this century, methane from manure has been used for power in European farm hamlets when natural gas was hard to obtain.

Figure 44.30 A manure-burning electric generator. *(Courtesy of Iowa State University)*

Methane is usable, like natural gas. There is nothing new or mysterious about this process. Sanitary engineers have long known that a family of bacteria produces methane when they ferment organic material under strictly anaerobic conditions. Because of the large outlay in capital and technical resources needed, for some time to come the commercial production of methane by anaerobic digestion will likely be limited. If all animal manure were converted to energy, it has been estimated that it could produce energy equal to 10 percent of the petroleum requirements or 12.5 percent of our natural gas requirements.

While the costs of constructing plants to produce energy from manure on a large-scale basis may be high, some energy specialists feel that a prolonged fuel shortage will make such plants economical (Figure 44.30). In addition, on-farm anaerobic digestors are available and are used on some farms. The high costs associated with the purchase and maintenance of these digestors has limited their use at the current time. The major advantage of anaerobic digestors may be in their ability to stabilize raw manure rapidly rather than their ability to generate biogases.

MANURE AS A FEED

Recycling manure as a livestock feed is the most promising of the nonfertilizer uses. Various processing methods are employed, and some manure is being fed without processing. More and more feedlot manure will be either incorporated in a grower ration or fed to breeding herds during periods when pasture supplementation is beneficial, with the residues distributed over grazing areas where they would have fertilizing value.

Animal wastes contain several nutrients that are capable of being utilized when the material is recycled by feeding. Nitrogen, present in both protein and nonprotein forms, is a major constituent. Available energy is rather low. Fiber and ash are generally high. The high ash content indicates that animal wastes are high in minerals; they are especially rich in phosphorus. They also contain certain vitamins synthesized in the digestive tract.

One characteristic of all animal wastes is variability in composition due to diet, kind and amount of bedding; length of time before collecting; and processing method. The main difference in composition between raw and processed wastes is in the moisture content; many of the processed wastes are low in moisture. The high fiber and considerable nonprotein nitrogen of animal wastes indicate that they are best suited for feeding to ruminants because they possess a digestive tract capable of utilizing high fiber and nonprotein nitrogen efficiently. Because of their low energy content, animal wastes are best adapted for use in maintenance and gestating rations, rather than in lactating and growing rations.

Animal wastes processed by ensiling, dehydration, and other methods can be fed successfully to a wide range of animals. The inclusion of excessive amounts of waste in dairy rations results in an excessive level of fiber and/or minerals, followed by lowered animal performance. Because of this limitation, not more than ten percent waste should be included in high-energy rations, such as in lactating cow rations.

NUTRIENT BUDGETING

Nutrient budgeting refers to the strategic feeding of animals to maximize the performance of the animal while simultaneously managing actively the environmental impact of those nutrients that escape the animal. Maintaining environmental accountability requires monitoring nutrient excretion, nutrient uptake by plants, and off-farm nutrient export. Currently, a major focus is on closely monitoring the phosphorus intake of animals to increase the usability of the resulting manure. Enhancing the protein quality of feeds or utilizing feeding practices that maximize nitrogen utilization in the rumen (enhancing nutrient synchrony) reduces nitrogen losses to the environment.

Another opportunity for balancing a nutrient budget includes intensifying cropping practices to remove more soil nutrients, especially phosphorus, thus allowing increased rates of application of animal wastes. Double- or triple-cropping practices can allow an operation to reduce the need to move manure off-site, thus improving profitability. Crops that remove the most phosphorus from the soil, such as corn silage, can be substituted for crops that are less efficient in removing phosphorus in a crop rotation.

SUMMARY

- Manure refers to a mixture of animal excrements.
- States may regulate the levels of nitrogen and/or phosphorus that can be applied to land by a dairy operation.
- The acceptable level of odors produced may also be regulated.
- The choice of bedding affects the handling characteristics and value of manure.
- Manure produced by a dairy farm can be handled effectively in several ways.
- Manure gases pose a significant hazard to both animal and human health.

QUESTIONS

1. What factors affect the amount of manure produced by cattle?
2. What is the total solids content of manure handled as a slurry?
3. Outline a system for handling low total solids content manure.
4. Describe methods of separating solids from water in a manure-handling system.
5. Describe the challenges in handling sand-laden manure and how these challenges are overcome.
6. What gases are produced by manure?
7. What is the nutritive value of manure?

ADDITIONAL RESOURCES

Book

MWPS-18. *Livestock Waste Facilities Handbook* (48824–1039). East Lansing: Michigan State University Extension, 1993.

Articles

Fulhage, C. D. "Sizing and Management Considerations for Settling Basins Receiving Sand-Laden Flushed Dairy Manure." Proceedings of the 9th International Symposium on Animal, Agricultural and Food Processing Wastes. Bleigh, NC: Research Triangle Park, 2003, pp. 456–462.

Stowell, R. R., and W. G. Bickert. "Storing and Handling Sand-Laden Dairy Manure" (Bulletin E-2561). East Lansing: Michigan State University Extension, 1995.

Pamphlet

ASAE D384.1. "Manure Production and Characteristics." St. Joseph, MI: ASAE Standards, 1998.

Internet

Composting Dairy Manure for the Commercial Markets: http://www.ctic.purdue.edu/Core4/Nutrient/ManureMgmt/Paper35.html.

Dairy Manure and Milk Center Wash Water Management: http://www.uaex.edu/Other_Areas/publications/HTML/MP359/introduction.asp.

Dairy Manure Management: http://edis.ifas.ufl.edu/DS096.

Fertilizing Cropland with Dairy Manure: http://www.extension.umn.edu/distribution/cropsystems/DC5880.html.

Fertilizer Nutrients in Dairy Manure: http://muextension.missouri.edu/explore/envqual/wq0307.htm.

45

Managing Preweaned Calves

OBJECTIVES

- To outline when and how to render assistance during the delivery process.
- To describe the importance of colostrum and passive immunity.
- To understand the feed and labor costs associated with preweaned calves.
- To stimulate rumen development via feedstuffs and water.

The first day of life is a critical period that can affect the lifetime profitability of an animal. Optimal management during this critical period can improve feed efficiency and rate of weight gain, improve health throughout the rearing period, and allow optimal expression of milk production potential during the first lactation. The detrimental effects of poor management during this period, therefore, cannot be overemphasized. Calf management during the first day of life is focused on reducing the stress of the birth process and maximizing passive immunity.

BIRTHWEIGHT ISSUES

The birthweights of calves vary by breed and parity. First-calf heifers have smaller calves than do multiparous animals, and different breeds have distinctly different size calves. The optimal birthweight of the calf depends on the size of the cow carrying that calf; the optimal birthweight of the calf should be roughly 6.5 percent of the dam weight. Jersey calves vary only slightly from that value, and consequently Jerseys have little difficulty with dystocia. Holsteins have the most difficulty regulating calf growth in utero, and birthweights may range from 5 to 12 percent of the dam's weight. It is easy to understand that a 1,200-pound heifer delivering a 150-pound calf will have problems at calving, and the calf is more likely

to be stillborn. It is surprising to most producers, however, that the 1,300-pound heifer delivering a 60-pound calf is also likely to experience dystocia, and the calf is more likely to be stillborn.

For Holsteins, calves with birthweights between 70 and 90 pounds are optimal for the future productivity and profitability of that animal. Birthweights above and below the optimal weight increase the risk of both dystocia and stillbirth. The average birthweight of Holstein calves is higher than this range (Table 45.1), suggesting that current trends in genetic selection need to be reevaluated. There is strong evidence to support this conclusion; the incidence of both dystocia and stillbirths has increased steadily since 1985. Currently, over 20 percent of all Holstein births receive some form of obstetrical assistance.

Despite the increased availability of calving-ease data for sire selection, there has been a decrease in the percentage of unassisted births and an increase in the percentage of births requiring considerable force or experiencing extreme difficulty over the last ten years. During the same period of time, the stillbirth rate has increased 10 percent, and there has been a 37 percent increase in calf mortality from birth to weaning over the last fifteen years (excluding stillbirths). Calf mortality is now estimated at nearly 20 percent of all calves born (compared to an approximately 15 percent mortality rate fifteen years ago).

TABLE 45.1   Breed Effects on Birthweight

Breed	Average Birthweight
Holstein	91 lb
Ayrshire	75 lb
Brown Swiss	96 lb
Guernsey	66 lb
Jersey	54 lb

How can you change calf size? There are several ways to produce optimally sized calves. The slow but sure method is through genetic selection. Selecting for stature independent of milk production is detrimental to calf survival and to herd profitability. When selecting a group of sires using Net Merit as your primary index, remove the cows siring the tallest cows rather than those siring the smallest cows! Certainly, the use of calving-ease sires is important when breeding heifers, and such use results in calves that meet more closely our optimal calf-weight:dam-weight ratio. But producers should also consider using calving-ease sires on older cows when they meet production goals.

Sire selection for calving ease tends to reduce the birthweight of the resulting calves, and many people in the dairy industry have voiced concerns that these small calves will become small cows at increased risk for dystocia. These concerns may have limited the use of calving-ease sires in problem herds. Evidence suggests, however, while small calves become small cows and large calves become large cows, that larger heifers are more likely to have difficulty delivering their first calf than their smaller, calving-ease sired herdmates.

As mentioned previously, breeds also differ in their ability to regulate calf growth. Although this may be surprising to most people considering cross-breeding as a method to improve "livability," it is not necessarily best to use Jersey sires on a group of Holstein cows. In fact, breeding a Holstein sire to a Jersey dam results in a lower incidence of dystocia. The dystocia rate of this cross is even lower relative to purebred Holsteins despite the smaller size of the dam and increased calf birthweight:dam body weight. Maternal or (presumably uterine) constraints on fetal growth are important, but they can be overridden at least partially by sire effects.

DYSTOCIA

Dystocia is defined as any abnormal or difficult delivery process. It can be a result of poor communication between the fetal calf and its dam (poor preparation for calving), malpresentation of the calf (Figure 45.1), or difficulties due to inappropriate assistance. The scoring system for calving ease currently in use by the dairy industry is a five-point system. A calving ease of 1 indicates no assistance was provided, and a score of 5 indicates that extreme difficulty was encountered during delivery. A calving-ease score of 1 indicates that no assistance was provided, but it doesn't necessarily mean it was not needed. For example, a cow may be in labor for many hours, struggling to deliver her calf. If the producer walks in just as she finally finishes the de-

Figure 45.1 Malpresentations, such as this backward presentation, are more common in multiparous than in primiparous cows. *(Courtesy of Howard Tyler)*

livery, it is scored the same as the cow that delivered her calf in a matter of minutes with minimal difficulty. Unobserved calvings are scored 1 by definition. A calving scored as a 5 may be due to a prolonged and difficult delivery, an uncorrected malpresentation of the calf, or inappropriate timing of assistance. Although the calving-ease score is the same in all cases, the problem that created the dystocia is not.

Between 70 to 98 percent of calf deaths are directly related to calving difficulties. Calf mortality increases steadily with increasing calving-ease scores. Nearly 50 percent of calves are stillborn following a birth with a calving-ease score of 5, with even more of these calves dying by forty-eight hours of age. It is important to note, however, that calf mortality is increased even with slight assistance at calving. This suggests that even slight assistance is detrimental to the survival of the calf and points out the need for extreme restraint when assisting at delivery.

Most calf deaths are associated with a lack of oxygen during delivery, although a surprisingly high number are associated with trauma during the delivery process. Almost all trauma inflicted on calves is a result of inappropriate timing of assistance or excessive force during assistance. A European study reported that 40 percent of veterinary-assisted deliveries resulted in fractured ribs and 10 percent resulted in fractured vertebra in calves, with heifers suffering more traumatic damage at the same force or calving difficulty than bulls. As little as 600 pounds of force may fracture the leg bone of a calf, and this force is exceeded easily when using calf jacks (Figure 45.2) or even during forceful manual extractions. Not surprisingly, most traumatic injuries remain undiagnosed.

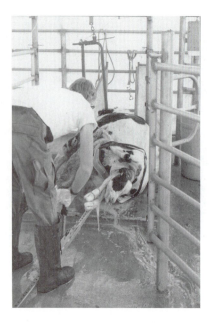

Figure 45.2 Mechanical calf jacks can generate over 1,700 pounds of force on the calf and should be utilized only in extreme situations and with extreme care. *(Courtesy of Iowa State University)*

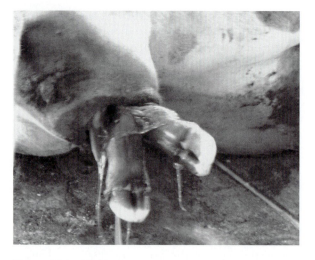

Figure 45.3 Visible hooves and nose indicate a calf in normal birth presentation. *(Courtesy of Howard Tyler)*

For example, calves with broken ribs are often suspected to have weak calf syndrome, but it is easy to understand their reluctance to stand and suckle. The fact that most broken ribs are a result of pulling the calf too early makes the cure easy enough to diagnose.

Following a difficult delivery, calves also have more difficulty in maintaining their body temperature. This inability to thermoregulate also increases death losses, especially during cold or windy conditions. The problems caused during a stressful birth continue to plague these calves for many months. More than 40 percent of calves that were highly stressed at birth die by three weeks of age, and the risk of illness is over twice as high during the first six weeks of life. These calves also often suffer from a failure of passive transfer of immunity. This is primarily due to a lack of maternal behavior by the dam and reduced suckling drive in the calf rather than a lack of ability to absorb colostral immunoglobulins. However, stressed calves that are force-fed colostrum via an esophageal feeder after birth can absorb immunoglobulins as well as unstressed calves.

PREVENTING PROBLEMS— WHEN AND HOW TO ASSIST

All calvings should be monitored carefully and the position of the calf should be determined and corrected early in labor. It is far easier to reposition a calf in the labor process than it is after several hours of uterine contractions. Even backward calves can be turned to a normal presentation in some cases if their position is discovered early enough in the calving process. With uncorrected posterior presentation (hind feet first), the delivery is almost certainly going to be difficult, and there is considerably more danger of having the calf suffocate through premature rupture of the umbilical cord or by strangulation.

If the position of the calf is normal, however, assistance should be provided only when the cow or calf appears to be in distress. But how can you tell when the cow or calf is having problems? This is the biggest challenge when delivering calves. It can be difficult to differentiate between a delivery that is progressing normally but slowly and a delivery where the calf is in imminent danger of dying if assistance is not provided immediately. Luckily, there are cues that the cow and the calf provide.

The easiest method for assessing the status of the calf is by observing the color and reflex responses of the tongue (Figure 45.3). Distress of the calf is indicated by a lack of reflex response to pain (easily checked by pinching the tongue) and by a darkening of the mucus membranes (indicated by a darkening of the tongue and gums that persists between uterine contractions). Normally, when the contractions are coming hard and fast, the tongue will darken because the umbilical cord is compressed by the contractions of the uterus. During the rest period between contractions, the color quickly returns to normal. If this does not happen, the calf needs to be delivered as quickly as possible. The same is true for a lack of response to pain reflexes; immediate assistance is required. Other visual cues that the calf should be delivered quickly are the appearance of either blood or the appearance of pieces of cotyledon during delivery. Both are indications that the placenta has been torn and damaged as the calf rotated and entered the birth canal. Keep in mind that the blood that is visible is the calf's blood, and the calf is slowly bleeding out in the uterus. The delivery of the calf solves this problem with rupture the umbilical cord and separation of the calf from the wounded placenta. Even in these extreme situations, excessive force is not warranted. With a steady force of approximately 150 pounds (about the force of one man pulling), it takes no more than three minutes for cervical dilation to occur and for the calf to be removed safely. The use of forces greater than 150 pounds does not speed cervical dilation; however, they force the calf through the cervix before it can safely fit through the cervical opening.

The use of extreme force during calf delivery is, in many cases, a self-fulfilling prophecy. The person assisting makes the decision that the calf needs immediate help. The calf is forced through a cervix that is not dilated to the proper size, and the calf is injured during the extraction. The calf hits the ground broken and damaged, and the person doing the assisting therefore feels the decision was validated—"Whew, we got that little fella out just in time, look how close to death he is!" The consensus is usually that earlier assistance would have been even more beneficial. The reality is that, in most cases, more patience would have resulted in a happier, healthier calf.

True cow exhaustion, or uterine inertia, is a rarity during deliveries. It is most often a result of hypocalcemia and is an early indication of subclinical milk fever. Providing intravenous calcium to the cow is more beneficial than forcefully delivering the calf. The provision of intravenous calcium immediately corrects the problem, and cervical dilation begins as soon as uterine contractions force the fetus back into position against the cervical opening. Providing assistance based on the length of time the cow has been in labor is wrong. Many cows, especially heifers, are easily distracted during the delivery process and may actually stop being in active labor for hours at a time. For example, it is common for labor to stop temporarily after the rupture of the chorioallantois or the amnion (the "water bags") (Figure 45.4). This occurs for two reasons: first, the pressure on the cervix (that drives active labor) is gone because the space created by the loss of fluids must be filled by the calf before labor can resume and second, the cow is often instinctively compelled to consume all of the fluids that are now soaking her bedding. This is probably a defensive instinct; removing evidence of the birth process is a natural protection from predators. In many cases, it can take up to forty-five minutes for labor to resume.

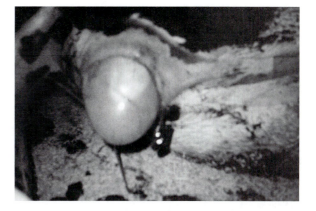

Figure 45.4 The calf is protected from the force of uterine contractions by the fluid-filled amniotic sac. *(Courtesy of Emily Barrick)*

The problem occurs when the progress of labor is monitored only at intervals. The cow is seen in hard labor with the tips of hooves and nose of the calf visible, and the person monitoring assumes labor is progressing according to schedule. They leave, the water bags break, the cow does not resume labor for forty-five minutes, and the person returns to find that the cow has made almost no progress in the last hour. Panicked, they hook up the calf jack and extract the calf as quickly as possible, even though cervical dilation has not had a chance to progress.

It is important that the person assisting the delivery can visualize the progress of cervical dilation. Early in the labor process, the cervix is too small to allow the head to fit into the cervical space. Typically, the cervical diameter must be greater than 12 inches before the head begins to enter the birth canal (cervix) and the nose becomes visible. Progress is obvious and apparent as the head slowly is pushed into the cervix. Once the head is fully engaged in the cervix (cervical diameter is about 16 to 18 inches), however, progress seems to stop until the cervix is dilated sufficiently to allow the shoulders to enter the birth canal (cervical diameter is at least 24 inches). This increase in dilation can easily take five to fifteen minutes, and it is the most common time for people to panic and assist prematurely. Progress is occurring, it is just not visible from outside the cow.

Trying to pull a 100-pound calf through an opening that was meant for 75 pounds naturally results in an injured calf. It helps to think of the cervix as a broad, thick elastic band. Although the shoulders can force the cervix to stretch to a small extent, it will snap back over the ribs. The ribs are fractured fairly easily at this point of development, and forcing the calf through this too-tight elastic band easily breaks ribs. Thus, contrary to common opinion,

Figure 45.5   Twin births occur in less than three of every 100 births and require special attention at calving. *(Courtesy of Mark Kirkpatrick)*

Figure 45.6   Individual twin calves weigh about 20 percent less than singletons. The combined weight of these twin Jerseys was barely over 70 pounds. *(Courtesy of Carolyn Hammer)*

producers should be more patient and assist later in the process for larger calves than for smaller calves. It takes longer for the cervix to dilate to the size of a 125-pound calf than it does to dilate for a 75-pound calf. Fortunately, hoof size is highly correlated to calf size, and it is fairly easy to gauge the need for patience early in the delivery process.

Twinning is a special exception and typically requires some assistance, especially in the case of the second twin (Figure 45.5). Twins occur in about two out of every 100 births (triplets occur in one out of 10,000 births). Although birthweights of twins are typically 20 to 30 percent lighter than single calves (Figure 45.6), their combined weight is much heavier, and they often experience complications during delivery. Occasionally both calves are presented simultaneously; one calf must be pushed back to allow the uncomplicated delivery of the other. The earlier this is done in the labor process, the less

stress is involved for both calves and for the dam. The firstborn calf rarely needs assistance. However, the second calf is often presented abnormally (backward is the most common malpresentation), and the risk of stillbirth is much higher in this calf. Identical twins are usually born in rapid succession, and frequently there is only one placenta. However, there may be a significant delay in the delivery of a second fraternal twin, but this is not necessarily just cause for forceful removal of this calf. The stretching of the uterus for a twin pregnancy followed by the delivery of the first twin results in a large, rather atonic uterus for the second twin. It typically takes five minutes to an hour or more for the second twin to assume normal presentation for delivery and for the uterus to close down enough to allow it to expel the second calf. Fortunately, the cervix is already dilated from the passage of the first twin, and delivery time is rapid if the calf is in a normal position.

When assistance is necessary during any delivery, gentle traction should be provided in concert with uterine contractions. No more than 150 pounds of force should be used in extracting Holstein calves, and no more than 75 pounds of force should be used in extracting Jersey calves. If chains or ropes are used, they should be double-half-hitched on both legs (above the fetlock on the cannon bone and below the dew claw) to distribute the force of the pull. Chains or ropes used for assisting deliveries should be sterilized after each use. Sterilization is easily accomplished by placing the chains or ropes in a small paper sack (sealed with masking tape) and placing them in an oven that has been preheated to 300°F for 1.5 minutes.

Lubrication also helps reduce the force of the pull but do *not* use soap and water inside the reproductive tract. Soap acts only to defat the birth canal and remove natural lubricants. Commercial lubricants are available for this purpose. Mineral oil or vaseline are also effective and inexpensive choices. The same lubricants used to lubricate your arm when artificially inseminating a cow are suitable for extracting or repositioning a calf.

Cleaning the perineal area (the area around the anus and vulva) is an important management consideration prior to a calf's birth. This practice is not just for appearance sake; it also minimizes the risk of fecal-oral transfer of any pathogens from the cow to her calf. For example, this step is important in reducing the incidence of cryptosporidia in young calves. Skipping this step results in nearly 100 percent of calves being exposed to the organism during the birth process. With all the organisms that can be transferred in this manner (Johne's organisms, for example), it certainly is worth the extra few minutes of effort.

Calves normally rotate from left to right as they are born; they also flex downward after delivery of the shoulders. These positional changes minimize the chances of hooking the hips of the calf on the pelvis of the cow. Calves should be pulled parallel to the angle of the rump of the cow until the head and shoulders are completely delivered; then the angle of delivery is closer to 45°. If hiplock does occur, the calf should be pushed back and rotated between 45° and 90°; traction at this point should be successful. If not, a fetotomy may be necessary. Regardless of the method of assistance, traction should stop when the last rib is delivered; the rest of the delivery should be allowed to proceed naturally to allow time for the transfer of blood from the placenta to the calf prior to umbilical rupture.

The umbilical cord must constrict tightly immediately after it ruptures to avoid excessive blood loss from the calf. This constriction is accomplished through two mechanisms. First, the stretching and tearing of the cord stimulates a reflex constriction in the umbilical arteries near the point where the rupture occurs. Second, compounds produced in the calf's lungs after the first breath is taken, called bradykinins, travel to the umbilical cord and cause a further constriction of the entire cord. Therefore, it is important that patience be exercised during the delivery process and that the calf be allowed to start breathing before the cord ruptures whenever possible. It is also important to allow the cord to break naturally, through stretching and tearing, rather then cutting the cord as it becomes exposed during birth. Calves born using these techniques have improved oxygenation compared to those delivered without allowing blood transfer to occur.

THE CALF IS BORN—NOW WHAT?

Following birth (Figure 45.7), calves should be removed immediately from their dams to a thermoneutral environment (between 60° and 75° is optimal) (Figures 45.8 through 45.10). Breathing should be stimulated if necessary. It is important that the first breath be a gasp. A gasp draws the pulmonary fluid across the lung and into the bloodstream, simultaneously coating the lungs with surfactant. A normal breath does not provide enough force to remove fluids from the lungs, and artificial respiration procedures cannot push the fluids out. Cold water applied to the face of the calf is most effective in inducing the gasp reflex, although rotating a little finger or a piece of thick straw in the nostril can also be effective. For the calf's sake, remember that this is not brain surgery; it is not necessary to ram a 10-inch length of stiff straw up the calf's nostril! Other common

Figure 45.7 The desired outcome of a successful reproductive program is a live, healthy calf. *(Courtesy of Mark Kirkpatrick)*

Figure 45.8 Heat lamps can help maintain body temperature while simultaneously drying the haircoat in newborn calves. *(Courtesy of Emily Barrick)*

Figure 45.9 Newborn calves should be kept in a clean, dry, and warm area until they are completely dry prior to being moved to cold-type housing. *(Courtesy of Iowa State University)*

Figure 45.10 During cold weather, young calves utilize brown fat deposits to generate heat effectively without shivering (nonshivering thermogenesis). *(Courtesy of Iowa State University)*

practices such as hanging the calf upside down or swinging the calf are not recommended. These procedures do not remove fluids from the lungs; however, they can compress the digestive organs against the diaphragm and make it more difficult for that first gasp to occur. Respiratory stimulants should be used only as a last resort; although they stimulate respiratory activity initially, they eventually act to depress respiration. After breathing has been initiated, calves should be dried vigorously using a rough absorbent towel, the umbilical cord should be dipped in a 7 percent iodine or chlorhexidine solution, and adequate quantities of high-quality colostrum should be provided.

GETTING PASSIVE IMMUNITY

The bovine placenta is impermeable to maternal antibodies; therefore, postnatal transmission via small intestinal absorption is the sole source of passive immunity for the newborn calf. The failure of this process results in mortality rates in excess of 50 percent and long-term impairments of health and productivity for the survivors. The mortality rate for calves is approximately 20 percent between birth and weaning (including stillbirths), with many deaths directly attributable to failure to attain adequate levels of maternal antibodies during the first day of life.

In addition to the direct economic loss this mortality represents to the dairy industry, hidden costs become apparent as the high mortality rate impedes genetic progress of the dairy herd. The average dairy cow is in production for four lactations; at this rate, 75 percent of all heifers must be raised as replacements to maintain herd size. If 20 percent of heifers die, then the voluntary culling rate cannot exceed 5 percent without purchasing replacements. Additional costs are associated with the increased morbidity and decreased productivity of survivors with low antibody titers. Calves with lower levels of passive immunity have decreased growth rates and increased health problems through the first six months of life and may even have decreased milk production in the first lactation.

Colostrum is the first milk secreted by mammals. The first six milkings from fresh cows are considered colostrum for milk-marketing purposes and cannot be sold. However, the most important colostrum for the newborn calf is the first milking. The transition from colostrum to milk is a rapid process, with dramatic composition changes occurring during the first few hours after parturition and more gradual changes occurring thereafter through the third day.

First milking colostrum contains a fiftyfold greater concentration of immunoglobulins than does normal milk. The high variation in this level can be attributed partially to the age of the cow because the concentration of immunoglobulins is correlated with the amount of exposure to pathogens. Pathogens are often farm-specific, and colostral antibodies are also farm-specific. Concentrations of immunoglobulins in colostrum decrease quickly following calving.

Colostrum provides more to the calf than passive immunity. The energy content is also critically important to the newborn, especially for the first day of life. Although the fat content is highly variable, it is easily digestible and efficiently used by the calf as an energy source. The lactose concentration is much lower than that of true milk. This characteristic is biologically important because lactase is not present in the small intestine during the first day of life and a high intake of lactose causes diahrrea in the calf. The low levels of lactose therefore allow abnormally high intakes of colostrum during the first day, which optimizes passive immunity.

Colostrum also provides a concentrated source of growth factors, including insulin-like growth factors I and II, epidermal growth factor, relaxin, nerve growth factor, and many other growth factors, cytokines, and hormones. In some cases, these peptides are actively transported into the mammary gland prepartum and may act to stimulate or regulate development postnatally.

The primary protective components of colostrum are the immunoglobulins. The primary immunoglobulins in colostrum are IgG, IgA, and IgM. IgG is the primary circulating antibody and is concentrated to the greatest extent in cows' colostrum. It is actively transported into the gland during the final few weeks prior to calving. The IgA in colostrum provides a protective coating for the small intestine during the first days of life, which prevents absorption of infectious organisms through the gut. The passive immunity attained in calves ingesting adequate colostrum is the primary protection for the calf through the first one to two months of life, after which the endogenous immune system of the calf is developed enough to assume this responsibility.

Immunoglobulins are absorbed nonspecifically through the wall of the small intestine, which means that the small intestinal cells transport immunoglobulins, albumins, and other proteins, as well as viruses and other pathogens equally effectively during the first twenty-four hours of life. It is important that colostrum be the first material presented to the small intestine after birth; the presence of antibacterial and antiviral substances in colostrum is protective to the calf and minimizes the risk of disease and death while the gut matures.

Absorption of immunoglobulins by the calf is a remarkably efficient process. About 35 percent of ingested immunoglobulins are absorbed for calves fed at birth, with this percentage decreasing rapidly thereafter. Less than 5 percent of the immunoglobulins fed after twenty hours of age appear in the circulation; however, they continue to play a protective role in the gut against enteric pathogens.

Closure refers to the point at which calves no longer absorb immunoglobulins from colostrum. As described, this process is a gradually accelerating one initiated at birth. In calves fed at birth, closure occurs near twenty hours after birth. It is highly critical, therefore, that calves be fed colostrum within two hours of birth to ensure maximal passive immunity. Calves can ingest large quantities of colostrum in the first twenty-four hours of life. Calves over 100 pounds at birth can be safely fed 4 quarts at birth, followed by 2 more quarts at twelve hours of age. Calves between 50 and 100 pounds should be fed 3 quarts at birth, followed by 2 quarts at twelve hours. Smaller calves (less than 50 pounds birthweight) can be safely fed 2 quarts at birth and 2 quarts at twelve hours of age. Although suckling does enhance absorption, 30 to 40 percent of calves left with their dam do not ingest enough colostrum to attain adequate levels of passive immunity. Therefore, colostrum needs to be hand-fed, either by nipple feeder (Figure 45.11) or esophageal feeders

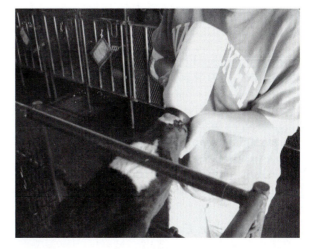

Figure 45.11 Feeding colostrum via a nipple bottle, although time-consuming, helps train calves to nurse from a bottle. *(Courtesy of Howard Tyler)*

Figure 45.12 Calves that won't voluntarily consume colostrum are force-fed with an esophageal feeder. *(Courtesy of Emily Barrick)*

(Figure 45.12), for less vigorous nursers. The total amount of immunoglobulins absorbed by any calf depends on the amount of ingested colostrum, the immunoglobulin concentration of the colostrum, and the time elapsed between birth and the first feeding.

AFTER THE FIRST DAY—ECONOMICS OF CALF REARING

Dairy calves are both an asset and an expense to a dairy operation. They generate no current income, and they compete for facilities, labor, capital, and feed. However, they determine the future profit potential for the producer and theoretically should represent the best genetics in the herd. It costs between

$1,100 and $1,600 to raise heifers to the time of first calving; this cost is highly dependent on feed costs, labor costs, and the age at first calving. The heifer-raising enterprise should be considered as a separate enterprise from the milking herd, with different strategies. Because heifers rarely generate income directly, this enterprise seldom receives the priority of management that it requires to operate at full efficiency. Management of calves after the first day of life is focused on three priorities: reducing feed and labor costs, optimizing rumen development, and maintaining animal health.

Feed and Labor Costs

Daily labor costs per animal and costs per unit of gain are higher for the preweaned calf than at any other time in the animal's life. Reducing these costs without compromising the health of the animal is a critical goal for the dairy producer.

Excess colostrum is a very nutritious feed, but the immunity benefit (antibody protection) to a calf beyond the first twenty-four hours of life is limited to enteric (within the gut) protection. Colostrum contains about one-third more solids than milk or reconstituted milk replacer and is highly digestible. Storage and subsequent feeding of excess colostrum can help reduce costs. It may be fed fresh; frozen, then thawed prior to feeding; or stored as sour (fermented) colostrum. Because it is higher in solids than normal milk, it should be diluted by 25 percent when fed to older calves to avoid overfeeding and scours. Naturally fermented, sour colostrum can easily putrify and become unfit for consumption, especially during the summer months. Producers can add acetic or propionic acid to acidify colostrum to a pH of 4.6 and extend its feeding life.

Physiologically, the newborn calf is not a functioning ruminant. The abomasum, which represents the largest portion of the stomach of the newborn calf (Table 45.2), is the primary functional unit of the gastric region. The sucking reflex (Figures 45.13 and 45.14) allows colostrum or milk to bypass the rudimentary rumen and reticulum via the reticular or esophageal groove and to proceed directly into the abomasum, where digestion is initiated. Milk is coagulated in the abomasum through the action of the enzyme rennin, and digestion and absorption proceed much like that in the nonruminant.

Milk or milk replacer should be viewed strictly as a nutritional supplement to dry feed. Milk replacer intake allows the calf to receive adequate nutrients for survival until rumen capacity and function allow enough nutrients to be derived from dry feed intake. The provision of 1 pound of a 20 percent protein, 20 percent fat milk replacer provides a 90-pound calf

TABLE 45.2   Relative Size of the Bovine Stomach Compartments (%)

	Birth	3–4 Months	Mature
Rumen	2.5	60	80
Reticulum	5	5	5
Omasum	10	10	7–8
Abomasum	60	25	7–8

Figure 45.13   Feeding positions where the head is extended on the neck help ensure complete closure of the esophageal groove. *(Courtesy of Iowa State University)*

Figure 45.14   Suckling behaviors are partly satisfied by feeding calves via nipple. *(Courtesy of Iowa State University)*

with about 2.5 Mcal of energy, which under ideal conditions meets the maintenance needs of the calf and allows 0.5 pound of weight gain per day. Maintenance needs increase under conditions of heat or cold stress, and this level of milk replacer intake may not support growth.

Milk replacer is generally fed at a rate of 1 pound (dry-matter basis) per head per day until the time of

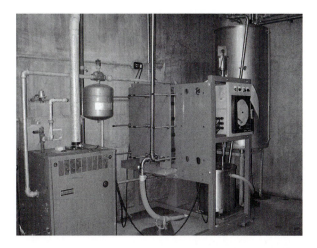

Figure 45.15 On-farm pasteurizers drastically decrease bacterial loads in milk and therefore enhance the performance of calves. *(Courtesy of Mark Kirkpatrick)*

weaning, although judicious use of waste milk and surplus colostrum can minimize expenses. Accumulated colostrum from an average cow provides enough feed for one calf for sixteen days. Waste (antibiotic-treated and mastitic) milk can be fed to calves with no loss of performance and without increasing the incidence of health disorders in these calves, as long as it is pasteurized prior to feeding (Figure 45.15). Pasteurization of waste milk increases the weight gain of calves, decreases mortality rates, and decreases sick days relative to calves fed unpasteurized waste milk. Mastitic milk should not be fed to calves less than two days old because the intestine is permeable to large protein molecules. Calves fed milk from cows treated with antibiotics should not be marketed for meat unless the required withdrawal period is observed prior to slaughter.

Milk replacers rather than whole milk are typically fed to calves to reduce costs and minimize the potential for disease transmission. Many vegetable protein sources have been utilized in milk replacers to reduce costs; however, whey- or skim milk-based milk replacers are the standard by which all other milk replacers are measured. Substitution of vegetable proteins for whey proteins in milk replacers results in decreased calf performance. Chemically modified soy protein, soy isolates, and soy concentrates are good, but as plant proteins, they are still less digestible and more allergenic than milk protein. Meat solubles, fish protein concentrate, distillers' dried solubles, brewers' dried yeast, oat flour, and wheat flour are inferior as protein sources in milk replacers.

Advances in the collection and processing of blood from the meat-packing industry have increased the availability of relatively inexpensive food-grade blood proteins. Spray-dried red blood cells and spray-dried plasma proteins both contain highly digestible proteins close to milk proteins in both quality and digestibility. Inclusion of these protein sources decreases the cost of milk replacers yet results in similar rates of weight gain when these replacers are fed to calves. If the cost of gain is less expensive for calves fed milk replacers containing blood proteins, then these protein sources should be an attractive alternative for dairy producers seeking to maximize profit.

A good milk replacer powder should contain a minimum of 15 percent fat, and it may contain more than 20 percent. The higher fat level tends to reduce the severity of diarrhea and produce additional energy for growth. Good-quality animal fats are preferable to most vegetable fats. However, soy lecithin, especially when homogenized, is an acceptable fat source and improves the mixing qualities of the replacer.

The calf can use two carbohydrate sources in milk replacer: lactose (milk sugar) and dextrose. However, two other carbohydrate sources, starch and sucrose (table sugar), are not satisfactory and should be excluded from milk replacers.

Milk replacers have traditionally contained 20 percent protein, however, this level is not based on the nutrient needs of the calf. Actually, higher levels of protein may more nearly meet the nutrient requirements of the milk-fed calf. Many companies are currently marketing *accelerated growth* programs based on 28 percent protein milk replacers. For calves not yet consuming appreciable amounts of calf starter, these higher protein milk replacers will more nearly meet the requirements of the calf and allow improved growth. This does not mean that all the programs utilizing these milk replacers are more profitable way to raise calves, however.

Most *accelerated growth* programs combine higher intakes of milk replacer and a higher protein intake of the milk replacer. The higher energy intake and improved protein:energy ratio from milk replacer in these programs does allow for a faster rate of gain; however, because of the increased energy intake from the milk replacer, such programs also reduce voluntary intake of calf starter. In general, early weaning (at 21 to 35 days of age) will reduce feed and labor costs associated with calf rearing (Table 45.3). Thus, the most effective means for reducing the costs of raising preweaned calves is to provide a feeding program that meets the nutrient requirements of the calf while simultaneously and rapidly stimulating rumen development. The provision of up to 2 percent of birthweight of a 28 percent protein, 20 percent fat milk replacer intake (2 pounds of milk replacer daily for a 100-pound calf) during the period prior to

TABLE 45.3   Performance variables and costs associated with raising calves using an early weaning program or an accelerated growth program compared with the national averages

Variable	National Average	Early Weaning	Accelerated Program
Birth weight, lb	95	87	95
Weaning Age, d	56	31	56
ADG, lb	.98	1.5	2.1
8-Wk Weight, lb	150	165	212
MR Intake, lb	64	30	121
Starter Intake, lb	74	130	45
Gain:Feed Ratio	.40	.49	.70
MR Cost, $	54	25	121
Starter Cost, $	13	24	9
Total Feed Cost, $	67	49	130
Feed $/lb Gain	1.20	0.63	1.12

Figure 45.17   The amount of calf starter fed daily should be limited to just slightly more than the calf consumes in the same period. *(Courtesy of Iowa State University)*

Figure 45.16   The labor associated with feeding calves individually makes cost per pound of weight gain higher than at any other time in the animal's life. *(Courtesy of Iowa State University)*

starter intake may be justified by the improved growth rates of these calves. After 7–10 days, reducing intake (to 1 percent of BW) of the same milk replacer should encourage starter intake. Weaning should then be determined by starter intake. The timing of all changes in the feeding program should be flexible based on calf health, environmental conditions, and economic factors.

In general, early weaning (at twenty-one to thirty-five days of age) reduces feed and labor costs (Figure 45.16). Thus, the most effective means of reducing the costs of raising preweaned calves is to feed them so that rumen development is stimulated.

Stimulating Rumen Development

Stimulation of rumen development depends on fermentation of ingested feedstuffs to volatile fatty acids (VFAs). The physical form of the feed is not important;

however, concentrates are more effective in stimulating papillae growth than are forages. All volatile fatty acids are mitogenic to the rumen epithelium, but butyrate is the most important VFA in this respect. Although forages are not necessary for rumen development during this period, a threshold level of abrasiveness is required to prevent abnormal papillae formation and excessive keratinization of rumen tissues. Thus, while concentrate intake drives rumen growth and development, a minimal level of forage inclusion in the ration is suggested to enhance the functional development of the rumen. As early as seven days of age, both bacteria and protozoa may be established in the rumen. Adult concentrations are reached by two weeks of age. By five weeks, fermentation can proceed on a body-weight basis at the same rate as in adults. These changes in rumen function are reflected in the growing calf by increasing blood urea (rumen ammonia is converted to urea in the liver), increasing blood acetate levels (from rumen acetate), and increasing blood ketone bodies (β-hydroxybutyrate and acetoacetate are produced in the rumen epithelium from rumenally produced butyrate).

Calf starter should be available to the calf during the first week and should be replaced daily (Figure 45.17). Because starter intake drives rumen development, the primary goal of the producer should be to manage the calf in the manner that best encourages starter intake. Starter intake is negatively correlated with the energy intake from milk or milk replacer, and increased provision of liquid feeds is not advisable for calves over ten days of age because limiting energy intake from milk replacer stimulates starter intake in older calves.

Calf starter should contain 18 to 20 percent crude protein and 80 percent total digestible nutrients and may include coccidiostats to decrease the incidence of

Figure 45.18 The innate drive to consume forage can be observed within a few days after birth. *(Courtesy of Iowa State University)*

subclinical coccidial infections. The physical form of the feed is more important at this stage of life than is the nutrient content. To encourage intake, calf starter must have a coarse texture with minimal fines to reduce dustiness. Pelleted complete feeds are often fed to encourage a balanced intake of all feed ingredients in a palatable form; however, pellets that are too hard are not as readily consumed as are softer pellets. Molasses (up to 7.5 percent) is often added to calf starters to increase palatability and decrease dustiness.

Hay intake should be limited for several reasons (Figure 45.18). Lowly digestible, overly mature legumes are not fermented as rapidly or masticated as effectively at this age; rumen retention is prolonged and dry-matter intake is decreased. In addition, the severe limitations in rumen capacity limit overall dry-matter intake, and even relatively low intakes of forage may limit starter intake and therefore slow rumen development. Therefore, the highest quality hay (preferably an immature grass hay or grass-legume mix) should be fed to calves.

Fresh water should always be available to the calf from birth. Young calves can dehydrate very rapidly, especially under conditions of heat stress. In addition, free water intake is crucial for maintaining a normal rumen environment because the water in milk or milk replacer bypasses the rumen. Calves offered water during the liquid feeding period (birth to four weeks) tend to consume more starter and perform better than calves fed liquid only. Thus, a lack of water availability limits dry feed intake and slows rumen development.

Because starter intake drives rumen development, intake can be monitored as an indirect reflection of readiness to wean. Weaning at a specific level of starter intake is the best weaning criterion. Calves with birthweights over 100 pounds at birth should be consuming at least 1.5 pounds of starter per day for several days prior to weaning. Calves with birthweights between 50 and 100 pounds should be consuming at least ¾ pound of starter for several days prior to weaning, and calves with birthweights under 50 pounds should be consuming 1 pound of starter for several days prior to weaning, Again, the calf must have a digestive system with adequate capacity so that the quantity of food eaten will sustain it.

Producers should continue to feed calf starter until intake is twice what was recommended at weaning and then switch to less expensive feed. In addition, maintaining calves in the same housing system for several weeks after weaning prevents sucking on other calves, reduces stress, and allows producers to note signs of stress or illness more easily.

OTHER MANAGEMENT PROCEDURES

Several other management procedures should be scheduled in the preweaning period. Immediately after birth, calves should be identified with an ear tattoo (permanent) and ear tag (more visible). Registered calves should be sketched or photographed.

Dehorning

Early dehorning, preferably before the calf is two months of age, is recommended. At this age, the horn bud is free-floating in the skin layer above the skull. At some point after two months of age, the horn bud attaches to the skull, and a small horn starts to grow. Young calves are easier to handle and lose less blood. Also, the danger of infection and screwworm trouble is minimized when calves are dehorned at an early age.

Several types of heated dehorners are available, including electric (12-, 24-, and 120-V) and gas models (Figure 45.19). These are used to burn the tissue surrounding the horn bud, cauterizing the vessels that supply blood to the growing horn. These are very effective as long as all the tissue surrounding the horn bud is burned all the way through. There is no blood loss using this technique; therefore, there is less chance of attracting screwworm flies. Producers may also choose to dehorn with caustic paste at two to three weeks; however, this option is less desirable and has more complications. Heifers may also be surgically dehorned later (Figure 45.20), but this method presents a much higher risk of infection.

Removing Supernumary Teats

Heifer calves may be born with more than four teats. The extra teats are usually located posterior to one or both rear teats, but they may be between the

Figure 45.19 Gas dehorners cauterize the blood supply to the horn bud and effectively dehorn without leaving an open wound. *(Courtesy of Leo Timms)*

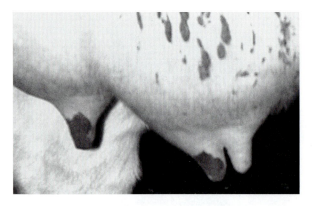

Figure 45.21 If extra teats are not removed during calfhood, extra glands develop and create significant management problems after freshening. *(Courtesy of Leo Timms)*

Figure 45.20 Barnes-type dehorners are most commonly used on older calves, and this technique requires removal of all horn-bud tissue for the procedure to be successful. *(Courtesy of Leo Timms)*

Figure 45.22 Oral vaccines are most effective in situations when the calf has been deprived of colostrum. *(Courtesy of Iowa State University)*

front and rear teats on one or both sides of the calf's udder. Because extra teats have no value, detract from the appearance of the udder, and may interfere with milking (Figure 45.21), they should be removed when the calf is one to two months of age. When extra teats are removed properly, there is usually little bleeding, and the scar is scarcely noticeable when the heifer freshens.

CALFHOOD HEALTH AND DISEASE MANAGEMENT

Because of the immaturity of their immune systems, young calves are extraordinarily susceptible to infectious disease. Maintaining health is critical to maintaining profitability during this period. Fortunately, the same biosecurity concepts can be

applied to calf management as are applied throughout the dairy operation. The most important components of a biosecurity plan can be summed up in two concepts:

Maximize host immunity: For the calf, the cornerstone of an effective biosecurity program is excellent colostrum management. Because of the immaturity of the neonatal immune system and the inhibiting effects of maternal antibodies on endogenous antibody production, vaccines are generally ineffective prior to four months of age (Figure 45.22). Colostrum-deprived calves can be intravenously infused with exogenous sources of bovine immunoglobulins to provide some support to the immune system. Preventing dehydration is important in allowing the immune system to function

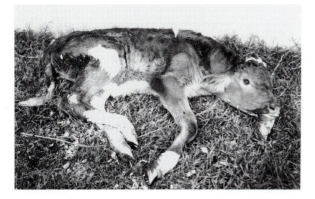

Figure 45.23 Calves with severe diarrhea dehydrate rapidly, and they often require oral rehydration therapy to restore fluid and electrolyte balance. *(Courtesy of Iowa State University)*

Figure 45.25 Pens without physical barriers facilitate good ventilation but also facilitate disease transmission via nose-to-nose contact. *(Courtesy of Iowa State University)*

Figure 45.24 Physical barriers help isolate calves and thus minimize the spread of disease, but they can impair the effectiveness of ventilation. *(Courtesy of Iowa State University)*

properly (Figure 45.23). Proper nutrition is also critical in allowing the immune system to respond optimally to a pathogenic challenge.

Minimize pathogen exposure: Sanitation and minimizing contacts are the keys to controlling pathogen exposure in young calves. Sanitation begins in the maternity stall; a calf born in a dirty environment is off to a compromised start. The choice of bedding and the adequacy of bedding throughout the preweaning period are critical to calf health. Calves should be housed in a well-ventilated, well-drained area. Calves can be raised successfully using many different types of housing systems; however, it is easier to control pathogen loads or exposure in individual housing systems. Well-ventilated housing systems are extremely important because calves are highly susceptible to respiratory pathogens (Figure 45.24). Adequate

space between calves (several feet) is also important because it minimizes calf-to-calf transmission of airborne pathogens (Figure 45.25). Raising calves in elevated stalls or on gravel without bedding allows fecal pathogens to be removed from the immediate environment of the calf frequently and thoroughly. Other types of housing that require bedding must be kept as dry, sanitary, and well-bedded as possible, especially when disease outbreaks occur.

Controlling exposure also requires that people interacting with the calves practice good hygiene. Calf care should be the first consideration of the herd veterinarian, prior to caring for the older stock, to prevent transmission of pathogens within the herd. Similarly, workers doing calf chores should avoid spreading pathogens to other areas of the farm or from calf to calf by developing a chore routine that keeps these principles in mind. Calf-raising areas should be isolated from other animals on the farm, and drainage should flow in a direction away from the calf-housing area. Concerns regarding the exposure of calves to extremely cold conditions are largely unfounded; winter temperatures are not dangerous as long as calves are kept dry. Calves have high levels of brown fat at birth, which allows nonshivering thermogenesis. However, keeping calves in a thermoneutral environment for several days prior to moving them to a cold environment is extremely dangerous because the levels of brown fat decrease rapidly postnatally if they are not being used. Calves should be moved to their new environment as soon as they are completely dried after birth, unless environmental conditions are extraordinarily extreme.

SUMMARY

- The first day of life is a critical period when optimal management can improve feed efficiency and the rate of weight gain, improve health throughout the rearing period, and allow optimal expression of milk production potential during the first lactation.
- Calf mortality increases with even slight assistance at calving.
- In the case of twins, the firstborn calf rarely needs assistance; however, the second calf is often presented abnormally.
- No more than 150 pounds of force should be used in extracting Holstein calves, and no more than 75 pounds should be used in extracting Jersey calves.
- The bovine placenta is impermeable to maternal antibodies.
- Colostrum should be the first material presented to the small intestine after birth.
- Pasteurization of waste milk increases the weight gain of calves, decreases mortality rates, and decreases sick days relative to calves fed unpasteurized waste milk.
- Stimulation of rumen development depends on fermentation of ingested feedstuffs to volatile fatty acids.
- Concentrate intake drives calf rumen growth and development.
- The physical form of the feed is more important at this stage of life than is the nutrient content.
- Weaning at a specific level of starter intake is the best weaning criterion.

QUESTIONS

1. Define *dystocia*.
2. What does a calving-ease score of 1 versus 5 indicate?
3. What is the percentage of twin births in Holsteins?
4. Why is it important that the first breath be a gasp?
5. What are the specific primary components of colostrum?
6. When should calves be fed colostrum to ensure maximal passive immunity?
7. How much colostrum can a 110-pound calf ingest in the first twenty-four hours of life?
8. What are three management priorities for calves after the first day of life?
9. Which enzyme is responsible for the coagulation of milk in the abomasum?
10. What are two carbohydrate sources that a calf can use in milk replacer?
11. When should calf starter be available to the calf? How often should it be replaced?
12. Which components of the bovine stomach remain relatively the same from birth to maturity?
13. Why should fresh water always be available to the calf?

ADDITIONAL RESOURCES

Books

Davis, C., and J. Drackley. *The Development, Nutrition, and Management of the Young Calf.* Ames: Iowa State University Press, 1998.

Phillips, C. *Cattle Behaviour and Welfare,* Second Edition. Oxford, UK: Blackwell Publishing, LTD, 2003.

Articles

Bøe, K., and O. Havrevoll. "Cold Housing and Computer-Controlled Milkfeeding for Dairy Calves: Behaviour and Performance." *Animal Production* 57 (January 1993): 183–191.

Faulkner, P., and D. M. Weary. "Reducing Pain after Dehorning in Dairy Calves." *Journal of Dairy Science* 83 (September 2000): 2037–2041.

Flower, F., and D. M. Weary. "Effects of Early Separation on the Dairy Cow and Calf: 2. Separation at 1 Day and 2 Weeks After Birth." *Applied Animal Behavioral Science* 70 (January 2001): 275–284.

Heinrichs, A. J., S. J. Wells, H. S. Hurd, G. W. Hill, and D. A. Dargatz. "The National Dairy Heifer Evaluation Project: A Profile of Heifer Management Practices in the United States." *Journal of Dairy Science* 77 (June 1994): 1548–1555.

Komisarek, J., and Z. Dorynek. "Genetic Aspects of Twinning in Cattle" (review). *Journal of Applied Genetics* 43 (January 2002): 55–68.

Place, N. T., A. J. Heinrichs, and H. N. Erb. "The Effects of Disease, Management, and Nutrition on Average Daily Gain of Dairy Heifers from Birth to Four Months." *Journal of Dairy Science* 81 (April 1998): 1004–1009.

Van Tassell, C. P., G. R. Wiggans, and I. Misztal. "Implementation of a Sire-Maternal Grandsire Model for Evaluation of Calving Ease in the United States." *Journal of Dairy Science* 86 (October 2003): 3366–3373.

Weary, D. M., and B. Chua. "Effects of Early Separation on the Dairy Cow and Calf: 1. Separation at 6 Hours, 1 Day and 4 Days After Birth." *Applied Animal Behavioral Science* 69 (October 2000): 177–178.

Pamphlet

Van Amburgh, M., C. Diaz, and J. Smith. "Nutrition and Management of the Milk Fed Calf." Proceedings of the 1999 Winter Dairy Management Schools, Cornell University, Ithaca, NY, pp. 54–63.

Internet

Absorption of Colostrol Lactoferrin in Newborn Calves: http://www.traill.uiuc.edu//dairynet/paperDisplay.cfm?ContentID=322.

Accelerated Growth for Dairy Calves: What's Behind the Controversy?: http://www.traill.uiuc.edu/dairynet/paperDisplay.cfm?ContentID=363.

Calf Care is Key: http://www.uwex.edu/ces/cty/marathon/ag/documents/DairyBriefs11_02.pdf.

Calf Management: Improving Calf Welfare and Production: http://www.agsci.ubc.ca/animalwelfare/publications/documents/calf_mgmt_weary.pdf.

Dairy Calf and Heifer Feeding and Management: http://www.das.psu.edu/dcn/CALFMGT/.

Raising Newborn Calves: http://wwwagcomm.ads.orst.edu/AgComWebFile/EdMat/EC1418.pdf.

46

Managing Replacement Heifers

OBJECTIVES

- To describe factors affecting heifer growth.
- To outline the development of the mammary gland during the heifer growth period.
- To describe rumen development during this period.

The average cow in the United States milks less than four years before she leaves the herd. To maintain a 100-cow dairy, therefore, at least twenty-five first-calf heifers must be available to replace these culled cows each year. But not all of today's heifers become tomorrow's cows. Approximately 20 percent of dairy calves die before reaching maturity, and still others may be culled.

To maintain the status quo in a milking herd, with no provision for expansion whatsoever, the average dairy producer replaces nearly one-third of the herd each year. This culling rate can be divided into culling for voluntary reasons (cows that are less profitable) and culling for involuntary reasons (health problems, reproductive failure, or death). Involuntary culling rates are much higher than voluntary culling rates on most farms. Increasing the ability to sell cows for purposes of increasing profits (voluntary culling) is the easiest way to increase the profit margin on most farms. The best way to accomplish this goal is to decrease involuntary culling rates, which translates into decreasing health problems, reproductive problems, and death losses. Why the high turnover in lactating cows, and why the high calf mortality? Of course, many factors are involved. Productive potential (growth, health, reproductive, or lactation potential) is determined by genetics, but management intensity either permits full expression of genetic potential or impairs it.

Management intensity varies throughout the lifetime of the dairy cow, depending on the needs of the animal. The growing heifer needs relatively little attention from weaning to calving when compared to milking cows or preweaned calves, but this does not mean that management of these animals is less important. Inattention or neglect during this period can have permanent effects on the productivity of the individual animal as well as farm profitability.

HEIFER GROWTH

The goal for a replacement heifer is to enter the milking herd as rapidly as possible without compromising health or lactational performance. Well-grown heifers can be bred at thirteen to fifteen months of age. Considerable data indicates that heifers calving at younger ages (around twenty-two to twenty-four months) are more productive and return more income during their lifetimes than do heifers that calve at an older age (thirty to thirty-six months). To achieve early calving goals, it is crucial that heifer growth rates be monitored and assessed on a regular basis. Both height and weight are important in assessing growth, and an accurate method of gathering these data is required. The most accurate method for measuring height is the use of an altitude stick, although quick "pen heights" can be estimated by creating a height scale on posts at fenceline feedbunks. Scales are the most accurate method for obtaining body weights, although this requires the availability of portable scales that can be moved from pen to pen.

If no scales are available, an ordinary tape can be used to measure the distance around the body immediately behind the front legs (heart girth) (Figure 46.1). Although this measurement is not precise, it is reasonably accurate and may be used for estimating body weights (Figure 46.2).

For large-breed dairy heifers, growth rates of approximately 1.75 to 2 pounds per day are necessary to reach optimal calving weight by twenty-two to twenty-four months of age (Figure 46.2). Preweaned calves typically average well under this target; if breeding size is to be attained by thirteen to fifteen months of age, nutrition and management of the growing heifer must be priorities. First estrus in heifers depends on size and weight, primarily weight. Heifers show their first estrus (puberty) at about 35 percent of their mature weight. Heifers fed high a plane of nutrition show estrus at an earlier age than heifers grown at recommended rates, and underfeeding of heifers delays estrus. Underfed or very slow growing heifers may ovulate, but estrus signs are often suppressed. Heifers in good condition and

gaining weight at breeding time generally show more definite signs of estrus and have improved conception rates over heifers in poor condition and/or that are losing weight. However, overconditioned heifers (heifers with a body condition score [BCS] greater than 3.0 at breeding) require more services per conception than heifers of normal size and weight. Heifers should be bred at 60 percent of mature weight so that they calve at 85 percent of mature weight. It is important to remember that type of gain is as important as rate of gain (Figure 46.3); heifers should be near a body condition score of 2.5 by six months of age (Figure 46.4), 3.0 at breeding, and 3.5 at calving (Figure 46.5). Overfeeding during the prepubertal period is associated with decreases in subsequent milk yield of approximately 10 percent.

Early breeding of heifers yields more profit than delayed breeding in several ways: it shortens the time from birth to lactation and lowers the cost of managing a nonproducing heifer, and lifetime production is greater. Early calving also permits faster genetic progress in the herd and helps maintain an established or desired seasonal calving schedule. Young animals make more economical weight gains than older ones because older animals require more feed for maintenance, and the added weight of young animals is high in water content and lower in energy (70 percent water in a calf at birth versus 45 percent in a fat two-year-old).

Mammary Development

Development of secretory tissue during the prepubertal period ultimately affects the profitability of the dairy cow. The most rapid increase in the number of milk secretory cells occurs between three months of age and the second or third estrous cycle following puberty. This allometric growth phase of

Figure 46.1 Heifer weight can be estimated accurately by measuring heart girth with a weight tape. *(Courtesy of Iowa State University)*

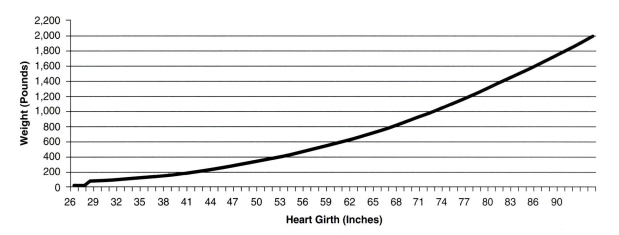

Figure 46.2 Estimated weight based on heart girth measurements.

TABLE 46.1 Relationship of Age, Weight, and Height of Dairy Heifers for Minimum Development for Calving at Twenty-four Months of Age

Age (Months)	Large* Weight (Pounds)	Large* Height (Inches)†	Medium* Weight (Pounds)	Medium* Height (Inches)†	Small* Weight (Pounds)	Small* Height (Inches)†
2	185	34	125	32	100	30
6	375	40	300	38	260	35
10	585	45	475	43	395	39
14	780	48	630	47	505	48
18	945	51	770	49	615	44
22	1,135	53	935	51	755	46
24	1,160	54	980	52	825	47

*Large = Holstein and Brown Swiss; medium = Ayrshire and Gurnsey; small = Jersey.
†Measured at withers.
Source: Stallings, C. C., and R. E. James. Feeding Heifers for Early Calving, *Virginia Cooperative Extension Publication number 404-108, 1998.*

Figure 46.3 The type of weight gain achieved by prepubertal heifers is more important than the rate of gain. *(Courtesy of Iowa State University)*

Figure 46.5 Bred heifers should be guided toward a body condition score of about 3.5 at calving. *(Courtesy of Iowa State University)*

Figure 46.4 By twelve months of age, a heifer should be at a body condition score (BCS) of about 2.75 to 3.0. *(Courtesy of Iowa State University)*

mammary development allows mammary growth to proceed at a rate three times that of somatic growth. Stimulating faster rates of weight gain during the prepubertal period by feeding high energy rations (140 percent of requirement) speeds up the onset of puberty; however, this early onset decreases growth hormone secretion, which decreases the development of mammary secretory tissue. Earlier onset of puberty also decreases the period of time allowed for proliferation of mammary secretory cells. This negative effect of high-energy rations in the prepubertal period on mammary development may be partially alleviated by increasing protein in the ration at the same time; protein:energy ratios need to be higher for heifers fed for faster rates of gain.

Although increased rates of gain in the post-pubertal period are not associated with impair-

Figure 46.6 By nine months of age, a calf's rumen capacity is adequate to meet nutritional needs almost completely by ingestion of high-quality forage alone. *(Courtesy of Iowa State University)*

Figure 46.7 High-quality forage provides the foundation for a good heifer ration. *(Courtesy of Iowa State University)*

ments in mammary development, prepubertal growth rates beyond 2.25 pounds per day cannot be recommended. Prepubertal growth rates determine age at puberty and age at first breeding, which ultimately determines age at calving. Therefore, goals for age at first calving should be maintained between twenty-two and twenty-four months of age to ensure optimal development of secretory tissue and maximal lifetime productivity of the dairy cow.

Rumen Development

The rumen of the prepubertal heifer undergoes changes during this period. Although much of the functional development of the rumen has occurred by the time the calf is weaned, rumen capacity is still extremely limited. At two months of age, the rumen comprises about 50 percent of the total four-compartment organ mass. By six months, it has expanded to about 60 percent of the same mass and is fully functional, although further increases in relative size will continue for several more months (Figure 46.6). This limited rumen capacity during the prepubertal growth period limits the ability of these animals to consume enough forage to meet nutrient requirements, especially if ensiled forages or pasture is fed. Because prepubertal heifers have a high energy requirement relative to their ability to consume feed, they should be maintained on a concentrate-based ration with enough fiber provided to maintain normal rumen function.

NUTRITION OF THE PREPUBERTAL HEIFER

Because the heifer has limited capacity for dry-matter intake, forage quality is critical if growth targets are to be achieved (Figure 46.7). The ability of the prepubertal heifer to digest overly mature forages is less than that of older heifers. Less digestible forages decrease the rate of passage, increase rumen fill, decrease total energy intake, and impair growth rates. Recommended energy intake for prepubertal heifers is approximately 0.72 Mcal per pound of dry matter, while protein-to-energy ratios should be maintained at approximately 6:1. Decreasing the protein-to-energy ratio below this level has detrimental effects on feed efficiency and growth rates. For most producers, the nutrient balance of the concentrate mix is determined by the quality of the forages available.

National Research Council (NRC) recommendations are based on certain assumptions (such as animals have free access to both feed and clean water). However, many heifers are not maintained under ideal conditions. For example, parasite loads can increase energy requirements by up to 10 percent. Therefore, it is not enough to provide a ration that meets the theoretical requirements of the heifer. We must work to adjust both the ration and the management strategies to match actual farm conditions so that optimal growth rates are achieved. The most recent NRC recommendations take into account many of the environmental variables that can affect the efficiency of nutrient utilization from feeds, including the effects of adverse weather conditions.

COMPENSATORY GROWTH

Compensatory growth refers to the ability of an animal to have an above normal growth rate following a period of growth restriction. Feed-restricted animals develop an increased energetic efficiency that is retained following return to full feed. This increased efficiency has been exploited by carefully alternating periods of feed restriction with periods of high energy intake in replacement heifers. The increased efficiency of lean growth also extends to improved mammary development in the prepubertal period and increased lactational efficiency after calving. Systems developed to induce compensatory gain are typically not implemented until about five months of age, and such systems are not practical to implement unless facilities allow heifers to be tightly grouped into small feeding groups with less than a three-month age spread.

VACCINATIONS

The effectiveness of vaccination programs for heifers prior to four months of age may be compromised by the presence of maternal antibodies from colostrum intake. Timing of initial vaccinations should be varied depending on the passive immunity status of the calves and the health status of the herd. Early initiation of a vaccination program leads to lower levels of protection for vaccinated animals.

The selection of vaccines should be made in consultation with the herd veterinarian using the following criteria:

1. Incidence of the disease.
2. Potential cost of the disease.
3. Cost of the vaccine.
4. Effectiveness of the vaccine.

The effectiveness of any vaccine may be compromised by several factors, including improper storage or administration and the use of disinfectants to clean syringes. In addition, agents in one vaccine may compromise the effectiveness of a second vaccine if they are given simultaneously or they are mixed together. Administration of antibiotics or other drugs at the same time as the vaccine can diminish the immune response. Vaccinating sick or unthrifty animals also results in an inadequate immune response. Timely booster vaccinations (typically three weeks after the initial injection) are necessary for optimal response to many vaccines, especially killed products.

PARASITE CONTROL

Control of parasites, both internal and external, is of special concern during this period. Susceptibility to parasites is highest at a young age and at first exposure, and an intensive control and treatment program is required to optimize heifer performance. Heifers with heavy parasite loads have a decreased feed efficiency and may develop diarrhea and anemia. Heifer facilities utilizing bedded manure packs should be cleaned often and thoroughly. Sanitation is the key to managing parasite loads within a facility, and treatment programs help decrease parasite problems within the individual animal.

Coccidiosis is a common internal parasite in dairy heifers (Figure 46.8). The oocytes from this protozoal organism are ingested and penetrate the gut of the animal. This condition results in impaired feed efficiency and diarrhea. As the life cycle progresses, the damage extends down the digestive tract, resulting in bloody diarrhea, anemia, and weakness in severely affected animals. Ionophores prevent and control coccidiosis and improve feed efficiency in growing heifers (no increase in intake but an increase in daily weight gain).

HOUSING CONSIDERATIONS

Replacement heifers can be raised successfully in many different types of housing as long as certain basic criteria are met. Both prepubertal and postpubertal heifers should be kept in facilities providing excellent ventilation, adequate space, and ready access to feed and water. Poorly designed or poorly managed facilities increase the risk of enteric and

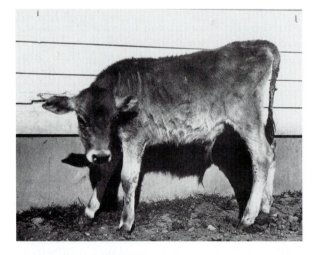

Figure 46.8 Coccidiosis is a common protozoal parasite affecting young heifers. Infected heifers appear listless and unthrifty. *(Courtesy of USDA)*

respiratory diseases and decrease feed efficiency and growth rates. Housing should allow heifers to be grouped in small groups of between six and ten heifers to minimize stress and competition at the feedbunk (Figure 46.9). Age spreads should not exceed one month within a group and weight spreads should not exceed 100 pounds (Figure 46.10).

BUSINESS ASPECTS OF HEIFER RAISING

Whether dairy producers are raising heifers for their own use or for sale to others, it should be done at a profit. The total cost of raising replacement heifers to twenty-four months of age is approximately $1,200.00. Although costs for individual inputs may vary from area to area and from year to year, the total cost has remained remarkably stable over time. Managers who focus their management attention on heifer rearing can often decrease rearing costs by $200 or more per heifer raised. On a 1,000-cow dairy, this reduction can mean an increase in total profit of approximately $50,000 per year. In addition, managers with low culling rates have the enviable option of either raising fewer heifers (thus lowering expenses associated with raising heifers) or marketing the excess heifers not needed as replacements (Table 46.2).

Figure 46.9 Postweaned calves should be housed in small groups of four or six calves per pen to minimize socialization difficulties. *(Courtesy of Iowa State University)*

Figure 46.10 Heifers should be housed in groups that closely limit age and weight differences. *(Courtesy of Mark Kirkpatrick)*

TABLE 46.2   Heifers Needed to Keep Herd Size at 100

				Age at First Calving				
	22	*24*	*26*	*28*	*30*	*32*	*34*	*36*
Cull Rate (%)				Number of Heifers				
20	40	44	48	51	55	59	62	66
22	44	48	52	56	61	65	69	73
24	48	53	57	62	66	70	75	79
26	52	57	62	67	72	76	81	86
28	56	62	67	72	77	82	87	92
30	61	66	72	77	82	88	94	99
32	65	70	76	82	88	94	100	106
34	69	75	81	87	94	100	106	112

Source: Modified from Hoards Dairyman, Brian L. Perkins, "What It Takes to Freshen Heifers Early," September 25, 1994, p. 647.

SUMMARY

- Approximately 20 percent of dairy calves die before reaching maturity.
- Culling can be divided into voluntary and involuntary culling.
- Heifers calving at younger ages are more productive and return more income during their lifetimes than heifers that calve at an older age.
- The type of weight gain is as important as the rate of gain.
- The level of development of secretory tissue during the prepubertal period ultimately affects the profitability of the dairy cow.
- Prepubertal heifers should be maintained on a concentrate-based ration, with enough fiber provided to maintain normal rumen function.
- Forage quality is critical if growth targets are to be achieved.
- Effectiveness of a vaccine may be compromised by several factors, including improper storage or administration and the use of disinfectants for cleaning syringes.
- Sanitation is the key to managing parasite loads within a facility.

QUESTIONS

1. What percentage of the herd does the average dairy producer replace each year?
2. When can well-grown heifers be bred?
3. Why is early breeding more profitable than delayed breeding?
4. At what age does the rumen comprise about 50 percent of the total four-compartment organ mass?
5. What is compensatory growth?
6. What criteria should be used for the selection of vaccines?
7. Name a common internal parasite in dairy heifers.
8. What is the purpose of ionophores?

ADDITIONAL RESOURCES

Articles

Gunn, S. W. "Vaccination." *Journal of World Surgery* 28 (April 2004): 420. Published on the Internet March 17, 2004.

Rice, L. E. "Nutrition and the Development of Replacement Heifers." *Veterinary Clinicians of North America Food Animal Practices* 7 (March 1991): 27–39.

Internet

Feeding and Managing Dairy Calves and Heifers: http://agbiopubs.sdstate.edu/articles/EXEX4020.pdf.

Management of Dairy Replacement Calves from Weaning to Calving: http://www.wcds.afns.ualberta.ca/Proceedings/1997/ch02-97.htm.

Raising Dairy Herd Replacements: http://www.oznet.ksu.edu/library/agec2/mf399.pdf.

Raising Dairy Replacement Heifers: http://edis.ifas.ufl.edu/DS150.

47

Managing Dry and Transition Cows

OBJECTIVES

- To define concepts in dry-cow management.
- To outline dry-off procedures in a dairy herd.
- To describe the feed and energy requirements of the dry cow.

Dry cows have several important tasks to complete prior to calving: regression of old milk-producing cells in the udder and regeneration of new milk-producing cells and developing the unborn calf. This necessitates that they be fed properly in late lactation and during the dry period. Dry-cow condition depends on the feeding and management practices during the last half to one-third of the lactation period, when body reserves should be replaced. At drying-off time, the cow's body condition score (BCS) should be 3.25 to 3.75; at calving, the body condition score ideally should not have changed appreciably (Figure 47.1). In cows with inadequate body condition at dry-off,

some gain in BCS is acceptable, although it should not exceed 0.5 BCS; gains greater than this figure can be detrimental to performance in the subsequent lactation. During the dry period, cows should be fed to gain only enough to support fetal development. Dry cows should not only be grouped separately from lactating cows (Figure 47.2), but they should also be grouped into several dry-cow groups. This grouping allows for specific management and feeding practices appropriate for the needs of the animal as she undergoes secretary cell renewal, or is preparing for parturition and the subsequent lactation.

Cows should be turned dry thirty to forty-five days before their expected calving date. Cows with no dry period have their production level lowered by as much as 25 percent. Cows with too short a dry period or that are not fed properly during the dry period are affected in proportion to the degree of the nutritional deficit. Conversely, extending the length of the dry period simply increases feed costs and management costs with no production benefit.

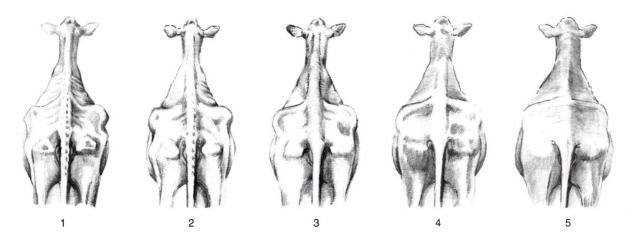

Figure 47.1   The body condition score is an effective technique for estimating the percentage of body fat in cattle. Images represent cows with body condition scores of 1 through 5. *(Courtesy of Dana Boeck)*

Figure 47.2 Dry cows gain weight through the dry period due to rapid fetal development, but their body condition scores should not increase prior to calving. *(Courtesy of John Smith)*

Figure 47.3 Quality forage provides most of the required nutrients for the dry cow. *(Courtesy of Iowa State University)*

Slightly longer dry periods are required as cow production increases, calving intervals decrease, and cow age decreases.

The dry period can be divided into two distinct phases. The first phase constitutes the cessation of lactogenesis in the mammary gland. The second phase is a period of colostrogenesis. At about three weeks prepartum, colostrogenesis is initiated. This period is marked by rapid differentiation of secretory epithelium; increased synthesis and secretion of fat, protein, and carbohydrates; and accumulation of immunoglobulins from plasma. The timing of initiation of colostrogenesis is determined by the same endocrine changes that prepare the cow for parturition. Throughout the dry period, secretory cells are renewed as the gland is prepared for the next lactation.

DRY-OFF PROCEDURE

Two strategies are typically employed to stop lactogenesis. Producers can simply discontinue milking or lengthen milking intervals. Lengthening milking intervals increases the risk of mastitis and is not recommended. For low- or moderate-producing cows, producers can simply discontinue milking; however, for higher-producing cows (over 45 pounds for large breeds, over 25 pounds for small breeds), they may need to decrease production prior to drying-off. Several strategies can serve to decrease milk production in a high-producing animal; they can be as simple as disrupting the routine or changing housing. In some facilities, water intake can be limited and feed quality can be reduced as well. The use of bovine somototropin should be discontinued for several weeks prior to dry-off. These practices re-

duce stress on the cow, decrease the risk of mastitis, and facilitate a successful drying-off. Following the final milking, cows should be treated with a long-lasting mastitis treatment (antibiotic) formulated for dry cows, and their teats should be dipped using a barrier-type teat dip. Cows should be observed closely for the first ten days to two weeks following dry-off for signs of mastitis. However, cows should not be milked out after having been dry for more than a few days; this practice can induce milk fever and increases the risk for mastitis.

FEEDING THE DRY COW

The dry period should be a time when the digestive tract has a chance to recondition itself following the rigors of fermenting large amounts of the high-concentrate feeds used during lactation. Roughage intake, especially long-stemmed hay, is extremely important during the dry period to stimulate muscle tone of the rumen (Figure 47.3). Ideally, the condition of the dry cow should be such that she will not have to gain much weight because weight can be put on much more efficiently during the latter stages of lactation. If forage quality is good, no concentrate may be required. The ration provided during this part of the dry period should be based on the body condition and size of the cow. Daily dry-matter intake should be near 2 percent of the cow's body weight, with forage intake a minimum of 1 percent of body weight and grain intake according to needs but not to exceed 1 percent of body weight. Dry cows should not be overconditioned and should not be fed to gain condition during the dry period. Feeding low-quality (bulky) forage, such as corn stalks or grass hay, is preferable to limit feeding. A protein level of 12 percent is sufficient during the first part of the dry period.

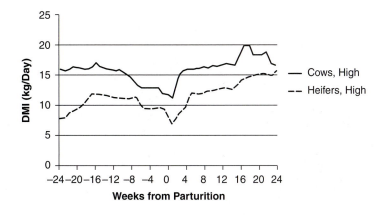

Figure 47.4 The dynamics of the changes in dry-matter intake (DMI) during the periparturient period for both heifers and cows.

Rumen bacterial populations during these first two stages of the dry period shift toward a higher proportion of forage digesters and fewer starch digesters. The decreased production of butyrate reduces the mitogenic capacity of the rumen epithelium; nearly half of the effective surface area of the rumen may be lost in the first seven weeks of the dry period as papillae shorten.

MANAGEMENT STRATEGIES FOR TRANSITION COWS

If possible, transition cows should be managed as a separate group. Housing facilities that allow individual cow management between two weeks prepartum and ten days postpartum facilitate a successful transition-cow program.

The goal in feeding during the last few weeks is to prepare the rumen and the rumen microbes and to maximize dry-matter intake. Typically, dry-matter intake is reduced by an average of 30 percent immediately prior to calving (Figure 47.4). Maintaining dry-matter intake at levels that allow the cow to meet her nutrient requirements not only reduces the incidence of metabolic disorders following calving, but also improves dry-matter intake following calving. This improved feed intake is reflected by increased peak milk production and reduced periods of negative energy balance (Figure 47.5). The objective is to allow each cow to reach peak production at or near her genetic potential. The advantage of achieving maximum peak production can be maintained throughout the remaining months of lactation. In addition, reproductive performance is enhanced.

The best strategy for avoiding most of the problems associated with the transition period is to maintain dry-matter intake during the final ten days prior to parturition. It is important to remember that the 30 percent decrease in dry-

matter intake triggered by the hormonal changes associated with preparation for parturition and lactation is simply an average. Strategies must be employed to stimulate intake in these cows and counteract their natural tendency to stop eating (Figure 47.6). To maintain dry-matter intake, cows should be fed smaller, more frequent meals. Palatability of feed ingredients needs to be a priority. Poor lighting can reduce feed intake; a 16:8 light:dark cycle is optimal for maintaining appetite. Both feed and fresh, clean water should be readily available. Cows nearing parturition are less likely to risk inconvenience or competition to gain access to feed or water. As always, cows should be kept clean, dry, and comfortable.

In facilities that allow cows to be fed individually over the last two weeks prepartum, ration changes to adapt the rumen and the rumen microbes to a lactating ration can be introduced gradually. This approach shifts rumen microbial populations to increase the numbers of starch-digesting organisms prior to calving. In addition, the butyrate produced by these organisms increases papillae surface area in the rumen, preparing the cow for liberal grain feeding immediately following freshening. It is also critical that numbers of lactate-utilizing bacteria be increased to reduce the risk of rumen acidosis in early lactation. Gradual ration changes also help maintain intake.

Palatability of ration ingredients becomes of paramount importance during this period. The most critical component of the ration in this respect is the forage source, but other ingredients of the concentrate supplement can also have dramatic effects. Anionic supplements need to be adjusted carefully as the ration formulation is changed to avoid negative effects on dry-matter intake. Intake of total mixed rations tends to be higher than when the ingredients are fed separately; thorough mixing

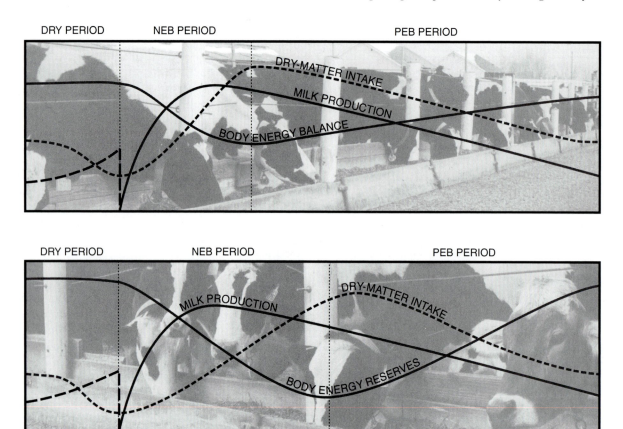

Figure 47.5   The effects of decreased feed intake in the transition period is felt throughout the entire lactation period. The top panel reflects a 30% decrease in feed intake while the bottom panel reflects a 70% decrease in feed intake. *(Courtesy of Dana Boeck)*

Figure 47.6   Dry-matter intake does not decrease significantly until the final week prior to calving. *(Courtesy of Monsanto)*

of feed ingredients during this period makes sense. If mixing equipment is not available and if the facilities allow individual feeding, rations can be hand-mixed as they are provided to the individual animal.

If milk fever is a problem, the cation-anion difference (CAD) of the ration should be monitored

closely. Vitamins A, D, and E, along with added selenium, may need to be provided. One commonly used procedure to adapt the ruminal microflora to the lactation ration is to feed the same ration, or a percentage of the same ingredients, that the cows will receive during lactation. It is critical, however, that dry cows not become overconditioned prior to calving.

Obviously, the success of a transition-cow program depends on management during the previous lactation. Decreases in feed intake are greater in cows with excessive body condition; cows that are overfed during late lactation and are overconditioned entering the dry period are more difficult to manage through the transition period (Figure 47.7). The strategies for success are similar, however, and simply become a greater priority in these problem herds.

If facilities do not permit cows to be grouped separately, electronic feeders in the dry pen allow the individual feeding of these cows. Providing electronic feeders to the dry-cow group is easier to justify in terms of return on investment than it is for any other management group, except possibly early lactation cows.

These extra efforts expended in managing transition cows are well worth the time and the expense.

Figure 47.7 The targeted body condition score for dry cows is between 3.5 and 3.75. *(Courtesy of Iowa State University)*

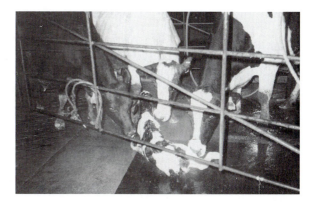

Figure 47.8 Maternity pens should be clean, dry, and well-bedded. *(Courtesy of Iowa State University)*

The returns will be manifested as decreases in metabolic disorders, health problems, and reproductive failures (with their associated costs) and increases in milk production and overall profitability.

MANAGING MATERNITY PENS

Preventing stress at calving starts with monitoring the cow during calving. If transition cows are handled as a group, maternity cows should be handled individually. Cows should be placed in individual, well-bedded maternity stalls one week prior to calving (Figure 47.8). Calving cows in group pens is a biosecurity threat for the calf and ultimately for the herd as well (Figure 47.9). Calving in tie stalls is detrimental for both the cow and calf and should be avoided if at all possible. Pastures are often touted as a perfect calving environment, but they also have many limitations that create a less-than-ideal environment for both cow and calf. Cows calving in a group environment, such as a pasture, try to isolate themselves during the calving process. Cows instinctively position themselves to limit access to predators during calving. As a result, they often calve in isolated corners of the pasture under unfavorable conditions. Pastures must be extremely clean and well-drained to even approach acceptable standards. However, cleanliness is the least of the problems encountered during pasture calvings. Many additional problems are a result of the stimulation of the maternal instincts of other cows in the group. The noncalving members of the group often congregate around the calving cow, and the labor process is extended as the cow in labor has to move continually to isolate herself. Other cows lick the calf as it is being delivered, and in many cases, calves are adapted and end up nursing cows other than their own dam. The threat of disease transfer dictates that a maternity pen is the only viable choice

Figure 47.9 The maternal instinct can be potentially detrimental in group calving environments because multiple cows can transmit pathogens to each calf. *(Courtesy of Emily Barrick)*

for a calving environment. Maternity pens allow the cow to eat and drink without competition, they allow the producer to monitor cows individually through this crucial period, and they protect the calf after delivery from exposure to herd pathogens.

Maternity pens should have a solid concrete base and be extremely well-bedded. They need to be approximately 200 square feet (with a minimum size of 120 square feet) to allow for both cow comfort and room for assistance if necessary. A ceiling hoist is a useful addition for cows that have difficulty in rising after calving. The solid base allows the pen to be cleaned and sanitized between calvings, which is crucial because both the cow and calf are immunocompromised and susceptible to infection at this point. Recent surveys suggest that less than half of all producers clean maternity pens after every calving and less than 10 percent of producers disinfect stalls in addition to stripping bedding. The costs of these practices (cleaning and disinfecting) are paid for easily by the reduced health problems of the calves and cows housed in these pens.

Bedding depth in maternity stalls should be greater than anywhere else on the farm. During calving, large volumes of fluids are expelled from the water bags surrounding the calf, and the cow defecates and urinates more frequently in the days leading up to calving. It is crucial to provide ample bedding so the calf is born in a clean environment despite all these challenges. Also, consider that the calf is born covered with sticky, viscous amniotic fluid; beddings such as shredded newspaper will stick to the calf and can result in papier-mâché calves (Figure 47.10). Deep straw bedding over a thick bed of wood shavings provides an excellent bed for both the cow and calf.

Figure 47.10   Sawdust, wood shavings, shredded newspaper, and seed hulls are poor choices for bedding in maternity pens. *(Courtesy of Monsanto)*

SUMMARY

- At drying-off time, the cow's body condition score should be 3.25 to 3.75.
- Following the final milking, cows should be treated with a long-lasting mastitis treatment, and their teats should be dipped using a barrier-type teat dip.
- Roughage intake, especially long-stemmed hay, is extremely important during the dry period to stimulate the muscle tone of the rumen.
- The amount of concentrates fed should be increased gradually over the last two to three weeks of the dry period.
- Palatability of ration ingredients becomes of paramount importance during the transition period.

QUESTIONS

1. What are some important tasks a dry cow must accomplish prior to calving?
2. What are two strategies to stop lactogenesis?
3. What is the optimal light:dark cycle for transition cows?

ADDITIONAL RESOURCES

Article

Gerloff, B. J. "Feeding the Dry Cow to Avoid Metabolic Disease." *Veterinary Clinicians of North America Food Animal Practices* 4 (July 1988): 379.

Internet

Dry Cows: http://agebb.missouri.edu/dairy/drycows/.

Feeding and Managing Dry Cows: http://pubs.cas.psu.edu/FreePubs/pdfs/ec372.pdf.

Feeding the Dry Cow: http://ianrpubs.unl.edu/dairy/g1201.htm.

Forages for Dry Cows: http://www.wisc.edu/dysci/uwex/nutritn/presentn/dcfor.pdf.

48

Managing Metabolic Disorders in Transition Cows

OBJECTIVES

- To describe the consequences of overconditioning.
- To identify the health-related disorders commonly associated with transition cows.

The transition period includes the final few weeks prior to parturition and the first few weeks following parturition. The level of management of dairy cattle during this critical period determines the potential for profitability throughout lactation. Transition cows, especially high-producing transition cows, have several major challenges to overcome. First, the hormonal changes associated with the transition from a pregnant state to a nonpregnant state tend to suppress immune function, which leads to an increased susceptibility to disease. Second, the hormonal and metabolic changes associated with the transition from a nonlactating to a lactating state create a susceptibility to metabolic disorders.

The sudden shift to an increased nutrient intake and nutrient outflow in the form of milk is a daunting challenge for the dairy cow. Although nutrient intake is greatly increased, energy demands are rarely met, and the cow exists in a state of negative energy balance for an extended period of time. The degree of susceptibility to periparturient disorders is highly influenced by management prior to calving. Most metabolic disorders can be viewed as a nutritional syndrome; minimizing hypocalcemia and overconditioning and maintaining dry-matter intake are critical prevention steps prior to calving. Maximizing intake of a properly balanced ration throughout the early lactation period is also critical to minimize the occurrence of most of these problems effectively. The occurrence of the first metabolic disorder at or near parturition greatly increases the risk for others. Figure 48.1 illustrates these relationships.

OVERCONDITIONING

Although overconditioning is not a metabolic disorder, it does predispose cows to an increased susceptibility to those disorders commonly occurring around calving. Body condition in dairy cattle is an effective way of monitoring the energy reserve status of the cow. In fact, changes in body reserves are more accurately estimated by assessment of changes in the body condition score (BCS) than by changes in actual body weight, especially in the transition cow. Cows with especially large changes in dry-matter intake following calving can mask changes in live weight; however, the mobilization of fat will be apparent during a body condition score assessment.

The scoring system most commonly used for dairy cattle uses a scale of 1 to 5, with a body condition score of 1 indicating an extremely emaciated cow and a score of 5 reserved for obese cattle. The optimal body condition score for a cow at calving is somewhere between 3.5 and 3.75. Cows carrying higher condition scores than recommended tend to have lower feed intake and lose more body condition following calving than those with optimal scores. Losses of more than one body condition score during the first sixty days of lactation is a strong indication of problems in the management of transition cows. The higher the condition score at calving, the greater the risk of both excessive loss of body condition and occurrence of metabolic disorders (Figure 48.2). The fat surrounding the reproductive tract creates difficulties during calving, and the fat imbedded in the internal organs decreases their efficiency of function. Overconditioned cows also tend to have lower milk yields and higher risk of difficulties in rebreeding. Overconditioned cows can be managed to avoid these problems. There is just a smaller margin for error in managing these cows; properly conditioned cows are simply a much lower risk option for the dairy producer.

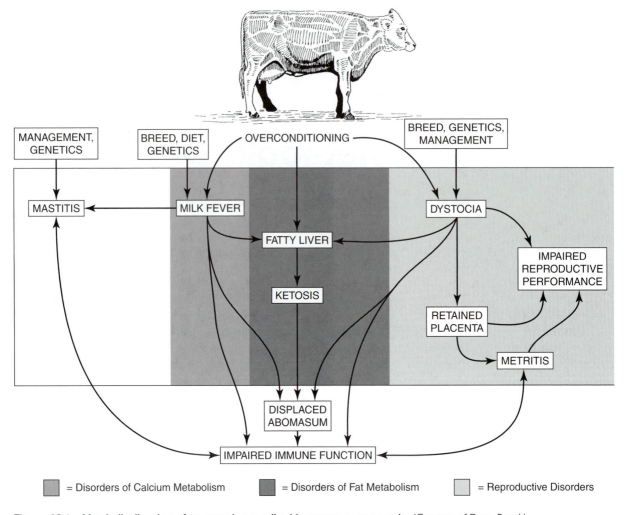

| | = Disorders of Calcium Metabolism | | = Disorders of Fat Metabolism | | = Reproductive Disorders |

Figure 48.1   Metabolic disorders often occur in a predictable sequence, or cascade. *(Courtesy of Dana Boeck)*

Figure 48.2   Cows that are overconditioned during the transition period are at greater risk for many metabolic disorders. *(Courtesy of Howard Tyler)*

FATTY LIVER SYNDROME (ALERT DOWNER COWS)

This syndrome commonly occurs within a few days of calving in excessively fat cows. It is most prevalent in cows receiving too much energy or not enough protein during late lactation and during the dry period. These cows accumulate excessive stores of internal fat. The liver fills with fat and is unable to function optimally. The problems become apparent when the cow enters negative energy balance, often during the final few weeks prior to parturition. The hormonal changes associated with preparation for parturition and lactation often affect appetite, and dry-matter intake decreases by about 30 percent on average. This decrease in feed intake is greater in cows with excessive body condition. To compensate for the decrease in energy intake, fat stores in the body are mobilized. Much of this fat accumulates in the liver, further impairing the ability of the liver to oxidize fats for energy. The cow mobilizes more and more body fat as the liver becomes increasingly filled with lipids and increasingly impaired in its ability to utilize these fats. This vicious cycle can result in a more than tenfold increase in liver lipids in severe cases. Partially oxidized fats (ketone bodies) are released from the liver into the bloodstream; these ketones further depress appetite in these animals. As a result, cows can lose several points in their body condition score within a short period of time during early lactation. Often these cows have an impaired immune function and increased susceptibility to other metabolic disorders.

Cows with fatty livers exhibit a decreased feed intake and drop in production. If the disorder becomes severe enough, cows go down but remain alert. Symptoms of ketosis are often apparent; fatty liver is considered to be a precursor to primary ketosis in most if not all cases. Fatty liver is also highly correlated with many other disorders: retained placenta, milk fever, displaced abomasum, metritis, and udder edema. In some cases, an elevated temperature may occur in these animals. Immune function is depressed, and the animals are more susceptible to mastitis as a result.

It is important to remember the cows that are the most likely candidates for fatty liver are those that are losing weight most rapidly. Cows nearing parturition with a body condition score of greater than 4 are at a high risk; also, cows losing more than 1.5 points in their body condition scores between calving and peak milk yield are at a greater risk for developing fatty liver. Treatments for fatty liver are limited to those that treat symptoms. The response is usually poor, and prevention is the key. Avoid overfeeding; balance rations for both protein and energy. Avoid feeding high-energy rations to dry cows. Exercise may help, but the amount of exercise needed may not be possible to achieve in many confinement systems. Choline chloride supplements improve lipoprotein transport and decrease fat levels in the liver, and they may be useful in high-risk animals.

KETOSIS (ACETONEMIA)

Ketosis usually affects high-producing cows ten days to six weeks following parturition. The incidence increases as the cow approaches peak lactation. It is one of the more common metabolic disorders encountered on dairy farms, with incidence estimates of approximately 15 percent. Ketosis is a metabolic disorder characterized by increased concentrations

of free fatty acids and ketones in the blood and urine and decreased concentrations of glucose. Ketosis can be characterized as either primary or secondary; the characteristics of the problem are similar in both types.

Ketones are formed from the incomplete oxidation of mobilized fat. Stores of body fat are mobilized in response to a negative energy balance in the animal; therefore, high-producing cows in early lactation are especially susceptible to ketosis because their feed intake is often unable to meet the energy demands of milk production. In ketotic cows, the liver cannot metabolize the high levels of free fatty acids quickly enough, forming ketone bodies faster than the rest of the body can metabolize them. Thus, ketones build up in the blood, milk, and urine. High concentrations of blood ketones tend to depress appetite further, which only exacerbates the problem. Primary ketosis is a metabolic disorder that occurs in cows mobilizing more fat than the liver can process; it is frequently, if not always, associated with the development of fatty liver syndrome in these animals. It is most commonly seen in cows that were overconditioned at calving because these cows are typically starting lactation with a lower feed intake and impaired liver function. Secondary ketosis, often referred to as nutritional ketosis, results from feeding low-energy rations or from decreased intake of normal rations. In addition, this type of ketosis can be induced when feeding a silage with high levels of butyric acid. (Butyric acid is converted to the keytone beta-hydroxy-butyrate.)

Symptoms of ketosis include an abrupt drop in milk production (although the percentage of fat in the milk increases), depressed appetite, excessive weight loss (gaunt, dull cow), dry feces and constipation, and drowsiness, although some cows appear to be more excitable. In addition, ketones are volatilized off through the respiratory system and pass into the milk, and affected cows have a sweet-smelling acetone breath and an acetone flavor in the milk. Ketone bodies in the blood and urine are increased, and blood glucose is decreased. In addition, liver glycogen levels are decreased in ketotic cows, while liver lipid levels increase from normal concentrations near 7 percent of liver weight to as high as 30 percent of liver weight. Clinical testing of these animals is via urine or milk ketones; these tests can be used prior to the appearance of physical symptoms.

Treatments for ketosis are designed to increase blood glucose. Intravenous glucose (40 percent dextrose) can be provided. However, it is used quickly by the cow, and the animal usually relapses. Feeding glucose is ineffective because rumen microbes convert glucose to volatile fatty acids (VFAs), but oral glucose precursors are often provided as a preventive measure and as a treatment for affected cows. Sodium propionate can be top-dressed at 0.5 pounds per cow per day or, more typically, propylene glycol (palatable, inexpensive, and easy to handle) is provided as an oral drench. These treatments provide a consistent source of glucose to the cow over an extended period of time.

Glucocorticoids stimulate the body to convert body protein to glucose and have been used to treat ketosis. The effect lasts several days, but prolonged use may deplete protein stores and glucocorticoid treatment definitely depresses the immune system. Niacin supplementation has also been provided in herds with a high incidence of ketosis. Niacin is a water-soluble B vitamin; rumenal microbes in high-producing cows may not produce enough niacin to meet demands for energy metabolism in feed-depressed, high-producing cows. Feeding supplemental niacin to preclinical ketotic cows decreases the levels of ketones in the blood, urine, and milk possibly because of increased efficiency in lipid metabolism. Some studies have also shown an increased rumen propionate and increased microbial protein synthesis with niacin supplementation, suggesting an improved rumen fermentation pattern. Niacin is fed 6 grams per day (preventive) up to 12 grams per day (treatment).

As with all metabolic disorders, the key to controlling ketosis is prevention. Producers should strive to keep cows in optimal condition at parturition (body condition scores from 3.25 to 3.75). Abrupt ration changes should be avoided, and dry-matter intake must be maximized. Producers can monitor problem animals with milk tests, and they can feed propylene glycol (125 to 250 grams per day) and/or niacin.

DISPLACED ABOMASUM (TWISTED STOMACH)

The abomasum of the dairy cow normally lies on the right ventral abdominal wall. The term *displaced abomasum* refers to a repositioning of the abomasal compartment; the twisting associated with this repositioning effectively slows or stops the flow of digesta through the gastrointestinal tract. The gas buildup (whether leading to this condition or as a result of this condition) leads to a characteristic bloat appearance. Pinging is detected with a stethoscope (although sometimes a stethoscope is not necessary) by thumping the cow near the last rib and listening on the left flank. Right-side displaced abomasums (DAs) are more serious (more complete torsion) and more difficult to treat because they almost always require surgical intervention. The exact causes of this condition are poorly understood. Displacements may be either left- or right-sided, although 85 to 90 percent are left-sided.

Symptoms of diplacement include a dramatically decreased feed intake (due to the inability of feed to leave the rumen) and a drastic decrease in production. Cows appear to be in pain and may stand with their back arched. Displacements are diagnosed most often before fourteen days postpartum, but nearly one-fourth of all cases appear after this time. Displacements rarely occur in heifers, and the incidence increases through the sixth lactation and then decreases again.

The incidence of displacements ranges up to 15 percent in problem herds. Many factors are associated with an increased risk of diplaced abomasum, but it appears to be a syndrome that can be induced through several pathways. Over two-thirds of cases are preceded by a different metabolic disorder or disease. Hypocalcemia decreases gut motility in general and may allow an increased production of gas in the abomasum. An increased risk is associated with high grain feeding or heavy corn silage feeding prior to parturition, and cows with high body condition scores at parturition are much more prone to displacements than are appropriately conditioned cows. High grain feeding increases VFA flow through the abomasum, which decreases abomasal contractility and may contribute to the incidence of this condition. Efficient absorption of the VFAs produced in the rumen requires adequate papillae development; this development is stimulated by concentrate feeding. Dramatic increases in concentrate feeding during early lactation without adequate rumen preparation during the transition period may increase ruminal outflow of volatile fatty acids. The incidence is also increased following sudden ration changes.

Many assume that anything that allows space in the body cavity for the abomasum to migrate increases the probability of clinical displacement. During late pregnancy, the omental attachments to the abomasum become stretched. The degree of stretching may be related to the risk of later displacement. The genetics of the animal also affects predisposition to this disorder; the heritability has been estimated at 24 percent.

Successful nonsurgical treatment of the displacement depends on removal of the gas from the abomasum. Strategies include rolling the cow over or taking her on a rough trailer ride. If successful, the affected cow accepts feed immediately; however, these approaches may alleviate the problem only temporarily. The surgical approach is to pull the abomasum into place and attach it surgically to the body wall. This technique provides a more certain and more permanent solution.

The resting position of the cow may affect the risk for displacement. Encouraging cows to lie on the left side has been hypothesized to reduce the risk of left displacement. In sloped barns, cows generally lay with their feet downhill, and high-risk cows lying in stalls that encourage them to lie on their right side may be at a greater risk for displacement. This hypothesis has not been tested; however, treatment protocols using this concept have been successful. Successful rolling followed by tethering of the right hind leg to force cows to lie on their left side allows a recovery rate of 87 percent compared to 92 percent for surgical intervention.

Prevention of this condition is accomplished most effectively by preventing other metabolic disorders. Cows should be in proper body condition at parturition (BCS of 3.25 to 3.75) and fed to encourage maximal intake of a well-balanced ration during the transition period. Prevent hypocalcemia prior to calving and utilize anionic salts in prepartum rations when appropriate.

MILK FEVER (PARTURIENT PARESIS, CLINICAL HYPOCALCEMIA)

Milk fever is caused by rapid drain of calcium from the blood to the milk. Calcium mobilization from bone cannot keep pace with the sudden demand, and hypocalcemia (low blood calcium) results. Blood changes associated with milk fever include a depressed blood calcium level, hypophosphatemia, hypermagnesemia, and sometimes hyperglycemia.

Milk fever is usually associated with parturition and the initiation of lactation. It most often affects high-producing cows during the first twenty-four hours following parturition (75 percent of cases), although it can occur at any time in lactation, even prior to calving. Milk fever seldom occurs in heifers and is most often observed between the third and seventh lactations. The first occurrence increases the probability of subsequent occurrences. It is a problem in all breeds; however, Jerseys are most susceptible.

Most of the symptoms of milk fever are related to the lack of blood calcium. Calcium is essential for muscle function, and the loss of this function (among other things) is the most apparent symptom in these animals. Three classic stages occur in the etiology of milk fever: initially, the cow staggers and is wobbly; she soon becomes sternally recumbent (down on her chest) and drowsy, with her head tucked back into her flank (Figure 48.3), and in the final stage, she becomes laterally recumbent (down on her side) and comatose. The severity of hypocalcemia associated with these stages varies dramatically from cow to cow, with some cows exhibiting no symptoms despite severe hypocalcemia (as low as 4 mg/dl), while others become laterally recumbent

Figure 48.3 A cow with milk fever typically tucks its head back against its flank. *(Courtesy of University of Illinois)*

with only moderate hypocalcemia (approximately 6.5 mg/dl). Despite the name, no fever is associated with milk fever. In fact, the ears and teats become hypothermic (cold), and the muzzle becomes dry in the early stages as the hypocalcemia induces a peripheral vasoconstriction that reduces blood flow to the extremities. There is a loss of appetite as feed is retained in the rumen. (Calcium is essential for gut motility.) Pulse rate and respiration increase, and the cow may exhibit labored breathing as the muscular diaphragm loses function.

Milk fever is 100 percent fatal if left untreated. Even after successful treatment, secondary problems may occur, including degeneration and necrosis of muscle and nerve paralysis if the cow is left down for too long. Cows often injure themselves and may experience a split pelvis if they fall on a slippery surface while staggering. The risk of ketosis is increased following milk fever if the depression in feed intake is long term. The incidence of mastitis is also greatly increased following milk fever; the muscles controlling the teat sphincter are less competent during the clinical phase of milk fever, thus allowing easier pathogen entry to the gland. In addition, hypocalcemia reduces the ability of neutrophils to destroy pathogens, allowing them to proliferate once they penetrate into the gland. Milk fever may occur prior to or during calving in rare instances; the loss of uterine contractions in these cases can prove fatal to the unborn calf. The risk of uterine prolapse is also greater in severely hypocalcemic cows.

Treatment is by infusion of calcium intravenously. Calcium borogluconate is typically the treatment of choice; 250 to 500 milliliters is given very slowly. Rapid delivery causes tachycardia and eventual heart seizure and death. Usually response to treatment is immediate and complete, although up to 30 percent of animals need retreatment eight to

Figure 48.4 Veterinary medical officer Jesse Goff administers calcium propionate. *(Courtesy of USDA-ARS)*

ten hours later. Oral calcium in the form of a calcium propionate gel is often administered at the same time as the intravenous treatment to prevent relapses. In some herds, oral calcium gels are provided routinely at calving to improve calcium status and reduce the risk of milk fever (Figure 48.4).

Prevention of milk fever is based on strategies to increase calcium absorption from the gut and increase the rate of resorption from bone. Incomplete milking may be effective in reducing the demand for calcium and the risk of milk fever; however, it also increases the risk for mastitis during a period of high susceptibility. Vitamin D, fed at the rate of 20 million units per day for five to seven days prepartum, increases the rate of absorption of calcium from the small intestine and increases resorption from bone. This treatment protocol needs several days to be effective and can be dangerous if continued too long. Therefore, it is difficult and impractical to implement.

Feeding Strategies to Prevent Milk Fever

Over the last twenty years, feeding recommendations for herds with a high incidence of milk fever have focused on limiting calcium intake during the dry period. More recent research suggests that limiting potassium intake in forages fed to dry cows is more effective. Both of these strategies are effective because they alter the cation-anion difference (CAD) in the ration. By increasing the concentration of anions (negatively charged ions) relative to the concentration of cations (positively charged ions), the diet becomes acidogenic, inducing decreases in blood pH in those cows fed such diets. Cows that are slightly acidotic are much less susceptible to milk fever. Cows are typically maintained on acidogenic diets for approximately three weeks prior to the anticipated date of calving for the most effective control of milk fever.

The primary cations present in most dry-cow rations are potassium, sodium, calcium, and magnesium, and the primary anions present are chloride, sulfur, and phosphorus. Absorption rates for sodium and potassium are much higher than for calcium and magnesium, and high concentrations of these cations in dry-cow rations are more problematic for producers attempting to maintain a CAD of approximately −30 mEq/pound of ration dry matter. Many forages contain over 3 percent potassium, making it difficult to utilize these forages in dry-cow rations balanced for herds in which milk fever is a concern. Corn silage contains about half this level of potassium, but the high-energy density of corn silage limits its inclusion rate in most dry-cow rations. Second-cutting hay crops tend to have less potassium than is present in the first cutting, and full-bloom alfalfa is lower in potassium than that harvested in early bud or prebud stages. With careful forage selection, it is possible to balance a dry-cow ration with a low CAD.

For dry-cow rations where the CAD is above +90 mEq per pound of ration dry matter, magnesium sulfate and calcium sulfate should be supplemented to bring ration magnesium and sulfur to 0.4 percent and 0.45 percent of ration dry matter, respectively. Dicalcium phosphate supplementation should be at a level that maintains daily phosphorus intake at 45 grams per day. Finally, calcium chloride is supplemented to create a CAD at approximately −30 mEq per day.

Monitoring the urine pH of close-up cows can be used to monitor diet effectiveness. Urine pH should be reduced to around 6.0 for the most effective control of milk fever. Additional supplementation of anionic salts continues to reduce urine pH, but it also results in a decrease in dry-matter intake because these salts are unpalatable. Excessive decreases in dry-matter intake prior to parturition are associated with decreased feed intake, increased risk of other metabolic disorders, and decreased milk production postpartum. Careful management of rations for close-up cows is therefore critical for both minimizing metabolic disorders in periparturient cows and for maximizing milk production and profitability of lactating cows.

Current research in this area focuses on developing more palatable alternatives to calcium chloride for dry cows. Hydrochloric acid is a potential acidifying agent for dry-cow rations that is more palatable than calcium chloride; however, handling and mixing problems currently prevent widespread acceptance of this alternative. The prudent use of these preventive feeding strategies, along with appropriate and timely treatment strategies, can minimize both the incidence of milk fever and the adverse effects of the cases that do occur.

DYSTOCIA

The carefully synchronized physiologic changes occurring in both the maternal and fetal systems during the final weeks prepartum are critical to ensure that labor and delivery proceeds with minimal stress for both the cow and her calf. Factors that disrupt either the fetal or maternal system may result in dystocia. Current strategies for minimizing dystocia have emphasized the use of calving-ease sires for high-risk females, especially nulliparous heifers. Despite the increased availability of calving-ease data for sire selection, there has been a decrease in the percentage of unassisted births and an increase in the percentage of births requiring considerable force or experiencing extreme difficulty over the last five years.

The impact of dystocia on cow performance has been well-documented. For the dam, increased calving difficulty is associated with decreased lactational performance (decreased yields of milk, fat, and protein), decreased reproductive performance (increased services per conception and days open), increased numbers of cow deaths, and increased risk of other periparturient metabolic disorders. Even slight assistance at calving results in significant effects on these parameters; however, the economic losses are greatest with extremely difficult deliveries.

Economic losses are incurred from several sources. Milk production decreases as calving-ease score increases. Losses exceed 1,700 pounds of milk per 305-day milk production with a calving-ease score of 5, although significant losses are apparent with even slight assistance (calving-ease score of 2). Changes in milk components are similar to those seen for milk volume. Reproductive performance of the dam following a dystocic delivery is also impaired. Services per conception increase with increasing calving-ease score, as does days open. Calving-ease scores of 5 are associated with increases in days open of over thirty days. As with lactational performance, significant decreases in reproductive performance are observed even at calving-ease scores of 2 (slight assistance). These estimates on cow performance may underestimate true losses because the effect of the increased culling rate of dystocic cows is not included. Culling rates for dystocic cows of up to 30 percent have been reported. Increased cow deaths are also seen with increasing calving-ease scores, with over 4 percent more cows dying after a calving-ease score of 5 than after unassisted calvings.

The incidence of metabolic disorders and other periparturient disorders is also increased in dystocic cows. Dystocia is a significant risk factor for both mastitis and lameness diagnosed before first service. Dystocia is associated with a twofold increase in the risk of milk fever, twofold to threefold increases in retained placenta and metritis, a threefold increase in cystic ovaries, and a greater than twofold increase in the risk for left-displaced abomasum. Again, the striking finding was that the increase in several of these metabolic disorders was present even following slight assistance. In most cases, the difference between an unassisted calving and a calving where slight assistance is rendered has more to do with the patience of the calving attendant than the needs of the cow. The impaired performance and increased incidence in metabolic disorders of cows following only slight assistance at calving may indicate that assistance itself induces the observed effects.

Events occurring as early as the first few days following conception may influence the parturition process. Several assisted reproductive technologies, including in vitro fertilization and nuclear transfer cloning, increase the incidence of dystocia in subsequent pregnancies. In many cases, the increased incidence of dystocia is associated with extended gestation lengths and abnormal fetoplacental development.

The sex of the fetus also affects the incidence of dystocia. Male calves have a higher incidence of dystocia than do female calves. The incidence of dystocia increases 1 percent for every pound that birthweight is increased. However, bull calves have more trouble calving even when birthweights are equivalent. Excessive birthweight is often suggested to be the primary cause of dystocia, although very small calves also have an increased incidence of dystocia relative to that seen in more moderately-sized calves (66 to 99 pounds). In addition, inbreeding decreases birthweight while increasing the incidence of dystocia.

Factors associated with the dam include body weight (extremes in body condition are associated with increased calving difficulty), parity (first-calf heifers are much more likely to have a prolonged or difficult parturition, while dystocia in multiparous cows is more often associated with malpresentation of the fetus), and age at first calving (young heifers have more difficulty). Although overfed, obese cattle have greater difficulty calving, lead feeding in dairy cattle is associated with a decreased incidence of dystocia. By contrast, long-term energy restriction decreases calf birthweight but has no effect on the rate of dystocia.

Maternal influences on dystocia are also apparent. Calving-ease scores of first and second parity cows are highly correlated with calving-ease scores of the same cows for their next calving, independent of sire effects. Although maternal effects are evident, breeding strategies for improving calving ease have focused on sire selection. Sire selection for calving ease tends to reduce the birthweight of the resulting calves, and many people in the industry have voiced concerns that these small calves will become small cows at increased risk for dystocia. These concerns may have limited the use of calving-ease sires in problem herds. While small calves become small cows and large calves become large cows, calves born from difficult deliveries tend to have a higher incidence of dystocia as cows than their unassisted counterparts. Daughter-dam heritability as a trait of the dam was estimated at 24 percent, suggesting that mature body size is less important than other maternal traits. Not surprisingly, calves born from sires with low calving-ease scores have less calving difficulty as cows than those born from sires with high scores. Despite the increasing availability of sire evaluations for calving ease, trends for both the incidence of dystocia and for stillbirth rates in calves suggest that losses to the industry are increasing rather than decreasing.

TWINNING

Twinning is similar to dystocia in terms of consequences, but it warrants a special discussion. The twinning rate has increased over the last ten to fifteen years, much like the dystocia rate and the stillbirth rate. The current rate of twinning is approximately 3 percent of all births. Suspected risk factors for twinning include milk production, ovulation rate, parity (increasing twinning with increasing parity), and genetics.

Ovulation rate is partially under genetic control and therefore can be increased or decreased by selection. In addition, the previous incidence of twinning is a risk factor for subsequent twinning. The incidence of double ovulation in dairy cattle is nearly 15 percent, much higher than in beef cattle. The incidence of double ovulations increases with parity, especially from the third pregnancy and after.

Twinning, much like dystocia, affects the cow in a negative manner. It is associated with an increased risk of most metabolic disorders, including dystocia, retained placenta, metritis, ketosis, and displaced abomasum. Cows delivering twins have impaired reproductive performance, with more services per conception prior to successful rebreeding and increased days open. Over 25 percent of cows carrying twins fail to carry them to term, compared to a 12 percent abortion rate in singleton pregnancies.

Milk production in the lactation after delivery of twins is decreased, although it is not different for the lactation during which twins are carried. However, peak milk is higher for the lactation during which twins are conceived, suggesting that higher-producing cows within a herd are at increased risk for twinning. In one study, cows averaging approximately 140 pounds of milk per day had a twinning rate nearly three times higher than cows milking an average of about 80 pounds per day.

Cows carrying twins can be managed differently during the transition period to minimize the adverse effects commonly observed. The plane of nutrition throughout the dry period should reflect the greater nutrient demand of two fetuses. Because the gestation length of twins is seven to ten days shorter than that for singletons, the dry period and the transition period should be adjusted accordingly. Appropriate assistance at calving minimizes some of the complications normally associated with these deliveries.

RETAINED PLACENTA

Retained placenta is defined as the failure of the fetal membranes to be expelled from the uterus. It is primarily due to failure of fetal villi in cotyledon to detach from maternal caruncle. Expulsion usually takes two to six hours in dairy cattle. Fetal membranes that are not expelled within twelve hours are arbitrarily defined as retained placentas. Cows delivering male calves retain the placenta for longer periods of time. Dystocia increases the incidence of retained placenta by two- to threefold. Although the physiological mechanism leading to this increase is unknown, similar prepartum hormonal defects have been observed in cows retaining fetal membranes as in dystocic cows. Cows with retained placentas typically have lower levels of progesterone, estrogen, and possibly prolactin. It is more common following abnormally short or abnormally long pregnancies, among older cows, and following twinning. A high incidence of retained afterbirth occurs when premature calving is induced by the administration of glucocorticoid drugs.

While infections such as brucellosis, vibriosis, and others have been associated with abortion and retained afterbirth, these are by no means the only causes. Its incidence appears to be related to the fat cow syndrome and increases dramatically with hypocalcemia. Hypocalcemia results in uterine inertia, and the expulsion of the placenta depends on continued uterine contractions following delivery of the calf. In these cases, treatment with intravenous calcium is warranted even if no clinical signs of milk fever are apparent. Deficiencies of vitamin A, selenium, copper, and iodine have also been implicated. The prepartum injection of selenium at low doses has been shown to reduce the incidence of retained placenta in some herds. Among cows that have previously retained the placenta, 20 percent are likely to do so again.

The estimated frequency of occurrence of retained placenta is 10 percent, but it can exceed

75 percent in some herds. Increases in occurrence rates have been linked to several pathological, physiological, environmental, and nutritional factors. Increases in the incidence of retained placenta are correlated with an increased risk of metritis, delayed uterine involution, and decreased reproductive efficiency. Retained placenta also typically decreases milk production over 550 pounds in affected cows.

Retained placentas are typically treated with intrauterine infusion of antibiotics or sulfonamides, although this treatment does little for enhancing the expulsion of the placenta. Such treatment does, however, significantly reduce the risk for metritis. Manual removal is not recommended and often aggravates the problem by leaving portions of the placenta in the uterus. This dead tissue then encourages pathogen growth in the uterus, again increasing the risk for metritis and subsequent reproductive difficulties. Infusion of warm water into the uterus triggers a release of prostaglandins and relaxin and may stimulate placental release.

Prevention of retained placenta is primarily achieved through nutritional means. Supplemental selenium may decrease incidence in areas with low concentrations of selenium in soils or in areas with acidic soils. When cows are fed ensiled feeds, both selenium and vitamin E are supplemented for best control. Other nutritional factors associated with an increased risk of retained placenta include vitamin A deficiency and fiber deficiency. Overconditioned cows have an increased incidence of retained placenta, and any factors that increase the incidence of an assisted delivery also increase the risk of placental retention.

UDDER EDEMA

Prior to parturition, hormonal changes associated with the onset of lactation initiate increased blood flow to the udder. In some cases, the lymph system is unable to accommodate the increased volume of fluids. Subsequent fluid accumulation leads to a swelling of the udder called udder edema (Figure 48.5). Udder edema, characterized by excessive accumulation of fluid in the intercellular spaces of the udder and forward of it, is sometimes of serious magnitude before calving. The cause is not well understood, but the reduction of blood proteins at calving time and increased blood flow without compensatory removal by the lymphatic system have been suggested as potential mitigating factors. The swelling associated with

Figure 48.5   Udder edema is a common precalving complication, especially in first-calf heifers. *(Courtesy of Mark Kirkpatrick)*

Figure 48.6   Udder edema can weaken the support structure of the mammary gland, which can lead to a pendulous udder. *(Courtesy of Howard Tyler)*

edema constricts many lymph vessels, leading to increased severity of this edema until the cow is milked and the pressure is reduced. It appears that high intakes of sodium chloride or potassium chloride increase the severity of udder edema and that restriction of salt intake reduces the severity in some cases. Severe edema may reduce milk production and may be one of the causes of pendulous udder (Figure 48.6). This condition is particularly prevalent in first-calf heifers and high-producing cows because of the extreme changes taking place in the body in preparation for the high demands of lactation.

SUMMARY

- The transition period includes the final few weeks prior to parturition and the first few weeks following parturition.
- Body condition in dairy cattle is an effective way of monitoring the energy reserve status of the cow.
- Overconditioned cows tend to have lower milk yields and a higher risk of difficulties in rebreeding.
- Fatty liver syndrome is a condition in which the liver fills with fat and cannot function optimally, thus impairing the ability of the liver to oxidize fats for energy. The condition occurs within a few days of calving. The cow goes down but remains alert.
- Ketosis is characterized by increased concentrations of free fatty acids and ketones in the blood and urine and decreased concentrations of glucose. It described as either primary or secondary.
- The abomasum of the dairy cow normally lies on the right ventral abdominal wall.
- Displacements of the abomasum may be either left- or right-sided, although 85 to 90 percent are left-sided.
- Hypocalcemia decreases gut motility in general and may allow an increased production of gas in the abomasum.
- Increased calving difficulty is associated with decreased lactation and reproductive performance, increased numbers of cow deaths, and increased risk of other periparturient metabolic disorders.
- Milk production in the lactation after delivery of twins is decreased.
- Retained placenta is primarily due to the failure of fetal villi in cotyledon to detach from maternal caruncle.
- Prevention of retained placenta is achieved primarily through nutritional means.

QUESTIONS

1. What does a body condition score of 1 indicate? What does a score of 5 indicate?
2. Fatty liver is correlated with what other disorders?
3. How can a dairy producer prevent fatty liver syndrome?
4. How are ketones formed?
5. What are the differences between primary and secondary ketosis?
6. What are some symptoms of ketosis?
7. What are some treatments for ketosis?
8. What are the symptoms of a displaced abomasum?
9. What is a nonsurgical, cost-efficient method of removing gas from the abomasum?
10. What are the causes of milk fever?
11. Which breed is most susceptible to milk fever?
12. What are the three classic stages in the etiology of milk fever?
13. How is milk fever treated?
14. What are some feeding strategies for preventing milk fever?
15. For the most effective control of milk fever, to what should the urinary pH be reduced?
16. Define *dystocia.*
17. What are some factors associated with an increased risk of dystocia?
18. Which breed has the highest incidence of dystocia?
19. Twinning is associated with what other metabolic disorders?
20. How much shorter is the gestation length of cows carrying twins?
21. Define *retained placenta.*
22. How long does it normally take a cow to expel the placenta?
23. What is the treatment for a retained placenta?

ADDITIONAL RESOURCES

Articles

Butler, W. R., R. W. Everett, and C. E. Coppock. "The Relationship Between Energy Balance, Milk Production and Ovulation in Postpartum Holstein Cows." *Journal of Animal Science 53* (September 1981): 742.

Overton, T. R. "Nutritional Management of Transition Dairy Cows: Strategies to Optimize Metabolic Health." *Journal of Dairy Science 87* (January 2004): E105–E119.

Internet

Feeding and Managing the Transition Dairy Cow: www.ext.nodak.edu/extpubs/ansci/dairy/as1203.pdf.

Feeding the Transition Dairy Cow: http://animalscience.tamu.edu/ansc/publications/dairypubs/L5197_feedingtransition.pdf.

Nutritional Management of Transition Dairy Cows: Strategies to Optimize Metabolic Health: http://www.adsa.org/jds/papers/2004/105.pdf.

49

Managing Lactating Cows

OBJECTIVES

- To describe the difference between a cow in negative energy balance and one in positive energy balance.
- To understand synchrony of nutrient availability.
- To outline the steps in the conversion of dietary components to milk components.
- To understand photoperiod effects and milking frequency effects on milk production.
- To become familiar with the use of bovine somatotropin.
- To describe the calving interval and its effects on profitability.

Figure 49.1 This well bedded, open pen for postcalving cows allows close observation of fresh cows. *(Courtesy of Howard Tyler)*

Maximizing profit in lactating cows is focused on optimizing milk production. Maximizing milk production throughout lactation does not necessarily maximize profit, especially during late lactation. During early lactation, however, the focus is clearly on maximizing milk production. Functionally, it is better to manage cows by their energy balance rather than their stage of lactation. As previously discussed in Topic 48, the period of negative energy balance (NEB) begins in the transition cow period.

COW ISSUES AFTER CALVING

Special efforts may be necessary to encourage the cow to rise and to move to the feed source, especially on the day of calving and the first two days following calving (Figure 49.1). After calving, the cow should be closely monitored for health problems (Figure 49.2). Rectal temperature and urinary ketone levels can be monitored daily, in addition to monitoring feed intake and milk production. Cows with health problems can be rapidly identified and

Figure 49.2 Marking crayons can be used to differentiate problem and normal cows in fresh cow pens. *(Courtesy of Mark Kirkpatrick)*

treated in this manner, and healthy cows should be returned to the general population at three to fourteen days postpartum, depending on their health status. The other area to monitor in the postpartum dairy cow is the return to normal reproductive function. Within three to four weeks, the resumption of the ovarian cycle should be noted. Because the first ovulation is typically not associated with behavioral signs of estrus, other methods must be used to assess ovarian function. The corpus luteum (CL) of pregnancy always regresses prior to calving, and the palpation of a CL in the postpartum cow is a sure indicator that ovarian function has resumed. During this palpation exam, the normalcy of uterine involution can also be assessed. Although further tissue changes must occur before the uterus is ready for another pregnancy, most of the palpable changes in involution are nearly complete by three weeks after calving. Palpable abnormalities at this time indicate a need for a professional veterinary evaluation and potential intervention. Early intervention in problem animals is the key to reducing delays in breeding.

MANAGING LACTATING COWS IN A NEGATIVE ENERGY BALANCE

Rations should be designed to maximize dry-matter intake (Figure 49.3) and to maximize nutrient intake during the NEB period. The best strategy for evaluating the effectiveness of these rations does not rely on milk production as the primary barometer for success. Because milk production during this period is determined by a combination of nutrient intake from the ration and mobilization of body reserves, monitoring milk production alone can mislead producers into a false sense of security regarding the quality of their NEB program.

Figure 49.3 To maximize dry-matter intake, feed should be accessible at least twenty hours per day for early lactation cows. *(Courtesy of Westfalia)*

Likewise, feed intake alone is not an accurate barometer for profitability. High dry-matter intake is not always linked to high milk production. Feed efficiency does vary between cows, and some cows may consume large amounts of feed without returning large quantities of milk. The correlation between high feed intake and high feed costs, however, is perfect. Feed efficiency for milk production is defined as the number of pounds of 3.5 percent fat-corrected milk produced per pound of dry matter consumed, and this number typically ranges from 1.3 to 1.7. This range is not insignificant. A cow producing 30,000 pounds of milk at a feed efficiency of 1.3 consumes roughly 3 tons more feed on an as-fed basis through that lactation than the same cow with a feed efficiency of 1.7.

Several factors should be included in a thorough evaluation of an NEB program, including milk production response, rate of change in body condition score (BCS), and incidence rate of metabolic disorders such as fatty liver, ketosis, and displaced abomasum. Obviously, the success of this program depends on the quality of the transition-cow management program, the dry-cow program that preceded it, and even the quality of management of the cows during the previous lactation. Poor early lactation performance cannot be viewed as a reflection of management or inappropriate rations during this period; more often than not, it reflects poor management in one or more of the previous stages.

The high-producing early lactation cow cannot meet the energy expenditures of milk production through feed intake alone; a negative energy balance is not only inevitable, but also desirable. One pound of mobilized body weight can support synthesis of 7 pounds of milk; without mobilization of body reserves, cows cannot hope to reach their genetic potential for milk production. The rate at which this condition is lost is what separates a successful NEB management program from an unsuccessful one. The scoring system most commonly used for dairy cattle uses a scale of 1 to 5, with a body condition score of 1 indicative of an extremely emaciated cow and a score of 5 reserved for obese cattle. Dairy cattle should lose no more than one point in their BCS during the early lactation period, regardless of the initial BCS at parturition. Ideally, cows should calve at a BCS of about 3.5 to 3.75; they should reenter a positive energy balance with a BCS of no less than 2.5. Cows with a higher BCS at calving should have a higher BCS at the end of the early lactation period. Rates of condition losses that exceed this recommendation are associated with higher risks of fatty liver, ketosis, and displaced

TABLE 49.1   Estimated Daily Dry-Matter Intake (DMI) for Milk Cows*

	Body Weight, lb			
	900	**1,100**	**1,300**	**1,500**
3.75% Fat Milk	*DMI, lb/day†*			
30	26	29	33	37
50	31	35	39	42
60	34	38	42	45
70	37	41	45	48
80	40	44	48	51
90	43	47	50	54
100		50	53	57
100+		52	56	60

*DMI (lb/day) = .0185 × body weight + .305 × 4% fat milk (lb/day)

4% fat milk (lb/day) = .4 × (lb/day) + 15 × fat (lb/day).

†Decrease DMI .2 percent per day for cows with less than ninety days in milk.

Source: Winfield J. Johnson, Using Quality Forages in the Dairy Cow Feeding Process, Northrup King Co., University of Minnesota Information-Chart C.

Figure 49.5 Total mixed rations improve feed intake by stabilizing the rumen environment. *(Courtesy of Iowa State University)*

Figure 49.4 Early lactation cows should never face an empty feedbunk. *(Courtesy of Monsanto)*

abomasum. Cows should have their BCSs evaluated every other week throughout lactation to allow an effective evaluation of rations.

The key factor controlling the rate of condition loss, other than milk production, is dry-matter intake of a nutrient-dense ration (Table 49.1 and Figure 49.4). The primary goal of the producer during this period should be maximizing the intake of the cows. Regulation of dry-matter intake in lactat-

ing cows is a complicated phenomenon, with many factors working apparently independently to either encourage or discourage intake. These factors include physical rumen fill and some poorly understood metabolic factors, including glucose, insulin, and the rate of oxygen consumption. Although the factors controlling dry-matter intake in lactating cows are not well understood, the practical management strategies that encourage cows to eat to their capacity are well-documented.

Strategies for increasing intake are focused on several goals: optimizing rumen fermentation patterns (Figure 49.5), stimulating appetite, and exploiting behavioral patterns of feed consumption in cows (Figure 49.6). Avoiding sudden ration changes and feeding more frequently tend to stabilize rumen pH, increase propionate production (which is converted to glucose in the liver and ultimately to lactose in the mammary gland), and increase digestibility of feeds by maximizing microbial numbers and activity. Feeding frequency also exploits cow behavior; cows tend to rise and eat when sounds or smells associated with feeding are present (turning on an auger, driving by with a mixer wagon). Keeping fresh feed in front of cows is critical (Figure 49.7). Early lactation cows consume about 10 percent of their daily intake per feeding bout; they therefore require at least ten eat-

Figure 49.6 Feed that exceeds the reach of these cows is moved closer, thus encouraging feeding behaviors and boosting feed intake. *(Courtesy of Monsanto)*

Figure 49.8 Cows spend up to five hours daily ingesting feed. *(Courtesy of Iowa State University)*

Figure 49.7 A portable mixing wagon delivers feed at a fenceline feedbunk. *(Courtesy of South Dakota State University)*

Figure 49.9 Cows eagerly await feed delivery time in this meal-fed herd. *(Courtesy of Iowa State University)*

ing periods (roughly thirty minutes per feeding period) for maximal intake. The difference between high-intake cows and low-intake cows is the amount of feed consumed per meal rather than the number of meals. High-producing cows consume one-third more in the same period of time than their low-intake herdmates. Although total eating time is not much more than five hours, feed needs to be present in front of the cow for longer than this amount of time (Figure 49.8). Meals are naturally spaced to allow adequate time for ruminal fermentation of the previous meal, cud-chewing activity, and finally for bulk to be eliminated from the rumen. Limiting access to feed reduces production, feed intake, and

profit. Fresh feed should be available to NEB cows almost twenty-four hours per day (Figure 49.9).

Therefore, factors that affect free access to feed reduce dry-matter intake. Because the timing of feeding bouts is so crucial, even relatively small amounts of time without access to feed can reduce total daily intake. Managers must develop strategies that reduce time spent without access to feed, including holding-pen time. Other factors that influence intake are empty bunks, poorly designed (Figures 49.10 through 49.12) or maintained feedbunks, or uneven distribution of feed at the bunk. Feed intake can be improved by increasing feeding frequency in many cases; even pushing up old feed can stimulate intake

Figure 49.10   Feedbunk barrier design affects the distance over which cows can access feed. *(Courtesy of Mark Kirkpatrick)*

Figure 49.13   Feed is pushed up to the feed barrier to make it more accessible for cows when they return from the milking parlor. *(Courtesy of Monsanto)*

Figure 49.11   These feedbunks were designed to minimize the barriers that might adversely affect feed-intake behaviors. *(Courtesy of Iowa State University)*

Figure 49.14   Cows waste more feed when it is fed in an elevated bunk than when it is fed at ground level. *(Courtesy of Iowa State University)*

Figure 49.12   The barrier forming the back of this feed manger eliminates the need to push feed back within reach of the cows. *(Courtesy of Iowa State University)*

(Figure 49.13). Elevated bunks increase feed wastage (Figure 49.14); in addition, saliva production is lowered in cows that eat from elevated bunks, thus increasing the risk of rumen acidosis.

The number of times fresh feed needs to be provided depends on the bunk life of the ration and labor availability. Feeds with a longer bunk life can be fed less frequently and simply pushed up more often (Figure 49.15). However, rations with a short bunk life require more frequent feedings to optimize palatability (Figure 49.16). This decision is not simple; the bunk life of the same ration may vary with seasonal weather changes. At minimum, feed should be pushed up three to five times daily, and fresh feed should be mixed and provided to cows two to three times per day. The amount mixed and

Figure 49.15   In regions of the country where the bunk life of feed is extended, once daily feeding can be implemented. *(Courtesy of Monsanto)*

Figure 49.17   Poor water quality can have a substantial impact on feed intake and lactation performance. *(Courtesy of Monsanto)*

Figure 49.16   Overfilling feedbunks leads to excessive feed losses. *(Courtesy of Mark Kirkpatrick)*

Figure 49.18   Clean water tanks providing fresh clear water encourage both water and feed intake. *(Courtesy of Monsanto)*

fed should be calculated based on feed refusals; 5 to 7 percent feed refusal is optimal for maximal intake.

Palatability of ration ingredients becomes of paramount importance during this period. The most critical component of the ration in this respect is the forage source, but other ingredients of the concentrate supplement can also have dramatic effects. Intake of total mixed rations tends to be higher than when the ingredients are fed separately. Feeding with a computer feeder allows similar intakes for mixed rations and ingredients fed separately, suggesting that feeding frequency and rumen stability are the critical factors for success.

Lighting can affect dry-matter intake; 70 percent of feed intake occurs during daylight hours. A 16:8 light:dark pattern appears to be optimal but is rarely achieved in dairy facilities.

Fresh, clean water should be readily available. Large-breed cows consume approximately 4 pounds of water for every pound of milk produced, while small breeds may consume close to 3 pounds of water per pound of milk. In hot, dry environments, this amount can increase dramatically. Limited water availability or water quality can dramatically decrease dry-matter intake (Figures 49.17 and 49.18). Producers need to account for both the amount of water needed by their early lactation cows and the timing of water consumption. High-producing cows drink about fifteen times per day and consume 1 to 1.5 gallons per drinking bout. They prefer warm water to cool water, and cold water reduces consumption significantly.

Cows also prefer to drink water immediately after being milked. Limited water availability at this time, as indicated by numerous cows fighting for access to the same water trough, discourages water consumption by all but the most dominant animals. In competitive situations, the dominant cow may even control access to the water source. If water tanks allow access by only a single cow, then multiple tanks need to be made available, no matter how acceptable the flow rate of the tank. At least one water source (or two linear feet of water trough space) should be available for every ten cows in the NEB group, although this number can be greatly reduced for all other lactation groups and age groups on the farm.

Optimizing Rumen Function

Optimizing rumen digestion starts with maintaining a rumen environment that maximizes microbe function. Several factors are needed for microbes to function: a warm, wet environment with a stable pH and a constant source of substrate to digest. Each rumen micobe maintains a symbiotic relationship not only with the host cow, but also with other types of bacteria and protozoa present in the rumen. The perfect environment maximizes function of the whole population rather than specific groups of rumen bacteria.

Thus, rumen pH in the NEB period should be maintained between 6.0 and 6.2 (rumenocentesis provides values about 0.2 points lower). This number is lower than that recommended at other stages of lactation, but NEB cows must be kept at the edge of acidosis if nutrient intake is to be maximized. This condition allows little room for error; acidosis is a common nutritional disorder in NEB lactation cows. If transition cows have been managed properly, the rumen should be functionally prepared for this type of high-energy ration.

Acidosis

If cows have not been exposed to a higher-energy ration through the transition period, the rumen is at a disadvantage in several ways. Papillae are shorter; the absorptive capacity for volatile fatty acids is greatly reduced; and microbial populations, especially starch digesters and lactate-utilizing bacteria, are poorly balanced. Such cows rapidly become acidotic and go off feed when exposed to a high-energy ration.

Lactate-producing bacteria, such as *Streptococcus bovis,* increase numbers more rapidly in response to a high-energy ration than lactate-utilizing bacteria. Following a rapid change to a high-energy ration, lactate concentrations in the rumen can increase dramatically. Lactate production decreases rumen pH more rapidly than volatile fatty acids; it is a stronger acid and is absorbed less efficiently. The short papillae and decreased absorptive capacity of the rumen contributes to all the rumen acids building up to higher concentrations. The lower the pH falls, the less efficiently these acids are absorbed. This acidosis often results in laminitis, which can provide a "visual cue" in herds with widespread acidosis problems. Keep in mind that a delay of one or two months occurs between the time of the acidosis problem and the appearance of hoof problems.

Acidic conditions also kill off many populations of rumen microbes. Decreased total numbers of microbes and altered populations dramatically affect forage digestibility; the optimal pH for forage-digesting (cellulolytic) bacteria is above 6.5. This decreased digestibility of forage slows rate of passage, increases rumen fill, and decreases appetite, especially for forage. If feeds are available separately, the animal preferentially consumes concentrates, further exacerbating the rumen acidosis. Total mixed rations are preferred during the NEB period for this reason.

Minimizing acidosis problems can be difficult when feeding wet silages, finely ground or pelleted concentrates, high-starch feeds, and/or too little long-stemmed hay. Added buffers can help maintain rumen pH above 6.2, as can proper feed sequencing (feeding forages prior to concentrate). Total mixed rations help maintain pH by minimizing selective eating, and computerized grain feeders can split concentrate into many feedings. In component-fed herds, forages (i.e., 5 to 10 pounds of hay) should be fed prior to concentrates (Figure 49.19) to initiate saliva flow prior to rapid carbohydrate fermentation.

Perhaps the most important consideration in maintaining a stable rumen environment is maintaining adequate salivation. The consumption of forage, especially long-stem hay (Figure 49.20), is critical for ensuring that adequate quantities of saliva reach the rumen. During rumination, the animal regurgitates and rechews a soft mass of coarse feed particles called a bolus. Each bolus is chewed for about a minute, then swallowed again. Ruminants may spend eight hours or more per day ruminating, depending on the nature of the diet. Coarse, fibrous diets result in more time ruminating.

Rechewing serves two important functions: reducing particle size of the forage and stimulating secretion of saliva, an important buffer for rumen

Figure 49.19   In some component-fed herds, silage is delivered directly to the feedbunk from the silo. *(Courtesy of Iowa State University)*

Figure 49.20   Portable round bale feeding rings allow flexibility in providing long-stem forage sources to cows. *(Courtesy of Iowa State University)*

pH. Thus, rumination has an important bearing on the amount of feed the animal can eat and utilize. Feed particle size must be reduced to allow passage of the material from the rumen. Because high-quality forages contain less fiber than low-quality forages, they require much less rechewing and pass out of the rumen at a faster rate; although they allow the cow to eat more, they also decrease saliva flow to the rumen and can increase the risk for rumen acidosis. The buffering capacity of saliva is critical; over 45 gallons of saliva can be secreted daily into the rumen. Supplemental buffers in the ration cannot fully replace the buffering capacity of saliva;

they also tend to be unpalatable when fed in high amounts. At least half and preferably two-thirds of all resting (lying down and not sleeping) cows should be chewing their cuds at any time.

Therefore, balance rations for NEB cows with careful attention to fiber levels. However, current systems for feed analysis have no direct measurement to indicate rumination time (and therefore saliva production). This balancing becomes part of the art of feeding NEB cows. The producer must monitor the cow's response to the diet to ensure that the diet maintains adequate rumination time, which depends on the amount, particle size, and maturity of the forage. However, producers can measure each of these characteristics independently.

The amount of forage is part of the ration-balancing procedure; it should be a known factor. Chemical analysis of the feed determines the fiber levels, which is indicative of maturity. Fiber contains a combination of cellulose, hemicellulose, and lignin. Acid detergent fiber (ADF) tests for cellulose and lignin primarily, while neutral detergent fiber (NDF) primarily isolates hemicellulose. The ADF value is a good predictor of the energy value of feed (higher ADF, lower energy value), while NDF (cell wall content) is inversely related to the quantity of forage consumed and increases with maturity. NDF can be a remarkably sensitive indicator of feed intake, and as little as a 1 percent increase in NDF decreases both milk yield and dry-matter intake (DMI). Ration ADF should be at 19 percent, while NDF should be at 28 percent for optimal rumen function in NEB cows. These numbers may vary slightly with the form of ration fed; total mixed rations permit NDF to drop to 27 percent, while rations with inadequate particle size may require 29 percent NDF to minimize the risk of acidosis.

The particle size of the ration is a crucial component to the equation. Penn State particle size separator boxes provide a useful estimator of the distribution of particle sizes in any ration (Figures 49.21 and 49.22). This tool consists of a series of three (or, less commonly, four) boxes that fit on top of one another. The top box has holes that are 0.75 inches in diameter, the middle box has holes that are 0.30 inches in diameter, and the bottom box has no holes and collects the fine particles. A known quantity of feed is placed in the top box and shaken vigorously; the proportions of feed that are contained in each box at the end define the particle size distribution of that feed. Ideally, 8 to 10 percent of a total mixed ration remains in the upper box, no more than 50 percent reaches the bottom box, and the remainder is found in the middle box. A similar tool is available to assess the particle size of grain mixes (Figures 49.23 and 49.24).

Figure 49.21 Forage shaker boxes allow for a rapid on-site estimation of particle-size distribution in forage mixes or total mixed rations. *(Courtesy of Leo Timms)*

Figure 49.23 Grain sieves allow the determination of particle-size distribution in a grain mix. *(Courtesy of Mark Kirkpatrick)*

Figure 49.22 A feed sample is placed in the top box and then shaken to allow smaller particles to drop to the lower boxes. *(Courtesy of Leo Timms)*

Figure 49.24 Each successive pan in a grain sieve retains smaller particles from the grain mix being tested. *(Courtesy of Mark Kirkpatrick)*

It is important to determine the particle size distribution of the feed actually consumed by the cow rather than the distribution of the feed presented to the cow. Occasionally, particle-size distribution should be determined at the time feed is first presented and at several other points prior to the next feeding. This approach allows the producer to determine both if and when sorting occurs. Sorting throughout the feeding period indicates that either feed was not thoroughly mixed or forage was not evenly chopped prior to feeding. Sorting that occurs primarily near the end of the feeding interval can be corrected by the addition of an extra feeding. Sorting of an excessively dry ration can be reduced by adding water to the ration prior to feeding; this practice has the added benefit of increasing intake.

Rations with insufficient particle size result in milk-fat depression as well. Milk-fat depression and protein:fat inversion (percentage protein exceeds percentage fat) are commonly seen in NEB cows. The de novo synthesis of fat in the mammary gland is reduced by the presence of trans fatty acids. These trans fatty acids are produced from polyunsaturated fats under acidic rumen conditions. Both milk-fat depression and protein:fat inversion can be prevented by preventing rumen acidosis or by eliminating vegetable sources of fats in the ration.

Producers seeking to maximize intake in NEB cows are therefore seeking to maximize rumen turnover rate (the speed at which feed passes through the rumen) and maximize microbial pro-

TABLE 49.2 Nutrition Guidelines for High-Producing Herds

Dry matter intake	4–5% of body weight
Neutral detergent fiber (NDF)	26–30% of DM
Forage NDF	20–22% of DM
Nonstructural carbohydrates	35–40% of DM
Fat	5–7% of DM
Crude protein	17–19% of DM
Degradable protein	60–65% of CP
Undegradable protein	35–40% of CP

Source: Adapted from Chase, L. E. 1993. Feeding Programs to Achieve 13,600 kg of Milk. In Advances in Dairy Technology 5:13–20.

TABLE 49.3 Ration Guidelines for Heat Stress

Nutrient	Level
Crude protein (% DM)	16–18
Undegradable (% of CP)	38–40
ADF (% DM)	19–21
NDF (% DM)	25–28
Effective NDF (% DM)	20–22
Fat (% DM)	
Unprotected	5–6
Unprotected and protected	7–8
NFC (% DM)	38–40
Sodium (% DM)	.4–.5
Potassium (% DM)	1.2–1.5
Magnesium (% DM)	.30–.35
Sodium bicarbonate (lb)	.3–.5
Niacin (g)	6
Aspergillus oryzea (g)	3

Source: Hutjens, Michael F., "Feed to Beat Hot Weather." Hoards Dairyman (July 1994): 497 .

TABLE 49.4 Lysine and Methionine Profiles of Protein Supplements and Microbial Protein Compared to Milk

Item	Lysine	Methionine	EAA
	% of total EEA		% of CP
Milk	16.4	5.1	38.4
Bacteria	15.9	5.2	33.1
Protein supplements			
Blood meal	17.5	2.5	49.4
Brewer's dried grains	6.7	4.5	46.3
Corn gluten meal	3.8	7.2	44.2
Corn DDG + solubles	5.9	5.9	37.7
DDG + solubles	6.5	3.7	43.3
Feather meal	3.9	2.1	31.4
Fish meal (menhaden)	16.9	6.5	44.8
Meat and bone meal, 45%	12.4	3.0	39.4
Meat and bone meal, 50%	14.2	3.7	36.6
Soybean meal (solvent)	13.8	3.1	47.6
Expeller soybean meal	13.0	2.9	49.6

Source: Adapted from Schwab (1994).

tein synthesis without unduly decreasing digestibility of feeds or inducing rumen acidosis. They should feed to encourage eating at least four to five times per day and avoid slug feeding (never feed over 5 pounds of concentrate [dry matter] at any one time). Nonstructural carbohydrates (NSC) should be in the range of 38 to 40 percent. This can be a difficult goal to meet when feeding rations with high levels of oil seeds or supplemental fats. The energy demands of the cow appear to be met on paper (Tables 49.2 and 49.3), but the rumen microbes are not receiving adequate fermentable carbohydrate. These fermentable carbohydrates are required for

maximal microbial growth and protein synthesis, although excess amounts at any single feeding can lead to acidosis.

Synchrony of Nutrient Availability

Adequate degradable protein (60 to 65 percent of total nitrogen or 12 percent of ration dry matter) and rumen soluble protein (30 percent of total nitrogen in component-fed rations, 45 percent in total mixed rations) are needed for optimal rumen ammonia (2 to 5 mg/dl) and rumen peptides to be used for the synthesis of microbial protein (Table 49.4). The goal of a properly balanced ration is to feed adequate amounts of cheap rumen degradable protein (RDP) and nonprotein nitrogen (NPN) to maximize microbial protein synthesis and subsequently maximize the amount of microbial proteins digested by the animal. The remaining needs of the animal must be met by higher quality (more expensive) rumen undegradable protein (RUP) sources.

The most difficult facet of designing a ration for a ruminant is to create one that maximizes microbial protein outflow from the rumen. The rumen microbes break down degradable proteins in the feed to ammonia and combine this ammonia with carbon backbones from carbohydrate fermentation to form amino acids and ultimately high-quality microbial proteins. The key is to ensure that the nitrogen source and the carbon source are available to the rumen microbes in the needed amounts at the same time. Soluble protein sources, such as soybean meal, are broken down rapidly to ammonia that is used primarily by cellulolytic bacteria. More insoluble protein sources, such as corn gluten meal, are broken down to peptides that are used more effectively by starch digesters. Rapid degradation of nonstructural carbohydrates (including pectins, starches, and sugars in concentrates) coupled with slowly degraded protein sources results in acidosis. Feeding NPN sources with slowly degradable carbohydrate sources (structural carbohydrates in forages) results in the excess ammonia being absorbed across the rumen wall and converted to urea by the liver and ultimately excreted rather than utilized.

Concentrations of urea in the blood are highly correlated with levels of milk urea nitrogen. These concentrations can be measured in the milk samples obtained by Dairy Herd Improvement Association personnel, and the values are becoming increasingly available to producers as a management tool used to monitor several aspects of herd nutrition. Milk urea nitrogen (MUN) averages around 14 mg/dl in most herds. Increasing crude protein in the ration or decreasing the nonfiber carbohydrate to crude protein ratio increases MUN. Excessively high MUN concentrations are predictive of reproductive problems and typically reflect rations that need excess protein removed or nonfiber carbohydrate increased. However, MUN also increases in any situation where cows are mobilizing significant amounts of body reserves. Under these conditions, muscle protein catabolism releases ammonia, the liver converts the free ammonia to urea, and urea concentrations increase in milk.

The importance of properly feeding rumen microbes cannot be overstated. Rumen microbes are over 50 percent protein, with a digestibility of 80 to 85 percent. Cows producing 100 pounds of milk per day can derive up to 80 percent of their protein needs from digestion of rumen bacteria. Microbial protein is closer in amino acid content to that of milk than any plant or animal protein supplement. Profitability of a ration is directly related to how cheaply and how well it meets the needs of the rumen microbes. The more postruminal nutrient supplementation required in a given ration, the greater the expense of that ration at any given level of milk production.

Postruminal Feeding

Feed selection for rumen-by-pass nutrients should complement the nutrients provided by rumen microbes. Rumen undegraded protein (typically 40 percent of total nitrogen) should deliver extra amino acids postruminally to the small intestine. Protein quality should be a primary consideration.

Balancing dairy rations for amino acid balance has not been a common practice in the past. Synthesis of essential amino acids by rumen microbes was thought to be sufficient to meet the needs of the cow. Even twenty-five years ago, however, researchers realized that methionine and lysine are often limiting in the rations of high-producing dairy cattle (Table 49.5). For nearly that long, companies have sought to develop a coating material for purified amino acids that would escape ruminal degradation yet provide a consistent and high rate of release postruminally. Lysine and methionine requirements can be met by feeding higher levels of undegraded rumen protein; however, the other amino acids provided in excess are degraded, thus increasing concentrations of urea and utilizing even more energy. The development of rumen-stable amino acids allows the opportunity for producers to provide a more balanced amino acid profile in the digesta reaching the small intestine. Continuing increases in milk production, especially in the yield of milk proteins, has created new interest in supplemental amino acids for dairy rations.

Supplementing rumen-stable methionine and lysine may increase milk production, especially in early lactation; however, the most predictable response to supplementation is a 3 to 5 percent increase in milk protein (equivalent to approximately 0.1 percent higher value for protein percentage than unsupplemented cows). Observed improvements in milk production are associated with improved fat metabolism, reflected in decreased liver fat levels and decreased ketosis, and subsequent increases in feed intake during this crucial period. The ratio of lysine:methionine should be maintained at approximately 3:1 for optimal response. This ratio is difficult to achieve without compromising least-cost principles or oversupplying total protein. However,

TABLE 49.5   Extended Chemical Scores of Protein Sources in Relation to Milk Protein

Protein Source	Histidine	Phenylaline	Leucine	Threonine	Methionine	Arginine	Valine	Isoleucine	Tryptophan	Lysine
Blood meal	100	100	93	86	45	33	70	10	76	91
Fish meal	77	69	58	68	100	59	59	47	71	80
Feather meal	11	59	66	59	23	32	38	32	29	13
Meat meal	67	65	46	59	49	76	51	36	39	58
Meat and bone meal	64	64	46	59	49	76	48	36	32	55
Corn gluten meal	67	100	100	60	100	36	48	40	30	18
Alfalfa meal, dehydrated	69	100	55	80	60	50	66	51	100	46
Brewer's grain	56	100	83	65	78	53	65	74	87	34
Distillers with solubles	74	84	72	63	81	42	53	38	45	24
Soybean meal	89	100	56	74	56	89	60	55	75	70
Microbes	90	97	54	100	97	79	66	61	99	100

Source: Adapted from Chandler (1989) and calculated as follows: %AA in feed − %AA in milk (× 100). A score of 100 is the maximum allowed for each value.

the high cost of rumen-stable amino acids currently favors the use of feeds high in by-pass proteins for practical management situations.

Unprotected fats can be added to NEB rations, and over 90 percent of the fat will pass down to the small intestine. Inclusion of fat increases the energy density of the ration, which allows a greater energy intake at the same dry-matter intake. However, fat suppresses appetite to some extent, offsetting any advantages that the increased density provides. Excessive fat in the rumen also adversely affects fiber digestion; consequently, fats should not exceed 3 percent in NEB rations. Protected fats are inert in the rumen, providing extra energy postruminally. Up to 2 pounds per day can be supplemented safely during this period.

Figure 49.25 Late lactation cows are fed to optimize milk production and to target proper body condition scores for the upcoming dry period. *(Courtesy of Iowa State University)*

MANAGING COWS IN A POSITIVE ENERGY BALANCE

Management strategies and goals shift when the cow reaches a positive energy balance (PEB). The reentry into a PEB signals readiness for reproduction. Reproductive success prior to this point is far less likely to be successful. This is also the period when the use of supplemental bovine somatotropin (bST) can be considered. In addition, the focus of the feeding program shifts. The goals in the NEB period are focused on maximizing profit from the current lactation, while the goals in the PEB period are focused on preparing the cow for the next lactation. However, the production of milk in the current lactation is not totally neglected.

Assessment of the program is still based on both milk production response and the rate of change in body condition score. In this case, however, it is the rate body condition score (BCS) gain that is critical. Cows in the PEB program should be fed to maintain

milk production as high as possible (persistency of lactation) without gaining excessive weight. The goal is to dry the cow off at a BCS of 3.5, and the rate of gain must be calibrated by the reproductive state of the animal (Figure 49.25). Cows with anticipated long calving intervals, whether planned or unplanned, should be fed to minimize rate of gain. Cows conceiving relatively early in lactation, with calving intervals of less than twelve months, need to be fed for a substantially higher rate of gain if BCS goals at dry-off are to be met. Thus, cows need to be grouped carefully based on level of production, BCS, and reproductive status (Figure 49.26).

Grain intake during this period should not exceed 2.3 percent of the cow's body weight (dry-matter basis). High-quality forage should be provided, with a minimum intake of 1.5 percent of the cow's body weight (dry-matter basis) to maintain rumen

Figure 49.26 These Brown Swiss cows have ample bunk space to allow maximal feed intake. *(Courtesy of Iowa State University)*

Figure 49.27 Feeding groups for PEB cows are based on a combination of production level, body condition score, and reproductive status. *(Courtesy of Iowa State University)*

function and normal fat test (Figure 49.27). The level of energy intake should be adequate to meet production requirements and to begin replacing body weight lost during the NEB period. Lactating cows require less feed to replace a pound of body tissue than dry cows do; hence, it is more efficient to have cows gain body weight during lactation than during the dry period. Young cows should receive additional nutrients for growth.

MAINTAINING LACTATIONAL PERSISTENCE

Strategies used to improve lacatational persistence include the use of bovine somatotropin, long-day lighting, and milking three times per day. The responses to these strategies are additive; the use of one strategy does not impair the response to the introduction of a second strategy.

Milking Frequency

Increasing the frequency of milking from two times daily to three times daily increases milk production about 6 to 8 pounds per day, regardless of the level of milk production or parity. Increased frequency of milking increases lactogenesis in the gland by maintaining a greater number of secretory cells throughout lactation (secretory cell numbers typically decline during the PEB period). In addition, the rate of synthesis of milk in each individual alveoli decreases with time after milking. By increasing milking frequency, the average rate of synthesis throughout the day is increased. Finally, removing milk from the gland more frequently also removes the pathogens present in that milk before they have the opportunity to multiply to concentrations higher than the immune system can control. Increasing milking frequency reduces the incidence of clinical and subclinical mastitis, thereby minimizing the damage to the secretory cells caused by this disease. This approach also reduces treatment costs and increases the percentage of saleable milk produced.

The economic gain resulting from improvements in milk production, increased saleable milk, and decreased mastitis treatment costs must be weighed against the increased labor costs and equipment maintenance costs associated with increasing the frequency of milking. In addition, increased milk production in these cows is associated with increased feed intake, and increased feed costs must also be factored into this decision.

Photoperiod Effects

Lighting is often neglected as a crucial component of cow management. Although improper lighting does not result in disease or injury, it does affect reproduction and lactational performance. The goals for facility lighting are to provide 20 foot-candles of lighting intensity and to simulate a 16:8 light:dark cycle. For smaller tie-stall or stanchion barns, this requirement can often be met using standard 40-watt fluorescent lights as long as mounting height is less than 10 feet. Separation distances should not exceed 15 feet, and fixtures should be in front of the stanchions or tie stalls to provide light for the cow, rather than in the alley providing light for the workers.

In larger freestall barns with higher sidewalls, metal halide fixtures are often required to provide enough intensity from the higher mounting sites. For example, mounting heights of over 20 feet require 400-watt metal halide or high-pressure sodium fixtures mounted no more than 30 feet apart to provide the required light intensity. Light fixtures should be mounted above stalls and above feedbunks to provide light where the cows spend most of their time.

The total lumens required is calculated by multiplying the square footage of the facility times the desired foot-candles of intensity times a constant. In tie-stall or stanchion barns, the area required in the equation is the area over the feed manger, and the constant used is 2. In freestall barns with fixed sidewalls, the area required for the equation is the total area of the facility, and the constant is 2. In freestall barns with open sidewalls or curtains, the area required is the total area of the facility, and the constant is 3. Metal halide fixtures are preferred because they mimic sunlight most closely for the human eye. High-pressure sodium lights are more efficient but provide less pleasing work conditions for producers. Mercury vapor lights are less efficient, are a potential environmental hazard, and are not recommended for use in dairy facilities.

Cows exposed to long-day lighting (sixteen hours light, eight hours dark) produce about 5 pounds more milk per day compared to those exposed to naturally varying day lengths. This response does not occur with the minimal amount of lighting available in many older facilities. The recommended target for light intensity is 20 foot-candles. This level of lighting intensity can be met by 32-watt fluorescent lights mounted at heights of less than 10 feet or by 250-watt metal halide or high-pressure sodium lights mounted at heights of less than 20 feet. Separation distances between fixtures should not exceed 1.5 times the mounting height for even light distribution. It is important that the fixtures are mounted in the resting areas and feeding areas of the facility; the light intensity must reach the eye of the cow for the response to occur (Figure 49.28). The critical period for exposing cows to high-intensity lighting occurs between thirteen and fifteen hours after dawn, and lighting regimens may not require continuous lighting for sixteen hours to achieve the desired production responses. A six-hour period of continuous exposure coupled with a two-hour exposure period during the critical exposure period may provide a response equal to the continuous sixteen-hour exposure period while reducing electricity usage by 50 percent.

When combining long-day lighting with milking three times daily, it is important that lighting in the housing facility not be continuous. This can be a challenge because light is required for farm workers to be able to move cows safely; however, the dark period is as important as the light period for the lactation response. Continuous lighting provides no milk production response and has the highest level of energy utilization. In situations where some level of lighting is required at all times, dim red bulbs can be used to provide light for workers. The red bulbs (approximately 4 watts) are not perceived as light by the cows.

Figure 49.28 Light fixtures must be spaced properly to provide adequate light intensity throughout the barn. *(Courtesy of Howard Tyler)*

The response to lighting is mediated through the pineal gland. High-intensity light reaching the eyes of the cows induces a decreased melatonin output, which ultimately increases insulinlike growth factor I (IGF-I). The increase in milk production drives an increased feed intake seen in these cows. It takes roughly one month to realize the full response of milk production to long-day lighting.

Bovine Somatotropin

Bovine somatotropin (bST) is a management tool available to dairy producers for improving persistency of lactating cows. Injections of bST act to supplement naturally released somatotropin in the circulation of the cow. This circulating bST binds to receptors in the liver of the cow and stimulates the production and release of insulinlike growth factor I (IGF-I); IGF-I mediates the nutrient partitioning that actually results in increased milk production in the cow. Injections of bST increase blood flow and nutrient availability to the mammary gland.

Recombinant bovine somatotropin (rbST) was approved by the U.S. Food and Drug Administration (FDA) for commercial sale in February 1995. The approval of this product was controversial to both producers and consumers. Consumer concerns included the use of recombinant DNA technology in the production of bST, animal health issues, and milk safety concerns. These concerns have been addressed and resolved through continued product research and public education efforts. Milk composition is not appreciably altered, and commercial milk sales have not suffered following the approval of rbST and its adoption by producers as a management tool.

Many producers had similar concerns. The potential for increased mastitis in rbST-treated cows and uncertainty over effects on milk composition left many producers reluctant to adopt this new technology. However, a two-year postapproval monitoring program by the FDA's Veterinary Medicine Advisory Committee followed twenty-eight commercial herds utilizing rbST. The program found a slightly lower incidence of mastitis than had been predicted in preapproval trials. Mastitis incidence is not higher in rbST-supplemented cows than in untreated cows at similar levels of milk production. There were also no increases in the amount of milk discarded because of antibiotic residues related to rbST use.

Many producers expressed concern that use of rbST in high-producing cattle would lead to excessive fat mobilization and subsequent increases in infertility, ketosis, fatty liver, and chronic wasting. However, it is critically important to understand that rbST has no real stimulatory effects on production in cows that are in a negative energy balance. Liver receptors greatly decrease in NEB cows in relation to the degree of negative energy balance. While this restricts some portions of the effective usefulness of the product as a management tool to those cows in a positive energy balance, it also protects the animal from the detrimental effects of excessive body fat mobilization.

The impact of rbST on socioeconomic issues in the dairy industry is still being resolved. The decision to use rbST does not depend on the size of the herd. Little capital investment is required, and dairy producers with smaller herds may have some advantage in their ability to monitor cows individually. The effect on total milk production nationally and milk prices received by producers has been negligible.

The most profitable use of rbST as a management tool is restricted to producers with excellent management skills. Increases in feed intake lag behind increases in milk production by several weeks. Therefore, nutritional management, especially for the first few weeks following initial administration, is critical for long-term success in adopting this technology. Increasing concentrate intake prior to this time can result in decreases in forage intake and subsequent digestive disorders. Most of the variation in fat-corrected milk response to rbST treatment is due to variation in ration undegradable protein (RUP). Additional RUP should be incorporated into rations based on forages containing high proportions of highly degradable proteins. Feeding strategies for rbST-treated cows are otherwise identical to untreated cows at the same level of production.

Effective herd health and reproductive management skills are equally important for success, and excellent records are essential for determining the appropriate use of rbST for both individual cows and the entire herd. Increasing numbers of dairy producers have adopted bST for use in their herds; however, the percentage of the herd being treated has decreased because producers have determined the most effective use of the product. Currently, approximately 20 percent of all producers utilize rbST on an average of 33 percent of all cows in their herds.

The current label recommendations for the use of rbST suggest initiating use at sixty-three days postpartum. In practical management situations, use of rbST to stimulate milk production should be restricted to PEB cows. Thus, the date of treatment initiation varies and is based on BCS results in individual cows. This strategy allows the most profitable use of the technology. Treatment should be discontinued several weeks prior to anticipated dry-off date.

MILK PRODUCTION AND REPRODUCTION

A question of interest for dairy producers is whether high levels of milk production affect reproduction negatively. This has been a challenging question to answer, and the debate is ongoing among experts in this field.

Analyzing DHIA records across farms with high and low milk production suggests that cows from higher-producing operations actually have a higher pregnancy rate than do cows from lower-producing facilities. Unfortunately, concluding that higher-producing cows are more fertile is not valid; the same level of management expertise that gets cows to produce more milk also gets more cows bred. Farms with lower levels of management intensity are expected to have both poorer reproductive performance and lower milk yields. In addition, for some farms, the improved reproduction may lead to higher levels of milk production rather than the other way around.

This might lead you to analyze the reproductive performance of cows on a single operation based on milk yield. When analyzing farm records for a single farm, however, it is also important to consider the records that are not there. Producers are much more forgiving of cows that are highly profitable. Therefore, low-producing cows are typically culled earlier for reproductive failure than are high-producing cows. When analyzing the past reproductive performance of cows, the cows that were removed from the herd are often not included. This omission makes it appear as though higher-producing cows take longer to breed simply because they are allowed more opportunities to breed.

The connecting link between high milk production and lowered reproduction may be feed intake. Higher feed intake results in increased blood flow to the liver and an increased hepatic metabolism of steroids. Because progesterone and estrogen clearance rates are both controlled by the rate of hepatic metabolism, both hormones are removed from the blood faster in higher-producing cows. The production rate of these hormones is unchanged in high-producing cows, but the rapid clearance rate results in lower circulating levels (approximately 40 percent lower than in lower-producing cows). Fertilization rates are as high in lactating cows as they are in nonlactating cows. Perhaps as a result of the altered steroid levels, however, embryo quality and embryo survival rates are both lower in lactating cows.

In the final analysis, it appears that high milk production affects reproduction indirectly. It may take a greater degree of management intensity to maintain optimal reproductive performance in extremely high-producing cows. Producers must balance this negative impact on reproduction against the increased income from higher milk production. Certainly, at this point in time, nobody is recommending that producers deliberately limit milk production to enhance reproductive performance.

CALVING INTERVAL

The calving interval of dairy cattle is defined as the period of time between the initiation of two subsequent lactation cycles. For decades, the most favorable calving interval was defined as the shortest possible calving interval. Following peak milk, production dropped so rapidly (low persistency of lactation) that it was not profitable to milk most cows for longer than 300 days. While genetic potential for peak milk yield has increased, the greatest genetic improvements have been in improving lacta-

tional persistency. High-producing cows are now profitable for a much longer period, and the difference between profitability at peak milk and in late lactation is not nearly as great. These genetic improvements and the potential to use technologies, such as bST to improve the persistency of lactation even more, have opened the possibilities of extending calving intervals without greatly decreasing profits.

Calving interval is determined by days open; the decision to breed a cow determines the calving interval for the animal irrevocably (Table 49.6). Days open should be determined by the desired calving interval. The producer should determine the most profitable calving interval and use that information to determine when to begin breeding cows. The most profitable calving interval for a herd (or an individual cow) is a complex decision that requires a thorough understanding and analysis of several factors:

Persistency: Highly persistent cows favor a longer calving interval (the use of growth hormone affects this decision).

Mastitis incidence: A high incidence of mastitis favors a short calving interval because damaged alveolar tissue is repaired only during the dry period.

Feed cost:milk price ratio: If this ratio is low, calving intervals should be long; if the ratio is high, calving intervals should be short. However, the relationship to consider is the relationship that will exist in the future.

Culling rate: A high involuntary culling rate requires a short calving interval; a low rate may or may not favor a long interval. (It depends on the value of the cows.)

Calf and heifer mortality: High mortality rates favor a short interval; low mortality rates may or may not favor a long interval. (It depends on the value of the heifers) (Table 49.7).

TABLE 49.6 Average Calving Intervals for Herds with Varying Heat Detection and Fertility Rates with a Fifty Day Rest Period

Herd Fertility Rates, %	Heat Detection Rates, %				
	40	50	60	70	80
40	456	427	405	390	379
50	439	406	391	378	369
60	421	395	382	370	362
70	412	388	375	365	357
80	404	382	370	360	353

Source: R. W. Touchberry, 1991 California Dairy Day Report.

TABLE 49.7   Effect of Calving Interval on Herd Replacements in 100-Cow Herd*

Calving Interval (months)	Average Number of Calves Born per Year	Bred Heifers Available for Herd Replacement per Year
12	100	38
13	92	35
14	84	32
15	76	29

*Assumes 75 percent of the female calves born survive to freshen as first-calf heifers.

Age at first calving: This factor is similar to the mortality rate in importance.

Cow BCS: Cows with a BCS above 2.75 or 3.0 at the time when milk production peaks are poor candidates for extended lactation simply because it becomes difficult to keep them from becoming overconditioned by the time of dry-off. Conversely, cows with a BCS below 2.0 are poor candidates for short calving intervals because it can be difficult to get them to the desired condition score by dry-off.

Other factors: Additional considerations may complicate this decision even further. Herds with poor transition cow management seem to favor a long calving interval to minimize the problems associated with parturition. However, it is difficult derive profit from an extended lactation unless the initiation of that lactation has been well-managed and trouble-free. In addition, it is difficult to manage cows with both long and short calving intervals appropriately in the same feeding group. Some facilities do not permit subgrouping of PEB cows; it is more practical in these situations to maintain a consistent length of calving interval for the whole herd. Finally, high-producing, highly persistent cows generally are more valuable and provide the most valuable contributions toward the genetic improvement of the herd. Extending the calving interval decreases their contribution to genetic improvement and reduces the total number of calves obtained from these cows (unless embryo transfer is used).

To summarize, factors that make a short calving interval more profitable include herds with low-persistence or low-producing cows, overconditioning problems, a high incidence of mastitis, high culling rates, high calf mortality, poor heifer management, and highly valuable animals. Factors that make a longer calving interval profitable are herds with highly persistent cows (either naturally or through the use of bST), outstanding mastitis control, anticipated periods of low feed costs and high milk prices, low culling rates, and tremendous calf and heifer management.

SUMMARY

- Functionally, it is better to manage cows by their energy balance rather than their stage of lactation.
- Rations should be designed to maximize dry-matter and nutrient intake during the NEB period.
- Dairy cattle should lose no more than 1 point on their BCS during the early lactation period, regardless of the initial BCS at parturition.
- Rations with a short bunk life require more frequent feedings to optimize palatability.
- Rations with insufficient particle size result in milk-fat depression.
- The most difficult facet of designing a ration for a ruminant is to create one that maximizes microbial protein outflow from the rumen.
- The nitrogen and carbon sources need to be available to the rumen microbes at the same time.
- Reentry into PEB signals readiness for reproduction and the use of supplemental bovine somatotropin (bST).
- The dark period is as important as the light period for the lactation response.
- Circulating bST binds to receptors in the liver and stimulates the production and release of IGF-I.
- For in cows that are in NEB, rbST has little stimulatory effect on production.

QUESTIONS

1. What is the ideal BCS for a cow at parturition?
2. At what BCS should a cow reenter a positive energy balance?
3. What is the key factor, other than milk production, controlling the rate of condition loss?
4. How many water sources should be available for a herd of 100 cows that are all in NEB?
5. What is the optimal pH for forage-digesting bacteria?
6. What two important functions does rechewing serve?
7. What happens to the ruminal pH as daily feeding frequency increases?
8. What is a useful estimator of the distribution of particle size in a ration?
9. How is protein in the diet converted to protein in milk?
10. Which two amino acids are limiting in the rations of high-producing dairy cattle?
11. What should the ratio of lysine:methionine be for optimal response?
12. How much fat can be supplemented safely during the NEB period?
13. Increasing the frequency of milking from two times daily to three times daily increases milk production by how much?
14. Which gland mediates the response to lighting?
15. When should the dairy producer initiate the use of rbST?
16. When should the use of rbST be discontinued?
17. Define *calving interval.*
18. What are some of the factors affecting the most profitable calving interval for a herd?

ADDITIONAL RESOURCES

Articles

Butler, W. R., R. W. Everett, and C. E. Coppock. "The Relationship Between Energy Balance, Milk Production, and Ovulation in Postpartum Holstein Cows." *Journal of Animal Science* 53 (September 1981): 742.

Caroll, D. J, B. A. Barton, G. W. Anderson, and R. D. Smith. "Influence of Protein Intake and Feeding Strategy on Reproductive Performance of Dairy Cows." *Journal of Dairy Science* 71 (December 1988): 3470.

Heuer, C., W. M. Van Straalen, Y. H. Schukken, A. Dirkzwager, and T. M. Noordhuizen. "Prediction of Energy Balance in High Yielding Dairy Cows with Test-Day Information." *Journal of Dairy Science* 84 (February 2001): 471–481.

Magliaro, A. L., R. S. Kensinger, S. A. Ford, M. L. O'Connor, L. D. Muller, and R. Graboski. "Induced Lactation in Nonpregnant Cows: Profitability and Response to Bovine Somatotropin." *Journal of Dairy Science* 87 (October 2004): 3290–3297.

Villa-Godoy, A., T. L. Huges, R. S. Emery, L. T. Chapin, and R. L. Fogwell. "Association Between Energy Balance and Luteal Function in Lactating Dairy Cows." *Journal of Dairy Science* 71 (April 1988): 1063.

Internet

Body Condition Scores: http://babcock.cals.wisc.edu/downloads/de_html/ch12.en.html.

Body Condition Scoring with Dairy Cattle: http://www.uaex.edu/Other_Areas/publications/HTML/FSA-4008.asp.

Fertility and Body Condition Score Negative Energy Balance: http://www.kt.iger.bbsrc.ac.uk/FACT%20sheet%20PDF%20files/kt17.pdf.

The Influence of Negative Energy Balance on Udder Health: http://www.nmconline.org/articles/ketosis.pdf.

INDEX